Plant Gene Research

Basic Knowledge and Application

Edited by

E. S. Dennis, Canberra

B. Hohn, Basel

Th. Hohn, Basel (Managing Editor)

P. J. King, Basel

J. Schell, Köln

D. P. S. Verma, Montreal

Springer-Verlag Wien New York

Genes Involved in Microbe-Plant Interactions

Edited by D. P. S. Verma and Th. Hohn

Springer-Verlag Wien New York

Dr. Desh Pal S. Verma
Department of Biology, McGill University, Montreal

Dr. Thomas Hohn
Friedrich Miescher-Institut, Basel

With 54 Figures

Printed in Austria

Library of Congress Cataloging in Publication Data

Main entry under title:

Genes involved in microbe-plant interactions.

(Plant gene research)
1. Plant genetics. 2. Microbial genetics. 3. Plants – Microbiology. I. Verma, D. P. S. (Desh Pal S.), 1944– . Hohn, Thomas, 1938– . III. Series.
QK981.G435 1984 581.1'5 84-13855

ISBN 0-387-81789-1 (U.S.)

ISSN 0175-2073
ISBN 3-211-81789-1 Springer-Verlag Wien – New York
ISBN 0-387-81789-1 Springer-Verlag New York – Wien

Introduction to the Series

Looking back over the last decades one can only marvel at the depth and the quality of the progress made in our understanding of living organisms in molecular terms. This is particularly true with regard to the structure, function and regulation of the genetic program which governs the biochemistry of living cells and which constitutes the basis of the evolutionary link between them. However Plants have been largely ignored in this scientific progress. Most of the modern concepts in Molecular Biology and Cell Biology as well as most of the available data regarding gene structure and function, are based on the study of microbial and animal organisms and their respective viruses. So much so that about a decade ago only very few, if any, people would have been able to claim the title of "Plant Molecular Biologist". This situation has now changed. Recombinant DNA techniques have contributed very much to molecular biology in general and to plant molecular biology in particular. The timely discovery that natural and very efficient gene vectors for plants exist in the form of the tumor-inducing plasmids (Ti-plasmids) of *Agrobacterium tumefaciens* and possibly also in the form of plant viruses, has paved the way for the study of plant genes via "reversed genetics", i.e. the in vitro mutation of isolated genes followed by transfer and integration of the modified genes in plants.

This experimental capacity combined with the concomitant development of plant tissue culture techniques, somatic embryogenesis, protoplast fusions, anther cultures etc., have already turned the previously neglected plants into very attractive subjects for molecular and cellular biologists. Possibly the strongest influence in making plant gene research a growing field of scientific interest and activity is the expectation that this research will open new and very effective ways for the breeding of agriculturally and biotechnologically important plants. The combination of the scientific excitement generated by the vastly increased possibilities to solve fundamental questions with the prospects of far-reaching applications in the fields of agriculture and biotechnology, has produced considerable support for this field of research and should ensure a sustained effort.

The present series on Plant Gene Research is therefore very timely and can be expected to play a major role. In view of the fact that the study of Microbe-Plant Interactions are playing a central role in the renewed scientific interest in fundamental plant genetics, it is appropriate that this series should devote its first volume to these studies. Future volumes are expected to cover such aspects as "Genetic Flux in Plants", "Biochemical Genetics of Higher Plants" and "Gene Transfer and Regulation in Plants".

Köln, May 1984 J. Schell

Preface

Interdependence between species is a law of nature. The degree of this interdependence is vividly evident in the plant-microbial world. Indeed, there is no axenic plant in nature and one finds various forms of interactions between these two kingdoms ranging from completely innocuous to obligate parasitic. Most of these interactions are poorly understood at the molecular and physiological levels. Only those few cases for which a molecular picture is emerging are discussed in this volume. With the advent of recombinant DNA technology and the realization that some of these interactions are very beneficial to the host plant, a spate of activity to understand and manipulate these processes is occurring.

Microbes interact with plants for nutrition. In spite of the large number of plant-microbe interactions, those microbes that cause harm to the plants (i.e., cause disease) are very few. It is thus obvious that plants have evolved various defense mechanisms to deal with the microbial world. The mechanisms for protection are highly diverse and poorly understood. Some pathogens have developed very sophisticated mechanisms to parasitize plants, an excellent example for this being crown gall caused by a soil bacterium, *Agrobacterium tumefaciens.* A remarkable ingenuity is exhibited by this bacterium to manipulate its host to provide nitrogenous compounds which only this bacterium can catabolize. This is carried out by a direct gene transfer mechanism from bacteria to plants. The transformed plant tissue differentiates and becomes autonomous with respect to plant growth hormones. In fact, cell proliferation appears to be an integral part of several diseases caused by bacteria and may be essential for the exchange of genetic material.

While most organisms interact on the cell surface or via some damaged tissue, in the case of *Rhizobium* a very specialized mechanism has evolved which allows endocytosis of bacteria into the host cell where the microbes reduce nitrogen for the plant. This microbe in its symbiotic state is behaving as an organelle since some of its functions have become physiologically dependent upon the host while the autonomy of *ex planta* growth is maintained. In this association as well one finds the involvement of plant-growth hormones.

The major problem associated with the study of plant-microbe interactions is that there is no synchronization in the infection process and only a few cells are initially affected. Biochemical and molecular studies on such early infection processes can best be performed at cellular levels. Current molecular biology is providing tools to explore new horizons with the possibility to eventually modify or maneuver these interactions for more benefit to mankind. This volume provides a glimpse of some aspects of plant-microbe interactions and various approaches by which these processes are being studied in more detail. For full treatment of each topic one needs a complete volume and several books have recently been written on some of the topics covered in this compendium. We hope that the present treatment of this wide subject would provide an overview of the problem along with some new ideas which could serve as a stimulus and challenge for researchers in this field.

Montreal, June 1984 D. P. S. Verma

Acknowledgements

The editors would like to express their thanks to J. Denarie for active interest and various suggestions, L. Dixon, D. Mohnen, C. Rowland, J. Schröder and the members of the genetic manipulation research group of the McGill University, Montreal, for critical readings of the manuscript, and H. Isler, Y. Mark and E. Robison for secretarial help.

Contents

Section I.
Recognition

Section II.
Symbiosis

Section III.
Plant Tumor Induction

Section IV.
Plant Pathogens and Defence Mechanisms

Section I.
Recognition

Chapter 1

Host Specificity in *Rhizobium*-Legume Interactions

F. B. Dazzo and A. E. Gardiol

Department of Microbiology and Public Health, Michigan State University, East Lansing, MI 48824, U. S. A.

With 8 Figures

Contents

I. Introduction

The process of cellular recognition between microorganisms and higher plants is receiving considerable attention in light of its effect on plant morphogenesis, nutrition, symbiosis, and protection against infectious disease. In general, these positive cellular recognitions are believed to arise from a specific union, reversible or irreversible, between chemical receptors on the surface of interacting cells. This hypothesis implies that communication occurs when cells that recognize one another come into contact, and therefore the complementary components of the cell surfaces have naturally been the focus for most biochemical studies. Such is the case for studies on the infection of legume roots by the bacterial symbiont, *Rhizobium*. This article deals with the analysis of lectins and their saccharide receptors in the infection process, their involvement in specific bacterial attachment to root hairs, their regulation, and a critical evaluation of the limitations of the present work relative to understanding the biochemical basis of host specificity.

Rhizobium is a genus of gram-negative soil bacteria which can infect and nodulate legume roots, forming a nitrogen fixing symbiosis. The specific infection of legume root hairs by *Rhizobium* involves many steps of cellular recogition which culminate in the formation of a root nodule that reduces atmospheric nitrogen into ammonia for the host plant. This symbiosis contributes significantly to the biogeochemical cycling of nitrogen and to the nitrogen economy of agricultural crops, since nitrogen is the nutrient most often limiting crop productivity. There is a high degree of host-specificity in this symbiosis, e.g., *R. meliloti* nodulates alfalfa but not clover, and *R. trifolii* nodulates clover but not alfalfa. An understanding of the recognition code for host specificity of the *Rhizobium*-legume symbiosis could provide ways to broaden the host range of agricultural crops which can enter efficient nitrogen-fixing associations. Therefore, this agronomically important symbiosis is ideal not only for studies of plant-microbe interactions but can serve as a model of cellular recognition.

II. The Infection Process

The infection process in the *Rhizobium*-legume symbiosis is an elegant sequence of cellular recognitions which can best be observed in enclosed slide cultures of small-seeded legumes, e.g., white clover, inoculated with pure cultures of the rhizobial symbiont (Fig. 1 a) (Fahraeus, 1957). Root hairs in various stages of development are the epidermal target cells through which the bacterial infection begins. When 10^7—10^8 fully encapsulated cells of *R. trifolii* are inoculated on clover seedling roots, they rapidly and specifically clump at the tips of the root hair cells. During the next several hours, other bacteria attach polarly along the sides of the root hair. A very marked curling of the root hair tip (the "shepherd's crook") is induced on a few root hairs along the root within 1—2 days later (Fig. 1 b). Occasionally, bacteria entrapped within the tight curl penetrate the root hair cell wall and form a tubular structure called the "infection thread" (Fig. 1c) which is an unambiguous sign of "successful" infection (Napoli and Hubbell, 1976; Callaham and Torrey, 1980). Host specificity is expressed at some event preceeding the formation of these infection threads within the root hair (Li and Hubbell, 1969). The growth of the infection thread follows the movement of the nucleus to the root hair base, where it then penetrates and enters the underlying root cortex. Within a few days, cortical cells in the root begin to proliferate in front of the invading bacteria and eventually emerge as a nodule (Fig. 1 d). These nodules contain the microbial symbionts within the plant cells (Fig. 1 e) and reduce atmospheric nitrogen into ammonia, making it readily available to the plant.

The infection of large-seeded legumes is less understood primarily because the large root size makes it difficult to use the Fahraeus slide technique. Recently, useful information has been obtained using cellophane growth pouches and spot-inoculation techniques (Bhuvaneswari *et al.*, 1980, 1981; Turgeon and Bauer, 1982). In this case, *Rhizobium* infections leading to nodule formation in soybean, alfalfa, and cowpea occur more frequently through infection threads in those primordial root hair cells that were located between the zone of root elongation and the zone of more mature root hairs at the time of inoculation. However, the worrisome limitation of the growth pouch technique for root hair infection studies is that it is open to microbial contamination throughout the growth of the host plant.

III. The Lectin Recognition Hypothesis

Over the last decade, much work has focused on the basic mechanisms which lead to the successful infection of legume roots by the nitrogen-fixing symbiont. A hallmark of this effort has been the discovery that certain legume lectins fulfill many criteria of a cell recognition molecule in this plant-bacterial symbiosis.

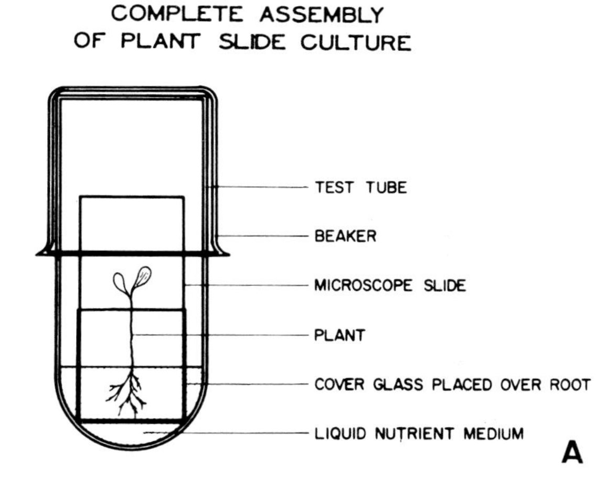

Fig. 1 A

Fig. 1. The infection of white clover roots by *Rhizobium trifolii* 0403 as studied by the Fahraeus slide technique

A. Schematic diagram of the Fahraeus slide culture, in which the bacterial and plant symbionts are grown between a cover slip and slide under bacteriologically controlled conditions. Drawing courtesy of David Hubbell.
B. Marked curling ("Shepherd's crook") of a root hair induced by the bacterial symbiont.
C. Root hair infected with the bacterial symbiont. Note the refractile infection thread within the host cell.
D. Infected root hair on top of a root nodule.
E. A cross-section of a nodule plant cell containing many cells of the bacterial symbiont

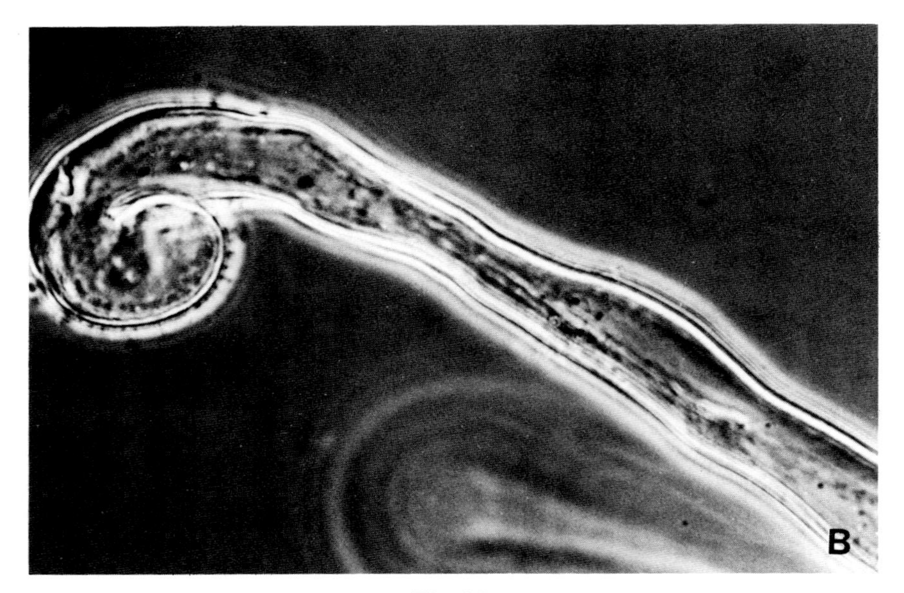

Fig. 1 B

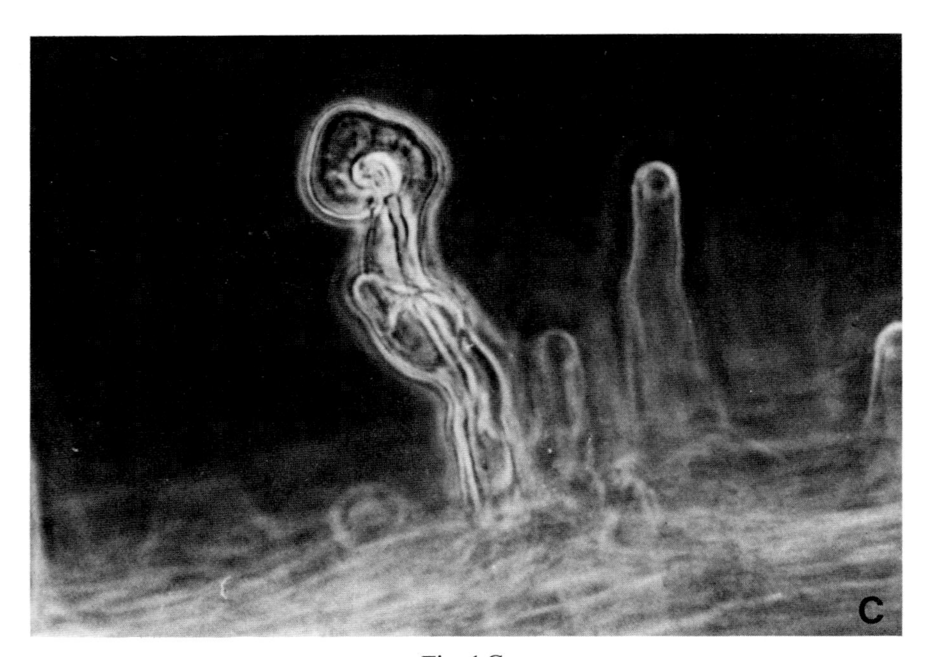

Fig. 1 C

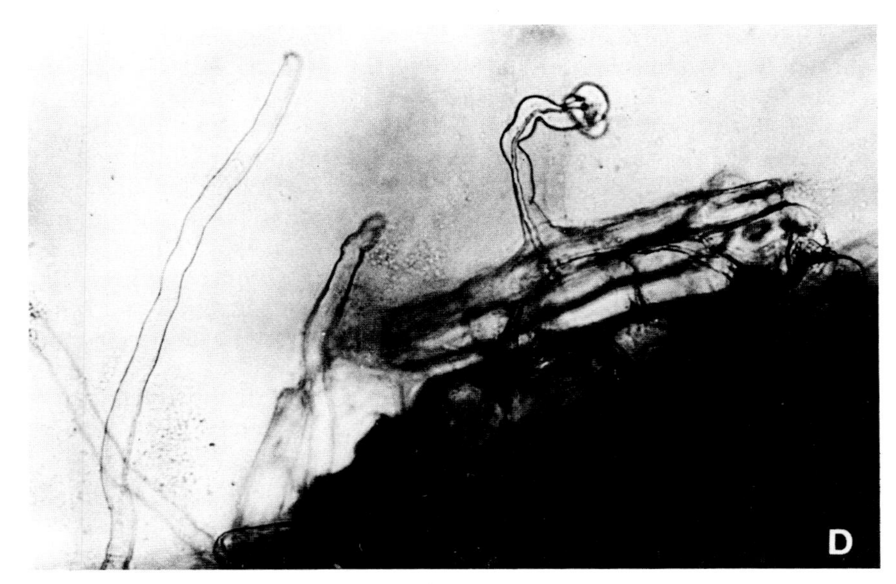

Fig. 1 D

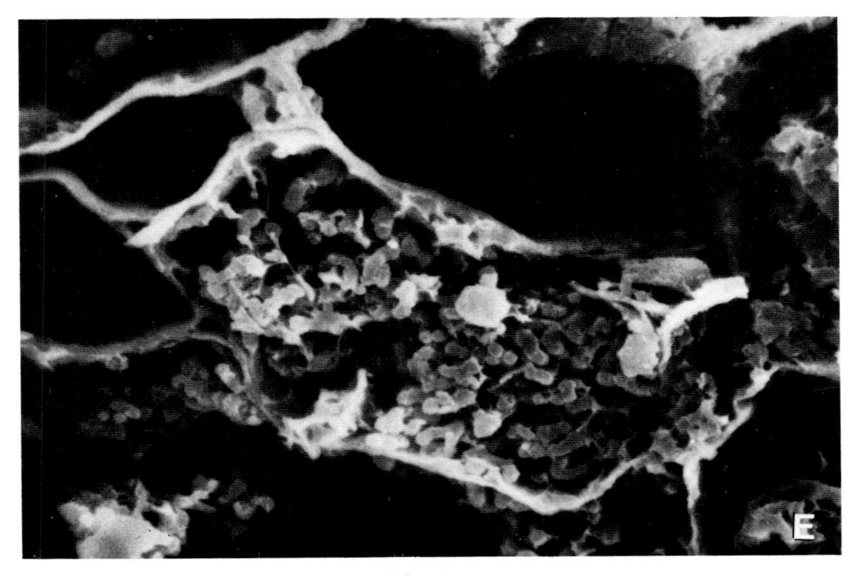

Fig. 1 E

The focus of lectin involvement in the *Rhizobium*-legume symbiosis began with the demonstration that bean lectin interacts with bean rhizobia (Hamblin, Kent, 1973). A year later, Bohlool and Schmidt (1974) reported that the soybean-nodulating rhizobia interact specifically with soybean lectin from seeds. This specific correlation was subsequently confirmed (Bhuvaneswari *et al.*, 1976) and extended to include several other legume-*Rhizobium* symbioses (Dazzo, Hubbell, 1975b; Wolpert; Albersheim, 1976; Kamberger, 1979a; Paau *et al.*, 1980; Kato *et al.*, 1980). This specificity became the basis for formulating the "lectin recognition hypothesis": that recognition between rhizobia and the legume host root involves a binding of the plant lectins to unique carbohydrates found exclusively on the bacterial symbiont as a prelude to infection. However, as early as 1975, it became clear that not all legume seed lectins specifically recognize the corresponding rhizobial symbiont. For example, concanavalin A from jackbean seeds bound in a hapten-reversible manner to rhizobia which do not nodulate jackbean roots (Dazzo and Hubbell, 1975a), and a soybean lectin from soybean seeds bound to some peanut rhizobia which do not nodulate soybean roots (Pueppke *et al.*, 1980). Law and Strijdom (1977) reported that lectins from *Lotononis* seeds non-specifically bound to rhizobia, but more recently found that lectin from the roots of this legume binds specifically to the bacterial symbiont (Strijdom, personal communication). These results emphasize that tests of the lectin recognition hypothesis should not be restricted to only consider the classical hemagglutinating lectins from legume seeds, but rather should focus on whether root lectins bind specifically to the rhizobial symbiont before infection.

We directed our attention to the *Rhizobium*-clover symbiosis to pursue the lectin recognition hypothesis. We discovered a new lectin (Dazzo *et al.*, 1978) using the agglutination of suspensions of the bacterial symbiont as the assay for the lectin. This lectin is a carbohydrate-binding glycoprotein with a subunit molecular weight of approximately 53,000 daltons (corrected from earlier estimates) and aggregates at neutral pH, near its isoelectric point, into particles which are approximately 10 nm in diameter. We have named this clover lectin "trifoliin A" (from *Trifolium*, the genus of clovers).

Trifoliin A has several unique properties which qualify it as a cell-recognition molecule in the *Rhizobium*-clover symbiosis (Dazzo *et al.*, 1978). Trifoliin A specifically binds to and agglutinates *R. trifolii* at very low concentrations (ug/ml). Its interaction with cells and their surface polysaccharides is specifically inhibited by the hapten 2-deoxyglucose (Dazzo and Hubbell, 1975; Dazzo and Brill, 1977; Dazzo *et al.*, 1978; Hrabak *et al.*, 1981). Based on Ouchterlony immunodiffusion, the lectin from the seed is antigenically related to a lectin from the root. Immunofluorescence microscopy detects this lectin on differentiated clover root hair cells (including root hair primordia) and not on undifferentiated epidermal cells. The distribution of trifoliin A immunofluorescence on maturing root hairs is a gradient which is greatest at the growing hair tip and least at the base of the root hair. Pea, alfalfa, and soybean lectins accessible for binding the corresponding rhizobial symbionts also seem to be located on the root in greatest quantity where root hairs develop (Gatehouse and Boulter, 1980; Kato *et al.*, 1980, 1981; Kijne *et al.*, 1980; Gade *et al.*, 1981; Stacey *et al.*, 1980; Paau *et al.*, 1981).

IV. Bacterial Attachment to Legume Root Hairs as an Early Recognition Step

Rhizobium attaches to the root hairs that it later infects. However, all attachments do not lead to infection. Specific bacterial attachment has been the step of cellular recognition studied in greatest detail. During early stages of the infection process, the bacteria attach via hapten-reversible interactions, and then later become irreversibly anchored to the host cell. Quantitative microscopic assays (Dazzo *et al.*, 1976; Dazzo, 1980a) and transmission electron microscopic studies (Dazzo and Hubbell, 1975b; Napoli and Hubbell, 1975; Kumarasinghe and Nutman, 1977) of the *Rhizobium*-clover symbiosis have revealed multiple mechanisms of bacterial attachment to the root hairs. A nonspecific mechanism allows all species of rhizobia to attach in low numbers (2—4 cells per 200 µm root hair length per 12 hr using low inoculum per seedling). In addition, a specific mechanism allows selective attachment of the rhizobial symbiont in significantly larger numbers (22—27 cells per 200 µm root hair length per 12 hr) under identical conditions (Dazzo *et al.*, 1976). Thus, the rhizobia which could infect clover attached in highest numbers to these root hair cells. Host-specific rhizobial attachment to root hairs has also been demonstrated in *R. japonicum*-soybean (Stacey *et al.*, 1980), and *R. leguminosarum*-pea (Kato *et al.*, 1980) root systems. However, specificity was not found in quantitative root attachment studies employing very high densities of radiolabelled rhizobia (10^9—10^{10} cells per seedling, Chen and Phillips, 1976), but many unattached bacteria which were not washed away could have contributed to this result. Similarly, studies based on plate-count data indicated nonspecific attachment of rhizobia to the epidermis below the zone of maturing root hairs of soybean (Vesper and Bauer, in preparation). However, these quantitative studies were not confirmed by direct microscopy. Some of these inconsistencies have been more clearly resolved by employing marble chips to dislodge bacteria attached to the root system before quantitative plating assays. In this case, "firm" attachment was found to be host-specific in *R. trifolii*-clover and *R. meliloti*-alfalfa systems, and "loose" attachment was nonspecific (Van Rensberg and Strijdom, 1982). It is therefore apparent that firm bacterial attachment to host roots is an early expression of cellular recognition in the *Rhizobium*-legume symbiosis.

V. Role of Legume Lectins in Attachment of Rhizobia

Since rhizobia which could infect clover attached in the highest numbers to clover root hair cells, we set out to identify the cell surface molecules which mediate this cellular recognition event. Our strategy has been to examine surface components of the bacterium and the host that interact with the same order of specificity as is observed with the adhesion of the bacterial cells.

The first clue that trifoliin A on the root may be involved in rhizobial attachment came from the observation that at 30 mM, the hapten 2-deoxyglucose (but not the analogue 2-deoxygalactose) specifically inhibited attachment of clover rhizobia to clover root hairs (Dazzo *et al.*, 1976; Zurkowski, 1980), reducing the high level of bacterial attachment to that characteristic of background. As a control, 30 mM 2-deoxyglucose did not inhibit attachment of alfalfa rhizobia to alfalfa root hairs. It is now known that an agglutinin on alfalfa root hairs specifically binds the alfalfa symbiont, *R. meliloti*, and 2-deoxyglucose is not an effective hapten for this alfalfa agglutinin (Paau *et al.*, 1981). Only about 90 % of the *R. trifolii* attachment to clover root hairs was inhibited by 2-deoxyglucose, providing further evidence for multiple mechanisms of attachment. Some mechanisms are host-symbiont specific and 2-deoxyglucose inhibitable, and others are non-specific at background levels. (The latter non-specific mechanism is 2-deoxyglucose resistant, mediated perhaps by plant polysaccharide-bacterial polysaccharide or bacterial pili-plant polysaccharide interactions.) Similarly, attachment of *R. japonicum* and *R. leguminosarum* to root hairs of their respective hosts includes a mechanism which is *Rhizobium*-specific and largely inhibited by corresponding haptens of the host root lectins (Stacey *et al.*, 1980; Kato *et al.*, 1980, 1981). These studies strongly suggest that a major mechanism of rhizobial attachment to host root hairs involves the symbiont-specific interaction of host lectins and the surface polymers of the bacterial symbiont.

We next studied the interactions between trifoliin A-binding polysaccharides of *R. trifolii* and the clover root surface to determine the location of receptor sites which recognize rhizobia (Dazzo and Brill, 1977). The bacterial capsular polysaccharide labelled with fluorescein isothiocyanate (which fluoresces green) specifically bound to the root hair cells, matching the distribution of trifoliin A (Fig. 2). Specificity of these receptor sites was demonstrated by the ability of unlabelled capsular polysaccharide from *R. trifolii* but not from *R. meliloti* to block the binding of the labelled polysaccharide to clover root hairs. This host-specific detection of receptor sites which specifically bind capsular polysaccharides of the rhizobial symbiont has been extended to include rhizobial symbioses with alfalfa, peas, and wild soybean (Dazzo and Brill, 1977; Hughes and Elkan, 1981; Kato *et al.*, 1980). In every case where the hapten is known, binding of the bacterial polysaccharide to the root hairs was inhibited specifically by the addition of the respective hapten sugars of the host lectin. This result strongly suggests that host lectin on the root hairs provides recognition sites for the bacterial symbiont. Hapten sugars selectively elute the clover, pea, and soybean lectin from off the corresponding intact root (Dazzo, Brill, 1977; Dazzo *et al.*, 1978; Kijne *et al.*, 1980; Gade *et al.*, 1981), suggesting that some of the lectin is reversibly bound to receptors on the root surface. The hapten could also block exposed sites of lectin irreversibly bound to the root hair surface.

In such hapten-facilitated elution techniques, it is believed that the sugar acts by combining specifically with the site on the lectin which is normally occupied by the natural saccharide receptor. This implies a close but not necessarily identical structure of the hapten and the native determinant. However, some lectins undergo conformational changes when associated with saccharide binding (Reeke *et al.*, 1975), and so this possibility must also be considered in the interpretation of hapten inhibition data.

The binding of FITC-capsular polysaccharides to legume roots has also highlighted the importance of epidermal cell differentiation in the development of receptor sites that recognize rhizobia. Close inspection of the photomicrographs revealed that undifferentiated epidermal cells in the clover root hair region did not bind the bacterial polysaccharide, whereas epidermal root hair primordia displayed this surface property (Fig. 2).

The results of immunochemical, physiological, and genetic studies support the hypothesis that the host lectin and its saccharide receptors are involved in this initial specific attachment of the bacterium to the root hairs. First, certain non-infective mutants of *R. trifolii* and heterologous rhizobia do not bind trifoliin A and only bind to clover root hairs at back-

Fig. 2. Specific binding of FITC-labelled capsular polysaccharide from *R. trifolii* 0403 to clover root hairs. The labelled bacterial polysaccharide bound to the root hairs and is illuminated by epifluorescence. From Dazzo and Brill (1977)

ground levels (Dazzo *et al.*, 1976). Second, competition assays using fluorescence microscopy indicated that the *R. trifolii* polysaccharides that bound to trifoliin A had the highest affinity for clover root hairs (Dazzo and Brill, 1979). Third, growth of the root in nitrate-supplemented media results in parallel reductions in levels of surface-associated trifoliin A and root hair attachment by clover rhizobia (Dazzo and Brill, 1978). Fourth, the transient level of these bacterial polysaccharide receptors directly correlates with the quantity and orientation of bacterial attachment to the clover root surface (Dazzo *et al.*, 1979; Sherwood *et al.*, submitted). Fifth, intergeneric transformations of *Azotobacter vinelandii* with *R. trifolii* DNA led to the formation of hybrid cells which make clover lectin receptors (Bishop *et al.*, 1977) and which specifically acquired the ability to attach to the clover root hairs in a hapten-inhibitable fashion (Dazzo and Brill, 1979). Sixth, Fab monovalent fragments against a cross-reactive clover root antigen (Dazzo and Hubbell, 1975a) compete specifically with trifoliin A for saccharide determinants on the bacterial surface and block the bacterial attachment to the clover root hairs (Dazzo and Brill, 1979). The hapten inhibition pattern of the cross-reactive antibody for the bacterial surface antigen matches that of trifoliin A (Dazzo and Hubbell, 1975a; Dazzo and Brill, 1979; Hrabak *et al.*, 1981). These observations and the facilitated hapten elution of lectin from roots described above suggest that the bacteria and clover root have similar receptors for the multivalent lectin. However, the definitive test of

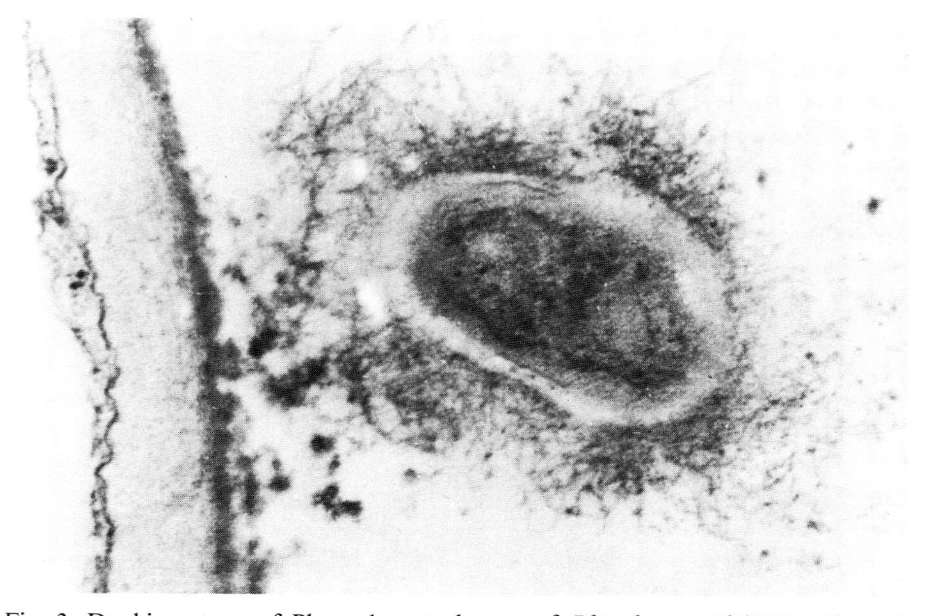

Fig. 3. Docking stage of Phase 1a attachment of *Rhizobium trifolii* NA-30 to the clover root hair cell wall. Note the fibrillar capsule which contacts electron-dense aggregates on the outer periphery of the root hair cell wall. From Dazzo and Hubbell (1975)

their identity as antigenically related structures will require knowledge of the minimal saccharide structure that binds trifoliin A.

Transmission electron microscopy (Dazzo and Hubbell, 1975a) disclosed that the initial bacterial attachment step consisted of contact between the bacterial capsule of *R. trifolii* and electron-dense globular aggregates which lie on the outer periphery of the clover root hair cell wall (Fig. 3). This "docking" stage is the first point of physical contact between the microbe and the host and occurs within minutes after inoculation of encapsulated cells of *R. trifolii* on the host clover. We have recently isolated these electron-dense particles which stain positively with uranyl acetate and accumulate on the root hair surface (Truchet *et al.*, in preparation). Fluorescence and electron microscopic studies indicate that these particles contain several proteins including trifoliin A, and they contain another component which binds this lectin. They can be dislodged from intact roots by gentle elution with the effective hapten, 2-deoxyglucose, consistent with previous hapten-elution data (Dazzo and Brill, 1977). In addition, we have found evidence for the *de novo* synthesis of trifoliin A in roots of axenically grown clover seedlings based on incorporation of labelled amino acids into a product which elutes from intact roots with 2-deoxyglucose, immunoprecipitates with anti-trifoliin A IgG, and comigrates with authentic trifoliin A in SDS-PAGE (Sherwood *et al.*, manuscript in preparation).

Based on all of the above evidence, we present a revision of the original lectin cross-bridging model (Dazzo and Hubbell, 1975a) on the specificity of attachment (Fig. 4). In summary, trifoliin A cross-bridges receptors on

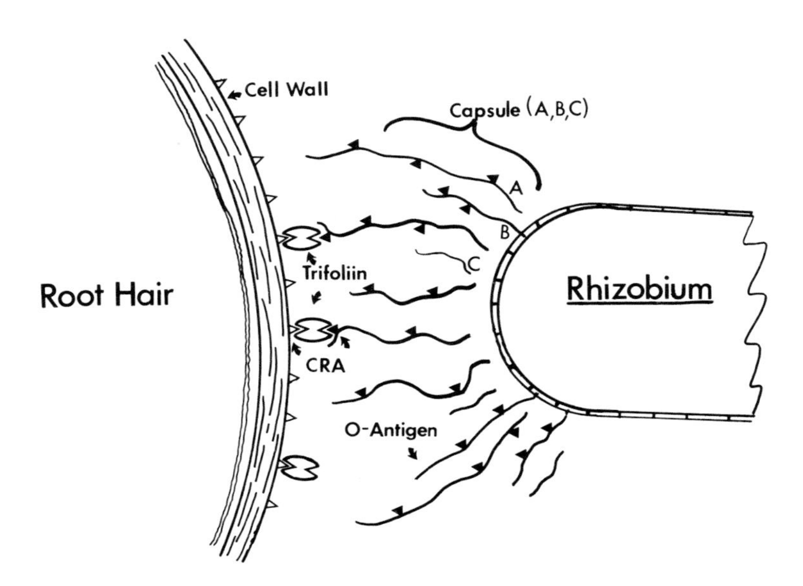

Fig. 4. Proposed cross-bridging of *Rhizobium trifolii* receptors to clover root hairs by the host lectin, trifoliin A. From Dazzo and Truchet (1983)

the root hair cell wall with bacterial capsular polysaccharide (primarily) and LPS (secondarily). Trifoliin A and these aggregated particles may literally play a central role in this initial attachment stage in that they are multivalent structures that can recognize and combine with unique saccharide sequences on both the microbe as well as the plant host. This revised model takes into account the growth phase-dependent LPS-trifoliin A interactions (Hrabak *et al.,* 1981) which are predicted to be involved in infection (Kamberger, 1979b), and the lectin involvement in the tip adhesion of unsaturated lectin receptors on adjacent root hairs of developing seedlings (Dazzo *et al.,* 1982b; Higashi and Abe, 1980).

Following this initial docking stage, the bacteria firmly anchor to the root hair surface (Dazzo *et al.,* in preparation). This adhesion may be important in maintaining the firm contact between the bacterium and the host root hair necessary for triggering the tight root hair curling (shepherd's crook) and successful penetration of the root hair cell wall during infection (Napoli *et al.,* 1975). During this firm adhesion, fibrillar materials, recognized by scanning electron microscopy, are characteristically found associated with the attached bacteria (Fig. 5, Dazzo *et al.,* submitted for publication). The nature of these microfibrils is unknown. One possibility is that they are bundles of cellulose microfibrils known to be produced by many rhizobia (Deinema and Zevenhuizen, 1971; Napoli *et al.,* 1975). Another possibility is that they are collections of pili, which have recently been demonstrated in *Rhizobium* (Stemmer and Sequeira, 1981; see also Kijne *et al.,* 1982). Future studies should be directed to isolate and charac-

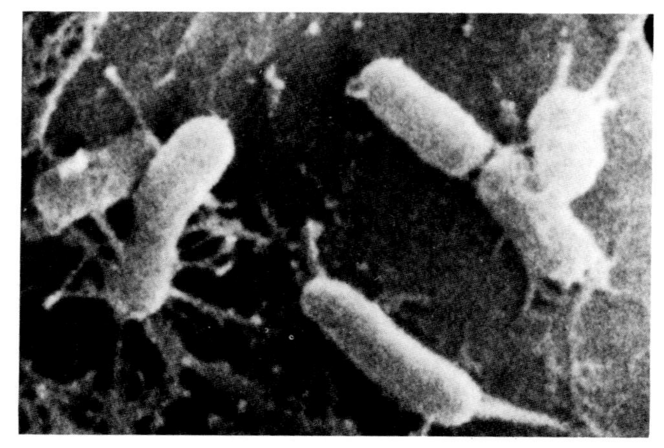

Fig. 5. Aggregated microfibrils associated with *Rhizobium trifolii* 0403 firmly attached to the clover root hair sufrace after prolonged incubation

terize these fibrils associated with the adherent bacteria in order to better understand the post-attachment adhesion process. This is particularly important in light of the recent demonstration that the degree of host-specific firm attachment of rhizobial strains to the root shows a significant positive correlation with the degree of their success in interstrain competition for nodule sites on the root (Van Rensberg and Strijdom, 1982).

VI. There Are Multiple Lectin Receptors on *Rhizobium*

Rhizobium produces several different polysaccharides in pure culture and a controversy of the past has been the question of which one(s) bind the host lectin. Current evidence indicates that *R. japonicum* binds soybean lectin through capsular and exopolysaccharides (Bal *et al.,* 1978; Calvert *et al.,* 1978; Mort and Bauer, 1980, 1982; Truchet *et al.,* submitted); *R. meliloti* binds alfalfa lectin through LPS (Kamberger, 1979a; J. Handlesman, personal communication); and the related species *R. trifolii* and *R. leguminosarum* specifically bind clover and pea lectin, respectively, through their capsular/polysaccharide and their LPS at certain culture ages (Kamberger, 1979a, Dazzo and Brill, 1979; Hrabak *et al.,* 1981; Kato *et al.,* 1979, 1980, 1981; Kijne *et al.,* 1980). Similarly, peanut lectin binds LPS and capsular polysaccharide of peanut rhizobia (Bhagwat and Thomas, 1980). Since the compositions and immunodominant structures of LPS vary widely among strains of a single *Rhizobium* species (Carlson *et al.,* 1978), the lectins of clover, pea, alfalfa, and peanut may be interacting specifically with a portion of the symbiont's LPS which is poorly immunogenic and common to different strains of the same *Rhizobium* species. The discoveries that the lectin-binding sites on the polysaccharides of *R. trifolii* are transient and are not immunodominant are very important since they suggest that this cell recognition process is regulated (Dazzo *et al.,* 1979; Dazzo and Brill, 1979; Hrabak *et al.,* 1981).

VII. Regulation of Lectin-*Rhizobium* Interactions

At least three components regulating lectin-*Rhizobium* interactions are known to exist in the *Rhizobium*-legume symbiosis. One is a plant component which is regulated by combined nitrogen, e.g., nitrate. A second is a bacterial component which is regulated by the growth phase (i.e., culture age) and has a profound effect on the ability of the bacterial polysaccharide to bind lectin and the bacterium to bind to the root. Finally, there is a less-well defined level of regulation which is detected only when the two symbionts are combined *in situ.*

Attachment of the rhizobial symbiont to the legume root is regulated by combined nitrogen (Dazzo and Brill, 1978; Truchet and Dazzo, 1982). The levels of trifoliin A on the root surface and attachment of *R. trifolii* to root hairs change dramatically when clover is grown with different levels of nitrate in the nutrient medium. During root growth in presence of 1 mM nitrate, the levels of lectin and bacterial attachment increase when compared to seedlings grown with no combined N. These levels decline as nitrate is increased to approximately 15 mM. Plants growing at 15 mM are healthy. The nitrate effect on the root is not due to the counterion or to a direct interaction between nitrate and trifoliin A, and it requires more than 1 hr exposure to affect bacterial attachment (Dazzo and Hrabak, 1982). Thus there must be some intervening mechanism with requires more than 1 hr to modulate this cell recognition process. In the very similar clover and pea systems, the *Rhizobium* binding lectin is synthesized in roots grown at high nitrate conditions, but does not accumulate on the root cell wall as much as when grown at low nitrate concentrations (Sherwood *et al.*, in preparation; Diaz *et al.*, in preparation). Similar control of lectin levels by nitrate supply has been shown in the *Azolla-Anabaena* nitrogen-fixing symbiosis (Mellor *et al.*, 1981).

Studies of the combined nitrogen effect have also revealed that clover and pea root cell walls change composition with nitrate supply (Dazzo *et al.*, 1981; Diaz *et al.*, 1981). For instance, nitrate supply increases the level of extensin, the hydroxyproline-rich glycoprotein in root cell walls (Dazzo *et al.*, 1981). Since rhizobia must penetrate the host cell wall, changes in the chemistry of the wall could have an important impact on the infection process. Detectable trifoliin A receptors on clover root cell walls are found to be reduced when the plant is grown with nitrate (Dazzo *et al.*, 1981). More studies are needed to determine how the accumulation of *Rhizobium*-binding lectins and their receptors on legume root surfaces is regulated by combined nitrogen.

The second mechanism of regulation, which involves the culture age-dependent changes in polysaccharide composition, is under bacterial control. In certain *R. japonicum* strains, galactose residues in the capsular polysaccharide become methylated at carbon-4 during stationary phase, resulting in a reduction in soybean lectin binding (Mort and Bauer, 1981). Soybean lectin binds certain strains of *R. japonicum* predominantly at one cell pole (Bohlool and Schmidt, 1976). This has been shown by transmission electron microscopy to reflect the polar location of the capsule (Bal *et al.*, 1978; Calvert *et al.*, 1979; Tsien and Schmidt, 1977, 1978; Truchet *et al.*, submitted for publication).

Shedding of capsular polysaccharide in early stationary phase (Truchet *et al.*, submitted for publication) explains why the culture as a whole retains soybean lectin binding activity (Tsien and Schmidt, 1979). Early models of soybean lectin-*R. japonicum* interactions were based on the inability of the lectin to agglutinate the bacteria despite binding of FITC-lectin (Broughton, 1976). This model has recently been refined with the finding that soybean lectin can agglutinate *R. japonicum* cells grown to a certain culture age on agar plates (Truchet *et al.*, 1983).

At a specific time in early stationary phase, both the capsular polysaccharide and the LPS of *R. trifolii* 0403 in broth culture exhibit binding to trifoliin A which is hapten reversible. The LPS undergoes profound immunochemical changes during normal batch culture growth (Hrabak *et al.*, 1981). Unique antigenic determinants of the LPS which bind trifoliin A are transiently exposed on the bacterial surface just as the culture leaves lag phase, and again as it enters stationary phase (Fig. 6). Cells in mid-exponential phase lack these LPS components and do not bind trifoliin A. Quantitative gas chromatography and combined GC-mass spectrometry showed several changes in the glycosyl composition of the LPS with culture age (Hrabak *et al.*, 1981). Quinovosamine (2-amino-2,6-dideoxyglucose) was one of the components of *R. trifolii* 0403 LPS which was found to increase as the culture aged. Quinovosamine in the β-anomeric configuration is a major

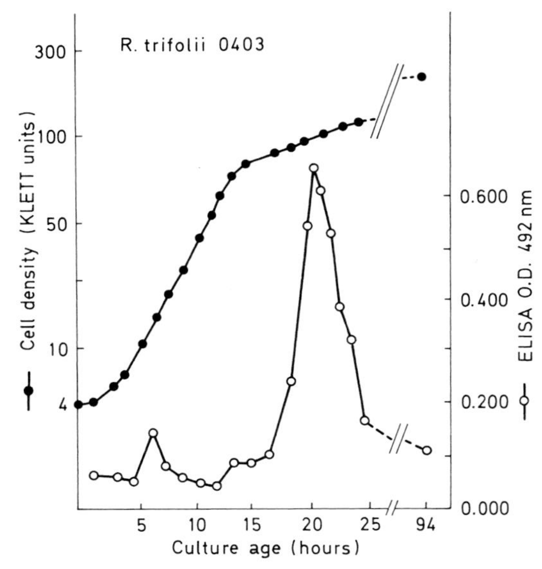

Fig. 6. Effect of broth culture age on appearance of unique determinants in the lipopolysaccharide of *R. trifolii* 0403 which bind trifoliin A. From Hrabak *et al.* (1981)

hapten of the unique LPS determinants at early stationary phase which bound trifoliin A (Hrabak *et al.*, 1981).

The potential importance of culture-age effects on lectin binding to the infection process was suggested by root hair infection studies using standardized inocula. White clover plants had more infected root hairs after incubation with an inoculum of cells in the early stationary phase than with cells in mid-exponential phase (Hrabak *et al.*, 1981). Similarly, changes in infectivity of *R. japonicum* for soybean nodulation were correlated with changes in the proportion of cells in a culture capable of binding soybean lectin (Bhuvaneswari *et al.*, 1983).

In plate culture, the development of the capsule on *R. trifolii* 0403 coincides with the appearance of trifoliin A receptors. These receptors are transient since the encapsulated cells partially lose their ability to bind the lectin as the culture ages (Dazzo *et al.*, 1979; Sherwood *et al.*, submitted). Transmission electron microscopy shows a dense, fibrillar capsule completely surrounding the cell in 5 day-old cultures (Fig. 7, Dazzo and Brill, 1979). Recent studies using a new method of specimen preparation (Mutaftschiev *et al.*, 1982) indicates that the capsule of *R. trifolii* 0403

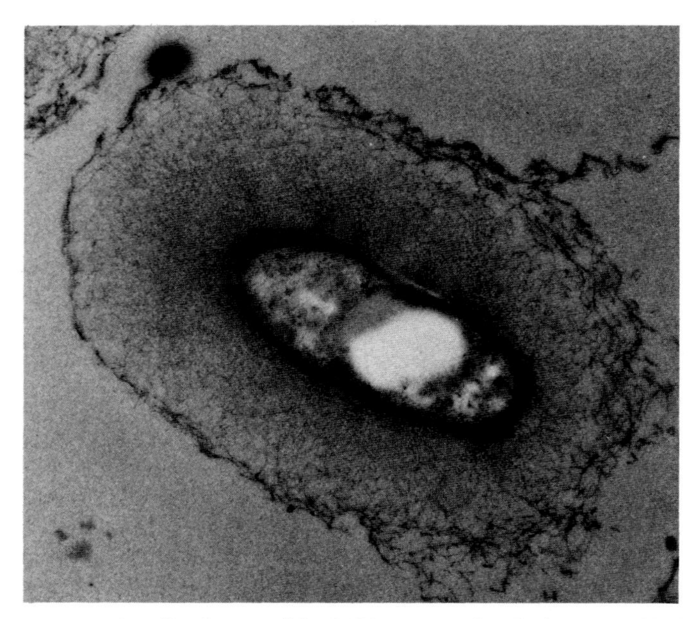

Fig. 7. Encapsulated cell of *R. trifolii* 0403 grown for 5 days on BIII plates and stained with ruthenium red. From Dazzo and Brill (1979)

develops first at one pole and then covers the whole cell (Sherwood *et al.*, submitted). The capsule consists of several polysaccharides, including both neutral glucans and acidic heteropolysaccharides (Dudman, in press). Use of bacteriophage-induced polysaccharide depolymerases has revealed chemical changes in the capsular polysaccharide of *R. trifolii* 0403 which occur with culture age (Abe *et al.*, in preparation). These growth-phase dependent chemical changes may reflect important mechanisms which regulate cellular recognition in the *Rhizobium*-legume symbiosis. They have already made an obvious impact on the validity of structural models of the repeating oligosaccharide units in *Rhizobium* polysaccharides, and represent a major reason for inconsistent results on studies of bacterial polysaccharide-lectin interactions.

A third level of regulation occurs when the bacterium encounters the root environment of the host. This is very important because it illustrates that an understanding of the biochemical basis of host specificity in the *Rhizobium*-legume symbiosis will require detailed studies of the microorganism in the rhizosphere of the host root as the normal case. For instance, some strains of *R. japonicum* bind soybean lectin in the root environment but not in pure culture (Bhuvaneswari and Bauer, 1978; Law *et al.*, 1982). This suggests that the host plays some role in expression of lectin-binding receptors on the rhizobial cell. Other strains of *R. japonicum* could bind soybean lectin better when grown in soil extract than in standard bacteriological media (Shantharam and Bal, 1981).

We have recently found another level of regulation in the rhizosphere of clover while studying the curious delay in polar attachment which occurs when fully encapsulated cells of *R. trifolii* are used as inoculum. While examining trifoliin A in clover root exudate of axenically grown plants (Dazzo and Hrabak, 1981), we also discovered an enzymatic activity in root exudate that alters and erodes the capsule in a way which favors polar attachment via polar-lectin receptors (Dazzo *et al.*, 1982a). Because it had been suggested that lectins are enzymes which degrade the rhizobial polysaccharides (Albersheim and Wolpert, 1977), we fractionated clover root exudate by immunoaffinity chromatography to purify trifoliin A and assayed fractions for this enzymatic activity (Dazzo *et al.*, 1982a). Reconstitution experiments showed that the enzymes responsible for this erosion of the capsule were antigenically unrelated to trifoliin A, and that trifoliin A did not fall into this special class of "enzymic lectins". Similarly, the N-acetylgalactosamine inhibitable soybean lectin (which specifically binds to the *R. japonicum* capsule) is not an enzymic lectin (Campello and Shannon, 1982).

The process of capsule erosion was documented by immunofluorescence and transmission electron microscopy (Dazzo *et al.*, 1982a). This alteration begins in an equatorial band arount the center of the bacterial cell and progresses towards both poles at unequal rates, resulting in an intermediate condition where cells have capsular fibrils only at one pole. If

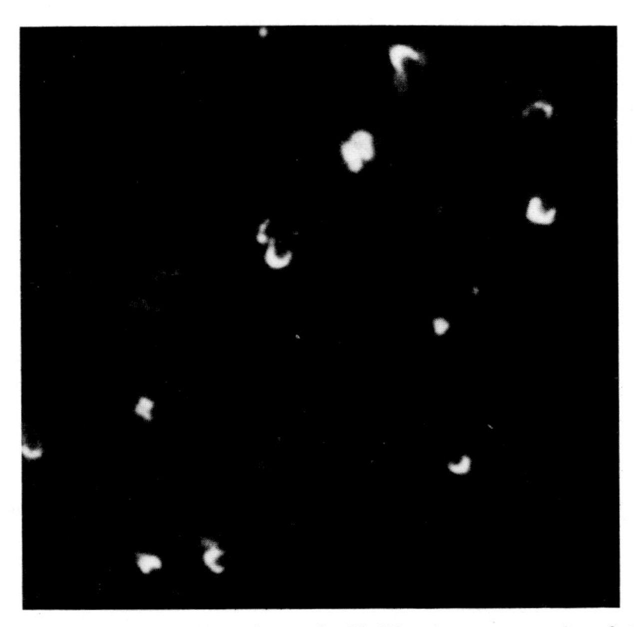

Fig. 8. Immunofluorescence detection of trifoliin A at one pole of *R. trifolii* 0403 cells growing in the root environment of white clover seedlings. From Dazzo *et al.* (1982)

the cells are heat-fixed to slides, all the capsule eventually erodes away. The kinetics of this enzymatic alteration are consistent with the time it takes for the transition of randomly oriented attachment to a predominantly polar attachment to clover root hairs. Interestingly, the equilibrium state of the nonadhering population of bacteria in the rhizosphere has trifoliin A bound *in situ* at one cell pole (Fig. 8). From these data, we have proposed a dynamic, multi-phase model which is fully consistent with the timing and orientation of attachment of *R. trifolii* to clover root hairs (Dazzo *et al.,* in preparation). At high inoculum of fully encapsulated, lectin-binding cells, the bacteria attach within minutes to root hair tips where trifoliin A accumulates in greatest quantity (Phase 1A). During the next 4—8 hr, the capsule is altered by host enzymes in equilibrium with *de novo* bacterial polysaccharide synthesis, resulting in a cell with predominantly polar fibrils that bind trifoliin A (Phase 1B transition). We speculate that these cells with trifoliin A-binding receptors at one pole later polarly attach along the sides of the root hair, either to trifoliin A already bound to the root hair, or by binding to lectin in root exudate and then to lectin receptors on root hairs (Phase 1C). Thus, polar attachment seems to be a manifestation of the location of the lectin receptor on the bacterium (Bohlool and Schmidt, 1976; Truchet *et al.,* in preparation).

Root exudate has been shown to affect the infection of legume roots by rhizobia. For instance, pretreatment of rhizobia with root exudate of clover (Nutman, 1957; Solheim, 1975; Napoli, 1976) or cowpea (Bhagwat and Thomas, 1982) significantly increases the rate of subsequent infection. Further studies are needed to identify the components in root exudate which are responsible for this phenomenon.

The presence of trifoliin A in clover root exudate which can bind to receptors on *R. trifolii* provides supporting evidence for a lectin recognition model proposed by Solheim (1975). According to this model, a glycoprotein lectin excreted from the legume root binds to the rhizobia. This active complex then combines with a receptor site on the root. Thus, both partners in the symbiosis could benefit from the discriminatory reaction of a cross-bridging lectin which could be either bound to a glycosylated receptor on the root hair cell wall (Dazzo and Hubbell, 1975b) or released from the root to bind to the rhizobial cell (Solheim, 1975). This event would help to ensure that only the symbiotic bacterium could establish the proper intimate contact with the host cell required to trigger other recognition events that lead to successful infection.

VIII. Rhizobial Attachment Is Only a Piece of the Puzzle in the Infection Process

Although attachment of infective rhizobia to target root hairs is one step of the infection process, several observations indicate that other undefined events must also occur to initiate root hair infection. First of all, very few root hairs to which infective rhizobia attach eventually become infected.

Thus, successful infection is a relatively rare event. This may be due to a transient susceptibility of the root hairs to infection by the rhizobial symbiont (Bhuvaneswari *et al.*, 1980; 1981). Under controlled conditions in Fahraeus slide cultures, the number of root hairs infected by a strain of clover rhizobia will be characteristic of that strain, and may be different for other strains. Secondly, genetic hybrids of *Azotobacter vinelandii* which carry the trifoliin A-binding saccharide receptor on their surface as a result of intergeneric transformation with DNA from *R. trifolii* (Bishop *et al.*, 1977), have acquired the ability to adhere specifically to clover root hairs (Dazzo and Brill, 1979) but do not infect them. Finally, although certain mutant strains which fail to bind the host lectin neither attach well to the host root hairs nor infect them (Dazzo *et al.*, 1976; Paau *et al.*, 1981; Stacey *et al.*, 1982), another class of non-infective mutant strains has been shown to bind the host lectin and attach to the host root hairs (Kamberger, 1979 b, Paau *et al.*, 1981). Each of these cases serves to illustrate the importance of lectin-mediated root hair attachment to the infection process, but makes it clear that other post-attachment events of cell recognition must occur to advance the infection process to the stage of root hair penetration.

Possible genes or gene products which may have not been expressed in the above situations include those controlling cell-wall hydrolytic enzymes (Hubbell *et al.*, 1978; Martinez-Molina *et al.*, 1979), inducers of host polygalacturonase (Ljunggren and Fahraeus, 1961; Palomares *et al.*, 1978), root hair deforming factors (Yao and Vincent, 1976; Solheim and Bhuvaneswari, in preparation), and periplasmic extrinsic substance ES-6000 which promotes root hair infection (Higashi and Abe, 1980 a) and has recently been identified as cyclic β-1,2-glucan (Abe and Higashi, 1982).

The presence of multiple lectin receptors on rhizobia raises the question of whether each one has a different role in root hair infection. Infection studies by Kamberger (1979 b) showed that mutants of *R. leguminosarum* with altered O-antigens could attach to pea root hairs but could not infect them. Their capsular polysaccharide still was able to bind pea lectin on the root hair. He suggested that the lectin cross-bridging hypothesis (Dazzo and Hubbell, 1975 b) needs to be modified. For example, the capsular polysaccharides could be responsible for attachment of high numbers of rhizobial cells to the target root hairs via cross-bridging lectins as an early recognition event. This would be followed by secondary recognition events requiring the host range-specific binding of lectin on localized sites of the root hair to LPS, which triggers subsequent invasive steps (Kamberger, 1979 b).

This hypothesis predicts a role of lectin-LPS interactions as an important post-attachment event for infection of legume root hairs by fast-growing rhizobia. Recent studies to test this hypothesis have shown that nanomolar concentrations of lectin-binding LPS from *R. trifolii* significantly enhances root hair infection of clover (Dazzo *et al.*, in preparation). This finding provides additional evidence that lectin-polysaccharide interactions contribute to successful infection.

Other analyses of symbiotically defective mutant strains of *Rhizobium* suggest that surface or extracellular components play a role in nodule formation. Mutant strains of *R. japonicum* 61A76 with an altered O-antigen (Maier and Brill, 1978) can attach to soybean root hairs but nodulate at a significantly lower rate (Stacey *et al.*, 1982). Other non-nodulating mutants of *R. japonicum* do not attach to soybean root hairs (Stacey *et al.*, 1982). Positive selection for non-nodulating mutants of *R. trifolii* (Barnet, 1980), *R. japonicum* (Stacey *et al.*, 1982), and *R. phaseoli* (Raleigh and Signer, 1982) has been accomplished by selection for resistance to suitable bacteriophage. The non-nodulating phage-resistant mutants of *R. japonicum* have an altered LPS antigen, and lack 128,000 and 37,500 dalton membrane proteins found in the wild type strain (Stacey *et al.*, 1982). The non-nodulating phage-resistant mutants of *R. phaseoli* do not produce detectable exopolysaccharide and lack the internal phosphoglucose isomerase activity (Raleigh and Signer, 1982) necessary for utilization of carbon sources via fructose-6-phosphate (Arias *et al.*, 1979). *R. japonicum* mutants selected for reduced capsule production in laboratory media attach 2 to 4-fold lower than does the encapsulated wild type strain to soybean root hairs (Law *et al.*, 1982). However, the lack of capsules on these mutants has not been confirmed by electron microscopy. These mutants nodulated soybean roots grown in cellophane pouches, albeit at a reduced level. Nodulation by several of the mutants was linearly proportional to the amount of acidic exopolysaccharide that they released into the culture medium during exponential growth phase, indicating that such polysaccharide synthesis is somehow important for nodulation. Two of the mutants appeared to synthesize normal lectin-binding capsules when cultured in association with host roots, but not when cultured in vitro. Nodulation by these mutants appeared to depend on how rapidly after inoculation they synthesized capsular polysaccharide. Selection for reduced capsule and exopolysaccharide production in *R. leguminosarum* has resulted in an "EXO–1" mutant strain which neither attaches to pea root hairs (C. Napoli, personal communication) nor nodulates pea roots (Saunders *et al.*, 1978; Napoli and Albersheim, 1980).

Analysis of symbiotically defective mutant strains of rhizobia is complicated by multiple pleiotropic effects when the mutated genes affect production of polysaccharides. For instance, Saunders *et al.* (1978) reported that 27% of the total LPS mass from *R. leguminosarum* mutant strain EXO–1 is anthrone-reactive carbohydrate, as compared to 63% of the total LPS mass from the wild type *R. leguminosarum* parent. The glycosyl and antigenic compositions of the O-antigen of the mutant and wild type strains do not seem to be different (Saunders *et al.*, 1979; Carlson and Lee, 1983), and both strains are lysed by the same collection of bacteriophages. The EXO-1 mutant does not seem to produce any of the typical acidic parental extracellular polysaccharides of the wild type, and its major excreted polysaccharides are an intact LPS and a fragment represented by the repeating O-antigen without R-core/lipid A which is not produced by the wild type strain (Carlson and Lee, 1983). Thus, there could be several defects in the synthesis, mobilization, and assembly of several outer membrane components of the mutant. For instance, a single mutation leading to a defective O-antigen polymerase would fail to polymerize in a block fashion the repeating O-antigen oligosaccharide on the polyisoprenoid acyl carrier lipid (Osborne *et al.*, 1972). When the O-antigen oligosaccharide is transferred to the R-core/lipid A acceptor via the translocase reaction, an LPS with reduced O-antigen polymerization would result. A mutation leading to a defective translocase, which regulates the polymerase reaction, would lead to the excreted O-antigen fragment without being attached to the R-core/lipid A acceptor. Again, an end result would be a shorter LPS on the outer membrane. All of these mutations which reduce the chain length of the O-antigen on intact LPS cause pleiotropic negative effects on the biogenesis and assembly of the outer membrane of the Gram negative cell. The reduction in chain length of this carbohydrate moiety of the LPS can cause a concomitant decrease and sometimes virtual loss of outer membrane proteins (Gmeiner and Schlecht, 1979). These pleiotropic negative effects complicate any direct interpretation of the significance of the symbiotically defective mutant strains which have less polysaccharide production to the infection process.

IX. Nodulation and Host Specificity Genes Are Plasmid Encoded

Nuti *et al.* (1982) have recently reviewed the subject of *Rhizobium* plasmids and symbiotic nitrogen fixation. Developments relevant to the topic of this chapter are summarized as follows (see also chapter 4, Rolfe and Shine).

(i) Genetic evidence for the role of extrachromosomal elements in determining host specificity in the *Rhizobium*-legume symbiosis has been provided (Higashi, 1967).

(ii) Non-nodulating mutants of *R. trifolii* have been isolated by using thermal shock to eliminate a large indigenous plasmid present in nodulating *R. trifolii* strains (Zurkowski and Lorkiewicz, 1979).

(iii) Host specificity traits of *R. leguminosarum* have been transferred into bean and clover by conjugation of a genetically marked plasmid conferring pea nodulation (Johnston *et al.*, 1980). Host range and nodulation genes have been recently located within a 10 kb region of the Sym plasmid of *R. leguminosarum* (Downie *et al.*, 1983).

(iv) It has been demonstrated (Zurkowski, 1980) that roa genes (for root hair attachment, Vincent, 1980, see also chapter 4, Rolfe and Shine) for the 2-deoxyglucose sensitive attachment of *R. trifolii* to clover root hairs are encoded on the nodulation plasmid of *R. trifolii.* Nod genes complementing a Nod$^-$ *R. trifolii* mutant strain have also been cloned (Lorkiewicz *et al.* in preparation) and the genetic determinants for nodule induction and development as well as for clover host specificity have been located within a 14 kb fragment of the Sym plasmid of *R. trifolii* (Schofield *et al.*, in preparation).

(v) Clover nodulation ability from *R. trifolii* has been transferred into a pTi-cured *Agrobacterium tumefaciens* by conjugation of a genetically marked Sym plasmid conferring clover nodulation (Hooykaas *et al.*, 1981).

Microscopic studies showed that moderate root hair curling was also transferred by conjugation of the *R. trifolii* Sym plasmid.

(vi) The "megaplasmid" of *R. meliloti* carries genetic information responsible for marked root hair curling and nodulation of alfalfa (Rosenberg *et al.*, 1981; Banfalvi *et al.*, 1981). Alfalfa specific nodulation ability has been transferred into *Agrobacterium tumefaciens* by conjugation of the Sym plasmid from *R. meliloti* (Truchet *et al.*, submitted). Nod gene(s) which complemented Nod$^-$, Hac$^-$ *R. meliloti* mutant strains have been cloned and located within 30 kb of the nif loci (Long *et al.*, 1982). This region also shows limited homology with slow growing *Rhizobium japonicum* and sequences able to confer root hair curling have been isolated from two strains (Sutton *et al.*, 1984).

(vii) The clover Sym plasmid of *R. trifolii* controls incorporation of the trifoliin A-binding aminodideoxysugar quinovosamine into its LPS (Russa *et al.*, 1982). Alterations in the surface polysaccharides of other symbiotic mutant strains of *R. trifolii* have also been reported. The LPS from a Nod$^-$, plasmid-less mutant strain was altered and no parental CPS was detected in a transposon generated Nod$^-$, Hac$^-$ mutant strain of *R. trifolii* (Carlson *et al.*, in preparation).

(viii) The conservation of root hair curling functions in a broad spectrum of *Rhizobium* strains has been reported. Heterologous information for clover nodulation (pBRIAN, pRt032), for pea nodulation (pJB5JI), for lucerne nodulation (pRmSL26), or for the nodulation of legumes and non-legumes (pNM4AN) was transferred to Nod$^-$, Hac$^-$ mutant strains of *R. trifolii, R. meliloti* and a *Rhizobium* strain that nodulates cowpea legumes and the non-legume *Parasponium*. Root hair curling capacity was restored and in most cases, nodulation capacity of the original plant host(s) (Rolfe *et al.*, in preparation).

(ix) The genetic information required for trifoliin A-binding, Phase 1 attachment, marked root hair curling (shepherd's crooks), root hair penetration, and infection thread formation is encoded on the Sym plasmid of *R. trifolii* 5035 which can be transferred and expressed in pTi-cured. *A. tumefaciens* (Dazzo *et al.*, in preparation). This opens a new approach for examining events involved in recognition at the root hair level which are encoded on the (Sym)biotic nodulation plasmid.

X. Concluding Remarks

The above discussion supports the validity of the lection recognition hypothesis as originally proposed by Bohlool and Schmidt (1974). However, the most important work which has countered this hypothesis is the finding that soybean lines which lack seed lectin nodulate normally with *R. japonicum* (Pull *et al.*, 1979). This latter discrepancy may be resolved by the recent detection of small quantities of lectin in seeds (Tsien *et al.*, 1983) and of lectin mRNA and lectin in seedling roots of these same soybean lines (Goldberg *et al.*, 1983; Sengupta-Gopalan *et al.*, in preparation). In addition, soybean seeds (including SBL-minus lines) contain a newly discovered 4-O-methyl-glucuronic acid specific lectin which specifically binds to the *R. japonicum* strains which were incapable of binding SBL (Dombrink-Kurtzman, *et al.* 1983) since they contained no galactose-like residues in their capsular polysaccharides. This study is also significant since 4-O-methylglucuronic acid residues are found in the xyloglucan polymers of plant cell walls. The reader is referred to recent reviews (Bauer, 1980; Schmidt, 1979; Dazzo, 1981; Dazzo and Truchet, 1983; Dudman, in press; Quispel *et al.*, in press) for lengthy discussions on other pros and cons of the lectin recognition hypothesis.

In summary, some lectins fulfill many criteria that are expected of a cell recognition molecule involved in the nitrogen-fixing *Rhizobium*-legume symbiosis. However, the research accomplishments of the last decade which have focused on testing the lectin-recognition hypothesis are only beginning to answer questions on the infection process. It is clear that lectin-mediated bacterial attachment is only one event necessary for successful infection, since root hair infections can also be genetically blocked at other steps of the process. The lectin-recognition hypothesis will never be universally accepted or rejected until more bacterial and plant mutants are available to determine the role of the *Rhizobium*-binding lectin and its saccharide receptors in the recognition code without affecting other events in this infection process.

Acknowledgments

Work in the author's laboratory was supported by NSF Grant 80-21906 and USDA-CRGO Grant 82-CRCR-1-1040. This is an issue of the Michigan Agricultural Experiment Station Journal.

XI. References

Abe, M., Higashi, S., 1982: Studies on cyclic β-1,2-glucan obtained from periplasmic space of *Rhizobium trifolii.* Plant Soil **64,** 315—324.

Albersheim, P., Wolpert, J., 1977: Molecular determinants of symbiont-host selectivity between nitrogen fixing bacteria and plants. In Solheim, B., Raa, J. (eds.) Cell Wall Biochemistry Related to Specificity in Host-Plant Pathogen Interactions, pp. 373—376, Oslo: Universitetsforlaget.

Arias, A., Cervenansky, C., Gardiol, A., Martinez-Drets, G., 1979: Phosphoglucose isomerase mutant of *Rhizobium meliloti.* J. Bacteriol. **137,** 409—414.

Bafalvi, Z., Sakanyan, V., Koncz, C., Kiss, A., Dusha, I., Kondorosi, A., 1981: Location of nodulation and nitrogen fixing genes on a high molecular weight plasmid of *R. meliloti.* Mol. Gen. Genet. **184,** 318—325.

Bal, A. K., Shantharam, S., Ratnam, S., 1978: Ultrastructure of *Rhizobium japonicum* in relation to its attachment to root hairs. J. Bacteriol. **133,** 1393—1400.

Barnet, Y., 1979: Properties of *Rhizobium trifolii* isolates surviving exposure to specific bacteriophage. Can. J. Microbiol. **25,** 979—986.

Bauer, W. D., 1981: Infection of legumes by rhizobia. Ann. Rev. Plant Physiol. **32,** 407—449.

Bhagwat, A., Thomas, J., 1980: Dual binding sites for peanut lectin on rhizobia. J. Gen. Microbiol. **117,** 119—125.

Bhagwat, A., Thomas, J., 1982: Legume-*Rhizobium* interactions: cowpea root exudate elicits faster nodulation response by *Rhizobium* species. Appl. Environ. Microbiol. **43,** 800—805.

Bhuvaneswari, T. V., Bauer, W. D., 1978: The role of lectins in plant-microorganism interactions. III. Influence of rhizosphere/rhizoplane culture conditions on the soybean lectin-binding properties of rhizobia. Plant Physiol. **62,** 71—74.

Bhuvaneswari, T. V., Bauer, W. D., 1983: Effect of culture age on root nodulation by *Rhizobium japonicum.* J. Bacteriol. **153,** 443—451.

Bhuvaneswari, T. V., Bhagwat, A. A., Bauer, W. D., 1981: Transient susceptibility of root cells in four common legumes to nodulation by rhizobia. Plant Physiol. **68,** 1144—1149.

Bhuvaneswari, T. V., Pueppke, S. G., Bauer, W. D., 1977: Role of lectins in plant-microorganism interactions. I. Binding of soybean lectin to rhizobia. Plant Physiol. **60,** 486—491.

Bhuvaneswari, T. V., Turgeon, B. G., Bauer, W. D., 1980: Early events in the infection of soybean (*Glycine max* L. Merr.) by *Rhizobium japonicum.* I. Location of infectible root cells. Plant Physiol. **66,** 1027—1031.

Bishop, P. E., Dazzo, F. B., Applebaum, E. R., Maier, R. J., Brill, W. J., 1977: Intergeneric transformation of genes involved in the *Rhizobium*-legume symbiosis. Science **198,** 938—939.

Bohlool, B. B., Schmidt, E. L., 1974: Lectins: a possible basis for specificity in the *Rhizobium*-legume root nodule symbiosis. Science **185,** 269—271.

Bohlool, B. B., Schmidt, E. L., 1976: Immunofluorescent polar tips of *Rhizobium japonicum:* possible site of attachment or lectin binding. J. Bacteriol. **125,** 1188—1194.

Broughton, W. J., 1978: A review: control of specificity in legume-*Rhizobium* associations. J. Appl. Bacteriol. **45,** 165—194.

Callaham, D. A., Torrey, J. G., 1981: The structural basis for infection of root hairs of *Trifolium repens* by *Rhizobium.* Can. J. Bot. **59,** 1647—1664.

Calvert, H. E., Lalonde, M., Bhuvaneswari, T. V., Bauer W. D., 1978; Role of lectins in plant-microorganism interactions. IV. Ultrastructure localisation of soybean lectin binding sites of *Rhizobium japonicum.* Can. J. Microbiol. **24,** 785—793.

Campello, E. D., Shannon, L. M., 1982: An α-galactosidase with hemagglutinin properties from soybean seeds. Plant Physiol. **69,** 628—632.

Carlson, R. W., Saunders, R. E., Napoli, C. A., Albersheim, P., 1978: Host-symbiont interactions III. Isolation and partial characterization of lipopolysaccharides from *Rhizobium.* Plant Physiol. **62,** 912—917.

Carlson, R. W., Lee R. P., 1983: A comparison of the surface polysaccharides from *Rhizobium leguminosarum* 128C53 sm^r rif^r with the surface polysaccharides from its Exo-1 mutant. Plant Physiol. **71,** 223—228.

Chen, A. P., Phillips, D. A., 1976: Attachment of *Rhizobium* to legume roots as the basis for specific associations. Physiol. Plant. **38,** 83—88.

Dazzo, F. B., 1980: Adsorption of microorganisms to roots and other plant surfaces. In: Bitton, G., Marshall, K. (eds.) Adsorption of Microorganisms to Surfaces, pp. 253—316. New York: Wiley.

Dazzo, F. B., 1981: Bacterial attachment as related to cellular recognition in the *Rhizobium*-legume symbiosis. J. Supramol. Struct. Cell. Biochem. **16,** 29—41.

Dazzo, F. B., Brill, W. J., 1977: Receptor site on clover and alfalfa roots for *Rhizobium.* Appl. Environ. Microbiol. **33,** 132—136.

Dazzo, F. B., Brill, W. J., 1978: Regulation by fixed nitrogen of host-symbiont recognition in the *Rhizobium*-clover symbiosis. Plant Physiol. **62,** 18—21.

Dazzo, F. B., Brill, W. J., 1979: Bacterial polysaccharide which binds *Rhizobium trifolii* to clover root hairs. J. Bacteriol. **137,** 1362—1373.

Dazzo, F. B., Hrabak, E. M., 1981: Presence of trifoliin A, a *Rhizobium*-binding lectin, in clover root exudate. J. Supramol. Struct. Cell. Biochem. **16,** 133—138.

Dazzo, F. B., Hrabak, E. M., 1982: Lack of a direct interaction between trifoliin A and nitrate as related to the *Rhizobium trifolii*-clover symbiosis. Plant Soil **69,** 259—264.

Dazzo, F. B., Hrabak, E. M., Urbano, M. R., Sherwood, J. E., Truchet, G. L., 1981: Regulation of recognition in the *Rhizobium*-clover symbiosis. In: Gibson, A. H., Newton, W. E. (eds.), Current Perspectives in Nitrogen Fixation pp. 292—295. Australian Academy of Science, Canberra.

Dazzo, F. B., Hubbell, D. H., 1975a: Cross-reactive antigens and lectin as determinants of symbiotic specificity in the *Rhizobium*-clover association. Appl. Microbiol. **30,** 1017—1033.

Dazzo, F. B., Hubbell, D. H., 1975b: Concanavalin A: lack of correlation between binding to *Rhizobium* and host specificity in the *Rhizobium*-legume symbiosis. Plant Soil **43,** 713—717.

Dazzo, F. B., Napoli, C. A., Hubbell, D. H., 1976: Adsorption of bacteria to roots as related to host specificity in the *Rhizobium*-clover association. Appl. Environ. Microbiol. **32,** 168—171.

Dazzo, F. B., Truchet, G. L., 1983: Interactions of lectins and their saccharide receptors in the *Rhizobium*-legume symbiosis. J. Membrane Biol. **73,** 1—16.

Dazzo, F. B., Truchet, G. L., Kijne, J. W., 1982: Lectin involvement in root hair tip adhesions as related to the *Rhizobium*-clover symbiosis. Physiol. Plant. **56,** 143—147.

Dazzo, F. B., Truchet, G. L., Sherwood, J. E., Hrabak, E. M., Gardiol, A. E., 1982: Alteration of the trifoliin A-binding capsule of *Rhizobium trifolii* 0403 by enzymes released from clover roots. Appl. Environ. Microbiol. **44,** 478—490.

Dazzo, F. B., Urbano, M. R., Brill, W. J., 1979: Transient appearance of lectin receptors on *Rhizobium trifolii.* Curr. Microbiol. **2,** 15—20.

Dazzo, F. B., Yanke, W. E., Brill, W. J., 1978: Trifoliin: a *Rhizobium* recognition protein from white clover. Biochim. Biophys. Acta **536,** 276—286.

Deinema, M., Zevenhuizen, L. P. T., 1971: Formation of cellulose fibrils by gram negative bacteria and their role in bacterial flocculation. Arch. Mikrobiol. **78,** 42—57.

Diaz, C., Kijne, J. W., Quispel, A., 1981: Influence of nitrate on pea root cell wall composition. In: Gibson, A. H., Newton, W. E. (eds.) current Perspectives in Nitrogen Fixation, p. 426. Australian Academy of Science, Canberra.

Dombrink-Kurtzman, M. A., Dick, W. E., Burton, K. A., Cadmus, M. C., Slodki, M. E., 1983: A soybean lectin having 4-O-methylglucuronic acid specificity. Biochem. Biophys. Res. Commun. **111,** 798—803.

Downie, J. A., Hombrecher, G., Ma, Q. S., Knight, C. D., Wells, B., Johnston, A. W. B., 1983: Cloned nodulation genes of *Rhizobium leguminosarum* determine host-range specificity. Mol. Gen. Genet. **190,** 359—365.

Fahraeus, G., 1957: The infection of clover root hairs by nodule bacteria studied by a simple glass slide technique. J. Gen. Microbiol. **16,** 374—381.

Gade, W., Jack, M. A., Dahl, J. B., Schmidt, E. L., Wolf, F., 1981: The isolation and characterization of a root lectin from soybean (*Glycine max* L.) cultivar Chippewa. J. Biol. Chem. **256,** 12905—12910.

Gatehouse, J. A., Boulter, D., 1980: Isolation and properties of a lectin from the roots of *Pisum sativum*. Physiol. Plant. **49,** 437—442.

Gmeiner, J., Schlecht, S., 1979: Molecular organization of the outer membrane of *Salmonella typhimurium*. Eur. J. Biochem. **93,** 609—620.

Hamblin, J., Kent, S. P., 1973: Possible role of phytohaemagglutinin in *Phaseolus vulgaris* L. Nature New Biol. **245,** 28—29.

Higashi, A., 1967: Transfer of clover infectivity of *Rhizobium trifolii* to *Rhizobium phaseoli* as mediated by an episomic factor. J. Gen. Appl. Microbiol. **13,** 391—403.

Higashi, S., Abe, M., 1980a: Promotion of infection thread formation by substances from *Rhizobium*. Appl. Environ. Microbiol. **39,** 297—301.

Higashi, S., Abe, M., 1980b: Scanning electron microscopy of *Rhizobium trifolii* infection sites on root hairs of white clover. Appl. Environ. Microbiol. **40,** 1094—1099.

Hooykaas, P. J., van Brussell, A. A. N., den Hulk-Ras, H., van Slogteren, G. M., Schilperoort, R. A., 1981: Sym plasmid of *Rhizobium trifolii* expressed in different rhizobia species and *Agrobacterium tumefaciens*. Nature **291,** 351—353.

Hrabak, E. M., Urbano, M. R., Dazzo, F. B., 1981: Growth-phase dependent immunodeterminants of *Rhizobium trifolii* lipopolysaccharide which bind trifoliin A, a white clover lectin, J. Bacteriol. **148,** 697—711.

Hubbell, D. H., Morales, V. M., and Umali-Garcia, M., 1978: Pectolytic enzymes in *Rhizobium*. Appl. Environ. Microbiol. **35,** 210—213.

Hughes, T. A., Elkan, G. H., 1981: Study of the *Rhizobium japonicum*-soybean symbiosis. Plant Soil **61,** 87—91.

Johnston, A. W. B., Beyon, I. L., Buchanan-Wollaston, A. V., Setchell, S. M., Hirsch, P. R., Beringer, J. E., 1978: High frequency transfer of nodulating ability between species and strains of *Rhizobium*. Nature **276,** 635—638.

Kamberger, W., 1979a: An Ouchterlony double diffusion study on the interaction between legume lectins and rhizobial cell surface antigens. Arch. Microbiol. **121,** 83—90.

Kamberger, W., 1979b: Role of cell surface polysaccharides in the *Rhizobium*-pea symbiosis. FEMS Microbiol. Lett. **6,** 361—365.

Kato, G., Maruyama, Y., Nakamura, M., 1979: Role of lectins and lipopolysaccharides in the recognition process of specific legume-*Rhizobium* symbiosis. Agric. Biol. Chem. **43,** 1085—1092.

Kato, G., Maruyama, Y., Nakamura, M., 1980: Role of bacterial polysaccharides in the adsorption process of the *Rhizobium*-pea symbiosis. Agric. Biol. Chem. **44,** 2843—2855.

Kato, G., Maruyama, Y., Nakamura, M., 1981: Involvement of lectins in *Rhizobium*-pea recognition. Plant Cell. Physiol. **22,** 759—771.

Kijne, J. W., van der Schaal, I. A. M., de Vries, G. E., 1980: Pea lectins and the recognition of *Rhizobium leguminosarum*. Plant Sci. Lett. **18,** 65—74.

Kijne, J. W., van der Schaal, I. A. M., Diaz, C. L., van Iren, F., 1982: Mannose-specific lectins and the recognition of pea roots by *Rhizobium leguminosarum*. In: Bog-Hansen, T. C., Spengler, G. A. (eds.) Lectins: Biology, Biochemistry, Clinical Biochemistry. Vol. III. pp. 521—529. Berlin: W. de Gruyter.

Kumarasinghe, M. K., Nutman, P. S., 1977: *Rhizobium*-stimulated callose formation in clover root hairs and its relation to infection. J. Exp. Bot. **28,** 961—976.

Law, I. J., Strijdom, B. W., 1977: Some observations on plant lectins and *Rhizobium* specificity. Soil Biol. Biochem. **9,** 79—84.

Law, I. J., Yamamoto, Y., Mort, A. J., Bauer, W. D., 1982: Nodulation of soybean by *Rhizobium japonicum* mutants with altered capsule synthesis. Planta **154,** 100—109.

Li, D., Hubbell, D. H., 1969: Infection thread formation as a basis for host specificity in *Rhizobium trifolii* – *Trifolium fragiferum* associations. Can. J. Microbiol. **15,** 1133—1138.

Ljunggren, H., Fahraeus, G., 1961: Role of polygalacturonase in root hair invasion by nodule bacteria. J. Gen. Microbiol. **26,** 521—528.

Long S. R., Buikema, W. J., Ausubel, F. M., 1982: Cloning of *Rhizobium meliloti* nodulation genes by direct complementation of Nod^- mutants. Nature **298,** 485—488.

Maier, R. J., Brill, W. J., 1978: Involvement of *Rhizobium japonicum* O-antigen in soybean nodulation. J. Bacteriol. **133,** 1295—1299.

Martinez-Molina, E., Morales, V. M., Hubbell, D. H., 1979: Hydrolytic enzyme production by *Rhizobium.* Appl. Environ. Microbiol. **38,** 1186—1188.

Mellor, R. B., Gadd, G. M., Rowell, P., Stewart, W. D., 1981: A phytohemagglutinin from the *Azolla-Anabaena* symbiosis. Biochem. Biophys. Res. Commun. **99,** 1348—1353.

Mort, A. J., Bauer, W. D., 1980: Composition of the capsular and extracellular polysaccharides of *Rhizobium japonicum:* changes with culture age and correlations with binding of soybean seed lectin to the bactria. Plant Physiol. **66,** 158—163.

Mort, A. J., Bauer, W. D., 1982: Structure of the capsular and extracellular polysaccharides of *Rhizobium japonicum* that bind soybean lectin. Application of two new methods for cleavage of polysaccharides into specific oligosaccharide fragments. J. Biol. Chem. **257,** 1870—1875.

Mutaftschiev, S., Vasse, J., Truchet, G., 1982: Exostructures of *Rhizobium meliloti.* FEMS Microbiol. Lett. **13,** 171—175.

Napoli, C. A., 1976: Ultrastructural and physiological aspects of the infection of clover *(Trifolium fragiferum)* by *Rhizobium trifolii* NA-30. Ph. D. Dissertation, University of Florida, Gainesville.

Napoli, C. A., Albersheim. P., 1980: *Rhizobium leguminosarum* mutants incapable of normal extracellular polysaccharide production. J. Bacteriol. **141,** 1454—1456.

Napoli, C. A., Dazzo, F. B., Hubbell, D. H., 1975: Production of cellulose microfibrils by *Rhizobium.* Appl. Microbiol. **30,** 123—131.

Napoli, C. A., Hubbell, D. H., 1975: Ultrastructure of *Rhizobium*-induced infection threads in clover root hairs. Appl. Microbiol. **30,** 1003—1009.

Nuti, M. P., Lepidi, A. A., Prakash, R. K., Hooykaas, P. J., and Schilperoort, R. A., 1982: The plasmids of *Rhizobium* and symbiotic nitrogen fixation. In: Schell, J. (ed.), Molecular Biology of Plant Tumors, pp. 561—588. New York: Academic Press, Inc.

Nutman, P. S., 1957: Studies on the physiology of nodule formation. V. Further experiments on the stimulation and inhibitory effects of root exudate. Ann. Bot. (London) **21,** 321—327.

Osborne, M. J., Cynkin, M. A., Gilbert, J. M., Mueller, L., Singh, M., 1972: Biosynthesis of lipopolysaccharide. Meth. Enzymol. **28,** 583—601.

Paau, A. S., Leps, W. T., Brill, W. J., 1981: Agglutinin from alfalfa necessary for binding and nodulation by *Rhizobium meliloti.* Science **213,** 1513—1515.

Palomares, A., Montega, E., Olivares, J., 1978: Induction of polygalacturonase production in legume roots as a consequence of extrachromosomal DNA carried by *Rhizobium meliloti.* Microbios **21,** 33—39.

Pueppke, S. G., Freund, T. G., Schulz, B. C., Friedman, H. P., 1980: Interaction of lectins from soybean and peanut with rhizobia that nodulate soybean, peanut, or both plants. Can. J. Microbiol. **26,** 1489—1497.

Pull, S. P., Pueppke, S. G., Hymowitz, T., Orf, J. H., 1978: Soybean lines lacking the 120,000 dalton seed lectin. Science **200,** 1277—1279.

Raleigh, E. A., Signer, E. R., 1982: Positive selection of nodulation-deficient *Rhizobium phaseoli.* J. Bacteriol. **151,** 83—88.

Reeke, J. N., Becker, J. W., Cunningham, B. A., Wang, J. C., Yahara, I., Edelman, G. M., 1975: Structure and function of concanavalin A. Adv. Expt. Med. Biol. **55,** 13—33.

Rosenberg, G., Boistard, P., Denarie, J., Casse-Delbart, F., 1981: Genes controlling early and late functions in symbiosis are located on a megaplasmid in *Rhizobium meliloti.* Mol. Gen. Genet. **184,** 326—333.

Russa, R., Urbanik, T., Kowalczuk, E., Lorkiewicz, Z., 1982: Correlation between occurrence of plasmid pUCS202 and lipopolysaccharide alterations in *Rhizobium.* FEMS Microbiol. Lett. **13,** 161—165.

Saunders, R. E., Carlson, R. W., Albersheim, P., 1978: A *Rhizobium* mutant incapable of nodulation and normal polysaccharide secretion. Nature (London) **271,** 240—242.

Schmidt, E. L., 1979: Initiation of plant root-microbe interactions. Ann. Rev. Microbiol. **33,** 335—376.

Shantharam, S., Bal, A. K., 1981: The effect of growth medium on lectin binding by *Rhizobium japonicum.* Plant Soil **62,** 327—330.

Solheim B., 1975: Possible role of lectin in the infection of legumes by *Rhizobium trifolii* and a model of the recognition reaction between *Rhizobium trifolii* and *Trifolium repens.* NATO Advanced Study Institute Symposium on Specificity in Plant Diseases, Sardinia.

Stacey, G., Paau, A. S., Brill, W. J., 1980: Host recognition in the *Rhizobium*-soybean symbiosis. Plant Physiol. **66,** 609—614.

Stacey, G., Paau, A. S., Noel, D., Maier, R. J., Silver, L. E., Brill, W. J., 1982: Mutants of *Rhizobium japonicum* defective in nodulation. Arch. Microbiol. **132,** 219—224.

Stemmer, P., Sequeira, L., 1981: Pili of plant pathogenic bacteria. Amer. Phytopathol. Soc. Abstr. 328.

Sutton, B. C. S., Stanley, J., Zelechowska, M. G., and Verma, D. P. S., 1984: Isolation and expression of cloned DNA from *Rhizobium japonicum* encoding an early soybean nodulation function. J. Bact. (in press).

Truchet, G. L., Dazzo, F. B., 1982: Morphogenesis of lucerne root nodules incited by *Rhizobium meliloti* in the presence of combined nitrogen. Planta **154,** 352—360.

Truchet, G. L., Dazzo, F. B., Vasse, J., 1983: Agglutination of *Rhizobium japonicum* 3I1b110 by soybean lectin. Plant Soil **75,** 265—268.

Tsien, H. C., Schmidt, E. L., 1977: Polarity in the exponential phase *Rhizobium japonicum* cells. Can. J. Microbiol. **23,** 1274—1284.

Tsien, H. C., Schmidt, E. L., 1980: Accumulation of soybean lectin-binding polysaccharide during growth of *Rhizobium japonicum* as determined by hemagglutination inhibition assay. Appl. Environ. Microbiol. **39,** 1100—1104.

Tsien, H. C., Schmidt, E. L., 1981: Localization and partial characterization of soybean lectin binding polysaccharide of *Rhizobium japonicum.* J. Bacteriol. **145,** 1063—1074.

Tsien, H. C., *et al.*, 1983: Lectin in five lines of soybean previously considered to be lectin-negative. Planta **158,** 128—133.

Turgeon, B. G., Bauer, W. D., 1982: Early events in the infection of soybean by *Rhizobium japonicum.* Time course and cytology of the initial infection process. Can. J. Bot. **60,** 152—161.

Van Rensberg, H. J., Strijdom, B., 1982: Root surface association in relation to nodulation of *Medicago sativa.* Appl. Environ. Microbiol. **44,** 93—97.

Vincent, J. M., 1980: Factors controlling the legume-*Rhizobium* symbiosis. In: Newton, W. E., Orme-Johnson, W. H. (eds.) Nitrogen Fixation, Vol. II, pp. 103—129, Baltimore: University Park Press.

Wolpert, J. S., and Albersheim, P., 1976: Host-symbiont interactions. I. The lectins of legumes interact with the O-antigen containing lipopolysaccharides of their symbiont rhizobia. Biochem. Biophys. Res. Commun. **170,** 729—737.

Yao, P. Y., Vincent, J. M., 1976: Factors responsible for the curling and branching of clover root hairs by *Rhizobium.* Plant Soil **45,** 1—16.

Zurkowski, W., 1980: Specific adsorption of bacteria to clover root hairs, related to the presence of the plasmid pWZ2 in cells of *Rhizobium trifolii.* Microbios **27,** 27—32.

Zurkowski, W., Lorkiewicz, Z., 1979: Plasmid-mediated control of nodulation in *Rhizobium trifolii.* Arch. Microbiol. **123,** 195—201.

Chapter 2

Interaction of *Agrobacterium tumefaciens* with the Plant Cell Surface

A. G. Matthysse

Department of Biology, University of North Carolina, Chapel Hill, N.C., U.S.A.

With 9 Figures

Contents

I. Introduction

Infection of a wound site on a dicotyledonous plant by *Agrobacterium tumefaciens* results in the formation of crown gall tumors. These tumor cells contain bacterial plasmid DNA sequences integrated in the plant chromosomal DNA. The transfer of plasmid DNA from the bacterium to the plant host cell is presumed to require intimate contact between the cells involved; however, there is little information available on the mechanism of this DNA transfer. The earlier steps in the interaction of *Agrobacterium* with the plant host cell prior to DNA transfer have received considerable attention. While not all of the details of this interaction are clear, it is still one of the best understood examples of the surface interaction and recognition between a phytopathogenic bacterium and a plant cell. This is due to the large amount of effort currently expended on research with *Agrobacterium*, to the availability of tissue culture model systems, and to the availability of some genetic information about the bacterium.

II. Evidence that Bacterial Attachment Is Required for Tumor Formation

That binding of *Agrobacterium* to plant cells is required for tumor formation was suggested before plasmid DNA transfer was known to occur. Prior or simultaneous application of avirulent or heatkilled agrobacteria to wound sites on bean leaves inhibits tumor formation by virulent *A. tumefaciens*. However, if the live virulent bacteria are inoculated first then there is little effect of the later addition of avirulent bacteria. The ability to inhibit tumor formation is a property of many avirulent strains of *A. tumefaciens* and of some strains of *A. radiobacter*. Other bacteria tested including *Escherichia coli, Pseudomonas auriginosa, P. savastonoi, Corynebacterium fascians* and *Bacillus megaterium* fail to inhibit tumor formation (Lippincott and Lippincott, 1969). These results have been interpreted as indicating that a site-specific attachment of *A. tumefaciens* to plant host cells is required for tumor formation. The failure of live virulent bacteria to induce tumors when inoculated subsequent to dead or avirulent bacteria is believed to be due to the occupation of a receptor site on the plant cell surface by the first bacteria inoculated. This receptor site is relatively specific for *A. tumefaciens* although one strain of *Rhizobium leguminosarum* apparently interfered with the binding of *Agrobacterium* at this site (Lippincott and Lippincott, 1976).

Similar results showing that prior but not subsequent inoculation of avirulent strains of *Agrobacterium* interferes with tumor formation by virulent strains have been obtained with pea by Manigault (1970), tomato by Kerr (1969), Jerusalem artichoke by Tanimoto *et al.* (1979) and potato by Glogowski and Galsky (1978). These experiments suggest but do not prove that a site-specific bacterial attachment is required for tumor formation by *A. tumefaciens*.

The existence of avirulent transposon mutants of the Agrobacterium which fail to bind to plant tissue culture cells and which do not inhibit tumor formation by virulent strain also strongly suggests that bacterial attachment is a necessary prerequisite for tumor formation (Douglas *et al.*, 1982; Matthysse, unpublished observation). That these mutants are in fact altered in a single gene was shown in one case by marker exchange of the DNA segment containing the transposon into a virulent *A. tumefaciens*. The transposon and the avirulent and non-attaching phenotypes are all transferred together suggesting that this mutant is indeed the result of the transposon insertion into an operon coding for attachment and virulence (Douglas *et al.*, 1982). While this result does not prove that lack of binding by the bacterium results in a lack of virulence (since the two phenotypes could result from two genes linked in the same operon whose expression was prevented by the polar effects of the transposon) the result does strongly suggest that the ability of the bacteria to bind to plant cells is required for bacterial virulence.

III. Methods Used to Assay Attachment of *Agrobacterium* to Plant Cells

A. Indirect Methods

Attachment of *A. tumefaciens* to plant cells was first assayed indirectly. These assays all depend on the ability of avirulent or non-viable (and therefore, non-tumorigenic) bacteria to interfere with tumor formation by virulent bacteria. The order of addition of the avirulent and virulent bacteria to the wounded plant or plant tissue affects the outcome of the experiment. Thus avirulent or heat-killed *A. tumefaciens* inhibit tumor formation by virulent bacteria if they are inoculated before the virulent organisms. However, the avirulent bacteria have no effect if inoculated after the virulent organisms. The dependence of the response on the order of addition of the bacteria has been assumed to be due to the irreversible occupation of available receptor sites by the bacteria inoculated first. This assumption implies that the inhibition should be dependent on the number of avirulent organisms used, and this has been shown to be the case for avirulent strain IIBNV6 and virulent B6 inoculated onto wounded pinto bean leaves (Lippincott and Lippincott, 1969). The data obtained fit a single particle hit curve where one IIBNV6 bacterium interferes with the binding of one B6 bacterium.

Most of the research using this type of assay has used wounded pinto bean leaves. The assay gives a linear increase in the number of tumors formed with the inoculation of increasing numbers of bacteria in the range of about 10^7 to 10^9 bacteria per ml (Lippincott and Heberlein, 1965). A similar assay has been developed using the inhibition of tumor formation on potato discs by avirulent or killed *A. tumefaciens* (Glogowski and Galsky, 1978). The number of tumors formed on the disc increases with increasing numbers of bacteria inoculated (Anand and Heberlein, 1977). Inhibition of tumor formation on Jerusalem artichoke discs has also been used as an indirect measure of bacterial attachment by Douglas *et al.* (1982).

These indirect assays of bacterial attachment have been extended to studies of the nature of the bacterial binding site by adding bacterial components potentially involved in binding to the wounded plant leaf or disc prior to the addition of the bacteria. Once again the effect of the order of addition of the substance and the bacteria was used to indicate that the inhibition of tumor formation was caused by inhibition of bacterial attachment rather than by interference with some other process required for the production of visible tumors (Lippincott and Lippincott, 1976; Whatley *et al.,* 1976a).

The nature of the plant receptor to which the bacteria bind has also been studied using this type of assay. The effect of the addition of various compounds to the wounded leaf or disc surface before and after the addition of the bacteria has been measured (Lippincott and Lippincott, 1976; Lippincott *et al.,* 1977; Lippincott and Lippincott, 1978; Pueppke and Benny, 1981; Pueppke *et al.,* 1982). In some studies the ability of prior incubation of avirulent or heatkilled *A. tumefaciens* with the putative receptor to neutralize its inhibitory effect has been used as an additional measure of the specificity of the interaction (Lippincott and Lippincott, 1976; Lippincott *et al.,* 1977; Lippincott and Lippincott, 1978).

These indirect assays have the advantage that processes which are important or required for tumor formation in the wounded plant leaf or cut disc are examined. The major disadvantage of the assays is that it is difficult to be certain that it is in fact bacterial attachment and not some other process required for tumor formation which has been inhibited.

B. Direct Assays

Direct assays of bacterial attachment of plant cells generally utilize tissue culture cells. These assays are of three types — bacterial viable cell count assays, radioactive assays, and microscopic assays. In general bacteria are inoculated into a suspension culture of plant cells. After varying incubation times the free bacteria and plant cells with attached bacteria are separated from each other by filtration through a material such as miracloth which retains plant cells but not bacteria (Matthysse *et al.,* 1978; Ohyama *et al.,* 1979; Douglas *et al.,* 1982) or by centrifugation through Ficoll (Smith and Hindley, 1978). If the bacteria were radioactively labelled then the radioactivity retained with the plant cells is used as a measure of the number of attached bacteria. The number of viable free and attached bacteria can also be determined by a simple viable cell count of the free bacteria and of a blended homogenate of the attached bacteria and plant cells (Matthysse *et al.,* 1978). The use of radioactively labelled bacteria allows the determination of all bound bacteria whether living or dead, but requires the use of rather high bacterial concentrations (greater than 10^6 bacteria per ml). The ratio of bacteria to plant cells or multiplicity of infection in these experiments ranges from 3 (Douglas *et al.,* 1982) to 10^5 (Smith and Hindley, 1978). The use of viable cell counts to determine the number of attached bacteria has the disadvantage that only live bacteria are measured. However, *A. tumefaciens* which are attached to plant cells appear to retain their viability and attached bacteria have been observed to divide using time-lapse microcinematography (Matthysse *et al.,* 1978; Gurlitz and Matthysse, unpublished observation). The advantage of the use of viable cell counts to determine the number of attached bacteria is that it allows measurements to be made at both low and high bacterial concentrations (10^3 to 10^8 bacteria per ml) and both low and high multiplicity of infection (10^{-2} to 10^5) (Gurlitz *et al.*, 1983).

Microscopic assays of bacterial attachment have the obvious advantage of being the most direct assay possible. Attached bacteria have been seen in the light microscope (Matthysse *et al.*, 1978; Ohyama *et al.*, 1978; Douglas *et al.*, 1982), the scanning electron microscope (Smith and Hindley, 1978; Matthysse *et al.*, 1978; Matthysse *et al.*, 1981; Douglas *et al.*, 1982), and in sectioned material in the transmission electron microscope (Ohyama *et al.*, 1978). In order to be able to find the attached bacteria in the microscope it is generally necessary to use a relatively high multiplicity of infection in these experiments (100 or greater).

Direct assays of bacterial attachment have the disadvantage that they employ tissue culture systems whose relevance to the infection process in the whole plant is uncertain. However, direct assays of attachment have several advantages — the attachment can be measured and visualized without inferences or assumptions as to whether it is in fact bacterial attachment which is being measured. The use of tissue culture systems also allows the study of the effects of alterations in the medium and the addition and removal of inhibitors.

In general the results of utilizing indirect and direct assays of bacterial attachment are in agreement suggesting that both assays are fundamentally measuring the same process. There are two cases in which these two assays have yielded different results: the question of the attachment or non-attachment of certain avirulent strains of *A. tumefaciens* and the identity of the plant receptor for bacterial attachment. The explanation of the first case appears to be rather straightforward and to involve the bacterial concentrations employed in the assays used. The explanation for the difference in results obtained in attempting to determine the identity of the plant cell receptor for attachment of *A. tumefaciens* is presently unclear, but it seems likely that further research will resolve this discrepancy in the near future.

IV. General Requirements for Attachment of *A. tumefaciens*

The effects of changes in temperature, pH, and medium composition on attachment have been examined using suspension culture cells and direct assays of bacterial attachment. The optimum pH for attachment is about 6 (Ohyama *et al.*, 1978). Divalent cations are not required; neither EDTA nor EGTA has an effect on binding of *A. tumefaciens*. In fact, Ca^{++} concentrations of 10 mM appear to be slightly inhibitory to attachment (Ohyama *et al.*, 1978; Gurlitz *et al.*, 1983). There is also no requirement for the other constituents of the plant tissue culture medium; attachment proceeds with unaltered kinetics in 4% sucrose. The ionic strength of the medium has little effect on bacterial binding to carrot cells which occurs in 4% sucrose, in Murashige and Skoog (1962) medium, or in 0.25M NaCl (Gurlitz *et al.*,

1983). The temperature optimum for binding to *Datura* cells is 28 degrees. This optimum is not sharp and binding is seen at 35 degrees but is much reduced at 47 degrees (Ohyama *et al.*, 1978). Similar results are seen measuring the attachment of *A. tumefaciens* strain K14 to tobacco cells (Fig. 1). Thus it seems unlikely that bacterial attachment is the step in tumor formation which is inhibited by temperatures above 32 degrees (Braun, 1962).

The attachment reaction in tissue culture is generally rather slow.

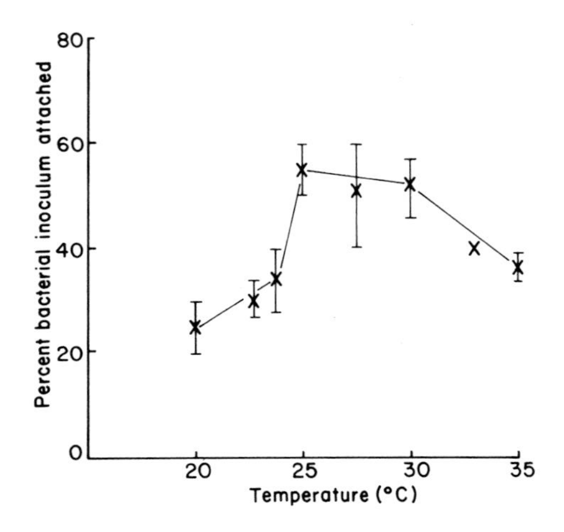

Fig. 1. The effect of temperature on the attachment of *A. tumefaciens* strain Kl4 to tobacco tissue culture cells. Bacterial attachment was measured after 120 min. incubation. Although attachment is decreased at 35 degrees this decrease is relatively minor. Bars indicate standard deviations of a minimum of 3 experiments

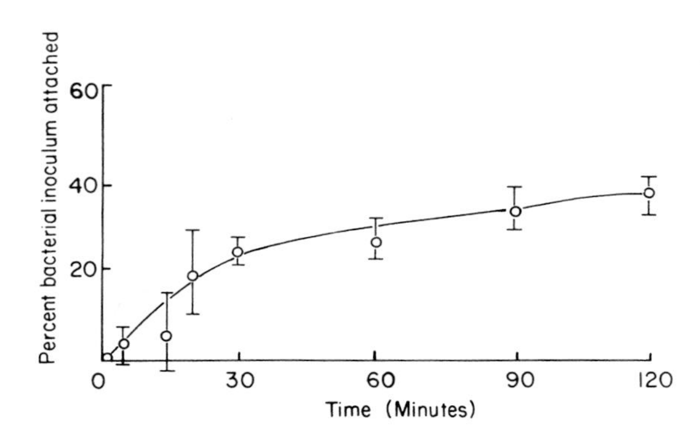

Fig. 2. Time course of binding of *A. tumefaciens* strain A6 to carrot suspension culture cells. Bars indicate standard deviations of a minimum of 3 experiments. Note the biphasic nature of the binding curve and the rather slow initial rate of binding

Ohyama *et al.* (1978) obtained half maximum binding of *A. tumefaciens* B6 to *Datura* cells at about 80 min of incubation. Binding of strain C58 to tobacco cells reached half maximum at about 60 min. (Matthysse *et al.*, 1978). Bacterial attachment to carrot (Fig. 2, see also Matthysse *et al.*, 1981) and zinnia (Douglas *et al.*, 1982) is slightly faster with half maximum attachment reached at between 15 and 30 min. In all of these cases only a fraction of the bacteria inoculated bound to the plant cells: 10% with B6 and *Datura*, 30% with C58 and tobacco, 50% with A6 and carrot, and 30% with A273 and zinnia. However, Smith and Hindley (1978) observed 90% binding of C58 to tobacco cells under their conditions. This represents a large number of bacteria attached to each tobacco cell — about 1×10^5 bacteria per plant cell. The reason for the increased percentage of bacteria bound in these experiments is unclear; it may be due to the particular conditions or assay used or to the fact that the tobacco cells used in these experiments were habituated cells.

Attachment of *A. tumefaciens* measured using indirect assays in wounded bean leaves or on potato discs appears to be faster than the reactions seen in tissue culture. No interference with tumor formation by virulent *A. tumefaciens* is observed when the avirulent bacteria are added 10 to 15 min. after the innoculation of the virulent strain (Lippincott and Lippincott, 1969; Glogowski and Galsky, 1978).

V. Plant Receptors for the Attachment of *A. tumefaciens*

A. The Nature of the Receptor

In the attachment of *A. tumefaciens* to plant tissue culture cells the bacterium appears to be an active participant while the plant cell is passive. Bacteria attach to carrot cells which were killed with heat or glutaraldehyde prior to the inoculation of the bacteria (Matthysse *et al.*, 1981). The kinetics of attachment to killed cells differ slightly from those of attachment to living cells (attachment to killed cells appears to be slightly more rapid). The bacterial strain specificity of attachment is unaffected by killing the plant cells suggesting that binding is to the same receptor on living and killed carrot cells. These experiments indicate that the receptor for attachment of *A. tumefaciens* is not made in response to the bacterial infection but preexists the introduction of the bacteria. Thus, it is feasible to examine the effects of treatments of suspension culture cells prior to the inoculation of the bacteria on subsequent bacterial attachment. The effects on tumor formation of the addition to the wound site of, or treatment of the bacteria with, various components of the plant cell surface have also been examined.

There are a number of compounds whose addition to tissue culture media has no effect on bacterial attachment. These substances include RNAse, DNAse, lipase, soybean lectin, concanavalin A, lecithin, lysolecithin, octopine, canavanine, arginine, galactose, arabinose, 2-deoxyglucose, glucosamine, mannose, α-methyl-D-mannoside, and 0.25M NaCl (Ohyama *et al.,* 1978; Gurlitz *et al.,* 1983). The only one of these substances which has been shown to affect tumor formation using the indirect assay is concanavalin A which inhibits tumor formation by strain B6M but not by other bacterial strains (Pueppke *et al.,* 1982; Lippincott and Lippincott, 1976). Protease and poly-L-lysine increase bacterial binding to *Datura* cells (Ohyama *et al.,* 1978). It is unclear whether this represents actual increased bacterial binding or an increase in the formation of large aggregates of bacteria which could be retained by the filter and thus included in the number of bacteria bound. Compounds which do not effect bacterial attachment using the indirect assay include galactose, galactan, cellex, gum arabic, locust bean gum, polyethylene glycol, wheat germ lectin, kidney bean lectin, potato tuber lectin, and glucuronic acid (Lippincott and Lippincott, 1976; Pueppke *et al.,* 1982; Pueppke and Benny, 1981). Pectin, polygalacturonic acid, sodium polygalacturonate, and arabinogalactan inhibit tumor formation by strain B6 in wounded bean leaves (Lippincott and Lippincott, 1976). Larch arabinogalactan does not inhibit tumor formation by B6 on potato discs, but citrus pectin, potato pectin, and polygalacturonic acid are inhibitory in this system as well (Pueppke and Benny, 1981). The addition of heparin resulted in a partial inhibition of tumor formation on potato discs.

The inhibition of tumor formation in bean leaves by the pectin fraction of plant cell walls has been studies extensively. The inhibition is dependent on the order of addition of the virulent bacteria and the pectin (Lippincott and Lippincott, 1980). Concentrations as low as 1 ng/ml of sodium polygalacturonate result in a measurable inhibition of tumor formation (Lippincott and Lippincott, 1976). Plant cell walls, but not membrane preparations, from pinto bean leaves inhibit tumor formation if added to wounded bean leaves prior to, or simultaneously with, virulent bacterial strain B6. Cell wall concentrations of 100 μg/ml are sufficient to result in a measurable inhibition of tumor formation. The inhibitory effect of these cell wall preparations could be reversed by pretreating the cell walls with heat-killed avirulent *A. tumefaciens* strain IIBNV6 but not by heat-killed *A. radiobacter* strains 6467 and S1005 (Lippincott *et al.,* 1977). Heat-killed *A. tumefaciens* IIBNV6 itself interferes with tumor formation by strain B6; the *A. radiobactor* strains have no effect on tumor formation (Lippincott and Lippincott, 1969). Lipopolysaccharide (LPS) from purified strain B6 but not from 6467 inhibits tumor formation (Whatley *et al.,* 1976a) and pretreatment of bean leaf cell wall preparations with LPS from B6 but not 6467 reverses the inhibitory effect of the cell walls (Lippincott *et al.,* 1977). Thus, the inhibition of tumor formation by plant cell walls appears to involve the same receptor site as the inhibition of tumor for *A. tumefaciens.*

Treatment of bean leaf cell walls with pectinase, macerase, cellulase, EDTA and Triton X-100, or sodium deoxycholate does not affect their inhibitory activity. However, treatment of cell walls with 0.05M H_2SO_4 for 90 min. at 100 degrees does neutralize their inhibitory activity (Lippincott and Lippincott, 1980). Treatment of cell walls with pectin methyl transferase also neutralizes their inhibitory activity; this reaction could be reversed by treatment with pectin esterase. Cell walls from crown gall tumors, monocots, or embryonic dicot tissues do not inhibit tumor formation. These cell wall preparations become inhibitory after treatment with pectin esterase (Lippincott and Lippincott, 1976).

These observations suggest that pectin may be the receptor (or a portion of the receptor) on the plant cell surface. Alternative explanations are possible, however. For example, Hahn *et al.* (1981) have shown that a plant cell wall fraction can act as an elicitor of phytoalexins. The effects of phytoalexins on *Agrobacterium* and on tumor formation are not known, but it is possible that plant cell walls could inhibit tumor formation by this mechanism or by some other mechanism involving interference with a process required for tumor formation. This process would not necessarily be bacterial attachment.

The nature of the plant receptor site for *A. tumefaciens* has also been studied using direct assays of bacterial attachment. The results obtained with carrot suspension culture cells are rather different than those obtained with indirect binding assays on bean leaves. In order to determine which

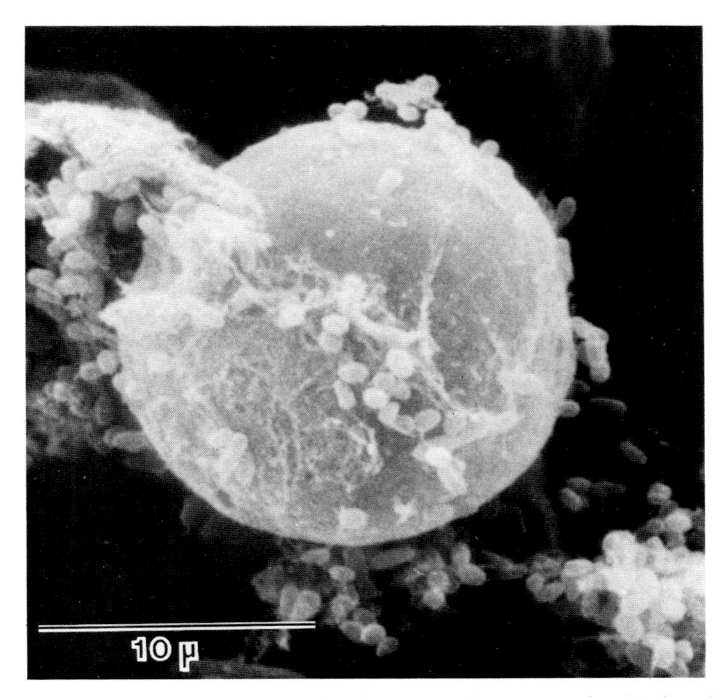

Fig. 3. Scanning electron micrograph of *A. tumefaciens* strain A6 incubated with killed carrot protoplasts for 19 hours. The bacteria are surrounded by fibrillar material. Large clusters of bacteria held together by fibrils are visible. Most of the bacteria in these clusters are only indirectly attached to the plant cell surface. Some individually attached bacteria are also visible

carrot cell wall components are required for bacterial attachment protoplasts were prepared with the intention of adding back cell wall constituents until bacterial binding occurred. However, *A. tumefaciens* binds to protoplasts from which all cell wall components visible in the scanning election microscope have been removed (Fig. 3). The bacteria bind to both living carrot cell protoplasts and protoplasts which were glutaraldehyde treated to prevent cell wall resynthesis. Neither cellulose, nor sodium polygalacturonate affects the kinetics of this binding reaction. Attachment to protoplasts shows the same bacterial strain specificity as attachment to intact cells (Matthysse *et al.,* 1982). At the present time it is not known whether the receptor sites to which the bacteria bind on the surface of carrot protoplasts are the same as the receptor sites on the surface of intact carrot cells. Microsopic observations of bacterial attachment to plasmalysed carrot cells show the bacteria attached to the cell wall surface in regions in which the plasmalemma has retracted from the cell wall suggesting that attachment does not occur preferentially at the points of emergence of plasmadesmata (Gurlitz *et al.,* 1983). Further experimentation is necessary to resolve these results with those obtained using the indirect assay of bacterial attachment.

The receptor for attachment of *A. tumefaciens* to the surface of intact carrot suspension culture cells can be removed or inactivated by treatment of the cells with the proteolytic enzymes, trypsin, chymotrypsin, or proteinase K (Fig. 4). These treatments do not irreversibly damage the carrot cells.

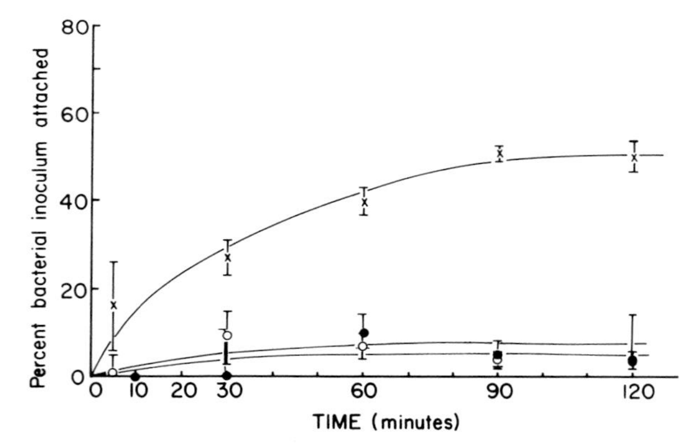

Fig. 4. Time course of attachment of *A. tumefaciens* strain A6 to carrot suspension culture cells. Symbols: x-untreated carrot cells; o-carrot cells treated with 0.1% trypsin for 60 min. prior to the addition of the bacteria; o-carrot cells treated with 0.1% chymotrypsin for 60 min. prior to the addition of the bacteria

If the enzymes are removed and the carrot cells returned to Murashige and Skoog (1962) medium they recover the ability to bind *A. tumefaciens* after about 6 hours (Gurlitz *et al.,* 1983). This recovery does not occur in the presence of cycloheximide indicating that protein synthesis is required for recovery. The receptor can also be extracted by Triton X-100 or by $CaCl_2$.

Once again these treatments do not irreversibly damage the carrot cells and receptor activity is recovered after incubation in tissue culture medium. When the proteins contained in the extracts are examined by SDS-polyacylamide gel electrophoresis more than 30 distinct polypeptides are found to be present. Calculations indicate that the more prevalent of these are present in 10^5 to 10^6 copies per carrot cell. The number of receptor sites for *A. tumefaciens* on the surface of a carrot cell has been estimated by measuring the binding of a cellulose-minus bacterial mutant to carrot cells. Unlike the wild type strains, cellulose-minus bacteria are capable only of attaching individually to the plant cell surface and large clusters containing both directly and indirectly attached bacteria are not formed. A maximum of between 200 and 500 cellulose-minus bacteria can be bound to each carrot cell suggesting that there may be only a few hundred receptor sites for the bacteria on each carrot cell (Gurlitz *et al.*, 1983). However, it is not clear how many protein molecules constitute a receptor site. If there is only one protein per receptor site, then the receptor protein would be a very minor constituent of the plant cell surface, and would be unlikely to be visible on SDS-polyacrylamide gels. If there are many protein molecules per receptor site, then the receptor protein might be visible on gel electrophoresis. Comparison of the polypeptide bands visible after gel electrophoresis of extracts of carrot suspension culture cells which bind *A. tumefaciens* and of the same cultures after the induction of embryos when the cells no longer bind the bacteria shows many differences in the polypeptides extracted. Any of the polypeptides present in extracts of cells which bind *A. tumefaciens* and absent in extracts of cells which do not bind the bacteria are candidates for involvement in the plant receptor site (Gurlitz *et al.*, 1983).

The addition of various substances to the medium in the attempt to inhibit bacterial attachment has provided only negative evidence about the plant receptor site. None of the sugars tested inhibits attachment nor do any of the lectins tested (Ohyama *et al.*, 1978; Gurlitz *et al.*, 1983). This suggests that the receptor protein may not be a lectin but it also is possible that some untested sugar or minor lectin fraction is involved. High ionic strength has no effect on binding of *A. tumefaciens* suggesting that proteins similar to the hydroxyproline-rich glycoproteins involved in the binding of *Pseudomonas solanacearum* to tobacco cells (Mellon and Helgeson, 1982) are not involved in the binding of *A. tumefaciens*. At the present time no other information on the nature of the plant surface protein involved in the binding of *A. tumefaciens* is available. Thus present information about the plant receptor to which *A. tumefaciens* binds suggests that it consists of a protein and, possibly, a pectin component, but these components have not been unambiguously identified.

B. Plant Host Range for Attachment of A. tumefaciens

Agrobacterium tumefaciens induces tumor formation on most dicots. However, the bacterium does not infect monocots and generally does not induce tumors on meristematic dicot tissues (Lippincott and Lippincott, 1975; Lippincott and Lippincott, 1976). The question of whether plant cells which can not be transformed to tumor cells by the bacterium lack the receptor for bacterial attachment has been investigated using both direct and indirect assays for bacterial attachment.

Cell walls from mature dicot cells inhibit tumor formation when added to wounded bean leaves either prior to or simultaneously with virulent *A. tumefaciens*. Cell walls from most monocots tested do not inhibit tumor formation. In those cases, principally barley, in which monocot cell walls did inhibit tumor formation this inhibition is clearly distinguishable from that of dicot cell walls since the inhibition is not reversed by pretreatment of the cell walls with heat-killed *A. tumefaciens*. Cell walls from embryonic dicot tissues also fail to inhibit tumor formation as do cell walls from dicot crown gall tumors (Lippincott and Lippincott, 1976; Lippincott and Lippincott, 1978). Although *Agrobacterium* does not induce tumors on monocots or meristematic dicot tissues, there is some evidence that crown gall tumor cells can be superinfected with a second strain of *Agrobacterium* (Otten *et al.*, 1981). It is not certain whether this is a general property of all crown gall tumors or is restricted to certain tumors or to tumor cells grown under particular conditions. In any case, it seems clear that the ability of plant cell walls to inhibit tumor formation by *A. tumefaciens* parallels the ability of the plant cells to become transformed into crown gall tumor cells by the bacterium. This suggests that the lack of the plant receptor for attachment of *A. tumefaciens* may be a major factor in determining the host range of the bacterium.

Results similar to those described for the indirect assay were obtained by Matthysse and Gurlitz (1982) using a direct binding assay. *A. tumefaciens* binds to all dicot tissue culture cells tested and to all crown gall tumor cells tested. The bacteria do not bind to monocot tissue culture cells. Carrot suspension culture cells bind *A. tumefaciens*. However, 24 hours after these cells have been induced to become embryos they no longer bind the bacteria. Although monocot cells and embryos do not bind the bacteria they do induce bacterial synthesis of cellulose and the resulting formation of bacterial aggregates (Matthysse and Gurlitz, 1982). It is possible that the binding of *A. tumefaciens* to monocots reported by Ohyama *et al.* (1978) is due to the formation of bacterial aggregates in their assays. Bacterial aggregate formation is dependent on bacterial concentration and these assays were carried out using more than 10^7 bacteria per ml.

Thus, the parallel between the binding of *A. tumefaciens* to plant cells and the ability of the bacterium to induce tumors in those plants is observed using both the direct and indirect assays for bacterial binding. The only exception to this correlation is soybeans to which the bacteria bind in culture but in which they do not induce tumors (Matthysse and

Gurlitz, 1982). It seems probable that the block to tumor formation in soybeans may be at some stage subsequent to bacterial attachment. The situation regarding the ability of crown gall tumor cells to be superinfected with *A. tumefaciens* is unclear, and it is therefore difficult to interpret the discrepancy in results obtained using direct and indirect assays of bacterial binding. However, it seems that one major factor determining the host range of *A. tumefaciens* is the lack of a receptor for bacterial attachment on most monocot and embryonic dicot cells.

VI. Binding Sites of *Agrobacterium tumefaciens* for Attachment to Plant Cells

A. The Role of the Bacterium in the Attachment Interaction

Although the plant cell is a passive partner in the binding of *A. tumefaciens* to tissue culture cells, the bacterium appears to play a more active role in this system. Live bacteria bind to dead carrot cells, but dead bacteria do not bind to either live or dead plant cells (Matthysse *et al.,* 1981, Fig. 5). However, bacterial protein synthesis is not required for attachment to carrot cells (Matthysse *et al.,* 1983).

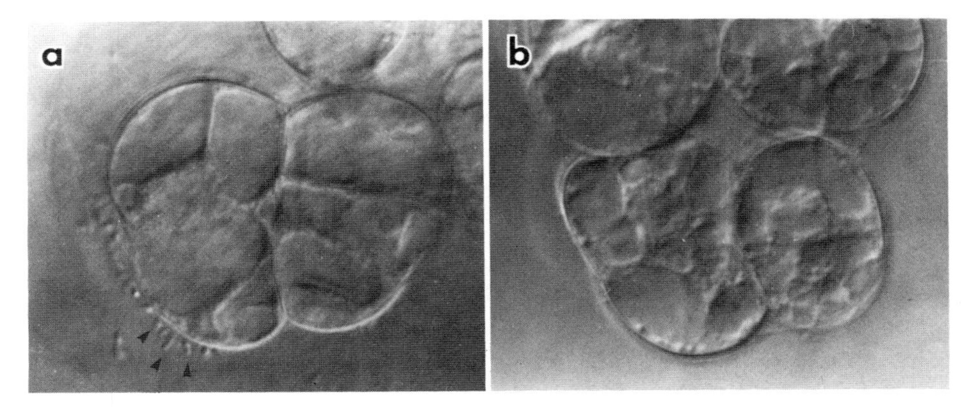

Fig. 5. Light micrographs of *A. tumefaciens* incubated with living carrot cells for 1 hour. a live bacteria. b heat-killed bacteria. About 10^7 bacteria per ml were present in both a and b. Note the attached live bacteria (a arrows). No attached bacteria were seen (b) when the bacteria were heat killed prior to their addition to the carrot suspension culture

In contrast to the failure of heat-killed bacteria to bind to tissue culture cells, heat-killed bacteria do interfere with tumor formation by live bacteria (Lippincott and Lippincott, 1969). Heat-killed *A. tumefaciens* also inhibit the interaction of live bacteria with isolated plant cell walls (Lippincott and Lippincott, 1978). The reasons for the requirement for live bacteria for attachment to tissue culture cells and the lack of such a requirement for the inhibition of tumor formation are unknown. Since bacterial protein synthesis is not required in either case it seems clear that the binding site for the attachment of *A. tumefaciens* to plant cells is made constitutively by the bacteria and is not induced by the presence of the plant cells.

B. The Nature of the Bacterial Binding Site

Using the indirect assay of bacterial attachment, Whatley *et al.* (1976a) found that lipopolysaccharide extracted from virulent strains of *A. tumefaciens* inhibits tumor formation. LPS preparations from *A. radiobacter* have no effect on tumor formation by virulent *A. tumefaciens* strain 15955. The order of addition of the LPS and the virulent bacteria to the wound site is important. LPS inhibits tumor formation if added prior but not subsequent to the inoculation of the virulent bacteria. When LPS is separated into lipid A and the polysaccharide O antigen, it is the polysaccharide portion which is inhibitory (Lippincott and Lippincott, 1976, Whatley *et al.*, 1976b).

The addition of lipopolysaccharide extracted from virulent *A. tumefaciens* to the culture medium also inhibits the binding of the bacteria to tobacco tissue culture cells (Matthysse *et al.*, 1978). LPS extracted from *A. radiobacter* 6467 (which does not bind to tissue culture cells) has no effect on the binding of virulent *A. tumefaciens*. Live spheroplasts of *A. tumefaciens* also fail to bind to carrot suspension culture cells until sufficient time has elapsed for the resynthesis of the bacterial cell wall and outer membrane (Matthysse *et al.*, 1983). Treatment of the bacteria with trypsin also delays their attachment, although for a shorter time than the total removal of the cell wall in the preparation of spheroplasts (Matthysse *et al.*, 1983). Avirulent mutants of *A. tumefaciens* which fail to attach to tissue culture cells show altered phage absorption, but the chemical basis of this alteration is not yet known (Douglas *et al.*, 1982). These results suggest that the bacterial binding site for attachment to plant cells contains lipopolysaccharide and possibly also a protein component.

C. The Role of Bacterial Cellulose Fibrils in the Attachment of A. tumefaciens

Observations on the binding of *A. tumefaciens* to tissue culture cells using the scanning electron microscope show the presence of fibrils surrounding the attached bacteria (Fig. 6). The fibrils are visible as early as 90 min. after the inoculation of the bacteria. By 19 hours of incubation the fibrils form networks that hold the bacteria to the plant cell surface and entrap and surround additional bacteria attaching them indirectly to the surface of the host cell. Eventually the mass of bacteria and fibrils forms a large network linking all of the plant cells in the culture together in a large aggregate (Matthysse *et al.*, 1981). Similar fibrils have been observed in transmission electron microscopy of *A. tumefaciens* attached to *Datura* cells by Ohyama *et al.* (1979) and in scanning electron microscopy of *A. tumefaciens* attached to zinnia cells by Douglas *et al.* (1982). These fibrils are synthesized by the

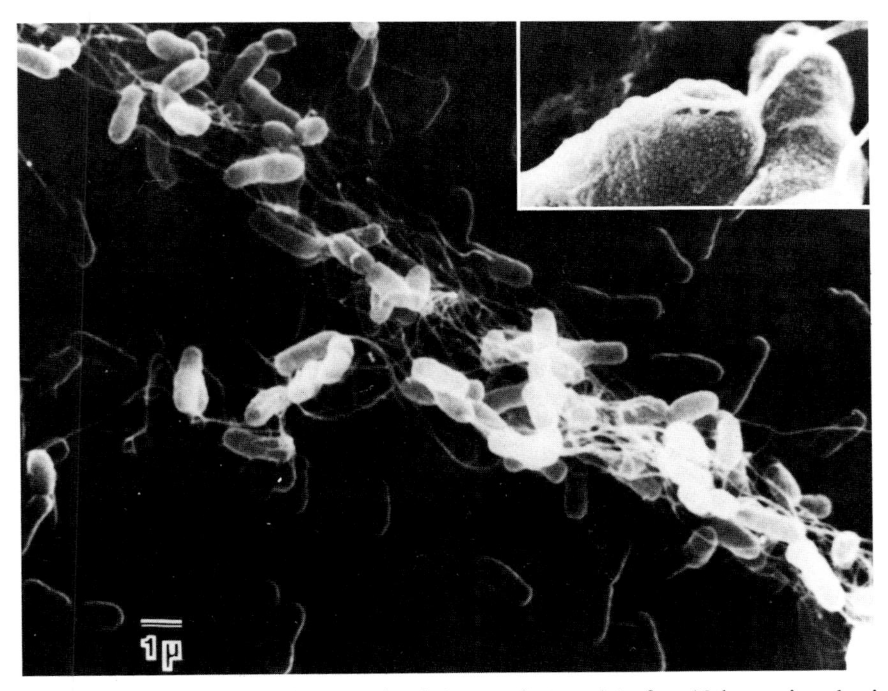

Fig. 6. Scanning electron micrograph of *A. tumefaciens* A6 after 19 hours incubation with heat-killed carrot cells. Note the abundant cellulose fibrils which emerge from the sides and ends of the bacteria

bacteria rather than the plant cell since they are made during the attachment of bacteria to heat-killed or glutaraldehyde-fixed carrot cells (Matthysse *et al.*, 1981, Fig. 7). Fibrils are also elaborated by the bacteria during attachment to glutaradehyde-fixed carrot protoplasts (Matthysse *et al.*, 1982). The bacteria can synthesize fibrils in the absence of plant cells. The amount of fibrillar material made by *A. tumefaciens* strain A6 is increased by the inclusion of soytone in the culture medium. Fibrils have been purified from these bacterial cultures. They are resistant to digestion with 2N trifluoroacetic acid at 121 degrees for 3 hours and to digestion with proteases. The fibrils show a fluorescent stain with Calcafluor which stains β-linked polysaccharide. The fibrils are susceptible to digestion with cellulase and with 6N HCl at 100 degrees. The products of cellulase digestion are cellobiose and glucose. The fibrillar material was purified and its infrared spectrum compared with that of cellulose. The two spectra are almost identical (Matthysse *et al.*, 1981). Thus, it appears that the fibrils synthesized by *Agrobacterium* are composed of cellulose.

In order to determine the role of bacterial cellulose fibrils in infections by *A. tumefaciens* transposon mutants which lack the ability to synthesize cellulose fibrils were constructed. These mutants synthesize no cellulose detectable using Calcafluor staining. The mutant bacteria attach to the sur-

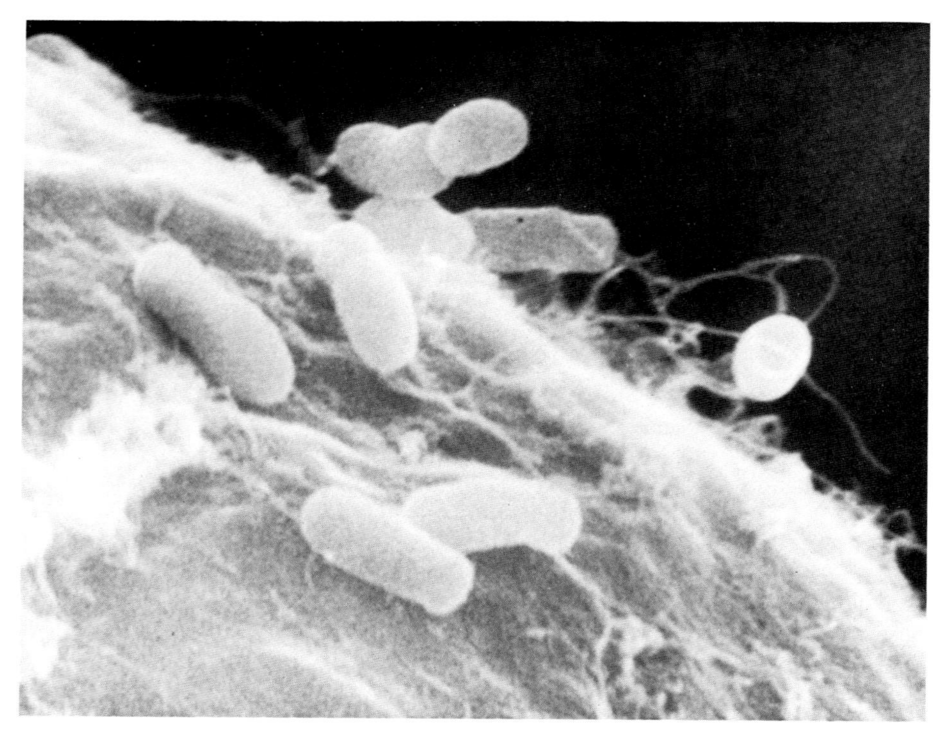

Fig. 7. Scanning electron micrograph of *A. tumefaciens* strain A6 incubated for 8 hours with glutaraldehyde-killed carrot cells. The bacteria have attached and have elaborated numerous fine fibrils

face of carrot suspension culture cells, but in contrast to the parent strain with which large aggregates of bacteria are seen on the plant cell surface, cellulose-minus mutant bacteria attach individually to the carrot cell surface. By 90 min. after attachment of the parent strain fibrils were visible surrounding the bacteria. No fibrils are visible around the cellulose-minus mutants (Fig. 8). Longer incubation of the parent bacteria with carrot cells results in the formation of large aggregates of bacteria, bacterial fibrils and plant cells. No such aggregates are formed by cellulose-minus bacteria. The absence of cellulose fibril synthesis also alters the kinetics of bacterial attachment. Revertants of the cellulose-minus mutants which regain the ability to synthesize detectable cellulose also regain all of the other properties of the parent strain at the same time (Matthysse, 1983 a).

Cellulose synthesis is not required for virulence; the cellulose-minus mutants induce tumors on tobacco and *Bryophyllum*. For most mutants only cellulose-minus bacteria could be recovered from the tumors indicating that the mutants themselves and not revertants caused the tumors. However, the ability of the parent strain to produce tumors on *Bryophyllum* leaves is unaffected by washing the inoculation site with water while the ability of the cellulose-minus bacteria to induce tumors is much reduced by water washing of the inoculation site. Thus, a major role of the cellulose fibrils synthesized by *A. tumefaciens* appears to be anchoring the bacteria to the surface of the host cells thereby aiding in the production of tumors (Matthysse, 1983a).

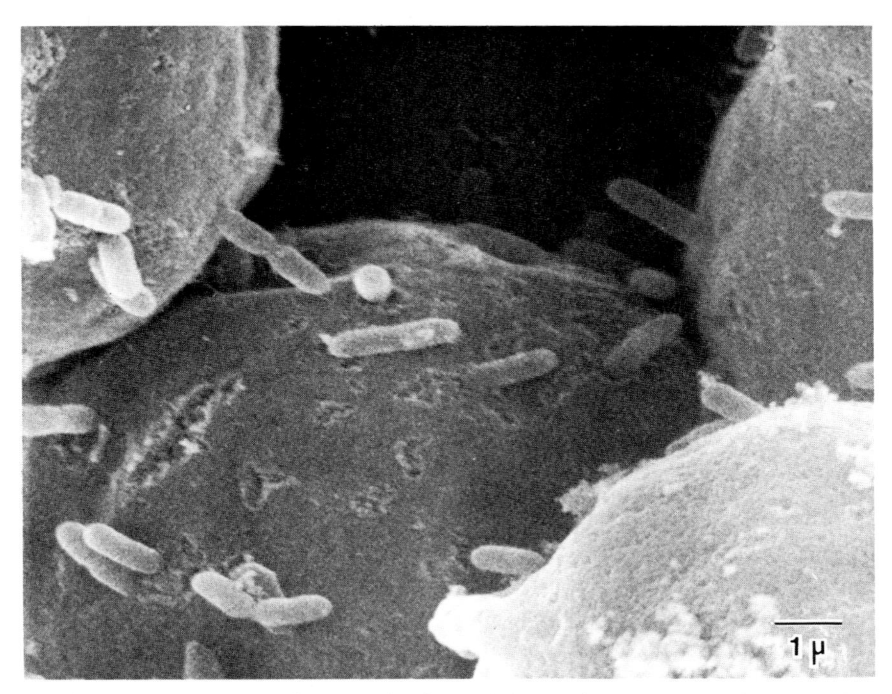

Fig. 8. Scanning electron micrograph of a cellulose-minus mutant of *A. tumefaciens* strain ACH_5C_3 incubated with killed carrot protoplasts for 19 hours. Note the absence of fibrillar material surrounding the bacteria

D. *The Genetics of Attachment of A. tumefaciens*

Using either direct or indirect assays of bacterial attachment virulent strains of *A. tumefaciens* containing the tumor-inducing (Ti) plasmid attach to plant cells. Avirulent strains of *A. radiobacter* which lack the Ti plasmid such as 6467, S1005, and TRI do not bind to plant cells (Lippincott and Lippincott, 1969; Matthysse *et al.*, 1978). Lipopolysaccharide (LPS) isolated from virulent strains of *A. tumefaciens* inhibits bacterial attachment. LPS isolated from *A. radiobacter* does not inhibit attachment of *A. tumefaciens* to wounded bean leaves or to tobacco tissue culture cells (Whatley *et al.*, 1976a; Matthysse *et al.*, 1978). The Ti plasmid can be transferred from *A. tumefaciens* to *A. radiobacter* by conjugation. Strains of *A. radiobacter* into which the Ti plasmid from either B6S3 or K84 has been introduced acquire the ability to bind to wounded bean leaves and to produce lipopolysaccharide which inhibits attachment of *A. tumefaciens*. These strains also become virulent and induce crown gall tumors (Whatley *et al.*, 1978).

These observations would suggest that at least some of the genes required for attachment of *Agrobacterium* to plant host cells are carried on the Ti plasmid. However, the results of experiments measuring the attachment to plant cells of strains of *A. tumefaciens* which have been cured of the Ti plasmid present a more confusing picture. Strains containing an octopine Ti plasmid generally appear to continue to attach to plant cells when cured of their Ti plasmid. Thus, strain ACH_5 cured of the Ti plasmid (called ACH_5A or ACH_5C_3) attaches to carrot tissue culture cells and to bean leaves (Whatley *et al.*, 1978, Matthysse, 1983 b), but apparently does not bind to Jerusalem artichoke tuber slices (Tanimoto *et al.*, 1979). Nopaline strains, particularly C58, cured of their Ti plasmids present a confusing picture. *A. tumefaciens* strain NT1 (which is an avirulent derivative of C58 cured of the Ti plasmid) binds to bean leaves (Whatley *et al.*, 1978), Jerusalem artichoke tuber slices (Tanimoto *et al.*, 1979), and zinnia cells (Douglas *et al.*, 1982). Strain NT1 binds only weakly (compared to its parent, C58) to habituated tobacco tissue culture cells (Smith and Hindley, 1978). These binding measurements were all made at bacterial concentrations in excess of 10^5 bacteria per ml. Binding of strain NT1 to tobacco and carrot suspension culture cells is dependent on the bacterial concentration used. Very few bacteria (less than 5% of the inoculum) bind at concentrations of 10^3 bacteria per ml. However, at concentrations of 10^7 bacteria per ml. 30% of the bacterial inoculum binds. The parent strain of NT1, C58, binds to the extent of about 40% of the inoculum at these high concentrations of bacteria (Matthysse and Lamb, 1981). The reason for these peculiarities in the binding of strain NT1 is unknown.

Avirulent transposon mutants of strain C58 which fail to bind to zinnia cells using a direct assay or to Jerusalem artichoke tubers using an indirect assay have been obtained by Douglas *et al.* (1982). These mutants are all chromosomal mutants.

The genetics of the attachment of *A. tumefaciens* to plant cells presents a very complex picture. Some genes required for attachment appear to be located on the Ti plasmid for both octopine and nopaline strains, since the introduction of the Ti plasmid into non-binding *A. radiobacter* strains gives them the ability to bind to plant cells. On the other hand, some strains such as ACH_5C_3 can bind in the absence of the Ti plasmid and for at least one strain, NT1, binding may be dependent on the conditions used to measure it. Non-binding mutants of one strain, C58, appear to be chromosomal mutants.

Thus, both chromosomal and Ti plasmid genes are apparently involved in attachment to plant cells. The location of these genes may differ in different bacterial strains or it is possible that in some strains the genes may be present in more than one copy and be located on both the Ti plasmid and the chromosome.

VII. A Model for the Attachment of *A. tumefaciens* to Plant Cells

On the basis of the observations described above we can construct a model for the attachment of *A. tumefaciens* to plant cells (Fig. 9). Of necessity, this model must be very incomplete. Some steps in the attachment process remain vague. There may also be other steps whose existence is not suspected at the present time.

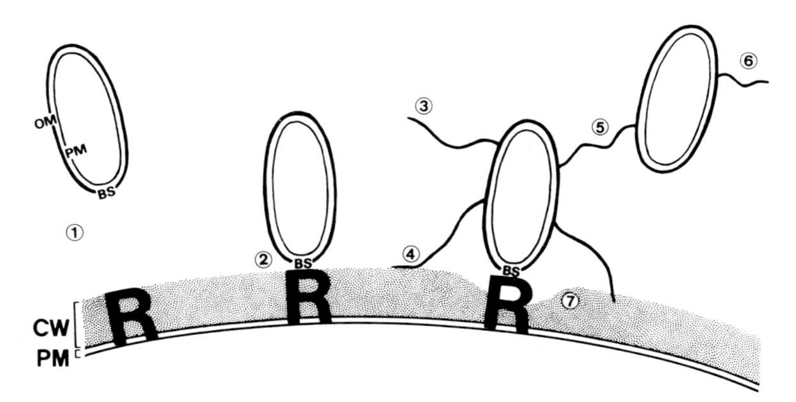

Fig. 9. A model for the attachment of *A. tumefaciens* to carrot cells. *A. tumefaciens* with a constitutive binding site on the bacterial outer membrane (OM) encounters a plant cell with a receptor site (R) on its surface (step 1). The bacterial binding site (BS) attaches to the plant receptor site (R) (step 2). The attached bacterium begins to elaborate cellulose fibrils (step 3). These fibrils anchor the bacteria to the surface of the plant cell wall (CW) (step 4) and entrap additional bacteria (step 5). These entrapped bacteria may themselves begin to synthesize cellulose fibrils (step 6) which may entrap additional bacteria. This facilitated attachment and multiplication of attached bacteria result in the development of a bacterial colony attached to the plant cell wall via binding sites and cellulose fibrils. The attached bacteria secrete enzymes such as pectinase which digest the plant cell wall (step 7) and may aid in establishing contact of the bacteria with the plant cell plasmalemma (PM) to facilitate transfer of the Ti plasmid from the bacteria to the plant host cell

Both the plant receptor for *Agrobacterium* and the bacterial binding site for plant host cells appear to exist prior to any contact between the bacterium and the plant. The bacterial binding site is composed of protein and lipopolysaccharide and is exposed on the outer membrane of the bacterium. The plant receptor site is composed of protein and possibly pectin. The receptor site is exposed on the surface of intact cells. In addition there is also a receptor site on the plasmalemma. It is not known if this second receptor site is identical to the receptor on the cell surface. The binding of the bacterium to the receptor on the plant cell surface does not require any active participation on the part of the plant cell. However, live bacteria are required for binding. The reason for this requirement is obscure. Once the bacterium has bound to the plant cell surface it begins to synthesize cellulose fibrils. Some of these fibrils anchor the bacteria to the surface of the host cell and are responsible for the extreme difficulty of removing attached *A. tumefaciens* from plant cells. Other cellulose fibrils are free in the medium and entrap unattached bacteria. These entrapped, indirectly-attached bacteria in turn elaborate cellulose fibrils which can entrap still more bacteria. In this manner a large aggregate of attached and entrapped bacteria and bacterial fibrils is built up on the plant cell surface. In addition the bacteria possess pectinase and possibly other enzymes which may modify the host cell surface in the region of the attached bacteria. The mechanism by which some or all of the attached bacteria then transfer Ti plasmid DNA to the plant cell is unknown.

Thus, even with only partial knowledge of the events following the encounter of *A. tumefaciens* with a plant host cell it is apparent that the surface interactions between the bacterium and the plant are a rather complex story. Further elucidation of these interactions seems likely to reveal a still more complicated sequence of events.

Acknowledgements

I thank Dr. Mary Deasey for helpful discussions. This research in the author's laboratory was supported by Public Health Service research grant CA18604 from the National Cancer Institute.

VIII. References

Anand, V. K., Heberlein, G. T., 1977: Crown gall tumorigenesis in potato tuber tissue. Amer. J. Bot. **64,** 153—158.

Braun, A. C., 1962: Tumor inception and development in the crown gall disease. Ann. Rev. Plant Physiol. **13,** 533—558.

Douglas, C. J., Halperin, W., Nester, E. W., 1982: *Agrobacterium tumefaciens* mutants affected in attachment to plant cells. J. Bacteriol. **152,** 1265—1275.

Glogowski, W., Galsky, A. G., 1978: *Agrobacterium tumefaciens* site attachment as a necessary prerequisite for crown gall tumor formation on potato discs. Plant Physiol. **61,** 1031—1033.

Gurlitz, R. H. G., Lamb, P. W., Matthysse, A. G., 1983: Involvement of carrot cell surface proteins in attachment of *Agrobacterium tumefaciens.* Submitted for publication.

Hahn, M. G., Darvill, A. G., Albersheim, P., 1981: Host-pathogen interactions. XIX. The endogeneous elicitor, a fragment of a plant cell wall polysaccharide that elicits phytoalexin accumulation in soybeans. Plant Physiol. **68,** 1161—1169.

Kerr, A., 1969: Crown gall of stone fruit I. Isolation of *Agrobacterium tumefaciens* and related species. Aust. J. Biol. Sci. **22,** 111—116.

Lippincott, B. B., Lippincott, J. A., 1969: Bacterial attachment to a specific wound site as an essential stage in tumor initiation by *Agrobacterium tumefaciens.* J. Bacteriol. **97,** 620—628.

Lippincott, B. B., Whatley, M. H., Lippincott, J. A., 1977: Tumor induction by

Agrobacterium involves attachment of the bacterium to a site on the host plant cell wall. Plant Physiol. **59,** 388—390.

Lippincott, J. A., Heberlein, G. T., 1965: The quantitative determination of the infectivity of *Agrobacterium tumefaciens.* Amer. J. Bot. **52,** 856—863.

Lippincott, J. A., Lippincott, B. B., 1975: The genus *Agrobacterium* and plant tumorigenesis. Ann. Rev. Microbiol. **29,** 377—407.

Lippincott, J. A., Lippincott, B. B., 1976: Nature and specificity of the bacterium-host attachment in *Agrobacterium* infection. In B. Solheim and J. Raa, eds. Cell Wall Biochemistry Related to Specificity in Host-plant pathogen Interactions. pp. 439—451. Trømso, Norway: Universitets-Forlaget.

Lippincott, J. A., Lippincott, B. B., 1978: Cell walls of crown-gall tumors and embryonic plant tissues lack *Agrobacterium* adherance sites. Science **199,** 1075—1078.

Lippincott, J. A., Lippincott, B. B., 1980: Microbial adherance in plants. In: E. H. Beachey, ed. Bacterial adherance, Receptors and Recognition. Series B, vol. 6, pp. 375—398, London: Chapman and Hall.

Manigault, P., 1970: Invervention dans la plaie d'inoculation de bacteries appartenant à differentes souches d'*Agrobacterium tumefaciens* (Smith et Town) Conn. Ann. Inst. Pasteur **119,** 347—359.

Matthysse, A. G., 1983a: The role of bacterial cellulose fibrils in infections by *Agrobacterium tumefaciens.* J. Bacteriol. **154,** 906—915.

Matthysse, A. G., 1983b: The use of tissue culture in the study of crown gall and other bacterial diseases. In: J. P. Helgeson and B. J. Deverall, eds., Use of Tissue Culture and Protoplasts in Plant Pathology, pp. 39—68, Canberra: Aust. Acad. Science.

Matthysse, A. G., Lamb, P. W., 1981: Soluble factor produced by *Agrobacterium* which promotes attachment to carrot cells. Amer. Soc. Microbiol. Ann. Meeting Abstracts, B95.

Matthysse, A. G., Wyman, P. M., Holmes, K. V., 1978: Plasmid-dependent attachment of *Agrobacterium tumefaciens* to plant tissue culture cells. Infection and Immunity **22,** 516—522.

Matthysse, A. G., Holmes, K. V., Gurlitz, R. H. G., 1981: Elaboration of cellulose fibrils by *Agrobacterium tumefaciens* during attachment to carrot cells. J. Bacteriol. **145,** 583—595.

Matthysse, A. G., Holmes, K. V., Gurlitz, R. H. G., 1982: Binding of *Agrobacterium tumefaciens* to carrot protoplasts. Physiol. Plant Path. **20,** 27—33.

Matthysse, A. G., Gurlitz, R. H. G., Lamb, P. W., Van Stee, K., 1983: Binding of *Agrobacterium tumefaciens* and of *Agrobacterium* proteins to carrot suspension cultures. Amer. Soc. Microbiol. Ann. Meeting Absts. B12.

Murashige, T., Skoog, F., 1962: A revised medium for rapid growth and bioassays with tobacco tissue cultures. Physiol. Plant. **15,** 473—497.

Mellon, J. E., Helgeson, J. P., 1982: Interaction of a hydroxyproline rich glycoprotein from tobacco callus with potential pathogens. Plant Physiol. **70,** 401—405.

Ohyama, K., Pelcher, L. E., Schaefer, A., Fowke, L. C., 1979: *In vitro* binding of *Agrobacterium tumefaciens* to plant cells from suspension culture. Plant Physiol. **63,** 382—387.

Otten, L., DeGreve, H., Hernalsteens, J. P., Van Montagu, M., Schieder, O., Staub, J., Schell, J., 1981: Mendelian transmission of genes introduced into plants by the Ti plasmids of *Agrobacterium tumefaciens.* Mol. Gen. Genet. **183,** 209—213.

Pueppke, S. G., Benney, U. K., 1981: Induction of tumors on *Solanum tuberosum* L.

by *Agrobacterium:* quantitative analysis, inhibition by carbohydrates, and virulence of selected strains. Physiol. Plant Path. **18,** 169—179.

Pueppke, S. G., Kluepfel, D. A., Anand, V. K., 1982: Interaction of *Agrobacterium* with potato lectin and concanavalin A and its effect on tumor induction in potato. Physiol. Plant Path. **20,** 35—42.

Smith, V. A., Hindley, J., 1978: Effect of agrocin 84 on attachment of *Agrobacterium tumefaciens* to cultured tobacco cells. Nature **276,** 498—500.

Tanimoto, E., Douglas, C., Halperin, W., 1979: Factors affecting crown gall tumorigenesis in tuber slices of Jerusalem artichoke (*Helianthus tuberosus,* L.). Plant Physiol. **63,** 989—994.

Whatley, M. H., Bodwin, J. S., Lippincott, B. B., Lippincott, J. A., 1976a: Role for *Agrobacterium* cell envelope lipopolysaccharide in infection site attachment. Infection and Immunity **13,** 1080—1083.

Whatley, M. H., Lippincott, B. B., Lippincott, J. A., 1976b: Site attachment in *Agrobacterium* infection involves bacterial "O-antigen". Amer. Soc. Microbiol. Absts. Ann. Meeting, B8.

Whatley, M. H., Margot, J. B., Schell, J., Lippincott, B. B., Lippincott, J. A., 1978: Plasmid and chromosomal determination of *Agrobacterium* adherance specificity. J. Gen. Microbiol. **107,** 395—398.

Section II.
Symbiosis

Chapter 3

Legume-*Rhizobium*-Symbiosis: Host's Point of View

D. P. S. Verma and K. Nadler*

Department of Biology, McGill University, Montreal, Canada

With 5 Figures

Contents

* Permanent Address: Department of Botany and Plant Pathology, Michigan State University, East Lansing, Mich., U.S.A.

I. Introduction

The legume-*Rhizobium* endosymbiosis may be the most highly evolved and perhaps ultimate association between a microbe and a plant in which the two partners can still grow independently. Undoubtedly the strong selective pressure on this association is the resulting nutritional complementation: the plant can be considerd a carbon-rich, nitrogen-deficient phototroph and the *Rhizobium* a carbon-deficient, nitrogen-fixing heterotroph. The resulting symbiosis which occurs in a specialized organ, the root nodule, makes the plant autotrophic with respect to the availability of reduced nitrogen, a limiting factor in plant nutrition. This unique intracellular association contributes significantly towards the yield of the agriculturally important legume crops.

The process of symbiotic nitrogen fixation and the steps leading to it are highly complex and are influenced by genetic factors of both the host and the endosymbiont. Other factors, e. g. nutrition, soil and climate also influence the development and efficiency of this association. However, these topics are not subject of this review and the reader is referred to the thorough discussions of Bergersen (1977) and Dart (1977). Moreover, a general outline of the molecular biology of the legume-*Rhizobium* association has recently been provided (Verma, 1982; Verma and Long, 1983) and the molecular genetics of nitrogen fixation has been extensively reviewed (Brill, 1980; Beringer *et al.*, 1980; Kondorosi and Johnston, 1981; Postgate, 1982). The early recognition events, including specificity and attachment leading to successful infection, have been enumerated by Dazzo and Gardiol in Chapter 1 (see also Bauer, 1981) and the recent developments on the specific role of *Rhizobium* in this process will be discussed by Rolfe and Shine in Chapter 4. Here we concentrate on the contribution of the host plant in this symbiotic process. We shall selectively review and reexamine the requirements of symbiosis and the possible origin of the symbiotic fixation of atmospheric nitrogen in legumes, the early responses of the host plant as it relates to the development of an effective infection, the specialized mechanism of controlled invasion of the plant as a possible means for regulation of the host defenses, and the role of plant hormones in the symbiotic process. These steps, carried out by specific host gene products are influenced or controlled by the activities of specific *Rhizobium* genes. Some of the plant gene products which may be involved in these processes have been identified and a few have been characterized in detail. With the rapid progress in molecular biology, particularly in recombinant DNA technology, it is now possible to explore the detailed structure and perhaps to deduct the possible functions of these genes including regulatory mechanisms which are operative in the symbiotic state. A detailed understanding of the legume-*Rhizobium* symbiosis at the molecular level may not only illuminate the shadowy route by which this process evolved but may provide ways to manipulate this very important association in nature.

II. Origin of Symbiotic Nitrogen-Fixing Association

Two basic requirements for the biological reduction of atmospheric nitrogen are the availability of metabolic energy and low pO_2. In spite of the fact that reduced nitrogen is vital for life, no eucaryotic organism has yet developed the capacity to reduce (fix) atmospheric nitrogen. This ability remains restricted to a few procaryotes. Considering the absolute requirement for low oxygen tension, it is unlikely that nitrogen fixation could occur efficiently in the aerobic metabolic environment of a plant cell. In fact, all vital gases such as oxygen, hydrogen and carbon dioxide are reduced inside specialized membrane compartments generally in eucaryotic cells. Organisms capable of directly reducing these gases existed before the evolution of eucaryotes; it is now widely believed that the present-day eucaryotic organelles originated from such organisms (Margulis, 1981). However, a "nitrogen-fixing organelle" (see Verma and Long, 1983) has not yet succeeded. The incompatabilities of nitrogen fixation with the aerobic environment of a eucaryotic cell and the availability of abundant reduced nitrogen in the atmosphere during early biological evolution may have prevented evolution of such an organelle.

Those nitrogen-fixing organisms which could not support their energy requirements in the free-living state formed associations with phototrophic organisms like plants. Thus, one finds in nature a variety of associations (Vose and Ruschel, 1981; Reisert, 1981; Verma and Long, 1983) ranging from colonization of rhizosphere and phyllosphere to the endosymbiotic nitrogen fixing state, as in *Rhizobium* and *Actinomycetes* associations with legumes and a few non-legumenous plants respectively. We will restrict our discussion here primarily to the legume-*Rhizobium* symbiosis.

A. Rhizobium as an Organelle

In the endosymbiotic state *Rhizobium* is transformed into a form known as the bacteroid, which performs various unique functions in close cooperation with the host and thus practically behaves like an organelle. Obviously none of the *Rhizobium* functions became obligately dependent upon the host genome like those of chloroplasts and mitochondria. However, in the endosymbiotic state *Rhizobium* has become physiologically dependent upon the host for several processes. *Rhizobium* bacteroids exhibit derepression not only of nitrogenase genes but also of those for hydrogenase and specific cytochromes, and genes responsible for changes in the outer membrane (see Sutton *et al.*, 1981). In addition, the metabolism of ammonia is altered (see Miflin and Cullimore, Chapter 5). Once inside the host cell, *Rhizobium* cell division and differentiation are influenced by the host. The two partners also share several structural components (e. g. peribacteroid membrane) and exchange a number of metabolites. Thus, it can be argued that *Rhizobium* is evolving towards a nitrogen-fixing organelle (Verma and Long, 1983), the ultimate association between the two organisms making them dependent on one another. A genetic exchange between the two organisms, in a manner similar to that with *Agrobacterium*, may facilitate this path.

B. Why Primarily Legumes?

Rhizobium does not generally fix nitrogen *ex planta* and thus requires both carbon and nitrogen for its survival. The nutritional environment provided by primitive legumes probably brought these organisms together, from where they coevolved into the modern symbiotic nitrogen-fixing forms. The fact that the host range of *Rhizobium* is very restricted, even within legumes (see Allen and Allen, 1981), suggests that, in addition to specificity, a tight relationship may exist with respect to the nutritional demands of the symbiont. The latter area of study has received little attention. In fact, the rhizosphere of a legume plant is very rich in nutrients and can generally support various auxotrophic mutations in *Rhizobium*. If recognition in *Rhizobium* to associate with legumes preceded their ability to reduce nitrogen, then this organism was a pathogen depending on the host for the supply of both carbon and nitrogen as a necrotroph or biotroph. The plant might respond to such a pathogen by developing restrictions to infection by keeping infection localized. Once such an organism acquired nitrogen fixation genes which could be expressed under some microaerophyllic conditions and make reduced nitrogen available to the plant, the host may have eliminated its defense mechanism against the *Rhizobium* or developed a special mode of infection. Controlled invasion (see below) which allows the host to localize infection and bring *Rhizobium* into close proximity may be a step in that direction.

On the other hand, nitrogen-fixing organsims existed before the origin of plants and nitrogen-fixing ancestors of *Rhizobium* lacking the ability to associate with the plant may have survived on hydrogen as a reductant for CO_2 and N_2 fixation, a capacity retained by some slow-growing *Rhizobium* (Lepo *et al.*, 1980). Since nitrogen fixation is an energy intensive process, 'proto-*Rhizobium*' may have developed an association with plants as a source of carbon and reductant.

Symbiotic nitrogen fixation only commences after a suitable association has developed; this argues that a first step towards the development of such an association was the removal or supression of host defense mechanism(s). This could be accomplished in two ways: 1) by actively suppressing the plant response(s) to invasion or 2) the removal from *Rhizobium* of the appropriate determinants which trigger the plant defense mechanism(s). Perhaps a combination of both allows *Rhizobium* to invade the host cell. Moreover, the host restricts the spread of this infection by compartmentalization and tissue specialization [root nodule formation]. This is accompanied by transformation of *Rhizobium* into a form which can function effectively in the endosymbiotic state. The fact that most legume-*Rhizobium* symbioses are very similar suggests the association evolved very early in evolution in an early ancestor of present day Leguminosae.

Considering the oxygen sensitivity of the nitrogenase enzyme system of diazotrophs, it is not surprising to find that only those plants which evolved a mechanism for the protection of this enzyme have developed effective symbotic associations. A direct correlation between the presence of leghemoglobin and effectiveness of symbiotic nitrogen fixation in legumes and the fact that a globin-like molecule also exists in non-leguminous (*Casvarina* and *Parasponia*) nodules (Appleby *et al.*, 1983), suggests that the evolution of an oxygen binding molecule and its induction only in infected nodule tissue is central to this process. In addition, a number of other specific host genes appear to be involved in the development of root nodules (see below). Perhaps the non-nodulating close relatives of symbiotic nitrogen-fixing legumes have all but a few of the genes necessary for root nodule development. Therefore, a detailed examination of these species at the molecular level may answer the question of why the *Rhizobium* association is so restricted. Since legumes produce protein-rich seeds there exists a sink for nitrogen which may provide further selection pressure on this association to evolve towards a symbiotic route. This helps a selected group of plants, primarily members of the family Leguminosae, to grow in nitrogen deficient soils.

III. Role of the Host Plant in Symbiosis

Involvement of the host plant in the symbiosis is apparent at all levels (see Nutman, 1981; Verma, 1982; Verma and Long, 1983), and it appears that both organisms coevolved to accomplish this highly efficient and mutually beneficial relationship. The host plant has not only to provide total energy, both for the process of nitrogen fixation and the survival of *Rhizobium*, and a mechanism for protecting nitrogenase system, but also must eliminate any pathogenic responses against the invader. The latter could be mediated via the development of a mechanism for controlled invasion and strict compartmentalization in the nodule structure. Any breakdown in this mode may render *Rhizobium* pathogenic on the host. In fact, the lack of development of an infection thread or premature release of *Rhizobium* from the infection thread (due to genetic incompatibilities or mutations in host or bacteria) aborts the process of nodule development and the *Rhizobium* infection becomes pathogenic (Verma *et al.*, 1978a; Bassett *et al.*, 1977b; Vincent, 1980). A number of specific *Rhizobium* mutants affecting various stages of nodule development and eliminating an effective association have been recently isolated (see Scott *et al.*, 1982 and Meade *et al.*, 1982). When mutant strains with different symbiotic defects are mixed with each other complementation may occur to produce an effective symbiosis (Rolfe and Greshoff, 1980). These results indicate a series of plant/*Rhizobium* interactions occurring at various steps e.g. infection, nodule initiation, bacteroid development and persistence of nodule function.

A. Early Responses of Host

The first step in the development of the root nodule symbiosis is the mutual recognition by the two partners — a process which is rather specific. The specificity of this recognition (an attachment leading to successful infection) has been discussed in Chapter 1 (see also Bauer, 1981; Reisart, 1981). The host plant responds rapidly to the challenge by *Rhizobium.* Within two to four hours, the infection that will give rise to nitrogen-fixing (fix^{+}) nodules apparently blocks nodule formation at adjacent sites on the root (Bauer, 1981). This indicates that effective nodulation is a negatively self-regulating phenomenon (Nutman, 1952) by which a plant controls the number of nodules and avoid excessive sink development with respect to the supply of carbon. While the nature of this regulatory control is completely unknown, it appears that the plant can distinguish the fate of an infection at a very early stage with respect to its potential in establishing the symbiotic nitrogen-fixing state.

An early structural response of the host is curling of the root hairs. This phenomenon is interpreted as inhibition of cell growth at the attachment site (see Bauer, 1981) and is found not only in most legume-*Rhizobium* associations but also in *Actinorrhiza* attachment to *Comptonia* and *Casuarina*, non-legume symbiotic nitrogen-fixers (Callaham and Torrey, 1977). The curled root hair forms a "pocket" entrapping the adhering *Rhizobium* and a hyaline spot appears at the attachment site. This pocket appears to concentrate nutrients for the entrapped bacteria and create a microenvironment conducive to the formation of wall loosening enzymes necessary for the appropriate invasion of the host (Callaham, 1979; see also Verma, 1982). In those legumes where no root hair curling is observed, the infection proceeds via breaks in the cell wall due to the emergence of lateral roots (see Dart, 1974, 1977).

The host plant appears to control not only the mode of infection but also the morphology and intracellular organization of the infected cell. A strain of *Rhizobium* (*e g. R. lupini* strain D-25) forms collar shaped nodules on *Lupinus*, containing rod shaped bacteroids enclosed singly in membrane envelopes, but cylindrical nodules on *Ornithopus*, in which several bacteroids are present in each membrane envelope (Kidby and Goodchild, 1966). Similar differences were observed when a strain of *Rhizobium* infected *Phaseolus vulgaris* and *Arachis hypogaea* (see Sutton *et al.*, 1981).

B. Controlled Invasion as a Means of Regulating Host Defense Response

The above observations suggest that following specific attachment of *Rhizobium* with a particular host, the plant recognizes the bacterium and allows infection to proceed in a controlled fashion. A unique feature of symbiotic nitrogen fixing associations (particularly between *Rhizobium* and legumes) is that bacterial invasion of the host is mediated by a tubular structure known as the infection thread. The infection thread grows inwards reaching and spreading to the cells that become meristematic as a result of or prior to the infection and restricts the invading microbe to an extracytoplasmic compartment. This situation also occurs in infection of non-leguminous plants by *Actinomycetes* (Lalande and Knowles, 1975) as well as in most ectendomycorrhiza infections (see Gianinazzi-Pearson, Chapter 8). The *Rhizobium* is finally released from the infection thread but remains enclosed in a membrane envelope (peribacteroid membrane) derived from the plasmalemma of the host (see Newcomb, 1981; Verma, 1982; and Verma and Long, 1983). Thus, the two organisms remain separated from each other throughout the endosymbiotic stage, keeping this association topologically extracellular. Such a controlled invasion may be considered to be a mechanism to avoid host defense responses against the invading "pathogen". Alternatively, deposition of infection thread material in the path of invading *Rhizobium* may itself be a defense mechanism of the host. As *Rhizobium* differentiates and is ready for release from the infection thread, the synthesis of cell wall material deposited as infection thread declines and wall-less regions appear at the tip of the infection thread (see Newcomb, 1981). Cell wall hydrolysing enzyme(s) of the host also appear to play a role in this process (Verma *et al.,* 1978b).

The composition of the infection thread wall is very similar to the primary plant cell wall. Indeed its biosynthesis is controlled by the host (Nutman, 1965; Sahlman and Fahraeus, 1963; Dart, 1974; Newcomb, 1976). Deposition of host cell wall material internal to the point of invasion has been observed (Callaham and Torrey, 1981). Recently, Robertson and Lyttleton (1982) have suggested that coated vesicles of the plant are involved in the biosynthesis of the infection thread wall material in a way similar to the normal cell wall synthesis. These vesicles appear to arise from Golgi bodies and transport primary cell wall material (Robertson *et al.,* 1978). In contrast to the invasion by other plant pathogens, the cell wall of the host is not completely dissolved at the site of *Rhizobium* invasion, and synthesis of host cell wall in the form of infection thread continues until the *Rhizobium* is "released" inside the host cell. The lumen of the infection thread is filled with mucopolysaccharide material apparently of bacterial origin (Dart, 1974). A similar situation is observed in the nitrogen-fixing associations between *Actinomycetes* and non-leguminous plants, although the composition and the origin of the matrix material appears to be different. The matrix is made up of pectic substances derived from the plant (Lalonde and Knowles, 1975). The production of copious amounts of this material may be a mechanism by which surface contact between the two symbiotic partners is restricted until the microbe has transformed its cell wall components and endocytosis occurs.

Virulent strains of phytopathogenic bacteria also produce large amounts of exopolysaccharides and thus avoid both the entrapment and the hypersensitive reactions of the host plant (see Sequeira, 1980). Avirulent derivatives lacking this exopolysaccharide interact with the plant cell wall and are agglutinated, which may result in a hypersensitive response. The interaction between bacterial lipopolysaccharides (LPS) and plant lectins apparently initiates the agglutination reaction. Bacterial production of exopolysaccharides (EPS) prevents this cognitive interaction, thus permitting invasion of the plant. It should be noted that LPS-lectin interaction results in avirulence in the *Pseudomonad*-plant system. By contrast, interaction of LPS with plant cell wall components is essential for virulence of *Agrobacterium* (see Sequiera, 1980). That *Rhizobium* is restricted to the infection thread and is embedded in a mucilaginous matrix suggests that rhizobial EPS prevents further direct interaction of the prospective symbionts, allowing the infection to proceed without evoking any defense in the host towards the invader.

More direct evidence of the involvement of exopolysaccharides in the above process is obtained from studies by Chakravorty *et al.* (1982). A transposon (Tn5)-induced mutant of *R. trifolii* which produces no water-soluble EPS causes infection of clover root hair, but the infected cell does not form proper infection threads and is filled by the bacteria. To some extent, this response is similar to the hypersensitive response of the plant. It appears that the plant reacts to the unencapsulated rhizobia and by inducing hypersensitive reaction(s) may prevent further spreading of infection. This mutation has been mapped to a region involved in exopolysaccharide biosynthesis.

If rhizobial EPS is elaborated in the infection to prevent *Rhizobium* -plant wall intractions then a corollory of this hypothesis is that as rhizobia are "released" from the infection thread they modify their outer wall components to be compatible with the host cellular environment and thus prevent any host responses. Such an alteration inadvertently may prevent specific synthesis of exo- and lipopolysaccharide material at this stage of bacterial transformation. In fact, when bacteria are released inside the membrane envelopes (peribacteroid membranes) no mucopolysaccharide is present in these vesicles (see Newcomb *et al.,* 1979a). The determinants which were present on the outer membrane during initial stages of infection may be removed and new membrane components developed and/or exposed. We have observed, in the case of *R. japonicum*, that the outer membrane is sloughed off during the early infection process and fragments of this material can be found inside the membrane envelopes enclosing the bacteroids (Fig. 1). Since the bacterial membrane is relatively insoluble in nonionic detergents, fragments of this membrane can be isolated from the nodule. The protein profile of these membrane fragments is very similar to the outer membrane of the free living *Rhizobium* (Bal *et al.*, 1980). In an incompatible association between soybean and *R. japonicum* (strain 61A24) some fibrous material is also found within the membrane envelopes (see Verma, 1982). Whether this material is of plant or bacterial origin remains

▷

Fig. 1. Endomembrane system of soybean root nodule, (A), cross section of a 2 week old infected cell; (B), cross section of an infected cell treated with Nonidet-P40 (0.5%) at 25°C. The endomembranes including the membrane envelope (peribacteroid membrane) are solubilized but the bacterial membrane fragments inside the membrane envelope (arrows) appear to remain intact. B, bacteroids; d, dictyosomes; er, endoplasmic reticulum; me, membrane envelope; cw, plant-cell wall; p, plasma membrane (from Bal *et al.*, 1980)

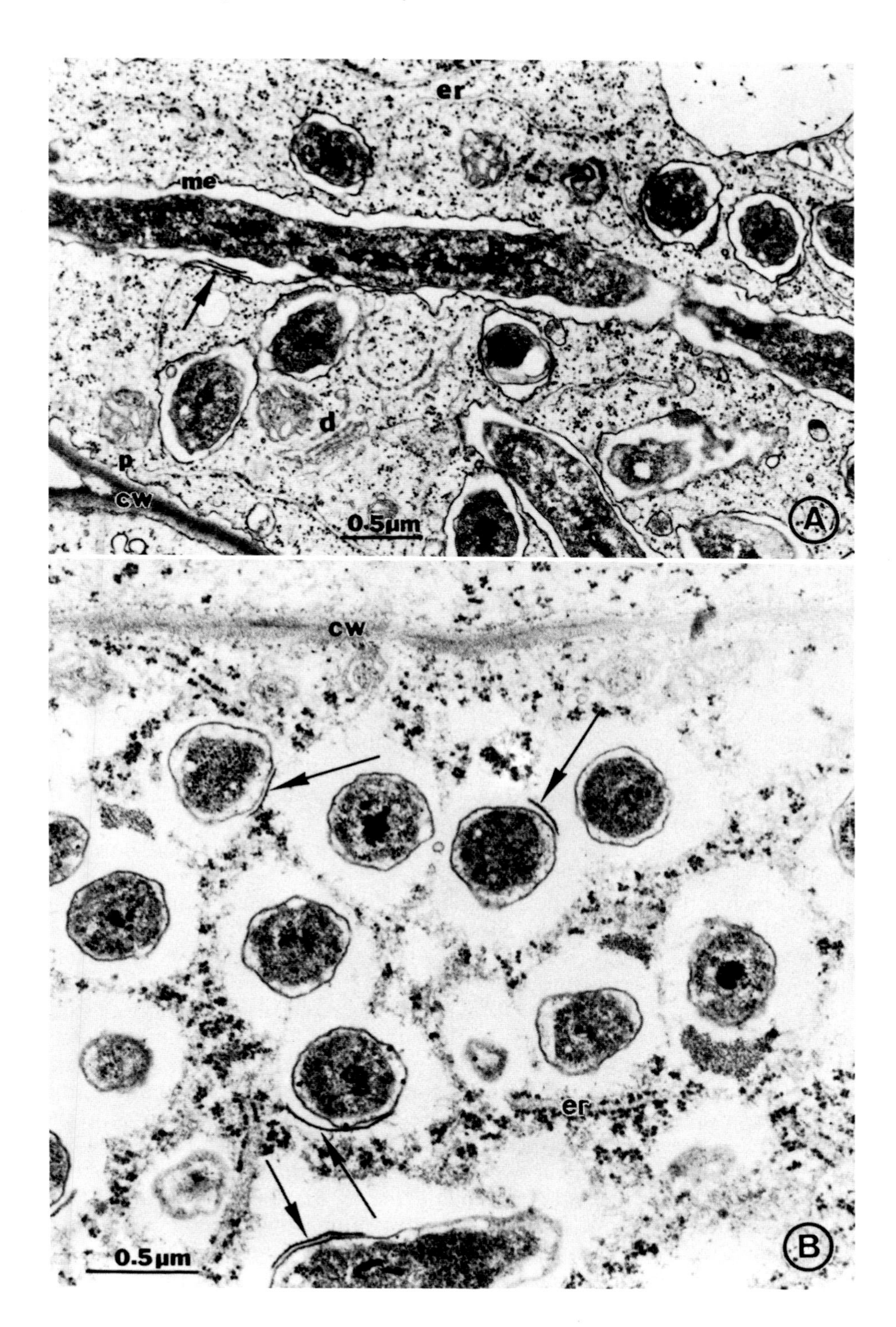
er
me
d
p
cw
0.5µm
A
cw
er
0.5µm
B

unclear. In the effective association, the infection thread wall is completely removed either by dissolution (Verma *et al.*, 1978 b) and phagocytosis (Bassett *et al.*, 1977 a) or by lack of synthesis (Newcomb, 1976), releasing bacteria enclosed in a membrane of host origin.

This bacteroid-enclosing membrane, known as the peribacteroid membrane, defines a cellular compartment where *Rhizobium* can function as an "organelle" (Verma and Long, 1983) and in which the host can provide the necessary requirements for symbiotic nitrogen-fixation. The peribacteroid membrane is a unique, subcellular structure elaborated by the host plant (Robertson *et al.*, 1978 a, b; Verma *et al.*, 1978 b). This membrane retains many of the properties of the plasmalemma while undergoing some specific changes (see Verma *et al.*, 1978 b). It is somewhat lighter than the plasmalemma (Robertson *et al.*, 1978 b), indicating that substantial host lipid synthesis is essential for bacterial vesiculation. Since all nutrients for the symbiotic rhizobia must traverse this membrane barrier, we would expect numerous plant genes for these transport systems to be expressed specifically during elaboration of this membrane system in nodules. These may be some of the products of nodule-specific plant genes (see below).

C. Phytohormones

i) Action at Distance: The involvement of plant growth hormones (e. g. indoleacetic acid, IAA) in root nodule symbiosis was suggested long ago (Thimann, 1936); however, to date there is no direct and conclusive evidence for their specific role. Since nodule development involves meristematic activity, plant hormones are believed to be integral to this process. A number of microorganisms that produce phytohormones cause deformation of root hairs. The culture filtrate of *Rhizobium* can induce curling or branching of host root hairs, but the complete curling characteristic of infection is only produced by the presence of live bacteria (Yao *et al.*, 1976). This may be due to specific attachment and a localized hormone action.

Following the invagination of root hair cells, an early response of the host is the elicitation of cell division in the cortex some distance away from the site of infection (Newcomb *et al.*, 1979a). A similar observation has been made in the early formation of *Actinomcyte* nodules (Calaham and Torry, 1977; Newcomb *et al.*, 1978). This suggests a diffusible substance stimulating cell division is produced by the endophyte. Growth promoting substances have been implicated not only in the growth of infection threads and in nodule morphogenesis (Kefford *et al.*, 1960) but may also be responsible for endoreduplication and the increase in ploidy level observed in the infected cells (see Libbenga *et al.*, 1973; and Libbenga and Torrey, 1973).

Extensive morphological evidence exists that diffusible factors play a role in nodule development (Libbenga *et al.*, 1973; Newcomb *et al.* (1979 a, b). In nodulating roots of soybeans and peas numerous cytological changes and cortical cell divisions are observed while the infection thread is still restricted to the root hair cell. In pea, cell divisions occur in the inner cortex, with the highest frequency noted close to the stele. In soybean, divisions occur in the outer cortex. Proliferation and hypertrophy of *Comptonia* root cortical cells is observed beneath the *Actinomycete*-infected root hair cell. These cell divisions lead to the formation of "multicellular colonies of dense cells" and initiate nodule growth. Wall in-growths in pericycle cells also are observed at this early stage of nodule formation (Newcomb, 1976).

Truchet *et al.* (1980) postulated the presence of a rhizobial "nodule organogenesis-inducing principle" (NOIP). In nodules induced by leucine-requiring mutant of *R. meliloti*, bacteria are not released from the infection threads and the nodules have a typical ineffective anatomy. When leucine is provided, these nodules develop a normal anatomy even though bacteria remain within the infection thread. Thus normal nodule morphogenesis can be triggered at a distance by the infecting rhizobia. The nature of NOIP is unknown although exogenous IAA and cytokinin induce cortical proliferation in pea root explants (Libbenga *et al.*, 1973).

ii) Phytohormones Produced by Rhizobium: Microbial production of plant growth hormones is a feature common to several plant-microbe interactions (see Kado, Chapter 12). It is now generally accepted that both *Rhizobium* and *Agrobacterium*, closely related members of Rhizobiaceae, produce IAA and cytokinins. These compounds have been identified unambiguously by GC/mass spectrophotometry in several species of fast-growing *Rhizobium* (Badenoch-Jones *et al.*, 1982; Wang *et al.*, 1982). Nodulating and non-nodulating strains of *R. trifolii* and *R. leguminosarum* both produce similar amounts of IAA. Thus IAA production in *Rhizobium* may not be related specifically to the virulence (nodulating) property. In the case of *Agrobacterium*, genes associated with IAA synthesis are located on the Ti-plasmid. The T-DNA region (the segment transferred to the host) of this plasmid influences the auxin/cytokinin ratio in the host (Liu *et al.*, 1982; Kao *et al.*, 1982). Some mutations in the T-DNA causing avirulence can be complemented by IAA, resulting in tumorigenesis (see Kado, Chapter 12 for details). Specific nutritional conditions are required (i. e. incubation in tyrosine rather than in tryptophan) far plasmid regulated IAA production in *Agrobacterium* to occur. Similar experiments have not been carried out yet with *Rhizobium.*

Formation of indeterminate nodules may be due to different responses of the host to IAA produced by *Rhizobium.* The level of IAA in the nodule may be further controlled by the pseudoperoxidase activity of leghemoglobin (Puppo and Rigaud, 1975). In addition to IAA, *Rhizobium* may produce other substances which regulate directly or indirectly the level of growth hormones normally produced by the host. There is preliminary evidence correlating IAA application and *Rhizobium* infection with respect to the appearance of specific host gene products (Verma *et al.*, 1983a). A group of low molecular weight proteins was found to be synthesized 3 days after the application of either IAA or *Rhizobium japonicum* to soybean seedlings.

D. Cell and Tissue Level Organization

The legume-*Rhizobium* symbiosis is restricted to a particular plant organ nodule (modified root). The cells in nodule tissue are highly organized and this organization is plant specific. A striking zonation within the root nodule is observed, with bacteria routinely excluded from the epidermal and meristematic zones. Two types of nodule growth habit, hence zonation, are observed (see Allen and Allen, 1981 and Newcomb *et al.*, 1979a). In determinate nodules, rhizobia are restricted to cells within the central, red-pigmented zone. This zone is generated by divisions of infected and some uninfected cells to form a mass of cytoplasmically-rich cells. In indeterminate (meristematic) nodules, the zones are arranged linearly with the meristem at the nodule tip followed sequentially by a zone of newly infected cells, of infected mature cells and lastly by a zone of senescing cells located at the base of the nodule. The infected cells enlarge as bacteria are released into them. There is a large generalized increase in numbers of host cell organelle profiles coincident with bacteroid formation. Presumably this concomitant hypertrophy is due to the metabolic demands of the intracellular symbiosis. In the zone of mature infected cells, the organelles of the host cell are forced to the periphery by the mass of membrane-enclosed bacteroids. The rough ER, ribosomes and starch deposits in the infected cells decline.

In a root nodule a maximum of about 50% of the total cells are infected by *Rhizobium* (see Dart, 1977; Newcomb, 1981 and Verma *et al.*, 1981) which are interspersed with uninfected cells. The DNA content of uninfected cells remains low (2 to 4C) (Truchet, 1978) while that of infected cells increases several fold (Libbenga and Bogers, 1974). Some cells in the pericycle of nodule vascular tissue develop "transfer cell" morphology (Pate *et al.*, 1979a) which may facilitate movement of photosynthates and/or nitrogen fixation products. In addition, outer cortical cells of the determinate nodules form aerenchyma and one to two layers of schlerenchyma. These cells perform special functions e. g. they facilitate diffusion of gases (Bergerson and Goodchild, 1973). Uninfected cortical cells are vacuolated and much smaller than their infected neighbours. The infected and uninfected nodule cells are not only different with respect to the presence or absence of bacteria but they also differ in metabolic functions. The distribution of organelles and associated enzymes between infected and uninfected cells was examined after separation of these cells on surcrose gradients (Hanks *et al.*, 1983). Allantoinase, an enzyme of the endoplasmic reticulum, and the peroxisomal enzymes uricase and catalase, required for the conversion of uric acid to allantoin, appear to be localized in the uninfected cells (see below). Moreover, several enzymes of purine biosynthesis are enriched in these cells. The data suggest that in the ureide exporting root nodules, e. g. soybean, much of ammonia assimilation occurs in uninfected cells. A vigorous intercellular transport system of nitrogenous compounds between infected and uninfected cells would thus be essential. The function of uninfected cells in indeterminate nodules is unknown but may be similarly specialized.

E. Nutritional Role of the Plant Host

Since *Rhizobium* does not generally fix nitrogen in the free-living state, it must depend on the host both for carbon as well as for nitrogen during the early stages of invasion. Nitrogen fixation commences at least 10 days after infection (Verma *et al.*, 1979). Any delay in the onset of nitrogen fixation may create a nitrogen imbalance deleterious to the host. Following the commencement of nitrogen fixation the nitrogen metabolism of both host and endosymbiont is altered (see Miflin and Cullimore, Chapter 5). Since the process of nitrogen-fixation is very energy intensive its activity is limited by the availability of photosynthates from the plant. How carbon and nitrogen compounds are metabolized in a specific association depends upon the two partners and may be specific for each association.

Nodulation is accompanied by qualitative and quantitative changes in the total carbon and nitrogen economy of the host plant. Pate *et al.* (1979a, b) have compared the fluxes of fixed carbon and N_2 in lupine with or without added nitrate. Although N_2- and nitrate-grown plants were essentially similar in their carbon and nitrogen nutrition, nodulated plants showed a lower conversion of photosynthate to dry matter and a high consumption of photosynthate by the below-ground parts. Interestingly, approximately 50% of the nitrogen content of root nodules themselves is derived from nitrogenous compounds cycled through the leaflets and translocated down to the nodules, the remainder being supplied by direct incorporation of recently fixed N_2. Thus the efficiency of these transfer processes is important in the N nutrition of the legume host.

With respect to nitrogen assimilation two major groups of plants exist, one metabolizing fixed nitrogen via the amide pathway and the other via the ureide pathway (for a detailed discussion on these pathways see Chapter 5). It now appears that although these pathways exist in the uninfected plants the specific enzymes involved in the nodule tissue may be synthesized *de novo* (see below and Chapter 5).

i) Carbohydrate Supply: A continous supply of photosynthates (as respiratory substrates for the symbionts as well as carbon skeletons for assimilation of nascent ammonia) is required for symbiotic nitrogen fixation. Hardy and Havelka (1976) summarized the evidence indicating that photosynthate availability is a major limiting factor in leguminous nitrogen fixation. Factors which tend to increase photosynthate supply such as increased illumination, CO_2 enrichment or the grafting of additional shoots onto a nodulated root system tend to increase nitrogen fixation. Similarly, defoliation or darkness, which decreases photosynthate supply, also decreases nitrogen fixation. Nodule activity at night is driven by nocturnal translocation and/or utilization of the reserve carbohydrates of the legume plant. Thus the length of time one species can continue fixing nitrogen in the absence of photosynthesis depends on the amount of carbon reserve it stores.

The nature of the respiratory substrate for each of the symbionts is the subject of some controversy. Whereas the carbon supply for the infected cells is assumed to be the major carbon transport form in each legume host, the substrate for the rhizobia might be reducing sugars (Trinchant *et al.*, 1981), sugar alcohols, or organic acids (Ronson *et al.*, 1981; Finan *et al.*, 1983; Glenn and Brewin, 1981). Organic acids are the best substrates for respiration and nitrogen fixation by bacteroid suspensions; sucrose is generally ineffective at substrate concentrations (see Bergersen, 1974). However, Rigaud and co-workers (Trinchant *et al.*, 1981) have found recently that sucrose and glucose support acetylene reduction by bacteroids maintained under very low oxygen tension by leghemoglobin and they propose that hexoses support nitrogen fixation in nodules. Thus failure of previous workers to demonstrate sugar-dependent nitrogen fixation by bacteroids may have been due to the major role played by oxygen in bacteroid uptake of sugars.

In contrast, several studies have indicated that bacteroids utilize dicarboxylic acids to support nitrogen fixation. *R. trifolii* mutants with defects in the uptake or metabolism of a variety of sugars induce effective root nodules on clover (Ronson and Primrose, 1979). Glenn and Brewin (1981) isolated a class of *R. leguminosarum* mutants which were resistant to high concentrations of succinate, fumarate or malate. These mutant strains were defective in uptake of organic acids and induced nodules with decreased nitrogenase activity. Ronson *et al.* (1981) also isolated mutants of *R. trifolii* which were defective in dicarboxylic acid transport and these mutants failed to grow on succinate, fumarate or malate and induced ineffective nodules on clover. Revertants regained the ability to utilize succinate and to symbiotically fix nitorgen. Since mutants induced nodules with morphologically normal bacteroids, the authors suggested that organic acids were the substrates supporting nitrogen fixation but that other carbon sources could be utilized during nodulation. Alternatively, as suggested by these authors, dicarboxylates may be essential as substrates for rhizobial heme synthesis.

ii) Nitrogen Metabolism: Nitrogen metabolism in nodules is clearly a shared activity of the symbionts (see Miflin and Cullimore, Chapter 5). The symbiotic bacteria are the source of ammonia exported to the plant cells as a stable product of nitrogen fixation (Bergersen, 1965; O'Gara and Shanmugam, 1976) but the host plant plays the major role in assimilation of this primary product. Joint assimilation of symbiotically-fixed nitrogen is the subject of chapter 5 and will not be discussed here at length.

It should be emphasized that the plant has specific functions in nitrogen metabolism other than that of a strong sink for fixed nitrogen. As with all other nutrients in this symbiosis, the plant is the source for a variety of nitrogenous compounds for *Rhizobium* during and after infection. In general, mutant strains of *Rhizobium* with requirements for amino acids induce effective nodules (Kondorosi *et al.* 1977; Dénarié *et al.*, 1976; Schwinghamer, 1977). Thus the plant provides the endosymbionts with many amino acids. Interesting exceptions to this generality are two types of amino acid auxotrophs of *Rhizobium* that are symbiotically defective. These require glutamine and leucine (Kondorosi *et al.*, 1977; Truchet *et al.*,

1980). Glutamate-requiring strains of *R. meliloti* were found to induce effective nodules but glutamine-requiring strains induced ineffective root nodules (Kondorosi *et al.*, 1977). The symbiotic defect of these latter mutants is believed to be a pleiotropic due to the impaired regulation of rhizobial glutamine synthetase. Leu^- mutants of *R. meliloti* are unable to leave the infection thread; provision of leucine or its precursors during infection restores effective symbiosis. It is likely that the host plant cannot supply leucine in the amount required for bacterial release from the infection thread (Dénarié *et al.*, 1977; Truchet *et al.*, 1980). Similarly, defects in symbiosis are associated with auxotrophy for purines (particularly adenine) and vitamins such as riboflavin. Provision of vitamins during infection restores the effective phenotype but adenine feeding does not (Schwinghamer, 1970; Pankhurst and Schwinghamer, 1974). The legume host clearly does not supply sufficient purines and flavins for formation of the symbiosis.

iii) Exchange of Other Nutrients: In addition to a vigorous exchange of carbon and nitrogen compounds, the symbionts also must engage in a complementary exchange of other essential nutrients. Among these are iron, sulfur, heme, cobalt and molybdenum (see Bergersen, 1974).

It is interesting to estimate the magnitude of these transport processes. Hardy and Havelka (1976; p. 433) have determined that the specific activity of nitrogen fixation of the root nodules of field-grown soybeans is approximately 330 mg N_2 fixed day^{-1} mg $nodule^{-1}$. Assuming that this fixation rate occurs over the course of a 12-hour day, that the bacteroids are about 20% of nodule by volume and the turnover number of nitrogenase is 3 moles/mole $enzyme^{-sec}$ (Orme-Johnson, 1977) we calculate that the concentration of nitrogenase in soybean nodule bacteroids is about 0.5 mM. Further assuming that the iron, sulfur and molybdenum content of *R. japonicum* nitrogenase is similar to other nitrogenases (Burris and Orme-Johnson, 1974), we calculate that the concentrations of iron, sulfur and molybdenum in soybean bacteroids are 14 mM, 20 mM and 1 mM respectively. When one includes the iron and sulfur importation required for synthesis of other iron and/or sulfur proteins, hemoproteins and the heme of leghemoglobin, it is clear that symbiosis demands a considerable flux of iron and sulfur into the bacteroids from the host plant.

Little is known about the mechanisms of iron and sulfur transport in nodules. Cystine-requiring strains of fast-growing rhizobia induce effective root nodules, indicating that the plant can provide organic sulfur to the bacteroids. Roessler and Nadler (1982) report that iron regulates rhizobial heme biosynthesis in laboratory-grown cells and suggest that iron released by the plant plays a regulatory role in bacteroid differentiation. Nothing is known about the uptake of cobalt and molybdenum or the release of heme to the plant by *Rhizobium.* Much more work is required on these interesting aspects of the root nodule symbiosis. If the "harmonics of symbiosis" (Bergersen, 1974) describe nutrient exchanges in the nodule, then these harmonics are polyphonic.

IV. Plant Genes Involved in Symbiosis

A. Genetics

A number of studies over half a century have conclusively demonstrated the involvement of the inheritable characteristics in the host plant which affect practically all aspects of symbiotic nitrogen fixation. These include number and size of nodules, timing of appearance as well as nodule morphogenesis and the rate of nitrogen fixation activity (for reviews, see Holl and LaRue, 1976; Caldwell and Vest, 1977; Vincent, 1980; Nutman, 1969, 1981; Verma and Long, 1983). It has been shown (see for details Brewin, Chapter 6) that the activity of hydrogenase, an enzyme involved in the reduction of hydrogen produced as a by-product of N_2 fixation, is also controlled by the host. In addition, several properties of the bacteroids are affected by a given host (see Sutton *et al.*, 1982).

Table 1 summarizes various host genes which have been identified via classical genetic experiments and selection of variants. Most of these genes are recessive and work singly. However, analysis of the ineffectiveness observed in *Medicago sativa* (Peterson and Barnes, 1981) suggests that more than one gene may be involved in a given mutation. The precise lesion caused by any of these mutations and the biochemical processes affected by them are completely unknown. However, by using current recombinant-DNA technology it may be possible to clone some of these genes and eventually identify the function of the products in symbiosis. Extensive work is needed on host mutants defective in symbiosis development of specific molecular probes for dissecting this complex process.

B. Molecular Studies

Two major groups of host gene products, leghemoglobins and nodulins are induced specifically during symbiotic nitrogen fixation. While the structure and function of leghemoglobins in root nodule symbiosis is relatively clear, little is known about the nodulins. The latter (with the exception of Nodulin-35, see below) have only been identified as mRNAs or protein products of the host detected following infection of the plant by *Rhizobium* and are either absent or present at very low levels in uninfected roots. In addition, another group of host genes whose modulation of expression can be mimiced by treatment with growth hormones has been detected (Verma *et al.*, 1983a) and two other sets of genes have been postulated to be involved in root nodule symbiosis (Fig. 2). Some members of these latter groups may be repressed following infection of the plant by *Rhizobium.* This repression may either be due to high levels of auxins produced in nodules as in auxin treated soybean hypocotyl tissue (Baulcombe and Key, 1980) or some other factors of *Rhizobium* origin. The auxins have no direct effect on the induction of leghemoglobins and nodulins (Auger, 1981).

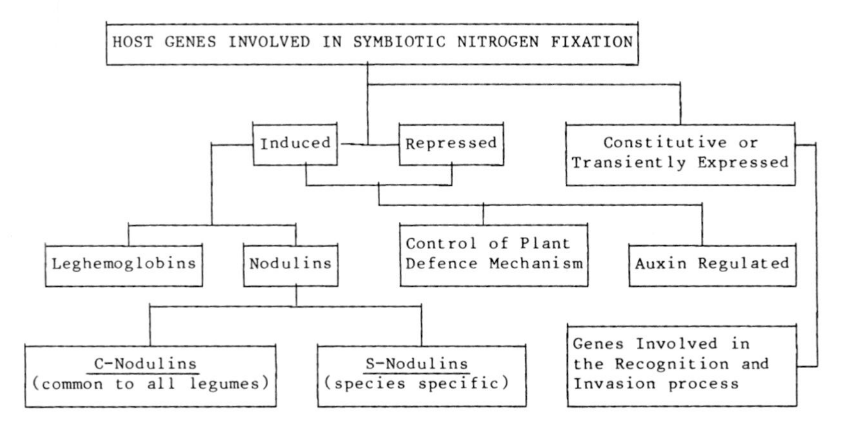

Fig. 2. General outline of our present understanding about various host genes which appear to be involved in the process of symbiotic nitrogen fixation

i) Leghemoglobin: This protein was first identified in 1939 by Kubo and is the best characterized plant-gene product in root nodules (see Appleby, 1974). The presence of leghemoglobin in all legumes and its direct correlation with the effectiveness of nodules suggest a very central role of this molecule in symbiotic nitrogen-fixation. This is also evident from the very high degree of conservation in its primary structure (and hence immuno-cross-reactivity) from various legumes (Hurrell *et al.*, 1979). On its initial identification, this molecule was proposed to be involved in nodule respiration, a role which is now fully established. The overall molecular structure of leghemoglobin is similar to that of animal myoglobins and hemoglobins but this molecule has a much higher affinity for oxygen. It releases oxygen for bacterial respiration only at low partial pressures (Wittenberg *et al.*, 1974; Appleby *et al.*, 1975a). By providing a sustained oxygen flux to bacteroids at low pO_2 it protects the nitrogenase enzyme from excess oxygen. Being located in the host cell cytoplasm (Verma and Ball, 1976), it may also provide a sufficiently high flux of oxygen to support the oxidative functions of the host cell and thus may serve a dual role in nodule tissue.

Leghemoglobin is only present in infected cells and its location is confined to the host cytoplasm (Verma and Bal, 1976; Robertson *et al.*, 1978b; Bisseling *et al.*, 1983) although a location within the bacteroid-containing envelopes has been suggested by previous work (see review by Dilworth, 1980). The protein chains of leghemoglobin are coded by plant genes and are synthesized on plant polysomes whereas the heme prosthetic group is produced by the bacteroids (Cutting and Schulman, 1977, see also Verma *et al.*, 1979). *Rhizobium meliloti* mutant strains with reduced levels of δ-aminolevulinic acid (δ-ALA) synthase, the first enzyme of heme biosynthesis, form small white nodules which do not fix nitrogen (Leong *et al.*, 1982). Symbiotic effectiveness is restored in the mutant by supplementing plants with δ-ALA or by complementing the mutation with the wild type δ-ALA synthase gene. A mutant strain of *R. leguminosarum* was isolated

Table 1. Host Genes Identified to Affect Nodulation and Nitrogen Fixation

Step in Symbiosis	Host	*Rhizobium* Strain	Host Gene(s)	Phenotype and Conditional Factors	Reference
No nodulation	*Glycine max*	*R. japonicum*	—	Non-nodulating	Williams and Lynch, 1954
			(*no*) rj_1	Prevent infection thread development, strain dependent Conditional nod-response rj_1/rj_1 plants will nodulate with some strains but not with others.	Caldwell, 1966
	Trifolium sp.	*R. trifoli*	r	Simple (rare) recessives fail to nodulate with homologous bacteria — (root hair curl but no infection). Associated with a maternally transmitted component.	Nutman, 1954
	Pisum sativum	*R. leguminosarum*	sym 1	No infection thread formed. Temperature dependent (Unknown dominance).	Lie, 1971; Holl, 1973a
			sym 2	Recessive, strain dependent	Degenhardt *et al.*, 1976
Appearance of nodules, their number and size	*Trifolium* sp.	*R. trifoli*	—	Inherited in a complex manner	(See Nutman, 1981)
	Desmodium intortum	*Rhizobium* sp.	—	Inherited in a complex manner	

Table 1. Continued

Step in Symbiosis	Host	*Rhizobium* Strain	Host Gene(s)	Phenotype and Conditional Factors	Reference
	Glycine wightii	*Rhizobium* sp.	—	Inherited in a complex manner	(See Nutman, 1981)
	Pisum sativum	*R. leguminosarum*	No and Nod	Abundant nodulation is recessive to sparse nodulation.	Gelin and Blixt, 1964
Nodule structure and morphology	*Lupinus* sp. *Ornithopus* sp.	*R. lupini*	Not known	Number and shape of bacteroids/membrane envelope is determined by host.	Kidby and Goodchild, 1966
	Phaseolus vulgaris *Arachis hypogea*	*Rhizobium* strain 2364	Not known	Form effective nodules but their shapes, size and the number of bacteroids/membrane envelope is different.	Sutton *et al.*, 1981
Ineffective nodules	*Trifolium pratense*	*R. trifolii*	i	Single recessive gene causing abnormal cell division in host cells and prevents bacteroid formation.	Nutman, 1957
			ie	Influenced by other host factors (a supressor, can restore effectiveness).	Nutman, 1954
	T. subterraneum		d	Small nodules containing a few Rhizobia and "empty" cells.	Bergersen and Nutman, 1957

Table 1. Continued

Step in Symbiosis	Host	*Rhizobium* Strain	Host Gene(s)	Phenotype and Conditional Factors	Reference
			t	Some bacteroid tissues form but remain ineffective.	Chandler *et al.*, 1973
			n	Small nodules failure of rhizobia to release from infection thread.	Gibson, 1964
	Glycine max	*R. japonicum*	Rj_2 and Rj_3	Small white nodules, independent dominants with particular strain of *Rhizobium*.	Caldwell, 1966 Vest, 1970
			Rj_4	Almost normal early development of nodules but remain ineffective.	Vest and Caldwell, 1972
	Pisum sativum	*R. leguminosarum*	sym 3	Ineffective nodules containing leghemoglobin but no fixation, supply of photosynthetase may be affected.	Holl, 1973b, 1975
	Medicago sativa	*R. meliloti*	'recessive'	Large nodules without leghemoglobin, containing large amounts of starch and bacteroids rapidly degenerate.	Gibson, 1962 Viandes *et al.*, 1979

(Nadler, 1981) which accumulated abnormal amounts of porphyrins, precursors of heme. This mutant induced ineffective nodules with greatly reduced levels of leghemoglobin. These results imply a vigorous transport of heme precursors from the host to the bacteroids as well as an equally vigorous efflux of heme from the endosymbiont to the host cell.

Leghemoglobins always occur in multiple forms in all legumes examined to date. Different roles for different leghemoglobin components have been proposed since they appear at different concentrations during the development of the nodule (Fuchsman and Appleby, 1979; Verma *et al.*, 1979; 1981). In soybean there are 4 major leghemoglobins, Lba, c_1, c_2 and c_3 which are post-translationally modified into 4 minor components, Lbb, d_1, d_2 and d_3 respectively (Whittaker *et al.*, 1981). The role of these minor components of leghemoglobins in root nodule symbiosis is completely unknown. The method of isolating the various leghemoglobin species and examining their physical-chemical properties has been well established (Appleby *et al.*, 1975b, Dilworth, 1980). To date no leghemoglobin mutant useful for studies of the specific role of each component in symbiosis has been isolated. Plants infected by a *Rhizobium* strain defective in nitrogenase may produce half as much leghemoglobin as the wild type (Fig. 3). On the other hand nodules formed by a wild type ineffecitve strain (61A24) produce very little leghemoglobin mRNA and protein (see also Verma *et al.*, 1981). Since the induction of leghemoglobin occurs prior to

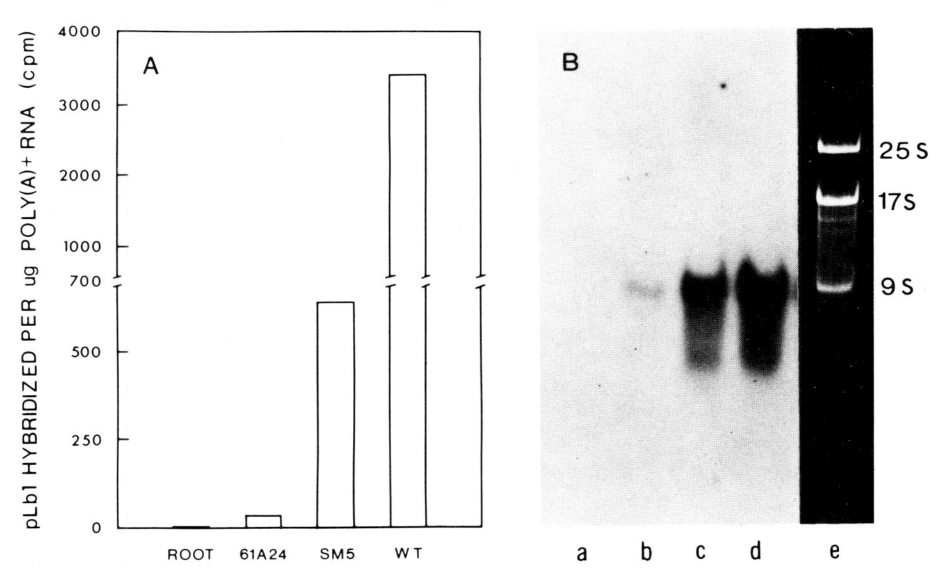

Fig. 3. Relative level of leghemoglobin mRNA in nodules formed by two different ineffective strains of *R. japonicum*. A leghemoglobin cDNA clone was hybridized to Northern blot of RNA following its electrophoresis on methyl mercuric hydroxid gel. A, histogram of the radioactive probe hybridized to Lb-mRNA; B, autoradiograph. Lane e in Panel B represents ethiduim bromide stained gel (from Verma *et al.*, 1981)

the appearance of nitrogenase activity in nodules (Verma *et al.*, 1979) and the ineffective nodules produce varying amounts of leghemoglobins, it appears that induction of this protein is completely independent of nitrogen fixation (Verma *et al.*, 1981; Bisseling *et al.*, 1978, 1983). Its induction may be one of the responses of the host to the invading "pathogen" for maintaining sufficient oxygen balance in the infected cell. However, functional protein can only be made if sufficient heme, synthesized by the bacteria, is available. Low oxygen tension and iron availability control bacterial heme synthesis (Avissar and Nadler, 1978). Perhaps these factors also control globin gene expression in the legume root.

Organization and structure of leghemoglobin genes: Leghemoglobins are encoded by a family of closely related sequences. While leghemoglobin proteins are very similar in a specific legume they differ greatly between legumes (see Hunt *et al.*, 1978). A decade of effort from a number of laboratories has culminated in a picture of soybean leghemoglobin genes and their organziation on the chromosome. Leghemoglobin mRNA was the first plant mRNA to be isolated (Verma *et al.*, 1974) and this led to the isolation of leghemoglobin genes (Sullivan *et al.*, 1981; Brisson and Verma, 1982) and the determination of their arrangement on the chromosome (Lee *et al.*, 1983; Lee and Verma, 1984). Other studies at both the protein (Ellfolk, 1971; Appleby *et al.*, 1975b; Hurrell and Leach, 1977) and DNA (Jensen *et al.*, 1981; Wiborg *et al.*, 1981) levels helped reach these goals.

Intriguingly, the structure of the leghemoglobin gene was found to be very similar to those of mammalian globin genes both with respect to the

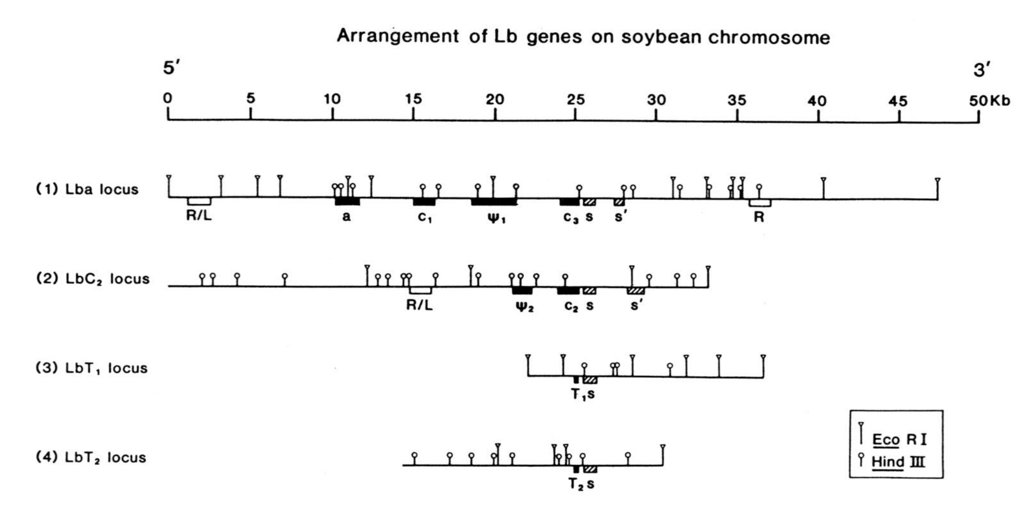

Fig. 4. Arrangement of soybean leghemoglobin genes on chromosome. Filled areas indicate leghemoglobin coding sequences; open areas indicate sequences in root (R) and root/leaf (R/L) tissue. Hatched areas show two types of repeat sequences (see Lee *et al.*, 1983; and Lee and Verma, 1984 for details)

position of two introns common to all globin genes as well as the presence of several regulatory sequences on the 5′ end of these genes (Brisson and Verma, 1982; Wiborg *et al.*, 1982, see also Verma *et al.*, 1983b; Lee and Verma, 1984). Furthermore, the arrangement of leghemoglobin genes on the chromosome is very similar to mammalian globin loci (Lee *et al.*, 1983). Also there occurs in the leghemoglobin gene family pseudo as well as truncated-genes (Brisson and Verma, 1982). Four regions containing leghemoglobin sequences have been found in soybean, two of which contain truncated sequences (Fig. 4). All leghemoglobin genes at the main locus are present in the same transcriptional orientation. The main characteristic of these regions is a repeat element found at the 3′ end of the main locus (Lee *et al.*, 1983). In addition, the main locus, which contains 3 normal (Lba, c_1 and c_3) and a pseudo gene (ψ_1), is flanked by genes preferentially expressed in leaf and root tissues. No linkage of the Lbc_2 gene is found to the main locus. It has been suggested that leghemoglobin genes may have been transferred horizontally from animals to legumes via a retrovirus-like vector. However, it seems unlikely in light of the structure of leghemoglobin genes (i. e., the presence of a third intron, Brisson and Verma, 1982) and the recent findings about the existence of similar genes in nitrogrn-fixing non-leguminous plants (Appleby *et al.*, 1983). If all nitrogen-fixing non-legumes have a globin-like gene but this gene is not present in non-nitrogen fixing plants, it makes the presence of this gene central to the symbiotic process. However, the high rate of sequence divergence in plants where these genes are not functional may make tracing the lineage of this gene very difficult.

ii) Discovery and Function of Nodulins: In an attempt to identify and eventually isolate host genes and their products (other than those for leghemoglobins), which play a role in the process of symbiotic nitrogen fixation, immunological and DNA-RNA hybridization techniques (Legocki and Verma, 1979, 1980; Auger and Verma, 1981) were employed. These studies detected the presence in soybean of several nodule-specific host gene products, termed nodulins. Recently, Bisseling *et al.* (1983) have also observed a number of nodule-specific products accumulated in pea nodules. It was observed (Verma *et al.*, 1981) that nodulin mRNA increases in parallel with that of leghemoglobin mRNA during nodule formation. Differences in the influence of mutations in *Rhizobium* and the apparent coordination in the induction of nodulins during effective nodule development (Verma *et al.*, 1981; Bisseling *et al.*, 1983) suggest that these host gene products play a role in root nodule symbiosis. Based upon their possible functions, these nodule-specific proteins may be divided into three categories: i) proteins responsible for the maintenance of nodule structure, ii) enzymes involved in specific nitrogen assimilation and carbon metabolism of nodules, iii) proteins that support bacteroid functions and thus facilitate nitrogen fixation (Fuller *et al.*, 1983). On the other hand, based upon the structural similarities (i. e. evolutionary conservation) they can be divided into two groups as follows:

C-Nodulins: Those nodule-specific host gene products which play a direct role in supporting the process of symbiotic nitrogen fixation are expected, like leghemoglobins, to be common to all nodules. This group of nodulins is termed "C-Nodulins" (common nodulins). The potential candidates for such host gene products are leghemoglobin reductase, some membrane proteins and products involved in energy coupling reactions. Since these proteins, like leghemoglobins, may play a basic role in the process of symbiotic nitrogen-fixation, their sequences are expected to be conserved. We have observed that a cDNA clone for soybean Nodulin-100 (Fuller *et al.*, 1983) cross-hybridizes with *Phaseolus* (Fuller and Verma, 1984), indicating that it may be a potential candidate for a C-nodulin. Furthermore, a novel form of glutamine synthetase has been isolated from *Phaseolus* nodules (Cullimore *et al.*, 1983; Lara *et al.*, 1983) and since glutamine synthetases from different plants may have significant homology (Cullimore, personal communication) and since this enzyme is essential for metabolism of NH_4^+, a primary product of N_2 fixation, it may be considered a C-type nodulin.

S-Nodulins: As indicated earlier (Verma *et al.*, 1983c), large numbers of nodulins are expected to be involved in carbon and nitrogen metabolism of nodules since most of the activity in this tissue is concentrated in maintaining this nutritional balance. Among the enzymes that show significant increases in activity (Table 2) those that are involved in nitrogen metabolism are most abundant. Since part of the nitrogen/carbon metabolism may be species specific, it appears that some nodulins may be present only in some species and not in others. The ineffective nodules formed due to mutations or incompatibilities either in the host or *Rhizobium* generally contain very low levels of nitrogen assimilating enzymes (Werner *et al.*, 1980; Groat and Vance, 1982). This activity could not be increased in plants supplied with reduced nitrogen, indicating that a new form of enzyme may be involved in assimilation of nitrogen fixed endosymbiotically. In fact, a new form of glutamine synthetase enzyme has been demonstrated in *Phaseolus* nodules as discussed above (Cullimore *et al.*, 1983: see also Miflin and Cullimore, Chapter 5). A nodule-specific uricase from soybean represented by nodulin-35 (subunit) has also been identified (Verma *et al.*, 1982; Jochimsen and Rasmussen, 1982; Bergmann *et al.*, 1983). Although the exact position of nodulin-35 is not clear, it may be an S-type nodulin (i. e. species specific). Nodulin-35 protein has been purified to homogeneity in native form and has been shown to have distinct properties which are not found in the leaf or root uricase enzymes. Its pH optima, subunit structure and stability are very different from the uricase enzyme present in other parts of the plant (Bergmann *et al.*, 1983, see also Tajima and Yamamato, 1975). The developmental data on the appearance of nodulin-35 suggest that it can be detected at about the same time as nitrogen fixation commences. Also this protein is present in ineffective nodules formed by strain 61A24 (an incompatible strain of *R. japonicum* (Legocki and Verma, 1979; Jochimsen and Rasmussen, 1982)). Using fluorescent-labelled antibodies this protein has been localized in uninfected cells of the nodule (Fig. 5) and thus it may represent the uricase enzyme suggested to

Table 2. Increase in Specific Host Enzymes and Proteins in Root Nodules

Enzyme/Protein	Approximate Increase* in Activity/Concentration (Nod./Root)	Plant Species	Reference
Allantoinase	3	*Glycine max*	Tajima and Yamamoto, 1975
		Vigna unguiculata	Shubert, 1981
			Herridge *et al.*, 1978
Asparagine synthetase	several hundred fold	*Lupinus angustifolius*	Scott *et al.*, 1976
		Glycine max	Reynolds *et al.*, 1982
Aspartate aminotransferase	(follows nitrogenase)	*Glycine max*	Ryan *et al.*, 1972
		Lupinus angustifolius	Reynolds and Farnden, 1979
Catalase	5—10	*Glycine max*	Frances and Alexander, 1972
Carbonic anhydrase	12.5	*Lupinus angustifolius*	Christeller *et al.*, 1977
Glutamine synthetase	300—500	*Lupinus angustifolius*	Robertson *et al.*, 1975
	(nodule specific)	*Phaseolus vulgaris*	Lara *et al.*, 1983
		Glycine max	Werner *et al.*, 1980
			Reynolds *et al.*, 1982
Glutamate synthase	10	*Lupinus angustifolius*	Robertson *et al.*, 1975
		Vigna unguiculata	Atkins *et al.*, 1980
Invertase	3—4	*Glycine max*	Streeter, 1977
Leghemoglobin	30—40% by mass/0	*Glycine max*	Nash and Schulman, 1976
			Verma *et al.*, 1981
Leghemoglobin reductase	—	*Glycine max*	Puppo *et al.*, 1980
Phosphoenolpyruvate carboxylase	3	*Lupinus angustifolius*	Christeller *et al.*, 1977
Uricase** (Nodulin-35)	(nodule specific)	*Glycine max*	Tajima and Yamamoto, 1975
			Jochimsen and Rasmussen, 1982
			Bergmann *et al.*, 1983
		Vigna unguiculata	Reynolds *et al.*, 1982
			Herridge *et al.*, 1978
Xanthine oxidase	0.014/0	*Glycine max*	Tajima and Yamamoto, 1975
Xanthine dehydrogenese	(nodule specific)	*Glycine max*	Shubert, 1981
			Reynolds *et al.*, 1982
Membrane envelope (peribacteroid membrane) proteins including the appearance of new proteins	several fold	*Vigna unguiculata*	Atkins *et al.*, 1980
		Glycine max	Verma *et al.*, 1978b

* Values are not directly comparable due to different methods used to measure concentration/activity.
** This enzyme is different from that present in leaf or uninfected root (see Bergmann *et al.*, 1983)

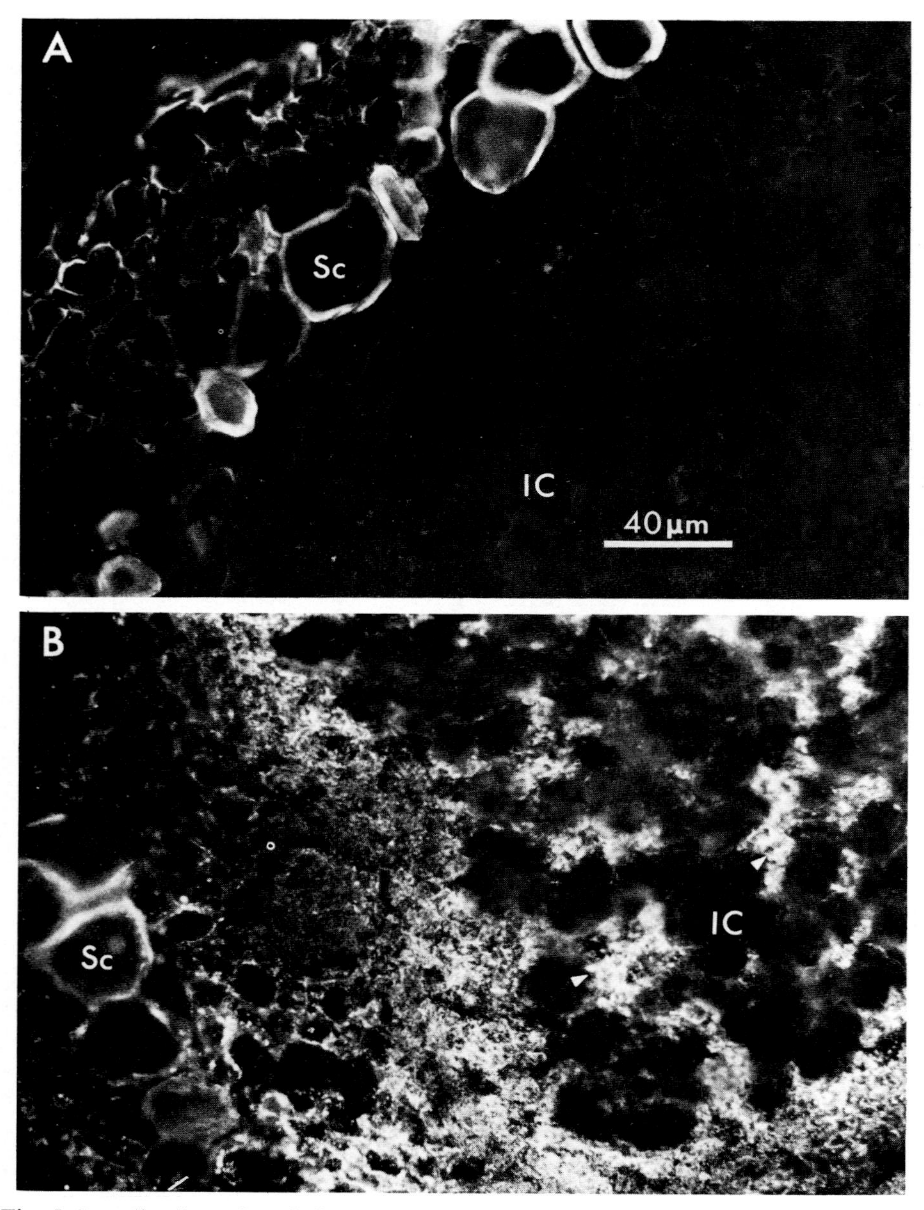

Fig. 5. Localization of nodulin-35, a subunit of nodule-specific uricase, using fluorescent labelled antibodies. Note the presence of specific fluorescence in only the uninfected cells (arrowhead). The autofluorescence in A is due to scalenchymatous tissue (SC). IC, infected cells (from Bergmann *et al.*, 1983)

be present in the peroxisomes of these cells (Newcomb and Tandon, 1981; Hanks *et al.*, 1983). This has been directly demonstrated by using immunohistochemistry with protein-A gold at the E. M. level (Verma *et al.*, unpublished data). A unique purine biosynthetic pathway appears to exist in some nodules (Atkins *et al.*, 1980; Atkins,1981; Schubert, 1981). Whether the other enzymes involved in this pathway, e. g. xanthine dehydrogenase (Triplett *et al.*, 1982; Boland *et al.*, 1983), as well as other nitrogen assimilating enzymes, e. g. aspartate aminotransferase (Reynolds and Farnden, 1979; Boland *et al.*, 1983), are S-type nodulins remains to be determined. These nodule-specific enzymes are either products of different genes which are not expressed in other tissues, or represent major modifications of the enzymes involved in similar functions and already present in other parts of the plant. Recently identified soybean nodulin sequences (Fuller *et al.*, 1983) may represent some of these enzymes. Further characterization of genes encoding these sequences and their regulation would facilitate our understanding about the molecular nature of this symbiosis.

V. Summary and Perspectives

Since plant-*Rhizobium* interactions occur throughout the life of nodule we can imagine these processes as a series of locked "plant gates", each of which must be opened by a different "rhizobial key". Mixed infections with different bacterial mutants (missing different keys) would be akin to reconstituting the complete set of rhizobial keys, permitting both mutant strains to pass through the gates and establish themselves in nodules. In this model, nodulation involves a series of subtle plant responses to bacterial invasion. One is tempted to speculate on the nature of these important "locks and keys". Lectins and lectin receptors have been implicated in rhizobial attachment to root hair cells. Determinants on the surface of the *Rhizobium* and on the infection thread wall may be the elusive locks and keys during nodulation. Perhaps rhizobia present different "keys" for different host "locks". The final bacterial key may be the removal of its outer membrane during release from the infection thread. Moreover, the host continously influences the *Rhizobium* as it passes through the infection thread, which may or may not allow expression of specific bacterial genes necessary for symbiotic nitrogen fixation. This is evidenced by the fact that one strain fully effective on one host only makes ineffective nodules on another host. These differences can occur even within the same species. The nature of these host signals which allow coordinate expression of *Rhizobium* genes is completely unknown. Some of these may be plant hydrolytic enzymes which affect bacterial cell surface and in turn alter it as it passes through the infection thread. Study at molecular level of the sequence of events during nodulation and the effect of various mutations, both in the plant and the *Rhizobium*, in this process is essential for our understanding of the ways in which these two organisms accomplish this highly complex symbiotic state. If we consider *Rhizobium* as a pathogen then comparison of this process with other pathogenic associations may shed some light on the ways in which this process evolved.

The host plant plays many important roles in the process of symbiotic nitrogen fixation. A number of host gene products which support the symbiotic state have been identified. Some of these host gene products are present in the infected cells while others are in unifected cells of the nodule, suggesting organ level cooperation in the nodules. Since legumes are essentially the only group of plants that can support endosymbioses with *Rhizobium*, genetic manipulations for nitrogen-fixation ability will have to deal with this problem which restricts these associations to a limited group of plants. Furthermore, incompatibility of the nitrogen reduction process with the aerobic environment of the plant cell has to be considered as a major problem for transfer of N_2-fixation genes directly to a plant. A strict compartmentalization (to provide high energy at low oxygen tension) has to be maintained in order to reduce this inert gas. Further identification and characterization of host genes obligatory for the development and maintenance of symbiosis may allow their eventual transfer to novel hosts enabling them to form associations with naturally occurring or new nitrogen-fixing microbes. Such a strategy may greatly enhance this resource for agricultural use.

VI. References

Allen, O. N., Allen E. K., 1981: The Leguminosae. Madison, Wisconsin: Univ. of Wisconsin Press.

Appleby, C. A., 1974: Leghemoglobin. In: The Biology of Nitrogen Fixation. Quispel, A. (Ed.), pp. 521—554. New York: Elsevier Pub. Co. Inc.

Appleby, C. A., Turner, G. L., MacNichol, P. K., 1975a: Involvement of oxyleghaemoglobin and cytochromome P-450 in an efficient oxidative phosphorylation pathway which support nitrogen fixation in *Rhizobium*. Biochim. Biophys. Acta **387**, 461—475.

Appleby, C. A., Nicola, N. A., Hurrell, J. G. R., Leach, S. J., 1975b: Characterization and improved separation of soybean leghaemoglobin. Biochemistry **14**, 4444—4450.

Appleby, C. A., Tjepkema, J. D., Trinick, M. J., 1983: Hemoglobin in a nonleguminous plant, *Parasponia*. Possible genetic origin and function in nitrogen fixation. Science **220**, 951—953.

Atkins, C. A., Rainbird, R. M., Pate, J. S., 1980: Evidence for a purine pathway of ureide synthesis in N_2-fixing nodules of cowpea (*Vigna unguiculata* L. Walp.). Z. Pflanzenphysiol. **97**, 249—260.

Atkins, C. A., 1981: Metabolism of purine-nucleotides to form ureides in nitrogen-fixing nodules of cowpea (*Vigna unguiculata* L. Walp). FEBS Letts. **125**, 89—93.

Auger, S., 1981: Ph. D. Thesis McGill University, Montreal, Canada.

Auger, S., Verma D. P. S., 1981: Induction and expression of nodule-specific host genes in effective and ineffective root nodules of soybean. Biochemistry **20**, 1300—1306.

Avissar, V. J., Nadler, K. D., 1978: Stimulation of tetrapyrrole formation in *Rhizobium japonicum* by restricted aeration. J. Bact. **135**, 782—789.

Badnenoch-Jones, Summons, R. E., Djordjevic, M. A., Shine, J., Letham, D. S., Rolfe, B. J., 1982: Mass-spectrometeric quantification of indole-3-acetic acid in culture supernatants; studies in relation to root hair curling and nodule initiation. Appl. Environ. Microbiol. **44**, 275—280.

Bal, A. K., Shantharam, S., Verma, D. P. S., 1980: Changes in the outer cell wall of *Rhizobium* during development of the root nodule symbiosis in soybean. Can. J. Microbiol. **26**, 1096—1103.

Bassett, B., Goodman, R. N., Novacky, A., 1977a: Ultrastructure of soybean nodules. I. Release of *Rhizobia* from the infection thread. Can. J. Microbiol. **23**, 573—582.

Bassett, B., Goodman, R. N., Novacky, A., 1977b: Ultrastructure of soybean nodules. II: deterioration of symbiosis in ineffective nodules. Can. J. Microbiol. **23**, 873—883.

Bauer, W. D., 1981: Infection of legumes by rhizobia. Ann. Rev. Plant Physiol. **32**, 407—449.

Baulcombe, D., Key, J., 1980: Polyadenylated RNA sequences which are reduced in concentration following auxin treatment of soybean hypocotyls. J. Biol. Chem. **225**, 8907—8913.

Bergersen, F. J., Goodchild, D. J., 1973: Aeration pathways in soybean root nodules. Anst. J. Biol. Sci. **26**, 729—740.

Bergersen, F. J., Nutman, P. S., 1957: Symbiotic effectiveness in nodulated red clover. The influence of the host factor *i*, i. e. upon nodule structure and cytology. Heredity **11**, 175—184.

Bergersen, F. J., 1977: Physiological chemistry of dinitrogen fixation by legumes: In: A Treatise in Dinitrogen Fixation, Hardy, R. W. F., Silver, W. S. (eds.), pp. 519—555, New York: John Wiley.

Bergersen, F. J., 1965: Ammonia an early stable product of nitrogen fixation by soybean root nodules. Aust. J. Biol. Sci. **18**, 1—9.

Bergersen, F. J., 1974): Formation and function of bacteroids. In: The Biology of Nitrogen Fixation. Quispel, A. (ed.), pp. 473—498, Amsterdam: North-Holland.

Bergmann H., Preddie, E., Verma, D. P. S., 1983: Nodulin 35: A subunit of specific uricase (uricase II) induced and localized in uninfected cells of nodules. The EMBO Jour. **2**, 2333—2339.

Beringer, J. E., Brewin, N., Johnston, A. W. B., Schulman, H. M., Hopwood, D. A., 1979: The *Rhizobium*-legume symbiosis. Proc. Roy. Soc. Lond. B **204**, 219—223.

Beringer, J. E., Brewin, N. J., Johnston, A. W. B., 1980: The genetic analysis of *Rhizobium* in relation to symbiotic nitrogen fixation. Heredity **45**, 161—186.

Bisseling, T., van den Bos, R. C., Weststrate, M. W., Hakkaart, M. J. J., van Kammen, A., 1978: The effect of ammonium nitrate on the synthesis of nitrogenase and the concentration of leghemoglobin in pea root nodules induced by *Rhizobium leguminosarum.* Biochim. Biophys. Acta **539**, 1—11.

Bisseling, T., Been, C., Klugkist, J., van Kammen, A., Nadler, K., 1983: Nodule specific host proteins in effective and ineffective root nodules of *Pisum sativum.* The EMBO Jour. **2**, 961—966.

Boland, M. J., Hanks, J. F., Reynolds, P. H. S., Blevins, D. G., Tolbert, N. E., Schubert, K. R., 1982: Subcellular organization of ureide biogenesis from glycolytic intermediates and ammonium in nitrogen fixing soybean nodules. Planta **155**, 45—51.

Boland, M. J., Blevins, D. G., Randall, D. D., 1983: Soybean xanthine dehydrogenase: A kinetic study. Arch. Biochem. Biophys. **222**, 2333—2339.

Brill, W., 1980: Biochemical genetics of nitrogen fixation. Microbiol. Rev. **44**, 449—467.

Brisson, N., Verma D. P. S., 1982: Soybean leghemoglobin gene family: Normal, pseudo, and truncated genes. Proc. Nat. Acad. Sci. **79**, 4055—4059.

Burris, R. H., Orme-Johnson, W. H., 1974: Survey of nitrogenase and its EPR properties. In: Microbial Iron Metabolism. Neilands, J. B. (ed.), p. 597. New York: Academic Press.

Caldwell, B. E., 1966: Inheritance of a strain-specific ineffective nodulation in soybeans. Crop. Sci **6**, 427—428.

Caldwell, B. E., Vest, H. G., 1977: Genetic aspects of nodulation and dinitrogen fixation by legumes: The macrosymbiont. In: A treatise on dinitrogen fixation. Sec. III, Hardy, R. W. F., Silver, W. S. (eds.) London: Wiley-Interscience, p. 557—575.

Callaham, D., Torrey; J. G., 1977: Prenodule formation and primary nodule development in roots of *Comptonia* (Myricaceae). Can. J. Bot. **55**, 2303—2318.

Callaham, D., 1979: M. Sc. thesis. University of Massachusetts, Amherst.

Callaham, D., Newcomb, W., Torrey, J. G., Peterson, R. L., 1979: Root hair infection in actinomycete-induced root nodule initiation in *Casuarina, Myrica,* and *Comptonia.* Botanical Gazette. 140 (Suppl.): S1—S9.

Callaham, D. A., Torrey, J. G., 1981: The structural basis for infection of root hairs of *Trifolium repens* by *Rhizobium.* Can. J. Bot. **59**, 1647—1664

Chakarvorty, A. K., Zurkowski W., Shine J., Rolfe, B. G., 1983: Symbiotic nitrogen fixation: molecular cloning of *Rhizobium* genes involved in exopolysaccharide synthesis and effective nodulation. J. Mol. Appl. Gen. **1**, 585—596.

Christeller, J. T., Laing, W. A., Sutton, W. D., 1977: Carbon dioxide fixation by lupine root nodules. 1. Characterization, association with phosphoenolpyruvate carboxylase, and correlation with nitrogen fixation during nodule development. Plant Physiol. **60**, 47—50.

Cullimore, J. V., Lara, M., Lea, P. J., Miflin, B. J., 1983: Purification and properties of the two forms of glutamine synthetase from the plant fraction of *Phaseolus* root nodules. Planta **157**, 245—253.

Cutting, J. A., Schulman, H. M., 1969: The site of heme synthesis in soybean root nodules. Biochim Biophys. Acta. **192**, 486—493.

Dart, P. J., 1974: The infection process. In: The Biology of Nitrogen Fixation. Quispel, A. (ed.), pp. 381—429, Amsterdam: Elsevier-North-Holland.

Dart, P. J., 1977: Infection and development of leguminous nodules. In: A treatise on dinitrogen fixation. Section III, biology, Hardy, R. W. F., Silver, W. S. (eds.), pp. 367—472, New York: John Wiley.

Degenhardt, T. L., LaRue, T. A., Paul, E. A., 1976: Investigation of a non-nodulating cultivar of *Pisum sativum.* Can. J. Bot. **54**, 1633—1636.

Dénarié, J., Truchet, G., Bergeron, B., 1976: Effects of some mutations on symbiotic properties of *Rhizobium.* In: Symbiotic Nitrogen Fixation in Plants. Nutman, P. S. (ed.), pp. 47—61, New York/London: Cambridge Univ. Press.

Dilworth, M. J., 1980: Leghemoglobin. In: Methods in Enzymology, Vol. 69c. San Pietro, A. (ed.), pp. 812—823, New York: Academic Press.

Dilworth, M. J., 1980: Symbiotic Associations and Cyanobacteria. In: Nitrogen Fixation II. Newton, W. E., Orme-Johnson, W. H. (eds.), p. 325, Baltimore: University Park Press.

Dullaart, J., 1967: Quantitative estimation of indoleacetic acid and indolecarboxylic acid in root nodules and roots of *Lupinus luteus* L. Acta Bot. Neerl. **16**, 222—230.

Ellfolk, N., Sievers, G., 1971: The primary stucture of soybean leghaemoglobin. Acta, Chem. Scand. **25**, 3532—3534.

Finan, T. M., Wood, J. M., Jordan, D. C., 1983: Symbiotic properties of C_4-dicarboxylic acid transport mutants of *Rhizobium leguminosarum*. J. Bacteriol. **154**, 1403—1413.

Fuchsman, W., Appleby, C. A., 1979: Separation and determination of relative concentrations of the homogenous components of soybean leghemoglobins by isoelectric focusing. Biochim. Biophys. Acta, **579**, 314—324.

Francis, A. J., Alexander, M., 1972: Catalase activity and nitrogen fixation in legume nodules. Can. J. Microbiol. **18**, 861—864.

Fuller, F., Künstner, P., Nguyen, T., Verma, D. P. S., 1983: Nodulin genes of soybean: Analysis of cDNA clones reveals several abundant sequences in nitrogen fixing root nodules. Proc. Nat. Acad. Sci. **80**, 2594—2599.

Fuller, F., Verma, D. P. S., 1984: Accumulation of nodulin mRNAS during the development of effective root nodules of soybean. Plant Mol. Biol. (in press).

Gelin, O., Blixt, S., 1964: Root nodulation in peas. Agr. Hort. Genet. **22**, 149—159.

Gibson, A. M., 1964: Genetic control of strain-specific ineffective nodulation in *Trifolium subterraneum* L. Aust. J. Agric. Res. **15**, 37—49.

Glenn, A. R., Brewin, N. J., 1981: Succinate resistant mutants of *Rhizobium leguminosarum*. J. Gen. Microbiol. **126**, 237—241.

Groat, R. G., Vance, C. P., 1982: Root and nodule enzymes of ammonia assimilation in two plant conditioned symbiotically ineffective genotypes of alfalfa (*Medicago sativa* L.). Plant Physiol. **69**, 614—618.

Hanks, J. F., Schubert, K., Olbert, N. E., 1983: Isolation and characterization of infected and uninfected cells from soybean nodules. Role of uninfected cells in ureide synthesis. Plant Physiol. **71**, 869—873.

Hardy, R. W. F., Havelka, U. D., 1976: Photosynthate as a major factor limiting nitrogen fixation by field-grown legumes with emphasis on soybeans. In: Symbiotic Nitrogen Fixation in Plants. Nutman, P. S. (ed.), pp. 421—439, New York/London: Cambridge Univ. Press.

Herridge, D. F., Atkin, C. A., Pate, J. S., Rainbird, R. M. 1978: Allantoin and allentoic acids in the nitrogen economy of the cowpea (*Vigna unguiculata*). Plant Physiol. **62**, 459—498.

Holl, F. B., 1973a: Host-determined genetic control of nitrogen fixation with *Pisum-Rhizobium* symbiosis. Can. J. Genet. Cytol. **15**, 659.

Holl, F. B., 1973b: A nodulating strain of *Pisum* unable to fix nitrogen. Plant Physiol. **51**, suppl., p. 35.

Holl, F. B., 1975: Host plant control of the inheritance of dinitrogen fixation with *Pisum-Rhizobium* symbiosis. Euphytica **24**, 767—770.

Holl, F. B., LaRue, T. A., 1976: Genetics of legume host plants. In: Proc. Ist International Symposium on Nitrogen Fixation. Vol. II. Newton, W. E., Nyman, C. J. (eds.), pp. 391—399, Pullman, Wash.: Washington State Univ. Press.

Hunt, L. T., Hurst-Caldrone, S., Doughhof, M. O., 1978: Globins. In: Atlas of protein sequences and structure. 5 supp. 3, 229—251.

Hurrell, J. G. R., Leach, S. J., 1977: The amino acid sequence of soybean leghaemoglobin c_2. FEBS letters **80**, 23—26.

Hurrell, J. G. R., Nicola, N. A., Leach, S. J., 1979: Evolutionary and structural relationship of leghemoglobins. In: Soil Microbiology and Plant Nutrition. Broughton, W. J., John, C. K. (eds.), pp. 253—281, Kuala Lumpur: University of Malaya Press.

Jensen, E. O., Palndon, K., Hyldig-Neilson, J. J., Jorgensen, P., Marker, K. A.,

1981: The structure of a chromosomal gene from soybean. Nature **291**, 677—679.

Jochimsen, B., Rasmussen, O., 1982: Appearance of nodule-specific proteins in effective compared to ineffective nodules of soybean. In: Molecular Genetics of the Bacteria-Plant Interaction. Pühler, A. (ed.), (Abs.). Symp., Univ. of Bielefeld.

Koa, J. L., Perry, K. L., Kado, C. I., 1982: Indoacetic acid complementation and its relation to host range specifying genes on the Ti plasmid of *Agrobacterium tumefaciens*. Mol. Gen. Gen. **188**, 425—432.

Kefford, N. P., Brockwell, J., Zwar, J. A., 1960: The symbiotic synthesis of auxin by legumes and nodule bacteria and its role in nodule development. Aust. J. Biol. Sci. **13**, 456—467.

Kidby, D. K., Goodchild, D. J., 1966: Host influence in the ultrastructure of root nodules of *Lupinus luteus* and *Ornithopus sativus*. J. Gen. Microbiol. **45**, 147—152.

Kondorosi, A., Svab. Z., Kiss, G. B., Dixon, R. A., 1977: Ammonium assimilation and nitrogen fixation in *Rhizobium meliloti*. Mol. Gen. Genet. **151**, 221—226.

Kondorosi, A., Johnston. A. W. B., 1981: The Genetics of *Rhizobium*. In: International Rev. of Cytol. suppl. 13. Giles, K. L., Atherly, A. G. (eds.), pp. 191—224, New York: Academic Press.

Kubo, H., 1939: Über das Hämoprotein aus den Wurzelknöllchen von Leguminosen. Acta. Phytochim. **11**, 195—200.

Kuykendall, L. D., 1981: Mutants of *Rhizobium* that are altered in legume interaction and nitrogen fixation. In: Int. Rev. Cytol. Suppl. 13, Giles, K. L., Atherly, A. G. (eds.), pp. 229—309, New York: Academic Press.

Lalonde, M., Knowles, R., 1975: Ultrastructure composition and biogenesis of the encapsulation material surrounding the endophyte in *Alnus cripa* var mollis root nodules. Can. J. Bot. **53**, 1951—1971.

Lara, N., Cullimore, J. V., Lea, P. J., Miflin, B. J., Johnston, A. W. B., Lamb, J. W., 1983: Appearance of a novel form of plant glutamine synthetase during nodule development in *Phaseolus vulgaris* L. Planta **157**, 254—258.

Lee, J., Brown, G. G., Verma, D. P. S., 1983: Arrangement of leghemoglobin genes in soybean chromosome. Nucl. Acid Res. **11**, 5541—5553.

Lee, J., Verma, D. P. S., 1984: An enigma of leghemoglobin genes. In: Genetic Engineering, Vol. 6. Setlow, J. K. (ed.), New York: Plenum Press (in press).

Legocki, R. P., Verma, D. P. S., 1979: A nodule-specific plant protein (Nodulin-35) from soybean. Science **205**, 190—193.

Legocki, R. P., Verma, D. P. S., 1980: Identification of "nodule-specific" host proteins (Nodulins) involved in the development of *Rhizobium*-legume symbiosis. Cell **20**, 153—163.

Leong, S. A., Ditta, G. S., Helinski, D. R., 1982: Heme biosynthesis in *Rhizobium*. J. Biol. Chem. **257**, 8724—8730.

Lepo, J. E., Hanus, F. J., Evans, H. J., 1980: Chemoautotrophic growth of hydrogen-uptake-positive strains of *Rhizobium japonicum*. J. Bact. **141**, 664—670.

Libbenga, K. R., Bogers, R. J., 1974: Root-nodule morphogenesis. In: The biology of nitrogen fixation. Quispel, A. (ed.), pp. 430—472, Amsterdam: North-Holland.

Libbenga, K. R., vanIren, F., Bogers, R. J. Shraag-Lamers, M. F., 1973: The role of hormones and gradients in the initiation of Cotex proliferation and nodule formation in *Pisum sativum L.* Planta **114**, 29—39.

Libbenga, K. R., Torrey, J. G., 1973: Hormone-induced endoreduplication prior to mitosis in cultured pea nodule cortex cells. Am. J. Bot. **60**, 293—299.

Lie, T. A., 1971: Symbiotic nitrogen fixation under stress conditions. Plant and Soil spec. vol. 117—127.

Liu, S.-T., Perry, K. L., Schardl, C. L., Kado, C. I., 1982: *Agrobacterium* Ti-plasmid indoleacetic acid genes is required for crown gall oncogenesis. Proc. Nat. Acad. Sci (U. S. A.) **79**, 2812—2816.

Margulis, L., 1981: Symbiosis in Cell Evolution, San Francisco: W. H. Freeman and Co.

Meade, H, Long, S. R., Ruvkun, G. B., Brown, S. E., Ausubel, F. M., 1982: Physical and genetic characterization of symbiotic and auxotrophic mutants of *Rhizobium meliloti* induced by transposon mutagenesis. J. Bact. **149**, 114—122.

Nadler, K. D., 1981: A mutant strain of *Rhizobium leguminosarum* with an abnormality in heme synthesis. In: Current perspectives in nitrogen fixation, Gibson, A. H., Newton, W. E. (eds.), p. 414 (Abs.), Canberra: Aust. Acad. Science.

Nash, D., Schulman, H., 1976: Leghemoglobin and nitrogenase activity during soybean root nodule development. Can. J. Bot. **54**, 2790—2797.

Newcomb, W., 1981: Nodule morphogenesis and differentiation. In: Int. Rev. Cytol. Supp. 13. Giles K. L., Atherly, A. G., (eds.). pp. 247—298, New York: Academic Press.

Newcomb, W., Peterson, R. L., Callaham, D., Torrey, J. G., 1978: Structure and host-actinomycete interactions in developing root nodules of *Comptonia peregrine*. Can. J. Bot. **56**, 502—531.

Newcomb, W., 1976: A correlated light and electron microscope study of symbiotic growth and differentiation in *Pisum sativum* root nodules. Can. J. Bot. **54**, 2163—2186.

Newcomb, W., Sippel, D., Peterson, R. L., 1979a: The early morphogenesis of *Glycine max* and *Pisum sativum* root nodules. Can. J. Bot. **57**, 2603—2616.

Newcomb, W., Peterson, R. L., 1979b: The occurrence and ontogeny of transfer cells associated with lateral roots and root nodules in Leguminoseae. Can. J. Bot. **57**, 2583—2602.

Newcomb, E. H., Tandon, S. R., 1981: Uninfected cells of soybean root nodules: Ultrastructure suggest key role in ureide production. Science **21**, 1394—1396.

Noel, K. D., Stacey, G., Tandon, S. R., Silver, L. E., Brill, W. J., 1982: *Rhizobium japonicum* mutants defective in symbiotic nitrogen fixation. J. Bact. **152**, 485—494.

Nutman, P. S., 1952: Studies on the physiology of nodule formation. III Experiments on the excision of root-tips and nodules. Annals of Botany (N. S.) **16**, 79—101.

Nutman, P. S., 1954: Symbiotic effectiveness in nodulated red clover. A major gene for ineffectiveness in the host. Heredity **8**, 47—60.

Nutman, P. S., 1956: The influence of the legume in root nodule symbiosis. A comparative study of host determinants and functions. Biol. Rev. Camb. Philos. Soc. **31**, 109—151.

Nutman, P. S., 1957: Symbiotic effectiveness in nodulated red clover. III. Further studies on inheritance of ineffectiveness in the host. Heredity II, 157—173.

Nutman, P. S., 1969: Genetics of symbiosis and nitrogen fixation in legumes. Proc. Roy. Soc. B. **172**, 417—437.

Nutman, P. S., 1981: Hereditary host factors affecting nodulation and nitrogen fixation. In: Current Perspectives in Nitrogen Fixation. Gibson, A. H., Newton, W. E. (eds.), pp. 194—204, Canberra: Aust. Academy of Science.

O'Gara, F., Shanmugam, K. T., 1976; Regulation of nitrogen fixation by rhizobia: Export of fixed N_2 as NH_4^+. Biochim. Biophys. Acta. **437**, 313—321.

Orme-Johnson, W. H., 1977: Biochemistry of Nitrogenase. In: Genetic Engineering for Nitrogen Fixation. Hollaender, A., Burris, R. H., Day, P. R., Hardy, R. W., Helinski, D. R., Lamborg, M. R., Owens, L., Vallentine, R. C. (eds.), p. 317—333, New York/London: Plenum Press.

Pankhurst, C. E., Schwinghamer, E., 1974: Adenine requirement for nodulation of pea by an auxotrophic mutant of *Rhizobium leguminosarum.* Arch. Microbiol. **100**, 219—238.

Pate, J. S., Layzell, D. B., McNeil, D. L., 1979a: Modeling the transport and utilization of carbon and nitrogen in a nodulated legume. Plant Physiol. **63**, 730—737.

Pate, J. S., Layzell, D. B., Atkins, C. A., 1979b: Economy of carbon and nitrogen in a nodulated and nonnodulated (NO_3-grown) legume. Plant Physiol. **64**, 1083—1088.

Peters, G. A., Ray, T. B., Mayne, B. C., Toia, R. E., 1980: *Azolla-Anabaena* association: Morphological and physiological strains. In: Nitrogen Fixation. Newton, W. E., Orme-Johnson, W. M., (eds.), Vol. 2, pp. 293—309, Baltimore: Univ. Park Press.

Peterson, M. A., Barns, D. K., 1981: Inheritance of ineffective nodulation and on-nodulation trait in alfalfa. Crop. Sci. **21**, 611—616.

Postgate, J. R., 1982: The Fundamentals of Nitrogen Fixation. London/New York: Cambridge University Press.

Puppo. A., Rigaud, J., Job, D., 1980: Leghemoglobin reduction by a nodule reductase. Plant Sci. Let. **20**, 1—6.

Puppo, A., Rigaud, J., 1975: Indole 3 acetic acid oxidation by leghemoglobin from soybean nodules. Physiol. Plant. **35**, 181—185.

Rawsthorne, S., Minchin, F. R., Summerfields, R. J., Cookson, C., Combes, J., 1980: Carbon and nitrogen metabolism in legume root nodules. Phytochem. **19**, 341—355.

Reisert, P. S., 1981: Plant cell surface structure and recognition phenomena with reference to symbiosis. In: Int. Rev. Cytol. Supp. 12. pp. 71—112.

Reynolds, P. H. S., Farnden, K. J. F., 1979: The involvement of aspartate amonotransferases in ammonium assimilation in lupine nodules. Phytochemistry **18**, 1625—1630.

Reynolds, P. H. S., Blevins, D. G., Boland, M. J., Schubert, K. R., Randal, D. D., 1982: Enzymes of ammonia assimilation in legume nodules: A comparison between ureide- and amide-transporting plants. Physiol. Plant. **55**, 255—260.

Robertson, J. G., Lyttleton, P., Bullivant, S., Graston, G. F., 1978a: Membranes in lupine root nodules I. The role of Golgi bodies in the biogenesis of infection threads and peribacteroid membranes. J. Cell Sci. **30**, 129—149.

Robertson, J. G., Warburton, M. P., Lyttleton, P., Fordyce, A. M., Bullivant, S., 1978b: Membranes in lupine root nodules II. Preparation and properties of peribacteroid membranes and bacteroid envelope inner membranes from developing root nodules. Ibiol. **30**, 151—174.

Robertson, J. G., Lyttleton, P., 1982: Coated and smoot vesicles in the biogenesis of cell walls, plasma membranes, infection threads and peribacteroid membranes in root hairs and nodules of white clover. J. Cell. Sci. **58**, 63—78.

Robertson, J. G., Warburton, M. P., Farnden, K. J. F., 1975b: Induction of glutamate synthase during nodule development in lupin: FEBS Letts. **55**, 33—37.

Roessler, P. G., Nadler, K. D., 1982: Effects of iron deficiency on heme biosynthesis in *Rhizobium japonicum* . J. Bacteriol. **149**, 1021—1026.

Rolfe, B. G., Gresshoff, P. M., 1980: *Rhizobium trifolii* mutant interactions during establishment of nodulation in white clover. Aust. J. Biol. Sci. **33**, 491—504.

Ronson, C. W., Lyttleton, P., Robertson, J. G., 1981: C_4-dicarboxylate transport mutants of *Rhizobium trifolii* from ineffective nodules of *Trifolium repens*. Proc. Nat. Acad. Sci. (U. S. A.) **78**, 4284—4288.

Ronson, C. W., Primrose, S. B., 1979: Carbohydrate metabolism in *Rhizobium trifolii*: identification and symbiotic properties of mutants. J. Gen. Microbiol. **112**, 77—88.

Ryan, E., Bodley, F., Fottrell, P. F., 1972: Purification and characterization of aspartate aminostransferases from soybean rot nodules and *Rhizobium japonicum*. Phytochem. **11**, 957—963.

Sahlmman, K., Fahraeus, G., 1963: An electron microscope study of root hair infection by *Rhizobium*. J. Gen Microbiol. **33**, 425—427.

Schwinghamer, E. A., 1970: Requriement for riboflavin for effective symbiosis on clover by an auxotrophic mutant strain of *Rhizobium trifolii*. Aust. J. Biol. Sci. **23**, 1187—1196.

Scott, D. B., Robertson, J. G., Farnden, K. J. F., 1976: Ammonia assimilation in lupine nodules. Nature **263**, 703—708.

Scott, K. F., Hughes, J. E., Gresshoff, P. M., Beringer, J. E., Rolfe, G. G., Shine, J., 1982: Molecular cloning of *Rhizobium trifolii* genes involved in symbiotic nitrogen fixation. J. Mol. App. Gen. **1**, 315—326.

Schwinghamer, E. A., 1977: Genetic aspects of nodulation and dinitrogen fixation by legumes: The microsymbiont. In: A Treatise on Dinitrogen Fixation III. Biology. Hardy, R. W. F., Silver, W. S. (eds.), p. 675, London: Wiley-Interscience.

Sequeira, 1980: Defenses triggered by the invader: Recognition and compatibility phenomena. In: Plant Disease, vol. V, New York: Acadmic Press.

Shubert, K. R., 1981: Enzymes of purine biosynthesis and catabolism in *Glycine max*. I. Comparison of activities with N_2 fixation and composition of xylem exudate during nodule development. Plant Physiol. **68**, 1115—1122.

Streeter, J. G., 1977: Asparaginase and asparagine transaminase in soybean leaves and root nodules. Plant Physiol. **60**, 235—239.

Sullivan, D., Brisson, N., Goodchild, B., Verma, D. P. S., Thomas, D., 1981: Molecular cloning and organization of two leghemoglobin genomic sequences of soybean. Nature **289**, 516—518.

Sutton, W. D., Pankhurst, C. E., Craig, A. S., 1981: The *Rhizobium* bacteroid state. In: Int. Rev. Cytol., Suppl. 13, Giles, K. L., Atherly, A. G. (eds.). pp. 149—177, New York: Academic Press.

Tajima, S., Yamamoto, Y., 1975: Enzymes of purine catabolism in soybean plants. Plant and Cell Physiol. **16**, 271—282.

Thimann, K. V., 1936: On The physiology of formation of nodules in legume roots. Proc. Nat. Acad. Sci. (U. S. A.) **22**, 511—514.

Trinchant, J. C., Birot A. M., Rigaud, J., 1981: Oxygen supply and energy-yielding substrates for nitrogen fixation (acetylene reduction) by bacteroid preparations. J. Gen. Microbiol. **125**, 159—165.

Triplett, E. W., Blevins, D. G., Randal, D. D., 1982: Purification and properties of soybean xanthine dehydrogenase. Arch. Biochem. Biophys. **219**, 39—46.

Truchet, G., 1978: Sur l'état diploide des cellules du méristème des nodules radiculaires des légumineruses. Ann. Sc. Nat. Bot. Biol. rég. **19**, 3—38.

Truchet, G., Michel M., Dénarié, J., 1980: Sequential analysis of the organogenesis of lucerne (*Medicago sativa*) root nodules using symbiotically-defective mutants of *Rhizobium* meliloti. Differentiation **16**, 163—172.

Verma, D. P. S., Bewley, J. D., Auger, S., Fuller, F., Purohit, J., Künstner, P., 1983a: Host genes involved in symbiosis with *Rhizobium*. In: Genetic Engineering. Application to Agriculture , Owens, L. D. (ed.), pp. 236—245. London/New Jersey: Granada Pub./Romman and Allanhead Pub.

Verma, D. P. S., Lee, J., Fuller, F., Brisson, N., 1983b: Leghemoglobin and nodulin genes of soybean: Organization and expression. In: Advances in gene technology: Molecular gentics of plants and animals, Ahmad *et al.* (eds.) New York: Academic Press (in press).

Verma, D. P. S., Bergmann, H., Fuller, F., Preddie, E., 1983c: The role of plant genes in soybean-*Rhizobium* interactions. In: Molecular genetics of the bacteria plant interaction, Pühler, A. (ed.), pp. 156—163, Berlin - Heidelberg - New York: Springer.

Verma, D. P. S., 1982: Host-*Rhizobium* interactions during symbiotic nitrogen fixation, In: The molecular biology of plant development. Smith, H., Grierson, D., (eds.), pp. 437—466, Oxford: Blackwell Pub.

Verma, D. P. S., Bal, A. K., 1976: Intracellular site of synthesis and localization of leghaemoglobin in soybean root nodules. Proc. Nat. Acad. Sci. (U. S. A.) **73**, 3843—3847.

Verma, D. P. S., Hunter, N., Bal, A. K., 1978a: Asymbiotic association of *Rhizobium* with pea epicotyls treated with plant hormone. Planta **138**, 107—110.

Verma, D. P. S., Kazazian, V., Zogbi, V., Bal, A. K., 1978b: Isolation and characterization of the membrane envelope enclosing the bacteroids in soybean root nodules. J. Cell Biol. **78**, 919—936.

Verma, D. P. S., Legocki, R., Auger, S., 1981b: Expression of nodule specific genes of soybean. In: Current perspectives in nitrogen fixation. Gibson, A. H., Newton, W. E., (eds.), pp. 205—208. Canberra: Aust. Acad. Sci.

Verma, D. P. S., Ball, S., Guérin, C., Wanamaker, L., 1979: Leghemoglobin biosynthesis in soybean root nodules. Characterization of nascent and released peptides and the relative rate of synthesis of major leghemoglobins. Biochem. **18**, 476—483.

Verma, D. P. S., Haugland, R., Brisson, N., Legocki, R., Lacroix, L., 1981: Regulation of the expression of leghemoglobin genes in effective and ineffective root nodules of soybean. Biochim. Biophys. Acta, **653**, 98—107.

Verma, D. P. S., Long, S., 1983: The molecular biology of *Rhizobium*-legume symbiosis. Int. Rev. Cytol. Suppl. 14, Jeon, K. (ed.), p. 211—245, New York: Academic Press.

Vest, G., 1970: Rj_3 — A gene conditioning ineffective nodulation in soybean. Crop Sci: **10**, 34—35.

Vest, G., Caldwell, B. F., 1972: Rj_4 — a gene conditioning ineffective nodulation in soybean. Crop. Sci. **12**, 692—693.

Vincent, 1980: Factors controlling the legume-*Rhizobium* symbiosis. In: Nitrogen Fixation Vol. II, Newton, W. E., Orme-Johnson, H. H. (eds.), pp. 103—129, Baltimore: Univ. Park Press.

Vose, P. B., Ruschel, A. P., 1981: Associative nitrogen fixation. Cleveland: CRC Press.

Wang, T. L., Wood, E. A., Brewin, N. J., 1982: Growth regulators, *Rhizobium* and nodulation in peas. Indole-3-acetic acid from the culture medium of nodulating and non-nodulating strains of *R. leguminosarum*. Planta **155**, 345—349.

Werner, D., Morschel, E., Striplf, R., Winchenbach, B., 1980: Development of nodules of *Glycine max* infected with an ineffective strain of *Rhizobium japonicum*. Planta **147**, 320—329.

Whittaker, R. G., Lennox, S., Appleby, C. A., 1981: Relationship of the minor soybean leghemoglobins *d*1, *d*2 and *d*3 to the major leghemoglobins c1, c2 and c3. Biochem. Int. **3**, 117—124.

Wiborg, O., Hyldig-Nielsen, J. J., Jensen, E. O., Palndan, K., Marcker, K. A., 1982: The nucleotide sequences of two leghemoglobin genes from soybean. Nucl. Acid. Res. **10**, 3487—3494.

Williams, L. F., Lynch, D. L., 1954: Inheritance of a non-nodulating character in soybean. Agronomy **46**, 28—29.

Wittenberg, J. B., Bergersen, F. J., Appleby, C. A., Turner, G. L., 1974: Facilitated oxygen diffusion. The role of leghemoglobin in nitrogen fixation by bacteroids isolated from soybean root nodules. J. Biol. Chem. **249**, 4057—4066.

Yao, P. Y., Vincent, J. M., 1976: Factors responsible for the curling and branching of clover root hairs by *Rhizobium*. Plant Soil **45**, 1—16.

Chapter 4

Rhizobium-Leguminosae Symbiosis: The Bacterial Point of View

B. G. Rolfe and J. Shine

Genetics Department, Research School of Biological Sciences, The Australian National University, Canberra City, A. C. T., Australia

With 7 Figures

Contents

I. Introduction

The group of soil bacteria classified as *Rhizobium* are characterized by their ability to successfully infect the roots of members of the plant familiy Leguminosae (Bergey, 1974). This infection causes an out-growth or "nodule" in which the invading bacteria fix atmospheric nitrogen into a form which the plant can assimilate and use for protein biosynthesis. As this association is normally of mutual benefit to both the invader (*Rhizobium* cells) and the invaded (plant host), it is called a symbiosis. However, if the infection fails to result in the fixation of nitrogen due either to a defect in the bacterium or the host, then the infected plants may become greatly debilitated in a fashion resembling a disease state. The symbiotic relationship between *Rhizobium* and plants can thus be viewed as that of a "controlled disease". Due to symbiotic nitrogen fixation with the various members of the *Rhizobium* genus, many pulse and pasture legumes can be grown without a requirement for nitrogenous fertilizer. On a global basis this association may reduce 20 million tonnes of atmospheric nitrogen to ammonia per annum which amounts to about 70—80% of all biologically fixed nitrogen per year (Burris, 1980).

Rhizobium species are defined on the basis of their host range with particular legumes being nodulated only by certain *Rhizobium* strains. Thus *R. trifolii* nodulates clovers (*Trifolium* species), *R. leguminosarum*-peas, (*Pisum sativum*), *R. meliloti*-lucerne (alfalfa) (*Medicago sativa*) and *R. japonicum*-soybeans (*Glycine max*) (Vincent, 1970). In contrast to the narrow specificity shown by these *Rhizobium* species on temperate legumes, another large group known as the "Cowpea Miscellany" contain those rhizobia which have a broad host range and can usually infect a variety of tropical legumes. This distinction between the two groups of rhizobia based on host range is paralleled by their different growth rates in the laboratory. The slow-growing rhizobia (now classified in a separate group known as the *Bradyrhizobium*) comprise most of the cowpea group of bacteria, whilst the narrow host range bacteria generally fall into the class of fast-growing rhizobia.

Several excellent reviews have recently been published which describe in detail some of the various aspects of the *Rhizobium*-legume interaction, including the infection of legumes (Bauer, 1981); genetic analysis of *Rhizobium* strains (Beringer *et al.*, 1979; 1980) and the biology of the Rhizobiaceae (Giles and Atherly, 1981). We shall attempt to describe here some of the very recent findings which illustrate some of the basic features of this *Rhizobium*-plant interaction from the bacterial point of view.

II. Strategies of Infection

Decades of research into biological nitrogen fixation as it occurs in *Rhizobium*-plant systems has shown that there are both a number of infection strategies used, as well as several different types of nodules formed (Dart, 1974a, b; 1975; 1977; Sprent, 1979; Allen and Allen, 1982). Nodule development in the temperate legumes (Fig. 1) is characterized by an initial bacterial colonization of the root and polar attachment by rhizobia to the root hair cells. This is then followed by a marked curling of the root hair which entraps the bacterium in a small pocket. An infection thread may form at this site and penetrate the root hair cell thus carrying the invading bacteria into the root cortex. The infection thread walls seem to be synthesised by the plant, although the matrix within the thread where the bacteria multiply is probably of bacterial origin. The infection thread passes through the outer cortical cells and then ramifies within the cortex (Fig. 2). Inner cortical cells begin to divide and initiate a focus of diving plant cells ahead of the advancing thread. The infection thread passes through the cortex and grows towards the stele. At the same time the proliferating cells which grow towards the infection threads become invaded by bacteria, which are

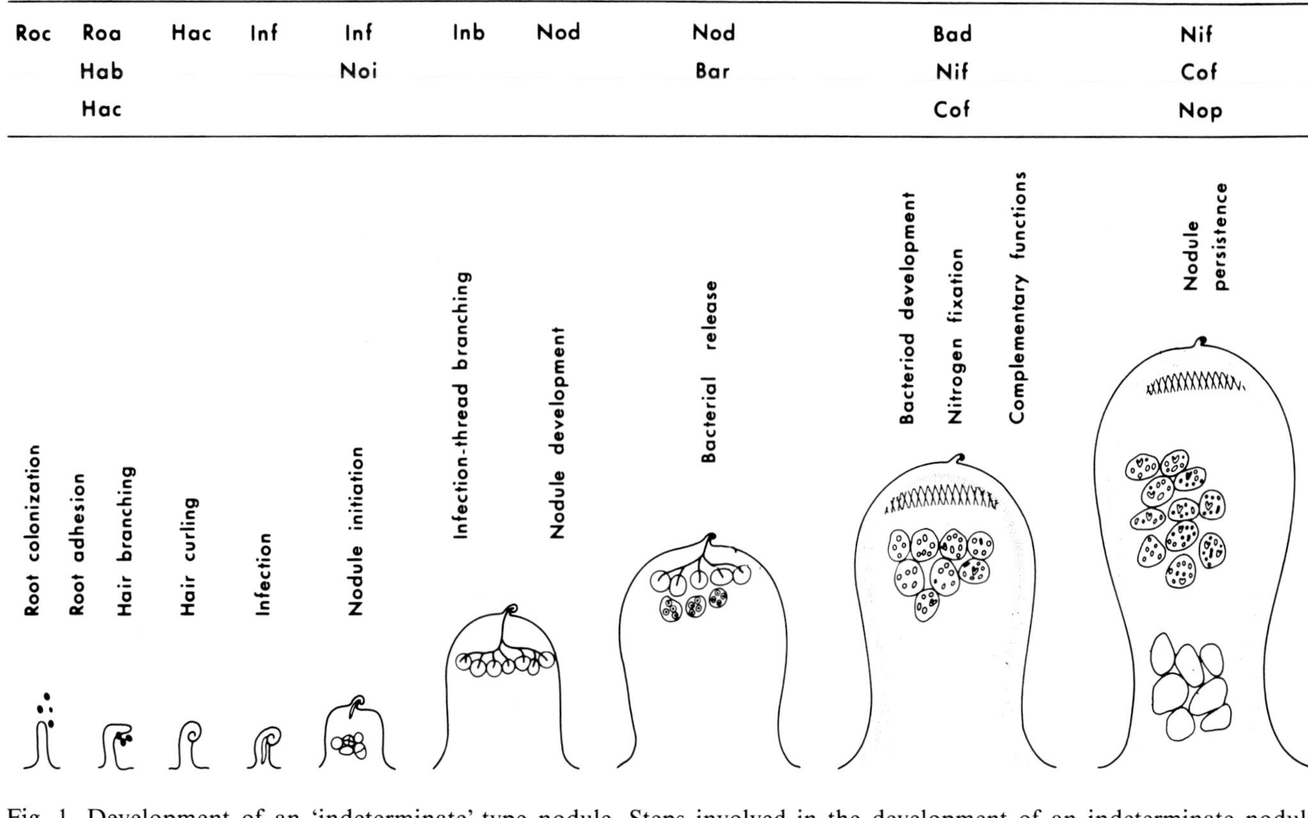

Fig. 1. Development of an 'indeterminate'-type nodule. Steps involved in the development of an indeterminate nodule showing the phenotype exhibited by bacterial mutants defective in particular stages of the symbiosis

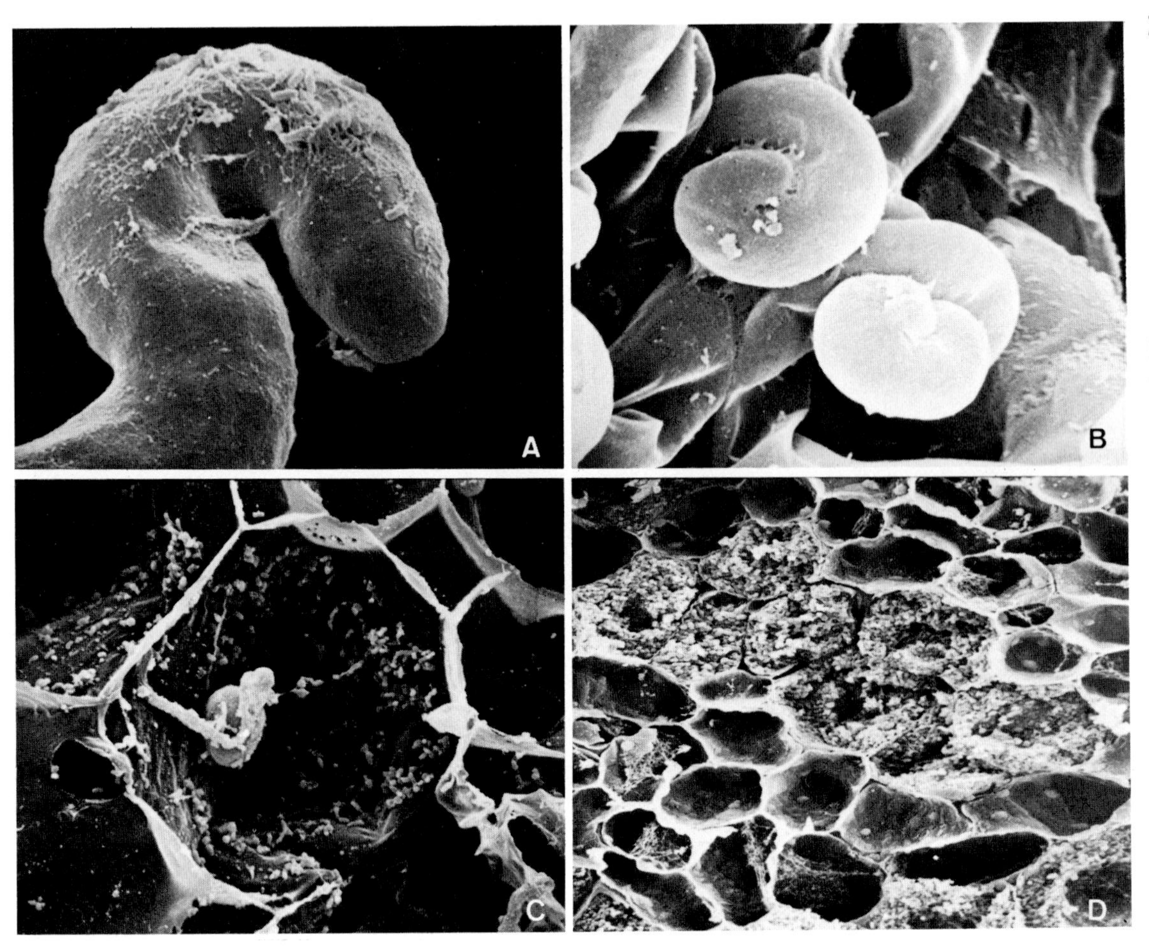

Fig. 2. Scanning electron micrographs of the *Rhizobium trifolii* — white clover symbiosis.

A and B: Curling of root hairs caused by attachment of bacteria

C and D: Nodule section showing infection thread and released bacteria and bacteroids.

Kindly provided by R. W. Ridge

released from the tips of these threads and become enveloped in plant cell membranes. These infected cells then cease to divide and start to swell (Libenga and Bogers, 1974; Newcomb, 1980). An apical meristem is formed by the adjacent uninvaded cells which continue to divide. Some of the cells produced from this meristematic activity are then also invaded by the growing threads; the other meristematic cells provide new cortex and vascular bundles. This activity continues until a morphologically defined root nodule is formed. As the nodule forms it receives its vascular supply from the stele (Dart, 1977). Examples of this type of nodule are found on clovers, peas, lucerne and some beans and are known as indeterminate nodules because of the maintenance of a continuous meristematic zone of dividing plant cells (Fig. 1).

During the development and maintenance of the nodule, specific bacterial and plant genes are expressed in a tightly coordinated and temporal fashion. Nodule-specific proteins (nodulins) are produced due to the derepression of certain plant genes including the gene encoding the protein moeity of the leghaemoglobin molecule (Legocki and Verma, 1980) — the haem moeity is produced by the bacterium (Dilworth and Appleby, 1979; Leony *et al.*, 1982). The leghaemoglobin molecule becomes important in the regulation of oxygen tension levels in the nodule and thus protects the oxygen-sensitive nitrogenase functions encoded by the bacterium.

A derepression of various bacterial genes also occurs — notably those encoding the proteins in the nitrogenase complex, as the bacteria undergo both biochemical and morphological changes. The bacteria stop dividing and take on swollen pleomorphic shapes at which stage they are called bacteroids. The bacteroids may then fix atmospheric nitrogen and utilize certain compounds provided by the plant as energy sources.

In summary, the nodulation process involves a series of complex, highly evolved interactions between a procaryote and its eucaryotic partner. Because of the complexity of this interaction it has been difficult to analyse the various steps involved in the process. Consequently, it has proven necessary to provide a framework by which different bacterial and plant mutants with recognisable phenotypic traits can be classified with respect to the various steps thought to be involved in the nodulation process (Vincent, 1980; Rolfe *et al.*, 1982). This scheme (summarised in Fig. 1) and the proposed phenotypic code aim to sub-divide various bacterial mutants into distinct groups in order to facilitate their further genetical and biochemical analysis.

Although there are many variations in the pattern of nodule development, two basic types can be recognized (Sprent, 1979). One consists of nodules where infection of new cells occurs by means of infection threads (as described above) whilst the other comprises those where the bacteria are mainly spread by division of preinfected cells (Sprent, 1979). This latter group have infection threads that do not branch and ramify extensively after entering the root cortex from the originally infected root hair cell. Generally the nodules without extensive infection threads are more spherical in shape, lack a persistent meristem and have a closed vascular system.

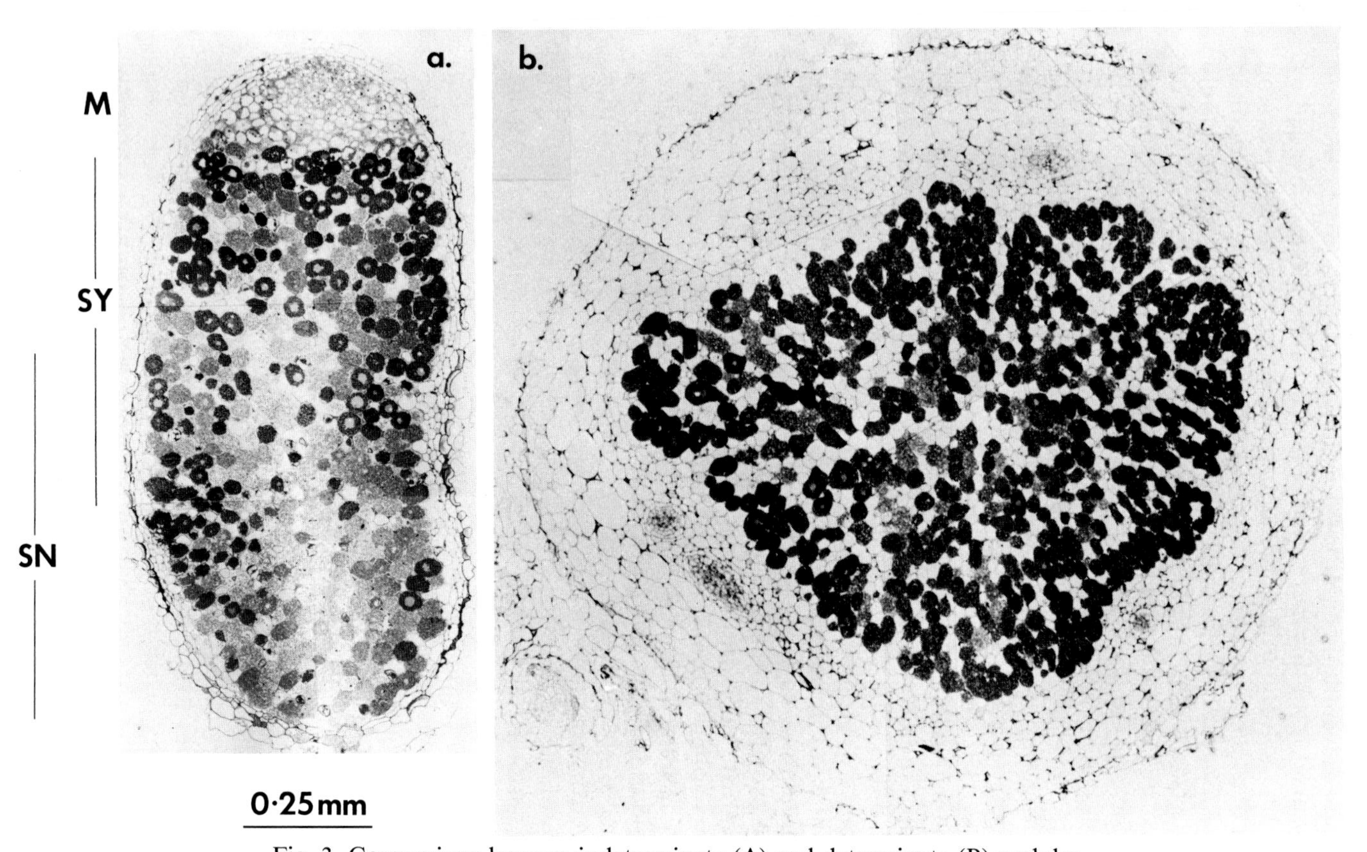

Fig. 3. Comparison beween indeterminate (A) and determinate (B) nodules

A: Light micrograph of a white clover nodule infected with the *R. trifolii* strain ANU843. M, meristem; SY, symbiotic zone; Sn, senescence zone.

B: Light micrograph of a siratro nodule infected with the cowpea strain CB756.
Sections embedded in Spurr's resin, sectioned at 0.5 micron, and stained with toluidine blue. Kindly provided by R. W. Ridge

COMPARISON OF TYPES OF LEGUME NODULES

Characteristic	Plant group	
	Temperate legumes Vicieae and Trifolieae (peas, beans, clovers, medics)	Tropical legumes Phaseoleae (soybeans, Vignas)
Nodule growth	Indeterminate persistent meristem	Determinate (3-5 weeks) non-persistent meristem
Vascular system	Open at apex 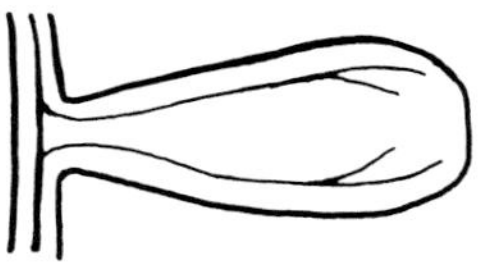	Closed continuous system  (Vascular flow rate ~5-10 times indeterminate system)
Vascular transfer cell	Present	Absent from nodule (perhaps in root vascular system at junction of nodule)
Infected cells	Vacuoles present	Non-vacuolated
Bacteroid morphology	Pleomorphic	Rods and swollen rods
Typical export products	Amides (glutamine) (asparagine)	Ureides (allantoin) (allantoic acid)

Fig. 4. Comparison of types of legume nodules

They are known as determinate nodules and examples are found in soybeans, phaseolus beans, cowpeas and siratro. A comparison of these two basic types of legume nodules is summarized in Figs. 3 and 4.

Legume infection via root hairs, whether leading to a determinate or indeterminate nodule, is most common but is not universal (Fig. 5). In peanuts and *Stylosanthes* root hairs are formed at the points of emergence of lateral roots and infection of these plants illustrates an invasion strategy that may be described as "crack entry". In these cases the *Rhizobium* cells

INFECTION SITES OF NODULE FORMING RHIZOBIA

	LEGUMES				NON – LEGUMES
	Clovers/Peas	Soybean	Siratro	Stylosanthes	Parasponia
					?*
Rhizobium					
Fast grower	+	+	+	+	+
Slow grower	–	+	+	+	+
ROOT HAIR DEVELOPMENT					
Present on tap root	+	+	+	occasional	+
Present on lateral roots	+	+	+	Whorls around lateral base	+
INFECTION SITE					
Root hairs	+	+	+	–	+
Between cells on lateral roots	–	–	–	+	–
NODULE LOCATION					
On tap and lateral roots	+	+	+	–	–
Axils of lateral roots	occasional	occasional	occasional	+	+

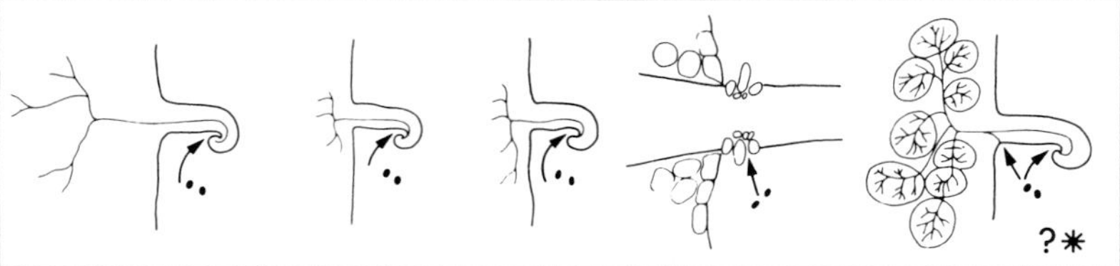

Fig. 5. Different strategies of *Rhizobium* infection

are thought to enter the plant tissue between cells at the base of the multicellular root hairs which occur on the emerging lateral roots (Fig. 5) (Chandler, 1978; Chandler, Date and Roughley, 1981). Although the actual process of entry has still not been resolved and no infection threads have been observed, the rhizobia are enclosed within plant membranes inside the cytoplasm of the host cells. In due course, masses of bacteria enter the cortical cells of the lateral root and both the host cells and the membrane-enveloped bacteria divide repeatedly to form a nodule.

As well as these different types of infection processes found within the legumes, some *Rhizobium* strains can also invade certain plants of the non-legume group of *Parasponia* (from the Ulmaceae) (Trinick and Galbraith, 1976; Trinick, 1979; Trinick and Galbraith, 1980). Although in these cases the initial mode of entry is still not clear, within the nodules ramifying infection threads are found which retain the invading bacterium (Fig. 5). The threads ramify between and within plant cells setting up an extensive network of invaded tissue. Furthermore, these nodules form at the points of emergence of the lateral roots.

Recently, nodulation along the stems of *Sesbania rostrata* has been examined and shown to involve a "crack entry" type system (Dreyfus and Dommergues, 1981; Tsien *et al.*, 1983; J. F. Olsson, personal communication). The rhizobia appear to enter via the sites from where the adventitious roots emerge from the stem and form a ramifying infection thread in this lateral root tissue. It appears that the rhizobia "take over" the adventitious root meristem and cause a termination of root growth with the concomitant formation of a nodule (J. F. Olsson, unpublished). If this is shown to be so, then the utilization of a pre-existing meristem contrasts with the induction of meristematic zones by invading bacteria, which is usually found in other legume nodules.

There are thus three essential types of infection processes (a) infection thread entry via root hairs; (b) crack entry with no observed infection thread formation, and (c) crack entry with infection thread formation. These infection processes can be observed with either fast or slow-growing rhizobia and while most rhizobia do not carry out all three potential strategies of infection, there are some strains where this does occur (Trinick and Galbraith, 1976; Trinick, 1979; Morrison *et al.*, 1983). Furthermore, using transmissible Sym plasmids it is possible to construct strains which can use each of these different strategies of infection (Morrison *et al.*, 1983) and which now have a broader host range (Djordjevic *et al.*, 1982; Djordjevic, 1983).

III. Localization of Infectible Root Cells

In a series of elegant experiments Bauer and his co-workers have identified which root cells are susceptible to nodulation by rhizobia (Bhuvaneswari *et al.*, 1980; 1981; Turgeon and Bauer, 1982; Bauer, 1981; Turgeon and Bauer, 1983). Using a combination of (a) plastic growth pouches, (b) precise recording of the growth of the root tip and emerging root hair zone, (c) inoculation with very small volumes (nanoliters) of *Rhizobium* cultures and (d) use of ion exchange beads (spot inoculation), they have shown that the infectibility of soybean epidermal cells is a transient property. Moreover, their techniques enable them to predict with an 80% probability the exact point of nodule formation after spot inoculation.

Infections in soybeans can occur through infection threads formed in emerged root hair cells. However, infections leading to nodule development appear to be initiated most frequently in the root zone where no emergent root hairs are present at the time of inoculation (Bauer, 1981). This zone of susceptible root cells moves acropetally as the root grows (Turgeon and Bauer, 1982). Furthermore, the susceptibility to nodulation by rhizobia is developmentally regulated and a transient property of root cells in soybeans, alfalfa, cowpea, peanut and white clovers. A similar finding has been made for siratro plants (R. W. Ridge, personal communication).

IV. Genetic Approach to the Analysis of Symbiosis

The isolation of various classes of bacterial mutants unable to carry out the symbiotic process has demonstrated that a number of different steps are involved in the establishment of effective nitrogen fixation (Vincent, 1980; Beringer, 1980). However, the genetic basis of this bacterial-plant interaction is poorly understood and the characterization and manipulation of symbiotic genes has been hampered by two main problems. The first has been the difficulty encountered in deciding whether a defective symbiotic phenotype resulting from a mutagenic treatment is the result of either single or multiple mutagenic events. The second concerns the technical difficulties of screening individual colonies where the only recognizable phenotype of symbiotic mutants is their inability to form fully functional nodules, thus requiring laborious plant assays.

Recently these problems have been largely overcome particularly in the fast growing strains by (a) the development of a rapid plant assay system to screen rhizobia for a defective symbiotic phenotype (Rolfe *et al.*, 1980; Rolfe *et al.*, 1981b; Cen *et al.*, 1982) and (b) the use of transposons (carrying drug-resistant genes) as mutagens (Beringer *et al.*, 1978; Meade *et al.*, 1982; Rolfe *et al.*, 1980; Forrai *et al.*, 1983). Transposons are small elements of DNA which can insert into different regions of the genome of a bacterium. Many of them carry a drug resistence gene which can be used to mark their location in the bacterial DNA. Thus, the recent advent of transposon mutagenesis as a general genetic tool has made available a mutagenesis system which can provide a "tag" for particular genes of interest. The transposon Tn5, which carries a kanamycin resistance marker (Kleckner, 1977), is a particularly useful mutagenic agent because of its ability to insert "randomly" into a genome. Current molecular techniques can be used to clone the transposon and associated *Rhizobium* DNA sequences. Coupling these techniques with a procedure for rapid screening for symbiotically defective associations provides a system for the isolation of transposon-induced mutants in the pathways involved in establishing a nitrogen-fixing nodule (Ruvkun *et al.*, 1982; Scott *et al.*, 1982).

V. Isolation and Cloning of Symbiotic Mutants

The isolation procedure used to obtain kanamycin-resistant rhizobia which are defective in their symbiotic properties is outlined in Table 1.

Table 1. Procedures in the Isolation of *Rhizobium* Symbiotic Genes

1.	Transfer of transposon Tn5 to chosen *Rhizobium* strain *via* bacterial conjugation.
2.	Detection of symbiotic mutants using the plate method to screen large numbers of plants rapidly.
3.	Examination of the DNA of presumptive symbiotic mutants to show that a single copy of the transposon is present for each mutant and that various mutants have different locations of the transposon in their DNA.
4.	Clone the transposon Tn5 and associated *Rhizobium* DNA sequences into amplifiable *E. coli* plasmids.
5.	Probe the wild-type *Rhizobium* DNA "clone bank" which is stored in phage λ with radioactively-labelled recombinant Tn5-containing plasmids to isolate the corresponding wild-type DNA sequence.
6.	Sub-clone the wild-type DNA sequences into a vector which can be transferred into the original mutant *Rhizobium* strain.
7.	Test the transconjugants for the correction of the symbiotic defect due to the introduction of the wild-type genes.

DNA is isolated from each of the presumptive symbiotic mutants and hybridized with a radioactively-labelled Tn5 probe to show that a single copy of the transposon is present in each mutant. To confirm that a true "symbiotic" gene has been identified by this approach, the corresponding wild-type *Rhizobium* DNA sequences are isolated and re-introduced into the appropriate symbiotic mutant and a correction of the symbiotic defect demonstrated. This approach is thus a type of "circuit experiment" and involves, (a) isolation of the original Tn5-induced mutant, (b) cloning of Tn5 and flanking *Rhizobium* sequences, (c) cloning of the wild-type gene sequences corresponding to those flanking the transposon and (d) transfer of these cloned sequences into the original mutant and demonstration of the correction of the original symbiotic defect (Scott *et al.*, 1982). The effectiveness of this type of approach has now been demonstrated for several Nod$^-$, Nod$^+$Fix$^-$ and Muc$^-$ mutants — in each case the original Tn5-induced mutation was shown to be the direct cause of the resultant symbiotic defect of the particular mutant. Although most of the "symbiotic" genes so far isolated by molecular cloning have been found to reside on large plasmids in the fast-growing rhizobia, other chromosomal genes are also clearly involved (Kondorosi and Johnston, 1981).

VI. *Rhizobium* Sym(biosis) Plasmids

Several excellent reviews are available which describe the indigenous plasmids of *Rhizobium* strains and the genetic analysis of such bacteria (Denarie *et al.*, 1981; Beringer *et al.*, 1980). We shall summarize here those properties which are relevant to *Rhizobium* Sym plasmids and the symbiotic process.

Although studies on the genetic basis of the *Rhizobium*-legume interaction had begun many years earlier, it was the findings of Higashi (1967) that indicated an extrachromosomal element might be involved in the control of nodulation and host specificity of *Rhizobium trifolii* strains. However, it was not until the studies of Zurkowski and Lorkiewicz 1976, Nuti *et al.*, 1977; Casse *et al.*, 1979; Prakash *et al.*, 1980, that plasmids of various sizes were definitely shown to exist in some *Rhizobium* strains. Zurkowski and Lorkiewicz (1978) subsequently found a method which was suitable for heat-curing many *R. trifolii* strains of their nodulation capacity. The mutant symbiotic phenotype was always correlated with the loss of a large plasmid. The role of plasmid encoded genes in symbiosis was confirmed when it was shown that the transfer of one of these large plasmids restored nodulation capacity to a plasmid-cured strain (Zurkowski, 1980).

A more precise equating of a particular *Rhizobium* plasmid with the encoding of both nodulation and fixation functions followed two developments. The first of these was the use of cloned nitrogenase genes (*nif* H, *nif* D) From *Klebsiella pneumoniae* to show the conservation of such DNA sequences in a variety of nitrogen-fixing bacterial strains (Ruvkun and Ausubel, 1980). Definitive evidence that a single large plasmid of a fast-growing *Rhizobium* strain encoded nitrogenase genes came from the hybridization of the cloned *Klebsiella* nitrogenase genes to plasmid DNA's resolved by gel electrophoresis. In addition, non-nodulating *Rhizobium* strains cured of one of their large plasmids, and strains with large deletions in one of their plasmid species, lacked the DNA sequence which hybridized to *Klebsiella nif* H, D probes. The second development was that of transposon mutagenesis using the transposon Tn5 (Beringer *et al.*, 1978) which was used to label different *Rhizobium* plasmids (Johnston *et al.*, 1982). This construction initially helped the genetic analysis of such plasmids by demonstrating that some plasmids (designated as Sym plasmids) encoded symbiotic genes and were self transmissible (Johnston *et al.*, 1978; Hooykaas *et al.*, 1981). More recently the use of Tn5-labelled Sym plasmids has permitted the construction of transmissible plasmids which code for both nodulation and fixation functions by co-integration of the Sym plasmids with other small mobilisable plasmids (Kondorosi *et al.*, 1982; Rolfe *et al.*, 1982; Bafalvi *et al.*, 1983; Morrison *et al.*, 1983; Morrison, unpublished). These techniques have enabled the transfer of the indigenous megaplasmid pRme41b of *R. meliloti* strain 41 into other *Rhizobium* strains and *Agrobacterium tumefaciens*, as well as the construction of a set of R-primes carrying nitrogen fixation (nif and fix) genes and nodulation (nod) genes of *R. meliloti* (Kondorosi *et al.*, 1982). A stable cointegrate plasmid (pNM4AN) has been constructed between the Sym plasmid of the broad host range *Rhizobium* NGR234 and pSUP1011 (Rolfe *et al.*, 1982). The latter plasmid carries the mobilization (mob_{RP4}) site of plasmid RP_4 and transposon Tn5. This cointegrate plasmid pNM4AN can also be mobilized between various *Rhizobium* and *Agrobacterium* strains at high frequencies (Morrison, personal communication).

Several important findings have emerged from all these studies on the *Rhizobium* plasmids:

(1) at least some of the genes involved in the initiation of root hair infection, nodulation and nitrogen fixation are located on large plasmids in most fast-growing *Rhizobium* strains.

(2) some *Rhizobium* strains can be cured of their Sym plasmid by incubation at higher growth temperatures.

(3) Sym plasmids can be transferred between symbiotic and non-symbiotic *Rhizobium* strains.

(4) some Sym plasmids also carry genes coding for bacteriocin and for pigment formation (Beynon *et al.*, 1980; Brewin *et al.*, 1980a).

(5) some chromosomally determined functions of *Rhizobium* are also involved in the establishment of a nitrogen-fixing nodule.

Most of the fast-growing *Rhizobium* strains contain variable numbers of plasmids of different sizes and, apart from several genes on the Sym plasmid, the information encoded on these plasmids is still unknown. Those biological functions that are known to be associated with plasmids in *Rhizobium*, particularly Sym plasmids, are listed in Table 2. Furthermore, studies of different heat-treated *Rhizobium* strains show that one and sometimes two of the plasmids (including the Sym plasmid) may be lost without greatly altering the performance of the strain under laboratory growth conditions (Djordjevic, personal communication). The cryptic information encoded by the Sym plasmid and other plasmids would thus appear to be not essential for the survival of a strain. However, *Rhizobium* can potentially have two life styles. One in the soil environment, often associated with plant roots, and the other within the infected plant root or nodule. The presence of the Sym plasmid certainly provides the opportunity for the latter alternative life style and perhaps as well the ability to catabolise unusual opine-like compounds synthesized in the nodule (Petit and Tempe, 1983). The possibility exists that these cryptic plasmids encode genes which enable survival in the soil.

The most intensively studied Sym plasmid so far is pRL1JI which encodes pea nodulation and fixation functions (the Tn5-marked derivative is known as pJB5JI) (Hirsch, 1979; Johnston *et al.*, 1978; Brewin *et al.*, 1980b; Benyon *et al.*, 1980; Ma *et al.*, 1982). This is a self transmissible plasmid which also encodes genes for the regulation of medium and small bacteriocin production. A transposon mutagenesis system was used to generate a variety of Nod^- and Fix^- mutants (Buchanan-Wollaston *et al.*, 1980) where the molecular lesion was mapped to the plasmid. Furthermore, characterization of these Fix^- mutants showed that they could be placed into two groups, one located within 4 kilobases (kb) of Nod^-

Table 2. Biological Functions Located on Sym Plasmids in *Rhizobium* Strains

Medium bacteriocin production (Mbp$^+$)	Beringer *et al.*, 1980; Brewin *et al.*, 1980a
Repression of small bacteriocin production (Rsp$^+$) production (Rsp$^+$)	Djordjevic *et al.*, 1983
Nodulation ability	
Root adhesion (Roa$^+$)	Rolfe *et al.*, 1981; Hooykaas *et al.*, 1981
Hair curling (Hac$^+$)	Johnston *et al.*, 1978; Rolfe *et al.*, 1981 Djordjevic *et al.*, 1983; Forrai *et al.*, 1983
Host range specificity	Truchet *et al.*, 1983; Rosenberg *et al.*, 1981
Nodule function (Fix$^+$)	Brewin *et al.*, 1980b
Nitrogenase enzyme complex	
*nif*H (Fe-protein)	Nuti *et al.*, 1979; Hombrecher *et al.*, 1981
*nif*D (-subunit of Mo, Fe-pro tein)	Morrison *et al.*, 1983
Hydrogenase (Hub$^+$) production	Brewin *et al.*, 1980c
Genes influencing cell surface polysaccharide synthesis	Carlson *et al.*, 1983; Kuempel *et al.*, 1983
Pigment production (Pig$^+$)	Beringer *et al.*, 1980; Lamb *et al.*, 1982
Transfer functions (Tra$^+$)	Beringer *et al.*, 1980
Incompatibility	Beringer *et al.*, 1980
Chromosome mobilization ability (Cma$^+$)	Beringer *et al.*, 1980

and Fix$^-$ mutants (Buchanan-Wollaston *et al.*, 1980) where the molecular lesion was mapped to the plasmid. Furthermore, characterization of these Fix$^-$ mutants showed that they could be placed into two groups, one located within 4 kilobases (kb) of the nitrogenase genes and the other located about 30 kb from this region (Ma *et al.*, 1982). Other studies have shown that it is possible to exchange about 35% of the pJB5JI plasmid molecule with a genetic region from a clover Sym plasmid and thus construct a transmissible, bacteriocinogenic plasmid encoding clover nodulation and fixation functions called pBR1AN (Rolfe *et al.*, 1982; Djordjevic *et al.*, 1983).

An investigation of the *R. leguminosarum* Sym plasmids pRL1JI and pRL6JI and plasmid incompatibility has enabled recombinant nodulation plasmids to be constructed. Such plasmid recombination experiments have made it possible to transfer the symbiotic determinants (Nod$^+$, Fix$^+$) and the hydrogen uptake (Hup$^+$) system between different replicons and thus construct *in vivo* new Sym plasmids (Brewin *et al.*, 1982). Furthermore, one particular recombinant plasmid pIJ1008, when transferred into different genetic backgrounds of *R. leguminosarum*, causes an increased nitrogen fixation in such strains (DeJong *et al.*, 1982). These results demonstrate a very important point for future strain construction studies, namely, "that both the quantity of plant biomass produced and the N content of that biomass can be increased by selective transfer of plasmids into strains of *R. leguminosarum" (DeJong et al.*, 1982).

VII. Chromosomal Location of Nitrogenase and Nodulation Genes

Studies of various slow-growing strains have had difficulty in demonstrating the presence of plasmids, although some slow-growing *R. japonicum* strains certainly do contain large plasmids (Russell and Atherly, 1982). In the slow-growers there does not appear to be plasmids of the Sym plasmid class as found in fast-growing rhizobia. For example, four different plasmid gel techniques have been used to examine the slow-growing cowpea rhizobia strains CB756 and 32H1 and the *Parasponia Rhizobium* strain ANU289 (J. Plazinski, personal communication). This study suggests that these strains do not contain any plasmids and that their nitrogenase genes are located on the chromosome. A similar conclusion was made for the fast-growing *Sesbania rostrata Rhizobium* strain ORS571 (J. Plazinski and R. Legocki, personal communication). Furthermore, one other fast-growing *Rhizobium* strain (*R. japonicum* strain PRC194), has been reported as having its nitrogenase genes located on the chromosome and not on a plasmid (Masterson *et al.*, 1982).

The Nif-Nod Region

There are two key regions located some 20—30 kilobases (kb) apart on the Sym-plasmids which are basic to the establishment of a nitrogen-fixing nodule (Fig. 6). The Hac (hair curling) region appears to code for genes involved in root hair branching and curling (Hab$^+$ and Hac$^+$ phenotypes) and may be composed of as few as 2 to 3 genes. Transfer of a DNA fragment of 14kb containing this Hac region from *R. trifolii* (clover bacterium) into a *R. trifolii* strain with no Sym plasmid enables this reconstructed strain to infect clover plants and form nodules with some bacteroids present (Schofield *et al.*, 1983a). Thus the infection steps such as infection thread formation (Inf$^+$), nodule initiation (Noi$^+$), infection thread branching (Inb$^+$), nodule development (Nod$^+$) and bacterial release (Bar$^+$) are all associated, directly or indirectly, with this transferred 14kb fragment. Furthermore, these sequences encode both host specificity and nodulation functions as demonstrated by conferring on related bacterial species such as *Agrobacterium* the ability to recognize and nodulate clover plants (Schofield *et al.*, 1983a). This indicates that there are a small number of closely linked bacterial genes essential for the highly specific initial (symbiotic) interaction with plants. Although it appears to vary with different species of *Trifolium*, in subterranean clovers some aspects of bacteroid development (Bad) are also influenced by genetic information associated with this 14kb DNA fragment (Shine *et al.*, 1983; Watson *et al.*, 1983).

In *R. leguminosarum*, random Tn5 mutagenesis of the Sym plasmid pRL1JI provided 12 symbiotically-defective mutants which are located within a span of about 45kb of plasmid DNA (Fig. 6) (Ma *et al.*, 1982; Downie *et al.*, 1983). The Fix$^-$ mutants occurred in two groups, one located within 4kb of the nitrogenase (*nif* HDK) structural genes and the other about 30 kilobases from this region. Within this 45kb region there are two

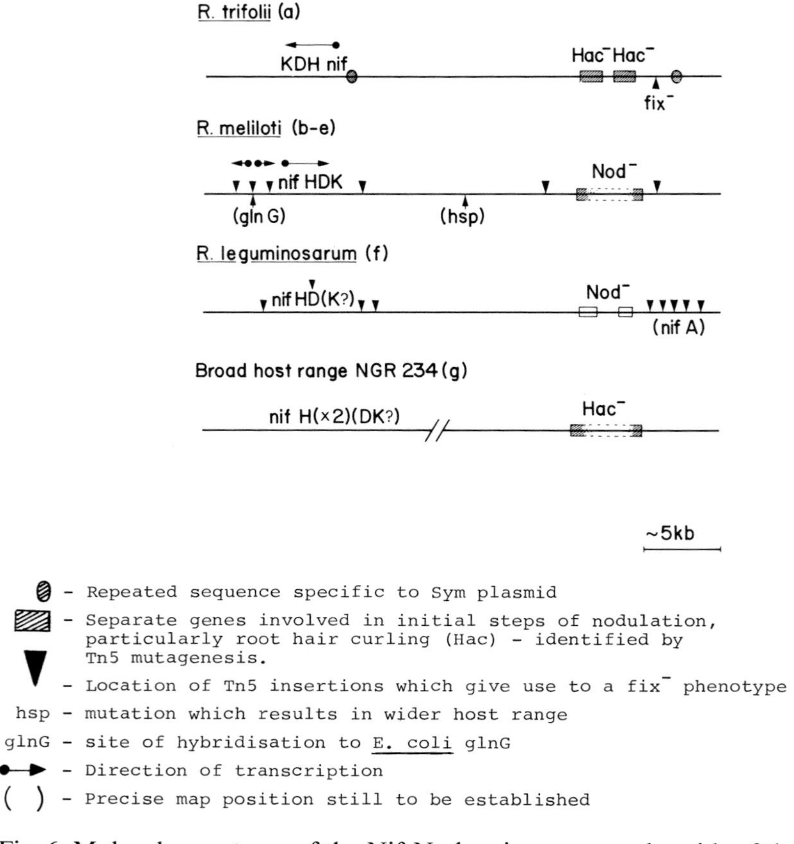

Fig. 6. Molecular anatomy of the Nif-Nod region on sym plasmids of the fast-growing rhizobia

areas, one about 20kb and the other about 6kb, which have no *fix* or *nod* mutations present. Such regions could contain genes which have an intermediate symbiotic phenotype (Downie *et al.*, 1983). In *R. trifolii* one Tn5-induced mutant which gave a level of acetylene reduction only about 30% of that of its parent strain was located within 2kb of the Hac genes (Fig. 6) (Schofield, personal communication).

The Nif region is centred around the location of the genes coding for the nitrogenase enzyme. In the fast-growing rhizobia these genes (the *nif* H [Fe-protein], *nif* D,K [Mo, Fe protein]) are closely linked and probably form part of a single operon (Shine *et al.*, 1983; Scott *et al.*, 1983).

The *R. trifolii* Fe-protein (*nif* H gene product), as predicted from the DNA sequence, is 297 amino acids in length and has a molecular weight of 31,903 daltons (Scott *et al.*, 1983a). It is now well established that the amino acid sequence of the Fe-protein subunit is strongly conserved among widely divergent organisms. The *R. trifolii* Fe-protein shares this sequence conservation and is 65% homologous to the *C. pasteurianum* (Tanaka *et al.*, 1977) sequence and 70% homologous to the sequences from *K. pneumoniae* (Scott *et al.*, 1981a; Sunderasan and Ausubel, 1981), *Anabaena* 7120 (Mevarech *et al.*, 1980), *Azotobacter vinelandii* (Hausinger and Howard, 1982) and *Parasponia Rhizobium* ANU289 (Scott *et al.*, 1983b). Furthermore, the *R. trifolii* Fe-protein has 90% identity with that of the related legume symbiont *R. meliloti.*

Interestingly, the Nif region in *R. trifolii* is associated with a repeat sequence (RS) which is present in 4 to 5 copies per cell but only on the Sym plasmid (Shine *et al.*; Schofield *et al.*, 1983 a) The fact that all copies of this sequence are located on the Sym plasmid suggests that they may have a functional role in controlling the expression of symbiotic genes. The presence of such repeated sequences may act as "hot-spots" for recombination. In *R. trifolii,* such a mechanism may account for the spontaneous loss of nitrogen-fixation functions, as is observed with some commercial inoculum strains (Djordjevic *et al.*, 1982). The Nod and Nif regions in *R. trifolii* are only 16kb apart and flanked by copies of this repeat sequence (Fig. 6), an organisation of *nif* or *nod* genes which is suggestive of a "transposon-like" structure.

The *nif* gene cluster in *R. meliloti* occupies only 14—15kb of which 12 to 13kb is indispensible for *nif* functions. Extensive Tn5-directed mutagenesis of this region has demonstrated the presence of several independent transcriptional units which are involved in the symbiotic phenotype (Corbin *et al.*, 1983; Ruvkun *et al.*, 1982) (Fig. 6). The regulatory region controlling this *nif* gene cluster, defined by hybridization with the *K. pneumoniae nif* A gene region (a positive gene regulator), has been tentatively located next to the Hac region in *R. leguminosarum* (Downie *et al.*, 1983). Similarly, sequences homologous to the *gln* G gene of *E. coli* are closely linked to the *nif* region in *R. meliloti* (Szeto *et al.*, 1983).

The successful demonstration of the homology of the *nif* A gene from *K. pneumoniae* to *R. leguminosarum* DNA was made with cloned fragments (Downie *et al.*, 1983) rather than with *Rhizobium* genomic DNA (Nuti *et al.*, 1979; Ruvkun and Ausuble, 1980). The *K. pneumoniae nif* A gene product is a positive regulator activating the initiation of transcription of all the *nif* promoters in that strain (Dixon *et al.*, 1980; Buchanan-Wollaston *et al.*, 1981) except that of *nif* A itself (Ow and Ausubel, 1983). Recently, it has been shown that positive transcriptional regulators of several genes involved in nitrogen assimilation are the products of the genes *gln* G (*ntr* C) and *gln* F (*ntr* A), whereas negative regulation involves *gln* G, *gln* L (*ntr* B) and *gln* B products (Ow and Ausubel, 1983). Furthermore, the *nif* A gene product can substitute for the *gln* G gene product in its regulator roles, including the activation of the *nif* LA operon (Ow and Ausubel,

1983). The *gln* F product is also involved in the *nif* A regulation of other genes. At present, studies of various *Rhizobium* strains suggest that a similar activation of the nitrogen fixation genes perhaps involves a *nif* A – *gln* G – *gln* F – like regulatory system.

In the slow-growing rhizobia the *nif* D (subunit) and *nif* K (subunit) genes are linked and appear to be expressed in one transcriptional unit in the order *nif* DK (Scott *et al.*, 1983b; Fuhrmann and Hennecke, 1982). However, the *nif* H gene is transcribed from a separate operon and is separated from *nif* DK by at least 13kb (Scott *et al.*, 1983b). Moreover, the amino acid sequence of the Fe-protein in *Parasponia Rhizobium* ANU289 is sufficiently divergent from that of the fast-growing rhizobia to suggest that there are at least two distinct classes of organisms which have evolved the capacity for symbiotic nitrogen fixation with legumes (Scott *et al.*, 1983b).

To date, the main studies examining which regions of the Sym plasmid are expressed in the nitrogen-fixing bacteroids of the fast-growing rhizobia have come from several laboratories investigating *R. leguminosarum* strains (Krol *et al.*, 1980; 1982; Prakash *et al.*, 1981; 1982). Hybridization with bacteroid RNA has shown that only Sym plasmid sequences and not other plasmid DNAs are strongly expressed in the nodule environment.

Furthermore, selective amplification of plasmid DNA does not occur during the formation of bacteroids. Several regions of the Sym plasmid are transcribed in bacteroids but none of these areas overlap with the region of the Sym plasmid that is expressed in stationary phase cultures (Prakash *et al.*, 1982). It is possible that two of the transcribed areas of the Sym plasmid equate with two Fix regions identified by the Tn5 mutants characterized by Ma *et al.* (1982) (see also Fig. 6).

VIII. Analysis of Host Specificity

The rhizobia which invade temperate legumes tend to exhibit a narrow plant host range (Vincent, 1980). In contrast, *Rhizobium* strains that infect tropical legumes tend to be less restricted and can invade a broader range of tropical legumes. However, with the recent construction of transmissible Sym plasmids, it has become possible to transfer a narrow host range capacity to normally tropical legume rhizobia and, *vice versa*, transfer a broad host range to normally temperate legume nodulating bacteria (Rolfe *et al.*, 1982; Morrison *et al.*, 1983; Morrison, personal communication).

The use of transmissible Sym plasmids has been very useful in examining the interaction of these plasmids and the contribution made by the background genome of the bacterial cell to the phenomenon of host specificity. The results in Table 3 summarize the findings made with a series of constructed strains testing the expression of symbiotic genes from various rhizobia in different bacteria. The general finding is that when a Sym plasmid is transferred to a different *Rhizobium* species, it can usually initiate nodulation but not nitrogen fixation on the new host plant. This implies that not all of the host specificity functions are encoded by the Sym plasmid. Exceptions occur where clover nodulation and fixation genes are

Table 3. Effect of Non-Homologous Sym plasmids on Symbiotic Nitrogen Fixation

Constructed strain	Plant response							
	Clovers							
	Red	White	Subterranean	Peas	Lucerne	Siratro	Desmodium	Reference
R. leguminosarum								
6015 (pJB5JI)		Hac$^+$	Nod$^+$Fix$^-$	Nod$^+$Fix$^-$				a
6015 (pBRIAN)	Nod$^+$Fix$^-$	Nod$^+$Fix$^+$	Nod$^+$Fix$^-$					a
6015 (pRMSL26)		Hac$^+$	Hac$^+$					a
6015 (pNM4AN)							Nod$^+$Fix$^-$	a
NZ 3841 (pPN1)	Nod$^+$Fix$^-$							b
R. trifolii								
ANU 845 (pJB5JI)		Hac$^+$	Nod$^+$Fix$^-$	Nod$^+$Fix$^+$				a
ANU 845 (pBRIAN)	Nod$^+$Fix$^-$	Nod$^+$Fix$^+$	Nod$^+$Fix$^-$		Hac$^-$			a
ANU 845 (pRMSL26)		Hac$^+$	Hac$^+$					a
ANU 845 (pNM4AN)						Nod$^+$Fix$^-$	Nod$^+$Fix$^-$	a
R. meliloti								
NZ 4015 (pPN1)		Hac$^+$			Nod$^+$Fix$^-$			b
41 (pBRIAN)		Hac$^+$			Nod$^+$Fix$^-$			a
41 (pNM4AN)						Nod$^+$Fix$^-$	Nod$^+$Fix$^-$	a
ZB 157 (pJB5JI)		Hac$^+$	Hac$^+$		Nod$^+$Fix$^-$			c
ZB 157 (pBRIAN)		Hac$^+$	Hac$^+$		Nod$^+$Fix$^-$			a
A. tumefaciens								
LBA 288 (pRtr5a::Tn5)	Nod$^+$Fix$^-$							d
C 58 (pBRIAN)		Nod$^+$Fix$^-$	Nod$^+$Fix$^-$					a
C 58 (pJB5JI)	Hac$^+$	Hac$^+$			Nod$^+$			a
C 58 (pRMSL26)					Nod$^+$Fix$^-$			a
C 58 (pNM4AN)						Nod$^+$Fix$^-$	Nod$^+$Fix$^-$	a
Broad host range strain								
NGR 234								
NGR 234 (pBRIAN)		Nod$^+$Fix$^-$	Nod$^+$Fix$^-$			Nod$^+$Fix$^+$		a
NGR 234 (pJB5JI)		Nod$^+$Fix$^-$	Hac$^+$	Nod$^+$Fix$^-$		Nod$^+$Fix$^+$		a
NGR 234 (pRme41b)					Nod$^+$Fix$^-$			e

Legend to Table 3

a, ANU laboratory, M. Djordjevic, N. Morrison, P. Schofield, J. Plazinski, G. Bender; b, Ronson *et al.*, 1983; c, Banfalvi *et al.*, 1981; d, Hooykaas *et al.*, 1981; e, Kondorosi *et al.*, 1982

pJB5JI, pea Sym plasmid; pBRIAN, clover Sym plasmid; pRMSL26, *R. meliloti* Hac$^+$ genes; pNM4AN, broad host range nodulation plasmid; pPN1, clover Sym plasmid; pRtr5a::Tn5, clover Sym plasmid; pRme41b, *R. meliloti* Sym plasmid.

transferred into *R. leguminosarum* strains, where pea nodulation and fixation genes are transferred into *R. trifolii* strains or where *R. phaseoli* Sym-plasmid pRP2JI is transferred into *R. leguminosarum.* In these cases a $Nod^{+}Fix^{+}$ phenotype results. These findings are important in that they indicate an involvement of host specificity in the "turn on" of the fixation (Fix^{+}) phenotype during nodule development. Moreover, the different fast-growing rhizobia vary in their ability to allow this broader host range interaction to successfully occur.

Transfer of clover or lucerne (alfalfa) nodulation and fixation genes into *Agrobacterium* strains enables clover and lucerne nodulation to occur, although again there is no fixation. This probably results from the fact that very few bacteria are released and little bacteroid material is formed in these Fix^{-} nodules. Interestingly, when plasmid pJB5JI, which encodes pea nodulation and fixation functions, is transferred into strain C58 of *Agrobacterium tumefaciens* it will cause marked root hair curling (Hac^{+}) on both white and subterranean clovers but will not cause nodules on peas. The block to infection is unknown but it may be the formation of the infection thread (Inf) stage, as no such threads are observed in sub-clovers. This is in contrast to the transconjugant C58 (pBR1AN) infection of both white and subterranean clovers where root hair curling and nodulation readily occur (Djordjevic, 1983). Another finding which also bears on the relationship of Sym plasmid interaction and host genome is that of transconjugant ANU845 (pRMSL26) which caused marked root hair curling on white clovers but not on lucerne (Djordjevic, *et al.*, 1983), even though the Hac genes of *R. meliloti* are carried on plasmid pRMSL26 (Long *et al.*, 1982). Again the apparent block on white clovers is at the Inf stage of the infection process as pRMSL26 can initiate nodules on lucerne if it is transferred into a Hac^{-} *R. meliloti* strain. Thus, it is not just the presence of the Sym plasmid within a fast-growing strain, but some combination of host background genes and the introduced hair curling region which will enable a nodule to be formed on a particular plant species.

When the plasmids pJB5JI (pea symbiotic genes) and pBR1AN (clover symbiotic genes) are transferred into slow-growing rhizobia of the cowpea group (which were shown to have no detectable endogenous plasmids), there are no discernible effects on plant nodulation by these strains (G. Bender, personal communication). The transconjugants are still able to nodulate tropical legumes and the non-legume *Parasponia* but not the temperate legumes (peas or clovers). However, the presence of these introduced Sym plasmids in the cowpea rhizobia results in a Fix^{-} phenotype on all tested tropical legumes. The nodules formed are defective in bacterial release into plant cells and exhibit very little bacteroid formation. The introduced Sym plasmid can be maintained in these slow-growing rhizobia and retransferred back to the original temperate legume rhizobia with no loss of nodulation and fixation functions. It seems that some of the symbiotic genes encoded on the introduced Sym plasmids can be expressed in the slow-growing rhizobia resulting in a defective nodule development.

Probably the most interesting recent findings are those which indicate

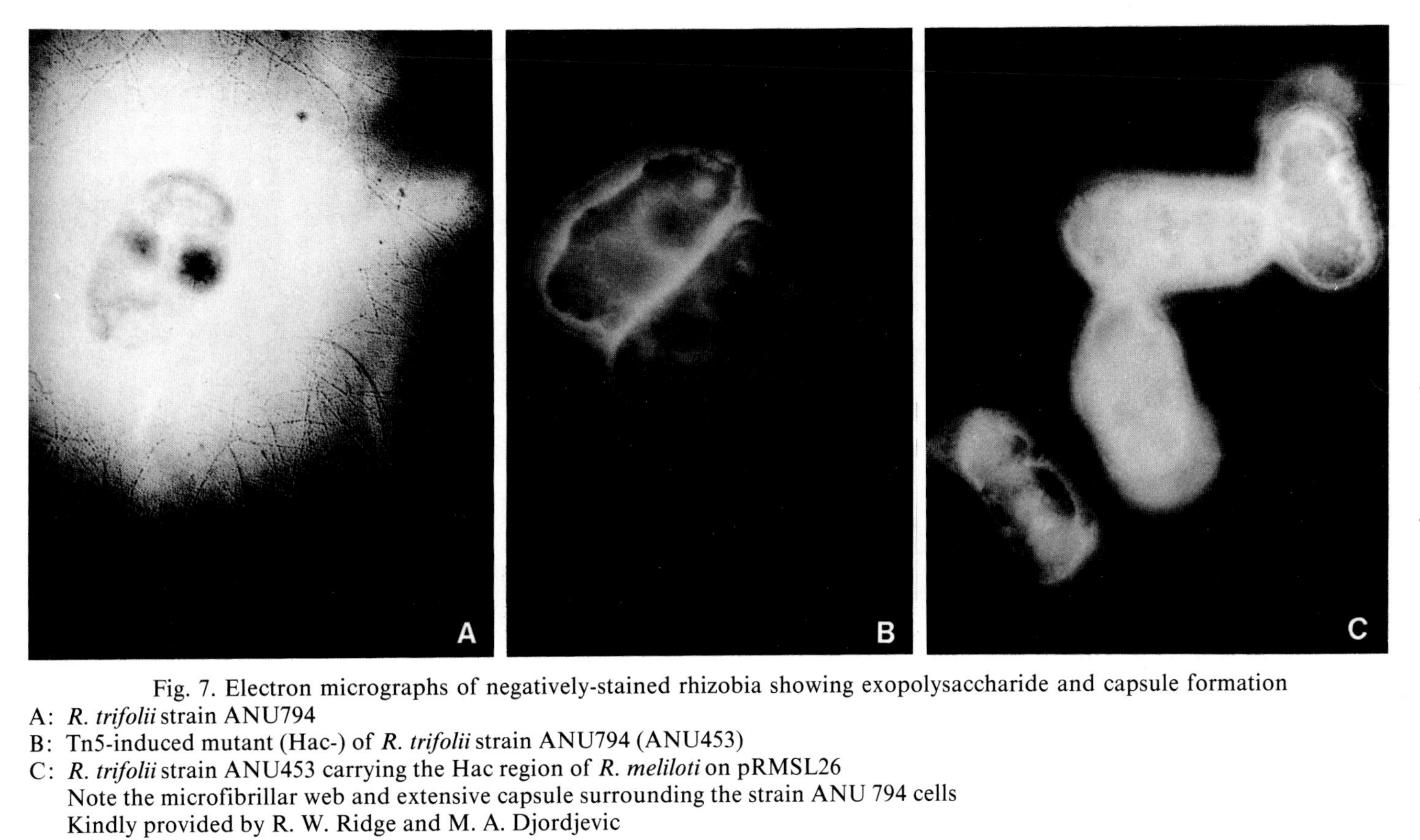

Fig. 7. Electron micrographs of negatively-stained rhizobia showing exopolysaccharide and capsule formation
A: *R. trifolii* strain ANU794
B: Tn5-induced mutant (Hac-) of *R. trifolii* strain ANU794 (ANU453)
C: *R. trifolii* strain ANU453 carrying the Hac region of *R. meliloti* on pRMSL26
Note the microfibrillar web and extensive capsule surrounding the strain ANU 794 cells
Kindly provided by R. W. Ridge and M. A. Djordjevic

that at least some of the Hac genes share some degree of homology in a broad spectrum of *Rhizobium* strains (Djordjevic *et al.*, 1983; Auger and Russell, 1983; Jarvis *et al.*, 1983). When different Hac$^-$ mutants from *R. trifolii, R. meliloti* and the broad host range strain NGR234 were examined, it was possible to suppress the defective nodulation phenotypes by the transfer of heterologous Sym plasmids to these mutants (Djordjevic *et al.*, 1983). Plasmids encoding nodulation for clover (pBRIAN; pRTO32), pea (pJB5JI), lucerne (pRMSL26), and the infection of legumes and non-legumes (pNM4AN) were able to restore root hair curling (Hac$^+$), and in most cases nodulation, on the original plant hosts. Furthermore, when plasmid pRMSL26 was transferred to the *R. trifolii* Hac$^-$ mutant (ANU453 (pRMSL26)), although there was a Hac$^+$Nod$^+$ phenotype on clovers, the presence of the plasmid did not restore a wild-type capsule structure, but did affect the production of some EPS structures on the cell surface as judged by electron microscopy (Fig. 7). In addition to this functional correction, there is a positive hybridization between the Hac region of *R. trifolii* and *R. meliloti* and other tested fast-growing rhizobia such as the fast-growing broad host range strain NGR234 and the *Sesbania rostrata* strain ORS571 (J. Plazinski and M. A. Djordjevic, personal communication).

IX. Role of the Bacterial Cell Surface

Wild-type *Rhizobium* cells produce large quantities of exopolysaccharide (EPS), and the colonies they form are mucoid (Muc$^+$) in appearance. It has been suggested that EPS has several important functions in the plant-*Rhizobium* symbiosis, including the requirement of EPS for bacterial adhesion to root surfaces (Dazzo and Brill, 1978) and determination of host specificity (Dudman, 1977). The *Rhizobium*-legume symbioses often have a high degree of host specificity, of which symbiont recognition could constitute one of the first in a series of necessary interactions between the bacterium and plant if symbiosis is to succeed (Broughton, 1978; Schmidt, 1979; Dazzo, 1980; 1981). The present evidence indicates that soybean and white clovers produce glycoproteins (lectins) which can interact specifically with particular cell surface carbohydrates synthesized by their *Rhizobium* symbionts (Bauer, 1981). The lectin-recognition hypothesis, currently under much review, states that recognition at infection sites involves the binding of specific legume lectins (carbohydrate-binding proteins) to unique carbohydrates found exclusively on the surface of the appropriate *Rhizobium* symbiont (Bohlool and Schmidt, 1974). Over the last 8 years Dazzo and co-workers have described a series of studies on a white clover lectin (trifoliin A) which binds specifically at low concentrations to surface receptors on *R. trifolii* strains. The properties of this trifoliin A are, (a) it can be isolated from seeds and is composed of subunits with a molecular weight of about 50,000 daltons; (b) it will combine specifically with an acidic heteropolysaccharide isolated from the capsule of *R. trifolii* strain 0403 and with the lipopolysaccharide from strain 0403 at a critical time in early stationary phase; (c) that the sugar 2-deoxy-D-glucose is an effective hapten inhibitor of the specific interaction between the polysaccharides from *R. trifolii* and trifoliin A; (d) that trifoliin A is also probably the same lectin isolated from seedling roots and root exudate and accumulates on clover root hairs; (e) that the trifoliin A-binding polysaccharides from *R. trifolii* specifically bind to the root hair surface (Dazzo and Hubbell, 1975; Dazzo and Brill, 1977; Dazzo *et al.*, 1978; Dazzo and Brill, 1979; Hrabak *et al.*, 1981).

These investigations have been interpreted as implying a host recognition of the bacterial polysaccharides on the root hair surface and that this interaction could contribute to host specificity. Other studies using quantitative microscopic assays argued for the existence of a multiphase sequence of events in the attachment of *Rhizobium* to legume root hairs. These consisted of a non-specific association and a selective adsorption which could be specifically inhibited by 2-deoxy glucose. Recent studies on the clover root recognition of an *Agrobacterium tumefaciens* strain carrying a Sym plasmid from *R. trifolii* has shown that the presence of the Sym plasmid results in these transconjugants being able to bind trifoliin A (Dazzo *et al.*, 1983).

The lectin recognition hypothesis and host specificity is an attractive but unproven model. Moreover, there are a number of problems which such a proposal must take into account including (a) some legume lectins do not bind specifically to only those rhizobia which infect that particular host (Chen and Phillips, 1976; Law and Strijdom, 1977); (b) some lectins do not bind to all strains of a particular *Rhizobium* species (Bohlool and Schmidt, 1974); (c) defects in Tn5-induced Hac$^-$ mutants of *R. trifolii* can be suppressed by the products from heterologous Hac genes from either *R. meliloti* or *R. legumionsarum* (Djordjevic *et al.*, 1983).

Another approach to the analysis of the role of the cell surface carbohydrates and the infection process has been the isolation of non-mucoid (Muc$^-$) exopolysaccharide-deficient mutants of different *Rhizobium* strains (Sanders *et al.*, 1978; Sanders *et al.*, 1981; Law *et al.*, 1982). In general, mutants that are deficient in EPS synthesis are able to form very few, if any, nitrogen-fixing nodules on their test plants (Sanders *et al.*, 1978). In fact, there appears to be a correlation between EPS production and the extent of root hair curling, infection and nodulation efficiency in various spontaneous mutants of *R. leguminosarum* (Napoli and Albersheim, 1980), although this potential correlation is disputed (Sanders *et al.*, 1981). Spontaneous mutants with an altered capsule synthesis, colony morphology and ability to bind host lectin have been isolated in *R. japonicum* strain 138 (Law *et al.*, 1982). Three of these mutants failed to form any detectable capsules and still nodulated and attached to soybean roots. The nodulation capacity of several mutants was proportional to the amount of acidic exopolysaccharide released into the culture medium during exponential growth. It was concluded from these studies that (a) the presence of a capsule physically surrounding the bacterium is not required for attachment or for infection and nodulation, and (b) CPS/EPS synthesis plays an important role in the infection and nodulation of soybeans by *R. japonicum.*

Most of these studies on exopolysaccharide mutants have been done with *Rhizobium* strains in which the genetic basis of the mutant phenotype is unclear. More recently, transposon Tn5 mutagenesis of *R. trifolii* strains has been used in an attempt to obtain specific, genetically well defined non-mucoid mutants (Chakravorty *et al.*, 1982). This approach has led to the isolation of several Muc$^-$ strains with greatly reduced levels of EPS synthesis but which can still nodulate clovers, although these nodules are poorly developed. One such mutant (ANU437) is drastically altered in its synthesis of both lipopolysaccharides (LPS) and EPS (Carlson, personal communication). These findings imply that substantial changes can occur in both the EPS and LPS while the *Rhizobium* strain can still infect and initiate a nodule.

A more detailed examination of various Muc$^-$ mutants from different fast-growing *Rhizobium* species has shown that they can form nodules, although these are non-nitrogen-fixing (Nod$^+$Fix$^-$) and usually small in size and many in number (Rolfe *et al.*, 1981b). Microscopic studies have also shown that while some bacteroid material is formed, this is readily broken down and the bacteria are usually not released from their infection threads (Bar$^-$ phenotype). Other studies (Rolfe and Djordjevic, unpublished) have shown that various Muc$^-$ mutants are more inhibited in their growth than their parent strains in a variety of environmental conditions such as higher pH, succinate and glycine additions. One possibility is that the EPS may not be important for the early stages of the infection process but be necessary as a "protective barrier" for the bacterial cell when it is present within the plant root environment. Alternatively, EPS may be necessary to elicit the correct plant response leading to the subsequent stages of nodulation. The sensitivity of the Muc$^-$ strains to different environmental conditions could cause a reduction in their growth rate within the infection thread and thus a loss of the stimulation necessary for plant cell division and nodule development.

Probably the most important set of transposon-induced mutants which cause a more subtle change in the surface properties of *Rhizobium* cells are the hair curling (Hac$^-$) mutants. Such mutants have been described for *R. meliloti* (Long *et al.*, 1982; Hirsch *et al.*, 1982; Forrai *et al.*, 1983) and *R. trifolii* (Rolfe *et al.*, 1981, Djordjevic *et al.*, 1983). Analysis of the surface polysaccharides of the *R. trifolii* strains has shown that while the parent and Hac$^-$ mutant strains produce identical EPS, the latter do not produce parental-type capsular polysaccharides (CPS) (Carlson *et al.*, 1983). In addition, the LPS core region from the Hac$^-$ strains appears to be altered while the O-antigen is identical in composition to the parent strain.

When individual cells from these Hac$^-$ mutants and their parent strain are examined by light and electron microscopy, no evidence of a capsule structure can be observed associated with the cells of Hac$^-$ strains. In contrast, the parent strain forms a distinct capsule structure and associated microfibrillar web (Fig. 6) (Kuempel *et al.*, 1983). Trifoliin A, which readily binds to the Hac^+ parent strains, is greatly reduced in its binding when tested with two different Tn5-induced Hac^- mutants (F. B. Dazzo, personal communication). Mapping data suggests that the Tn5 insertions map in different Hac genes of *R. trifolii* (P. R. Schofield and M. A. Djordjevic, personal communication). Hac^- mutants can therefore still produce trifoliin A receptor sites although probably at greatly reduced levels. In addition, trifoliin A was found not to bind to strain SU 847 (Rolfe *et al.*, 1980) which has a defective root adhesion (Roa^-) phenotype (listed as strain 2L) (Dazzo and Hubbell, 1975). In this strain the Hac region is deleted in the Sym plasmid (Djordjevic, unpublished). It thus appears that the Sym plasmid of *R. trifolii* encodes genes which influence capsule formation and that the Hac region of this plasmid may be either directly or indirectly involved in this process and in trifoliin A binding. Such an explanation is consistent with the observation that several of the capsule-negative mutants of *R. japonicum*, which were originally reported as failing to bind soybean lectins, are able to nodulate soybeans at reasonable efficiencies (Law *et al.*, 1982). Subsequent studies have shown that these mutants do actually bind soybean lectin at much reduced levels (Bauer, personal communication). Therefore, it might be expected that the Hac region of *R. japonicum* also codes for functions which influence the bacterial cell surface and thus soybean lectin binding to *R. japonicum* cells.

X. Future Studies

The combination of transposon-induced mutagenesis and molecular cloning has permitted the isolation of genes involved in some of the many steps leading to symbiotic nitrogen fixation. The use of cloned genes from the Hac-Nif region of a Sym plasmid, coupled with the isolation of mRNA from both culture-grown rhizobia and nodule bacteroids, should enable some analysis of the regulation of plasmid gene expression. Questions such as: "Are the Hac genes expressed in culture, or only in the presence of plant root hairs?"; and, "Are the Hac genes expressed in bacteroids?" can now be examined. Coupled *in vitro* transcription/translation systems could be used to synthesize the gene products from the Hac region in a similar fashion to that observed with the *R. meliloti nif* H,D,K genes (Puhler *et al.*, 1981). This would enable the production of antiserum or monoclonal antibodies to such products which could then be used to determine a precise location within the nodule or the bacterial cell. An alternative approach to the latter question will be to use the cloned genes directly as probes for a hybridization histochemistry analysis (Hudson *et al.*, 1981).

The possible conservation of the Hac region between a broad spectrum of *Rhizobium* strains raises a number of intriguing issues regarding the evolution of *Rhizobium* infection and the nature of the mechanisms involved. One particular Tn5-induced mutation of strain NGR 234 causes a Hac^- phenotype on siratro and a concomitant abolition of the different strategies of infection (root hair invasion and crack entry) and, in addition, legume and non-legume nodulation (Rolfe *et al.*, 1983; N. A. Morrison, personal communication). One possible explanation for the observation that a single mutation can abolish both types of infection strategies is to suggest that the Hac region encodes functions which cause a perturbation of normal membrane-cell wall biogenesis, whether it be of the dome-shaped tip of a root hair or the induction of endocytosis of the basal cells involved in crack entry infections. If the phenotype of root hair curling (Hac) is a symptom of an induced plant response due to a bacterial product released from the bacterium, this product may be released as a result of the action of plant products on the bacterium. Normal root hair enzymes responsible for the processing of the root hair cell wall polysaccharides could fortuitously process the polysaccharides of bound *Rhizobium* cells and thus release bioactive fractions. These might act to slow down the dome-shaped growth (Green, 1983) on one side of the tip of a growing root hair by perhaps disrupting the crystallization of the cellulose microfibrils and thus causing a bowing and curling of the root hair (Green, 1983). Alternatively, such released bacterial fractions may act as extracellular elicitors much like those found for *Phytophthora megasperma* (Ayers *et al.*, 1976; Albersheim and Anderson-Prouty, 1975). Similar types of enzymes would be expected on different root hairs where they may form part of the general nodulation of the cell walls of many plant root cells (for example, on the basal cells in the crack entry infection of *Stylosanthes*). The precise determination of the site of action of the Hac gene products for one *Rhizobium* system should provide an insight to many *Rhizobium*-plant interactions.

The outcome of much of the current effort on the molecular analysis of *Rhizobium* strains will be in the construction of bacterial strains which can more successfully infect chosen legumes and contain the desired properties of promptness of nodulation, persistance in some different soil types and an overall greater ability to compete with indigenous soil microorganisms. The re-examination of the phenomenon of stem nodulation, which is more broadly spread amongst legumes than previously thought, may lead to a manipulation of this characteristic to avoid some soil competition situations. It is also possible that one of the more important contributions that *Rhizobium* studies may make to agricultural microbiology lies in the construction of particular strains which can be used as probes of the plant response to infection (legume and non-legume) and thus generate new insights into the concepts of plant pathology.

Acknowledgments

We would like to express our deep appreciation to our colleagues for many helpful discussions and their dedicated efforts in the laboratory; and to R. W. Ridge and M. A. Djordjevic for help in the preparation of this manuscript.

XI. References

Albersheim, P., Anderson-Prouty, A. J., 1975: Carbohydrates, proteins, cell surfaces, and the biochemistry of pathogenesis, Ann. Rev. Plant Physiol. **26**, 31—52.

Allen, O. N., Allen, E. K., 1981: The Leguminosae, A Source book of characteristics, Uses and Nodulation. London: Macmillan.

Auger, S. G., Russell, P. F., 1983: Comparative organization of Nif and Nod genes in diverse strains of *Rhizobium japonicum*, 9th North American *Rhizobium* conference, p. 29. Boyce Thompson Institute, Cornell University, Ithaca, N. Y., U. S. A.

Ayers, A. R., Valent, B., Ebel, J., Albersheim, P., 1976: Host pathogen interactions. XI. Composition and structure of wall-released elicitor fractions. Pl. Physiol., **57**, 766—774.

Banfalvi, Z., Sakanyan, V., Koncz, C., Kiss, A., Dusha, I., Kondorosi, A., 1981: Location of nodulation and nitrogen fixation genes on a high molecular weight plasmid in *R. meliloti.* Mol. Gen. Genet. **184**, 318—325.

Bauer, W. D., 1981: Infection of legumes by rhizobia. Ann. Rev. Pl. Physiol. **32**, 407—449.

Beringer, J. E., Beynon, J. L., Buchanan-Wollaston, A. V., Johnston, A. W. B., 1978b: Transfer of the drug-resistance transposon Tn5 to *Rhizobium.* Nature **276**, 633—634.

Beringer, J. E., Brewin, N. J., Johnston, AW. B., Schulman, H. M., Hopwood, D. A., 1979: The *Rhizobium*-legume symbiosis. Proc. Roy. Soc. London B., **204**, 219—233.

Beringer, J. E., 1980: The development of *Rhizobium* genetics. J. Gen. Microbiol. **116**, 1—7.

Beringer, J. E., Brewin, N. J., Johnston, A. W. B., 1980: The genetic analysis of *Rhizobium* in relation to symbiotic nitrogen fixation. Hered. **45**, 161—186.

Bergey's Manual of Determinative Bacteriology, 1974: Baltimore: Williams and Wilkins Co., pp. 261—267.

Beynon, J. L., Beringer, J. E., Johnston, A. W. B., 1980: Plasmids and host-range in *Rhizobium leguminosarum* and *Rhizobium phaseoli.* J. Gen. Microbiol. **120**, 421—429.

Bhuvaneswari T. V., Turgeon, B. G., Bauer, W. D., 1980: Early events in the infection of soybean. (*Glycine max* L. Merr) by *Rhizobium japonicum* I. Localization of infectible root cells. Pl. Physiol. **66**, 1027—1031.

Bhuvaneswari, T. V., Bhagwat, A. A., Bauer, W. D., 1981: Transient Susceptibility of Root Cells in Four Common Legumes to Nodulation by Rhizobia. Pl. Physiol. **68**, 144—1149.

Brewin, N. J., Beringer, J. E., Buchanan-Wollaston, A. V., Johnston, A. W. B., Hirsch, P. R., 1980a: Transfer of Symbiotic genes with bacteriocinogenic plasmids in *Rhizobium leguminosarum.* J. Gen. Microbiol. **116**, 261—270.

Brewin, N. J., Beringer, J. E., Johnston, A. W. B., 1980b: Plasmid-mediated transfer of host-range specificity. J. Gen. Microbiol. **120**, 413—420.

Brewing, N. J., DeJong, T. M., Phillips, D. A., Johnston, A. W. B., 1980c: Co-transfer of determinants for hydrogenase activity and nodulation ability in *Rhizobium leguminosarum* Nature, **288**, 77—79.

Brewin, N. J., Wood, E. A., Johnston, A. W. B., Dibb, N. J., Hombrecher, G., 1982: Recombinant nodulation plasmids. J. Gen. Microbiol. in *Rhizobium leguminosarum* **128**, 15—194.

Broughton, W. J., 1978: Control of specificity in legume-*Rhizobium* associations. J. Appl. Bacteriol. **45**, 165—194.

Buchanan-Wollaston, A. V., Beringer, J. E., Brewin, N. J., Hirsch, P. R., Johnston, A. W. B., 1980: Isolation of symbiotically defective mutants in *Rhizobium leguminosarum* by insertion of the transposon Tn5 into a transmissible plasmid. Mol. Gen. Genet. **178**, 185—190.

Buchanan-Wollaston, V., Cannon, M. C., Beynon, J. L., Cannon, F. C., 1981: Role of the *nif* A gene product in the regulation of *nif* expression in *Klebsiella pneumoniae.* Nature, **294**, 776—778.

Burris, R. H., 1980: The global nitrogen budget: science of seance? In: Newton WE, Orme-Johnson, WH. (eds.) Free-living systems and chemical models. Nitrogen fixation. Vol. 1. pp. 7—16, Baltimore: University Park Press.

Carlson, R. W., Turnbull, E., Duh, J., 1983: Analysis of the surface polysaccharides of *R. trifolii* ANU 843 and its transposon generated symbiotic mutants. 9th North American *Rhizobium* Conference, P. 16, Boyce Thompson Institute, Ithaca, N. Y.

Casse, F., Boucher, C., Julliot, J. S., Michel, M., Dénarié, J., 1979: Identification and characterization of large plasmids in *Rhizobium meliloti* using agarose gel electrophoresis. J. Gen. Microbiol. **113**, 229—242.

Chakravorty, A. K., Zurkowski, W., Shine, J., Rolfe, B. G., 1982: Symbiotic nitrogen fixation: Molecular cloning of *Rhizobium* genes involved in exopolysaccharide synthesis and effective nodulation. J. Mol. Appl. Genet. **1**, 585—596.

Chandler, M. R., 1978: Some observations on infection of *Arachis hypogaea* L. by *Rhizobium.* J. Exp. Bot. **29**, 749—755.

Chandler, M. R., Date, R. A., and Roughley, R. J., 1982: Infection and Root-Nodule Development in *Stylosanthes* Species by *Rhizobium.* J. Exp. Bot. **33**, 47—57.

Corbin, D., Barran, L., Ditta, G., 1983: Organization and expression of *Rhizobium meliloti* nitrogen fixation genes. Proc. Natl. Acad. Sci. U. S. A. (in press).

Dart, J., 1974a: The infection process. In: The Biology of Nitrogen Fixation, Quispel, A. (ed.), pp. 381—429, Amsterdam: North Holland.

Dart, P. J., 1974b: Infection and development of leguminous nodules. In: Treatise on Dinitrogen Fixation. Section III. Biology. Hardy. R. W. F., Silver, W. S. (eds.), pp. 367—472, New York: John Wiley.

Dart, P. J., 1975: Legume root nodule initiation and development. In: The Development and Function of Roots, Torrey, J. G., Clarkson, D. T. (ed.), London: Academic Press, pp. 467—506.

Dart, P. J., 1977: Infection and development of Leguminous nodules. In: A treatise on Dinitrogen Fixation. Hardy, R. W. F., Silver, W. S. (eds.), pp. 367—472, New York: John Wiley.

Dazzo, F. B., Hubbel, D. H., 1975: Cross-reactive antigens and lectin as determinants of symbiotic specificity in the *Rhizobium*—clover association. Appl. Microbiol. **30**, 1017—1033.

Dazzo, F. B., Brill, W. J., 1977: Receptor sites on clover and alfalfa roots for *Rhizobium.* Appl. Env. Microbiol. **33**, 132—136.

Dazzo, F. B., Brill, W. J., 1978: Bacterial polysaccharide which binds *Rhizobium trifolii* to clover root hairs. J. Bacteriol. **137**, 1362—1373.

Dazzo, F. B., Yanke, W. E., Brill, W. J., 1978: Trifoliin: a *Rhizobium* recognition protein from white clover. Biochim. Biophys. Acta **539**, 276—286.

Dazzo, F. B., Brill, W. J., 1979: Bacterial polysaccharide which binds *Rhizobium trifolii* to clover root hairs. J. Bacteriol. **137**, 1362—1373.

Dazzo, F. B., 1980: Lectins and their saccharide receptors as determinants of specificity in the *Rhizobium*—legume symbiosis. In S. Subtelny and N. Wessells (eds.), The cell surface. Mediators of developmental processes, pp. 277—304.

Dazzo, F. B., 1981: Bacterial attachment as related to cellular recognition in the *Rhizobium*—legume symbiosis. J. Supramol. Stuct. Cell. Bichem. **16**, 29—41.

Dazzo, F. B., Truchet, G., Hooykaas. P., 1983: Clover root-recognition of *Agrobacterium tumefaciens* carrying the Sym plasmid of *Rhizobium trifolii.* 9th North American *Rhizobium* Conference, P. 52, Boyce Thompson Institute, Ithaca, N.Y.

Denarie, J., Biostard, P., Casse-Delbart, F., 1981: Indigenous Plasmids of *Rhizobium.* Int. Rev. Cytol. Sup. 13, Biology of the Rhizobiaceae, 225—246. Eds., Giles, K. L., Atherly, A. G., New York: Academic Press.

De Jong, T. M., Brewin, N. J., Johnston, A. W. B., Phillips, D. A., 1982: Improvement of symbiotic properties in *Rhizobium leguminosarum* by plasmid tansfer. J. Gen. Microbiol. **128**, 1829—1838.

Dilworth, M. J., Appleby, C. A., 1979: Leghaemoglobin and *Rhizobium* hemoproteins. pp. 691—764. In: A Treatise on Dinitrogen Fixation. Hardy, R. W. F., Bottomley, F., Burns, R. C. (eds.). New York: John Wiley.

Dixon, R., Eady, R. R., Espin, G., Hill, S., Iaccarino, M., Kahn, D., Merrick, M., 1980: Analysis of regulation of *Klebsiella pneumoniae* nitrogen fixation (*nif*) gene cluster with gene fusions. Nature **286**, 128—132.

Downie, J. A., Ma, Q. S., Knight, C. D., Hombrecher, G., Johnston, A. W. B., 1983: Cloning of the symbiotic region of *Rhizobium leguminosarum: the nodulation genes are between the nitrogenase genes and a nif* A-like gene. EMBO J. **2**, 947—952.

Djordjevic, M. A., Zurkowski, W., Rolfe, B. G., 1982: Plasmids and stability of symbiotic properties of *Rhizobium trifolii.* J. Bacteriol. **151**, 560—568.

Djordjevic, M. A., Ridge, R. W., Morrison, N. A., Schofield, P. B., Bassam, B. J., Watson, J. M., Shine J., Rolfe, B. G., 1983: Conservation of early nodulation (Nod)functions in different fast-growing rhizobia. 9th North American *Rhizobium* Conference, P. 33, Boyce Thompson Institute, Ithaca, N. Y.

Djordjevic, M. A., 1983: A genetic and molecular characterization of plasmid genes involved in the early nodulation steps of *Rhizobium trifolii.* PhD. thesis, Australian National University, Canberra.

Dreyfus, B. L., Dommergues, Y. R., 1981: Stem nodules on the tropical legume *Sesbania rostrata.* In: Current perspectives in nitrogen fixation. Gibson, A. H., Newton, W. E. (eds.), p. 471 (Aust. Acad. of Sci.).

Dudman, W. F., 1977: The role of surface polysaccharides in natural environments. In: Surface carbohydrates of the prokaryotic cell, Sutherland, I. (ed.), p. 357—414, New York: Academic Press.

Forrai, T., Vincze, E., Banfalvi, Z., Kiss, G. B., Randhawa, G. S., Kondorosi, A., 1983: Localisation of symbiotic mutations in *Rhizobium meliloti.* J. Bacteriol. **153**, 635—643.

Fuhrmann, M., Hennecke, H., 1982: Coding properties of cloned nitrogenase structural genes form *Rhizobium japonicum.* Mol. Gen. Genet., **187**, 419—425.

Giles, K. L., Atherly, A. G. (eds.) 1981: Biology of the Rhizobiaceae. Int. Rev. Cyto., Suppl. 13. New York: Academic Press.

Green, P. B., 1983: Analysis of axis extension. In: Positional controls in plant development, Banlow, P., Carr, D. (eds.), pp. 53—82, Cambridge: University Press.

Hausinger, R. P. Howard, J. B., 1982: The amino acid sequence of the nitrogenase iron protein from *Azotobacter vinelandii.* J. Biol. Chem. **257**, 2483—2490.

Higashi, S., 1967: Transfer of clover infectivity of *Rhizobium trifolii* to *Rhizobium phaseoli* as mediated by an episomic factor. J. Gen. Appl. Microbiol. **13**, 391—403.

Hirsch, R. P., 1979: Plasmid-determined bacteriocin production by *Rhizobium leguminosarum* J. Gen. Microbiol. **113**, 219—228.

Hirsch, P. R., van Montagu, M., Johnston, A. W. B., Brewin, N. J., Schell, J., 1980: Physical identification of bacteriocinogenic, nodulation and other plasmids in strains of *Rhizobium leguminosarum* J. Gen. Microbiol. **120**, 403—412.

Hirsch, A. M., Long, S. R., Bang, M., Hoskins, N., Ausubel, F. M., 1982: Structural studies on alfalfa roots infected with nodulation mutants of *Rhizobium meliloti.* J. Bacteriol. **151**, 411—419.

Hombrecher, G., Brewin, G., Johnston, A. W. B., 1981: Linkage of genes for nitrogenase and nodulation ability on plasmids in *Rhizobium leguminosarum* and *R. phaseoli.* Mold. Gen. Genet. **182**, 133—136.

Hooykaas, P. J. J., van Brussel, A. A. N., den Dulk-Ras, H., van Slogteren, G. M. S., Schilperoort, R. A., 1981: Sym plasmid of *Rhizobium trifolii* expressed in different rhizobial species and *Agrobacterium tumefaciens,* Nature **219**, 351—353.

Hooykaas, P. J. J., Snijdewint, F. G. M., Schilperoort, R. A., 1982: Identification of the Sym plasmid of *Rhizobium leguminosarum* strain 1001 and its transfer to and expression in other rhizobia and *Agrobacterium tumefaciens.* Plasmid. **8**, 73—82.

Hrabak, E. M., Urbano, M., Dazzo, F. B., 1981: Growth-phase dependent immunodeterminants of *Rhizobium trifolii* lipopolysaccharide which bind trifoliin A, a white clover letin. J. Bacteriol. **148**, 697—771.

Hudson, P., Penschow, J., Schine, J., Ryan, G., Niall, H., Goyhlan, J. Y., 1981: Hybridization histochemistry: Use of recombinant DNA as a "homing probe" for tissue localization of specific mRNA populations. Endocrinology **108**, 353—356.

Jarvis, B. D. W., Scott, K. F., Hughes, J. E., Djordjevic, M., Rolfe, B. G., Shine, J., 1983: Conservation of genetic information between different *Rhizobium* species. Can. J. Microbiol. **29**, 200—209.

Johnston, A. W. B., Beynon, J. L. Buchanan-Wollaston, A. V., Setchell, S. M., Hirsch, P. R., Beringer, J. W., 1978: High frequency transfer of nodulating ability between strains and species of *Rhizobium.* Nature **276**, 634—636.

Johnston, A. W. B., Hombrecher, G., Brewin, N. J., Cooper, M. C., 1982: Two transmissible plasmids in *Rhizobium leguminosarum* strain 300. J. Gen. Microbiol., **128**, 85—93.

Jorgensen, R. A., Rothstein, S. J., Reznikoff, W. S., 1979: A restriction enzyme cleavage map of Tn5 and location of region encoding neomycin resistance. Mol. Gen. Genet. **177**, 65—72.

Kleckner, N., 1977: Translocatable elements in Procaryotes. Cell, **11**, 11—23.

Kondorosi, A., Johnston, A. W. B., 1981: The genetics of *Rhizobium.* Int. Rev. Cyto., Sup. 13, Biology of the Rhizobiaceae, 191—224. Eds., Giles, K. L., Atherly, A. G., New York: Academic Press.

Kondorosi, A., Kondorosi, W., Pankhurst, C. E., Broughton, W. J., Banfalvi, Z., 1982: Mobilization of a *Rhizobium meliloti* megaplasmid carrying nodulation and nitrogen fixation genes into other rhizobia and *Agrobacterium.* Mol. Gen. Genet. **188**, 433—439.

Krol, A. J. M., Hontelez, J. G. L., Van Den Bos, R. C., Van Kammen, A., 1981: Expression of large plasmids in the endosymbiotic form of *Rhizobium leguminosarum.*, Nucleic Acids Res. **8**, 4337—4347.

Krol, A. J. M., Hontelez, J. G. J., van Kammen, A., 1982: Only one of the large plasmids in *Rhizobium leguminosarum* strain PRE is strongly expressed in the endosymbiotic state. J. Gen. Microbiol. **128**, 1839—1847.

Kuempel, P. B., Jones, A. J., Djordjevic, M. A., Ridge, R. W., Shine, J., Schofield, P. B., Watson, J. M., Rolfe, B. G., 1983: Genetic and Biochemical analysis of nodulating and non-nodulating strains of *R. trifolii*. 9th North American *Rhizobium* Conference, P. 21, Boyce Thompson Institute, Ithaca, N. Y.

Law, I. J., Yamamoto. Y., Mort, A. J., Bauer, W. D., 1982: Nodulation of soybean by *Rhizobium japonicum* mutants with altered capsule synthesis. Planta **154**, 100—109.

Legocki, R. P., Verma, D. P., 1980: Identification of nodule-specific host proteins (nodulins) involved in the development of *Rhizobium*-legume symbiosis. Cell, **20**, 153—163.

Leong, S. A., Ditta, G. S., Helinski, D. R., 1982: Heme Biosynthesis in *Rhizobium*. J. Biol. Chem. **257**, 8724—8730.

Libbenga, K. R., Bogers, R. J., 1974: Root-nodule morphogenesis. In "The Biology of Nitrogen Fixation" ed. A. Quispel, p. 430—471, Amsterdam: North Holland Publ. Co.

Long, S., Buikema, W. J., Ausubel, F. M., 1982: Cloning of *Rhizobium meliloti* nodulation genes by direct complementation of Nod^- mutants. Nature **298**, 485—488.

Ma, Q. S., Johnston, A. W. B., Hombrecher, G., Downie, J. A., 1982: Molecular genetics of mutants of *Rhizobium leguminosarum* which fail to fix nitrogen. Mol. Gen. Genet. **187**, 166—171.

Masterson, R. V., Russell, P. R., Atherly, A. G., 1982: Nitrogen fixation (nif) genes and large plasmids of *Rhizobium japonicum*, J. Bacteriol. **152**, 928—931.

Meade, H. M., Long, S. R., Ruvkun, G. B., Brown, S. E., Ausubel, F. M., 1982: Physical and genetic characterisation of symbiotic and auxotrophic mutants of *Rhizobium meliloti* induced by transposon Tn5 mutagenesis. J. Bacteriol. **149**, 114.

Mevarech, M., Rice, D., Haselkorn, R., 1980: Nucleotide sequence of a cyanobacterial *nif*H gene coding for nitrogenase reductase. Proc Natl. Acad. Sci. **77**, 6476—6480.

Morrison, N. A., Cen, H. Y., Trinick, Shine, J., and Rolfe, B. G., 1983: Heat curing of a Sym Plasmid in a Fast-Growing *Rhizobium* sp. That is able to Nodule legumes and the Nonlegume *Parasporia* sp. Journal of Bacteriology. **153**, 527—531.

Napoli, C., Albersheim, P., 1980: *Rhizobium leguminosarum* mutants incapable of normal extracellular polysaccharide production. J. Bacteriol. **141**, 1451—1456.

Newcomb, W., 1980: Control of morphogenesis and differentiation of pea root nodules. In Nitrogen Fixation Vol. 2, Newton, W. E., Orme-Johnson, W. H., eds., p. 87—102. Baltimore: University Park Press.

Nuti, M. P., Ledeboer, A. M., Lepidi, A. A., Schilperoort, R. A., Cannon, F. C., 1979: Evidence for nitrogen fixation (*nif*) genes on indigenous *Rhizobium* plasmids. Nature **282**, 533—535.

Ow, D. A., Ausubel, F. M., 1983: Regulation of nitrogen metabolism genes by *nif* A gene product in *Klebsiella pneumoniae*. Nature, **301**, 307—313.

Petit, A., Tempe, J., 1983: The opine trail (from crown gall to hairy root, from hairy root to legume nodules). P. II. Combined meeting, Eighth nitrogen fixation workshop and Fourth biennial meeting of the New Zealand Branch of the

International Association for plant tissue culture, February, Palmerston North, N. Z.

Prakash, R. K., Schilperoort, R. K., Nuti, M. P., 1981: Large plasmids of fast-growing rhizobia: Homology studies and location of structural nitrogen fixation (*nif*) genes. J. Bacteriol. **145**, 1129—1136.

Prakash, R. H., Hooykass, P. J. J., Ledeboer, A. M. Kijne, J. W., Schilperoot, R. A., Nuti, M. P., Lepide, A. A., Casse, F., Boucher, C., Julliot, J. S., Denarie J., 1980: Detection, isolation and characterization of large plasmids in *Rhizobium.* Proc. 3rd Inter. Sym. Nitrogen Fixation, Newton, W. E., Orme-Johnson, W. H., eds., Baltimore: University Park Press.

Prakash, R. K., Van Brussel, A. A. N., Quint, A., Mennes, A. M., Schilperoor, R. A., 1982: The map position of Sym plasmid regions expressed in the bacterial and endosymbiotic form of *Rhizobium leguminosarum.* Plasmid **7**, 281—286.

Rolfe, B. G., Gresshoff, P. M., Shine, J., 1980: Rapid screening for symbiotic mutants of *Rhizobium* and white clover. Pl. Sci. Let., **19**, 277—284.

Rolfe, B., Gresshoff, P. M., Shine, J., Vincent, J. M., 1980: Interaction between a non-nodulating and an ineffective mutant of *Rhizobium trifolii* resulting in effective (nitrogen-fixing) nodulation. App. Envir. Microbiol. **39**, 449—452.

Rolfe, B. G., Djordjevic, M., Scott, K. F., Hughes, J. E., Badenoch-Jones, J., Gresshoff, P. M., Cen, Y., Dudman, W. F., Zurkowski, W., Shine J., 1981 a: Analysis of nodule-forming ability of fast-growing *Rhizobium* strains. In: Current Perspectives in Nitrogen Fixation. Gibson, A. H., and Newton, W. E. (eds.) (Aust. Acad. of Sci.) pp. 142—145.

Rolfe, B. G., Shine, Gresshoff, P. M., Scott, K. F., Djordjevic, M., Cen, Y., Hughes, J. E., Bender, G. L., Chakravorty, A., Zurkowski, W., Watson, J. M., Badenoch-Jones, J., Morrison, N. A., Trinick, M. J., 1981 b. Transposon induced mutants of fast and slow-growing rhizobia and the molecular cloning of symbiotic genes. In: Proceedings of the 8th Nth. Amercan Rhiobium conference. K. W. Clark, J. H. G. Stephens (eds.). University of Manitoba printing Service, p. 34—54.

Rolfe, B. G., Djordjevic, M. A., Morrison, N. A., Plazinski, J., Bender, G. L., Gresshoff, P. M., Trinick, M. J., Shine, J., Genetic analysis of the symbiotic regions in *Rhizobium trifolii* and *Rhizobium parasponia.* In: ed. Pühler, Proc. of the 1st Int. Symposium, Molecular Genetics of the Bacteria-plant Interaction, Bielefeld. (in press).

Ronson, C. W., Scott, D. B., 1982: Identification, broad host range mobilization and mutagenesis of a *Rhizobium trifolii* Sym: R68.45 cointegrate plasmid. In: ed. Pühler, Proc. of the 1st Int. Symposium, Molecular Genetics of the Bacteria-plant Interaction, Bielefeld. (in press).

Ronson. C. W., Riddiford, A. H. H., Scott, D. B., Transfer of a *Rhizobium trifolii* symbiotic plasmid: R68.45 cointegrate to strains of *Escherichia coli, Pseudomonas aeruginosa* and fast-growing *Rhizobium* species. J. Bacteriol. (in press).

Rosenberg, C., Boistard, P., Denarie, J., Casse-Delbart, F. L., 1981: Genes controlling early and late functions in symbiosis are located on a megaplasmid in *Rhizobium meliloti.* Mol. Gen. Genet. **184**, 326—333.

Ruvkun, G. B., Ausubel, F. M., 1980: Interspecies homology of nitrogenase genes. Proc. Natl. Acad. Sci. U. S. A., **77**, 191—195.

Ruvkun, G. B., Long, S. R., Meade, H. M., van den Bos, R. C., Ausubel, F. M., 1982: ISRml: a *Rhizobium meliloti* insertion sequence that transposes preferentially into nitrogen fixation genes. J. Mol. Appl. Genet. **1**, 405—418.

Ruvkun, G. B., Sundaresan, V., Ausubel, F. M., 1982a: Site directed Transposon

Tn5 mutagenesis and complementation analysis of the *Rhizobium meliloti* symbiotic nitrogen fixation (*nif*) genes. Cell **29**, 551—559.

Sanders, R. E., Carlson, R. W., Albersheim, P., 1978: A *Rhizobium* mutant incapable of nodulation and normal polysaccharide secretion. Nature **271**, 240—242.

Sanders, R., Raleigh, E., Singer, E., 1981: Lack of correlation between extracellular polysaccharide and nodulation ability in *Rhizobium*. Nature **292**, 241—266.

Schofield, P. R., Ridge, R. W., Djordjevic, M. A., Rolfe, B. G., Shine, J., Watson, J. M., 1983a: Host-specific nodulation is encoded on a 14kb DNA fragment in *Rhizobium trifolii*. EMBO Journal (manuscript submitted).

Schofield, P. R., Djordjevic, M. A., Rolfe, B. G., Shine, J., Watson, J. M., 1983b: A molecular linkage map of nitrogenase and nodulation genes in *Rhizobium trifolii*. Mol. Gen. Genet. (manuscript submitted).

Schmidt, E. L., 1979: Initiation of plant root-microbe interactions. Ann. Rev. Microbiol. **33**, 355—376.

Scott, D. B., Ronson, C. W., 1982: Identification and mobilization by cointegrate formation of a nodulation plasmid in *Rhizobium trifolii*. J. Bacteriol. **151**, 36—43.

Scott, K. F., Hughes, J. E., Gresshoff, P. M., Beringer, J. E., Rolfe, B. G., Shine J., 1982: Molecular cloning of *Rhizobium trifolii* genes involved in symbiotic nitrogen fixation. J. Mol. Appl. Genet. **1**, 315—326.

Scott, K. F., Rolfe, B. G., Shine, J., 1983a: Biological nitrogen fixation: Primary structure of the *Rhizobium trifolii* iron protein gene. DNA **2**, 149—155.

Scott, K. F., Rolfe, B. G., Shine, J., 1983b: Nitrogenase structural genes are unlinked in the non-legume symbiont *Parasponia Rhizobium*. DNA **2**, 141—148.

Shine, J., Scott, K. F., Fellows, F., Djordjevic, M. A., Schofield, P. R., Watson, J. M., and Rolfe, B. G. Molecular anatomy of the symbiotic region in *R. trifolii* and *R. parasponia*. In: Puhler, A. (ed.), Proc. of 1st Int. Symposium. Molecular Genetics of the Bacteria-plant Interaction, Bielefeld (in press).

Sprent, J. I., 1979: The biology of nitrogen-fixing organisms. Maidenhead, England: McGraw-Hill Ltd.

Sunderasan, V., Ausubel, F. M., 1981: Nucleotide sequence of the nitrogenase iron protein from *Klebsiella pneumoniae*. J. Bio. Chem. **256**, 2808—2812.

Szeto, W. Zimmermann, L., Sundaresan, V., Ausubel, F., 1983: A *Rhizobium meliloti* locus regulating the transcription of nitrogenase and other symbiotic genes. 9th North American *Rhizobium* Conference, p. 27, Boyce Thompson Institute, Ithaca, N. Y.

Tanaka, M., Haniu, M., Yasunobu, K. T., Mortenson, L. E., 1977: The amino acid sequence of *Clostridium pasteurianum* iron-protein, a component of nitrogenase. J. Biol. Chem. **252**, 7013—7100.

Trinick, M. J., 1973: Symbiosis between *Rhizobium* and the non-legume *Trema aspera*. Nature **244**, 459—460.

Trinick, M. J., Galbraith, J., 1976: Structure of root nodules formed by *Rhizobium* on the non-legume Trema cannabina var. scabra. Arch. Microbiol., **108**, 159—166.

Trinick, M. J., 1979: Structure of nitrogen-fixing nodules formed by *Rhizobium* or roots of Parasponia andersonii Planch. Can. J. Microbiol. **25**, 565—578.

Trinick, M. J., Galbraith, J., 1980: The *Rhizobium* requirements of the non-legume *Parasponia* in relationship to the cross-inoculation group concept of legumes. New Phytol. **86**, 17—26.

Truchet, G., Rosenberg, C., Vasse, J., Julliot, J. S., Camut, S., Denarie J., 1983:

Transfer of *Rhizobium meliloti* pSym genes into *Agrobacterium tumefaciens.* 9th North American *Rhizobium* Conference, p. 31, Boyce Thompson Institute, Ithaca, N. Y.

Tsien, H. C., Dreyfus, B. L., Schmidt, E. L., 1983: Morphogenesis of stem nodules of *Sesbania rostrata.* 9th North American *Rhizobium* Conference, p. 3, Boyce Thomson Institute, Ithaca, N. Y.

Vincent, J. M., 1970: A manual for the practical study of root-nodule bacteria. (IBP handbook no. 15). Oxford: Blackwell Scientific Publications.

Vincent, J. M., 1980: Factors controlling the legume — *Rhizobium* symbiosis. In "Nitrogen Fixation" Vol. 2, W. E. Newton, W. H. Orme-Johnson, eds., p. 103—129. Baltimore: University Park Press.

Watson, J. M., Schofield, P. R., Ridge, R. W., Djordjevic, M. A., Rolfe B. G., Shine, J., 1983: Molecular cloning and analysis of a region of the Sym plasmid of *Rhizobium trifolii* encoding clover nodulation functions. In "Plant Molecular Biology" UCLA Symposia on Molecular and Cellular Biology, Vol. 12, R. B. Goldberg (ed.). New York: A. R. Liss.

Zurkowski, W., 1980: Specific adsorption of bacteria to clover root hairs, related to the presence of the plasmid pWZ2 in cells of *Rhizobium trifolii.* Microbios. **27**, 27—32.

Zurkowski, W., 1981: Conjugational transfer of the nodulation-conferring plasmid pWZ2 in *Rhizobium trifolii.* Mol. Gen. Genet. **181**, 522—524.

Zurkowski, W., 1982: Molecular mechanism for the loss of nodulation properties of *Rhizobium trifolii.* J. Bacteriol. **150**, 999—1007.

Zurkowski, W., Lorkiewicz, Z., 1976: Plasmid deoxyribonucleic acid in *Rhizobium trifolii.* J. Bacteriol. **128**, 481—484.

Zurkowski, W., Lorkiewicz, Z., 1978: Effective method for the isolation of non-nodulating mutants of *Rhizobium trifolii.* Genetic Research **32**, 311—314.

Zurkowski, W., Lorkiewicz, Z., 1979: Plasmid-mediated control of nodulation in *Rhizobium trifolii.* Archiv. Microbiol. **123**, 195—201.

Chapter 5

Nitrogen Assimilation in the Legume-*Rhizobium* Symbiosis: A Joint Endeavour

B. J. Miflin and J. V. Cullimore

Biochemistry Department, Rothamsted Experimental Station,
Harpenden, Herts, AL5 2JQ, U. K.

With 9 Figures

Contents

I. Introduction

The biological fixation of dinitrogen is the major source of renewable combined nitrogen available to the biosphere and is believed to be carried out solely by prokaryotic organisms (Postgate, 1982). The majority of this combined nitrogen is delivered directly to eukaryotic plants via a symbiotic relationship with nitrogen fixing bacteria. Such symbiotic relationships have been described for several genera of plants (see Broughton, 1982). By far the most important in an agricultural context is that between members of the Leguminoseae and *Rhizobium* sp., not surprisingly this has received the most detailed study and will be the subject of this review. As a general principle the prokaryotic partner contains the machinery for nitrogen fixation and the eukaryotic partner assimilates the ammonia produced into an organic form which is then used for the nutrition of the whole plant and also of the prokaryote. In return the eukaryote provides a suitable environment and a source of energy to enable the prokaryote to fix dinitrogen. A generalized scheme, which largely applies to the legume root nodule, is shown in Fig. 1.

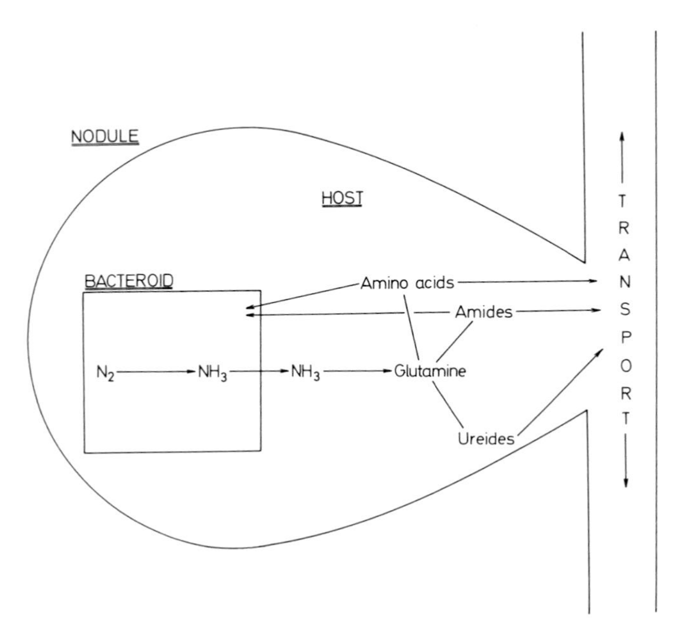

Fig. 1. Nitrogen flow during symbiotic nitrogen fixation

In this article we describe the present state of our knowledge of the steps involved in the assimilation of fixed nitrogen. Although the title of this book emphasizes gene expression, our current information is not always sufficient for us to describe the processes at this level particularly those that are encoded in the eukaryotic genome. Rather, therefore, we will identify the key enzymic steps in the hope that this will indicate some of the genes which should be studied in the future.

Beyond describing the process there is also a need to understand how the joint endeavour is orchestrated and how the two genomes communicate one to the other. Such communication must take place at many stages during the association. We will concentrate on the parallel initiation of nitrogenase activity and the export of ammonia from the bacterium, and the mechanism for assimilating that ammonia into a suitable form of organic nitrogen in the plant.

II. Rhizobial Metabolism

A. Nitrogenase

Our knowledge of nitrogenase and the genes involved in its synthesis has largely come from studies on free-living diazotrophs (for reviews see Brill, 1980, Kennedy *et al.*, 1981). Fig. 2 summarises the current knowledge of the

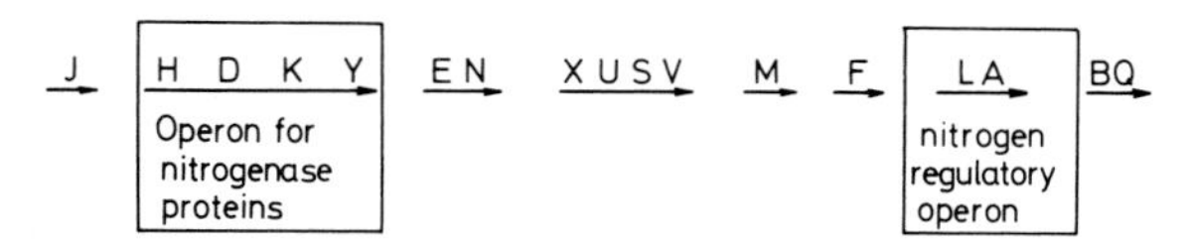

Fig. 2. The *nif* cluster of *Klebsiella pneumoniae*
After Kennedy *et al.* (1981) and Ow and Ausubel (1983)

nif cluster in *Klebsiella pneumoniae* and illustrates the complexity of the system. Nitrogenase, which is a composite protein, is coded for by genes *nifHDK*. The *nifH* gene product is the subunit of the nitrogenase Fe-protein and *nifD* and *nifK* code for the α and β subunits of the iron molybdenum (FeMo) nitrogenase protein. The production of the FeMo cofactor for this latter component of nitrogenase appears to be dependent on the genes *nifB, nifE, nifN* and *nifV; nifM* and *nifS* have been implicated in the processing of the *nifH* gene product. The *nifLA* operon appears to be concerned solely with the control of *nif* gene expression (see below) whereas the roles of the *nifQ, nifU, nifX* and *nifY* products are not known. The nitrogenase protein produced by this battery of genes has been subject to much study (see Yates, 1980, for a review); the properties are briefly summarized in Fig. 3.

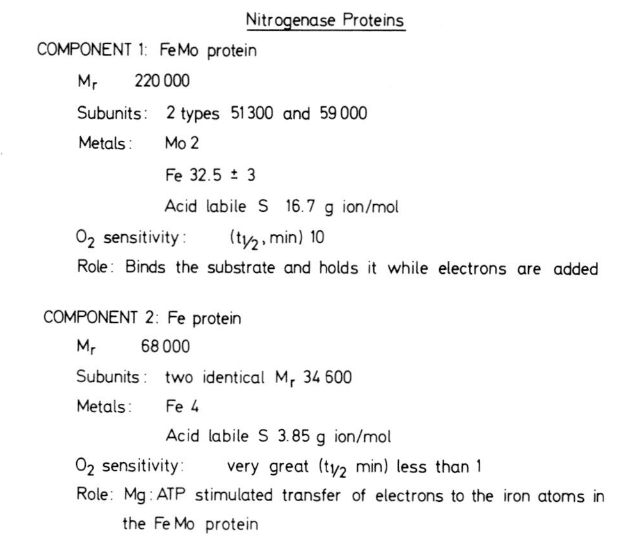

Nitrogenase Proteins

COMPONENT 1: FeMo protein

M_r 220 000

Subunits: 2 types 51 300 and 59 000

Metals: Mo 2

Fe 32.5 ± 3

Acid labile S 16.7 g ion/mol

O_2 sensitivity: ($t_{1/2}$, min) 10

Role: Binds the substrate and holds it while electrons are added

COMPONENT 2: Fe protein

M_r 68 000

Subunits: two identical M_r 34 600

Metals: Fe 4

Acid labile S 3.85 g ion/mol

O_2 sensitivity: very great ($t_{1/2}$ min) less than 1

Role: Mg:ATP stimulated transfer of electrons to the iron atoms in the FeMo protein

Fig. 3. Characteristics of nitrogenase proteins

The reaction catalysed by nitrogenase is

$$N_2 + 12\ ATP + 6e^- + 6H^+ \rightarrow 2NH_3 + 12\ ADP + 12\ Pi$$

Key aspects of this reaction are:
(i) The enzyme is highly sensitive to O_2, thus all diazotrophs have evolved systems to protect the enzyme from O_2 damage; some organisms only fix N_2 under anaerobic conditions others have developed ways of protecting the enzyme while fixing N_2 in a generally aerobic environment.
(ii) The enzyme is not highly specific and it can reduce a range of substrates. Of most practical importance is the reduction of acetylene ($C_2H_2 + 2H^+ + 2e^- \rightarrow C_2H_4$) which provides an alternative simple and sensitive assay for nitrogenase activity.
(iii) The enzyme evolves H_2 when fixing N_2 and this process is wasteful of energy (see chapter 6).

$$2H^+ + 2e^- + 4\ ATP \rightarrow H_2 + 4\ ADP + 4Pi$$

The ratio of hydrogen evolved to N_2 reduced is not constant although it always appears to be greater than or equal to one. In some nitrogen fixing organisms, including some strains of rhizobia (Schubert and Evans, 1976), the hydrogen may be reoxidized with a consequent recuperation of energy by means of a hydrogen uptake system.
(iv) The overall energy requirements of the biological system of dinitrogen fixation, even in the presence of an uptake hydrogenase, are considerable even though the overall thermodynamic equilibrium is in favour of reduction. This energy cost is a drain on the organism.

The genetics of nitrogenase in *Rhizobium* species is being worked out rapidly. The brief summary that follows is likely to be soon out of date. Perhaps the most important difference between *Rhizobium* sp. and *K. pneumoniae* is that the *nif* genes are on the bacterial chromosome in the latter whereas they are carried on a plasmid particularly in fast-growing rhizobia (see Beringer *et al.*, 1980, Brill, 1980, Denarie *et al.*, 1981, for reviews). The location of the *nif* genes in slow growing rhizobia (e.g. *R. japonicum* and cowpea *Rhizobium* sp.) is not yet clear, so far there is no well documented example of their location on a plasmid.

Study of the *nif* genes has been aided by the strong homology between nitrogenase from different species (Ruvkun and Ausubel, 1980). Sequencing of *nif* genes has given further information on the homology (see Scott *et al.*, 1981). Nucleic acid hybridization studies have shown that the genes *nifH, nifD* and *nifK* are present on large plasmids in *R. meliloti* (Ruvkun and Ausubel, 1980, Banfalvi *et al.*, 1981, Rosenberg *et al.*, 1981), *R. phaseoli* (Prakash *et al.*, 1981), *R. leguminosarum* (Nuti *et al.*, 1979, Krol *et al.*, 1980, Prakash *et al.*, 1981) and *R. trifolii* (Ruvkun and Ausubel, 1980, Prakash *et al.*, 1981). Furthermore, in some of the above species the *nifH, D* and *K* genes appear to be arranged in the same order and to be transcribed in a single operon. Transferring plasmids between strains or loss of plasmids from certain strains have also shown that the genes for nitrogen fixation (Fix) and nodulation (Nod) and host range genes are linked and on the same plasmid (Buchanan-Wollaston *et al.*, 1980, Banfalvi *et al.*, 1983). However the exact number of genes in the Fix and Nod systems in rhizobia is not known and current work suggests that the organization of them may be complex and vary between different species (e.g. see Quinto *et al.*, 1982). However the availability of cloned probes for *nif* genes should mean that the extent and position of these should be known before long.

B. Ammonia Assimilation in Free-living Rhizobia

Prior to 1970 ammonia assimilation in bacteria was considered to occur only via glutamate dehydrogenase (GDH). However in that year Tempest, Meers and Brown showed that another assimilatory route, which was of particular importance at low concentrations of ammonia, existed in *Klebsiella aerogenes* (Tempest *et al.*, 1970, 1973). Subsequent work has shown that this route operates widely in bacteria and in higher plants (see Brown *et al.*, 1974, Miflin and Lea, 1980, for reviews). This alternative assimilatory pathway, which we shall refer to as the 'glutamate synthase cycle', (Rhodes *et al.*, 1980) depends upon the linked action of glutamine synthetase (GS) and glutamate synthase to assimilate ammonia into the α-amino position of glutamate (Fig. 4). The same end is achieved by GDH but there are two critical differences between the pathways.

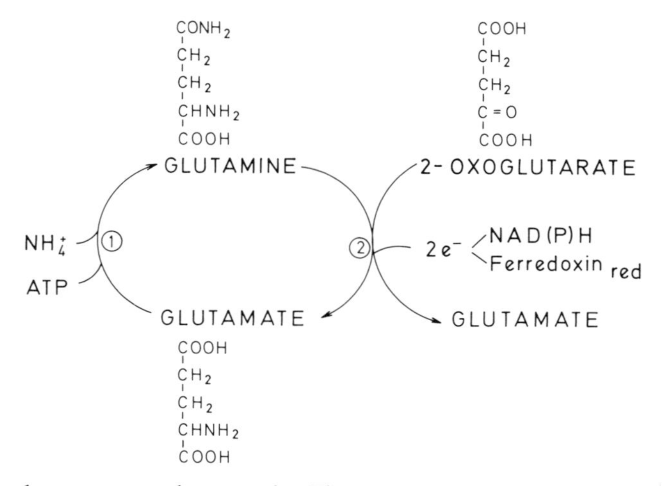

Fig. 4. The glutamate synthase cycle. The component enzymes are (1) glutamine synthetase EC.6.3.1.2 and (2) glutamate synthase EC.1.4.1.13 (pyridine nucleotide linked) EC.1.4.7.1 (ferredoxin linked)

(i) GS has a much lower K_m for ammonia than GDH and thus the glutamate synthase cycle can function at and maintain very low concentrations of ammonia in the tissue.

(ii) GS requires ATP, which is hydrolyzed to ADP, and the cycle thus uses 1 ATP and 1 NAD(P)H; GDH uses only 1 NAD(P)H. This difference in energy usage may be significant in certain circumstances. In most organisms GDH exists in two forms, one active with $NADP^+/NADPH$ and one with $NAD^+/NADH$; it is generally considered that the former is anabolic and the latter catabolic in function.

Brown and Dilworth (1975) attempted to establish the mode of ammonia assimilation in free-living rhizobia growing in chemostats. They showed that the bacteria contained GS, glutamate synthase, and both NAD- and NADP-dependent GDH. The relative K_m values for ammonia of GDH and for hydroxylamine (an analogue of ammonia used in the biosynthetic assay of GS) of GS were 13.9 mM and 0.4 mM respectively thus emphasizing the usual differences found between these two enzymes. Nitrogen-limited cultures of *R. leguminosarum* contained high activities of GS, glutamate synthase and NADP-GDH; a change to excess nitrate or ammonia caused the first two activities but not NADP-GDH to decrease virtually to nothing. Growing rhizobia on an excess of glutamate did not affect GS, but decreased glutamate synthase activity. In *R. trifolii* and *R. japonicum* GS activity responded in a similar manner but there was not such a marked decline in glutamate synthase activity in the presence of excess ammonia. From their studies Brown and Dilworth suggested that both pathways were operating in *Rhizobium.*

Subsequent work showed that the picture was more complicated. Firstly, there are isoenzymic forms of GS present in all species of *Rhizobium* tested so far (Darrow and Knotts, 1977, Fuchs and Keister, 1980a, b, Ludwig, 1980a, Darrow *et al.,* 1981). Secondly, mutant evidence, generally, but not universally, suggests that GDH does not function in assimilation (Ludwig, 1978, Osburne and Signer, 1980, Ali *et al.,* 1981, Osburne, 1982, see also sections II C and IV and Table 1).

Multiple forms of GS in rhizobia were first described in *R. japonicum* by Darrow and Knotts (1977). Subsequently they have been found in all common species of *Rhizobium* and in four species of *Agrobacterium* (Fuchs and Keister, 1980a, b). Isoforms of GS do not appear to be present in bacteria outside of the Rhizobiaceae although multiple forms of GS are common in higher plants (see Section C. I. b, Miflin *et al.,* 1981, and Mc Nally and Hirel, 1983, for reviews). The two forms of the rhizobial enzyme differ in a number of properties:

(i) GS I but not GS II is reversibly adenylylated in a manner similar to GS from the enterobacteria.
(ii) GS II is relatively unstable to heat.
(iii) GS I is larger with a S_{20w} of about 20 and with 12 subunits of M_r about 59,000 whereas GS II has a S_{20w} of about 11 and subunits of M_r 36,000; the subunit structure of the GS II holoenzyme is not known.
(iv) GS I binds DNA whereas GS II does not.

The general conclusion from physiological and biochemical studies, and from mutant evidence, is that the glutamate synthase cycle is the predominant functional pathway for ammonia assimilation in *Rhizobium;*

some strains, in the presence of high concentrations of ammonia, can use GDH others cannot. Further evidence for the crucial role of GS is the way in which a complex nitrogen regulatory system has evolved around this enzyme. The details of this are discussed in the next section.

C. Regulatory Controls on Ammonia Assimilation

a) General Aspects

Several recent results have led to the conclusion that, firstly, there are parallels between the general nitrogen regulatory genes in enteric bacteria and those controlling the *nif* operons within *K. pneumoniae* and, secondly, that the nitrogen regulatory gene products can control the expression of nitrogenase genes (e.g. see Merrick, 1982, 1983, Ow and Ausubel, 1983, Drummond *et al.*, 1983). It was originally proposed that in enteric bacteria GS controlled its own expression and that of other nitrogen utilization operons (see Tyler, 1978, and Magasanik, 1982, for reviews). This hypothesis provoked considerable experimentation and subsequently Kustu and colleagues (Bancroft *et al.*, 1978, Kustu *et al.*, 1979, Mc Farland *et al.*, 1981) showed that the regulatory element was in fact a gene which was closely linked to *glnA* (the structural locus for GS). Merrick (1982) has recently summarized the current position and this is sketched in Fig. 5a. From the work in *E. coli* (Pahel *et al.*, 1982, Ueno-Nishio *et al.*, 1983), *K. aerogenes* (Foor *et al.*, 1980, Reuveny *et al.*, 1981), *K. pneumoniae* (Espin *et al.*, 1981) and *Salmonella typhimurum* (Bancroft *et al.*, 1978, Kustu *et al.*, 1979, Mc Farland *et al.*, 1981) a consensus model has emerged. In this the control of expression of *glnA* and the other operons under nitrogen control (e.g. genes governing the utilization of histidine *(hut)*, proline *(put)* and arginine *(aut)*) is regulated by the products of three genes *ntrA (glnF)*, *ntrB (glnL)* and *ntrC (glnG)*. Repression requires the presence of both the *ntrB* and *ntrC* gene products, probably acting together as a complex. Activation is achieved by the *ntrC* gene product interacting together with the *ntrA* product. Mutants lacking the *ntrB* product constitutively express GS whereas mutants lacking *ntrA* cannot derepress genes under nitrogen control.

Superimposed on the regulation of the expression of GS is a control of the activity of the enzyme by reversible adenylylation of the GS protein (Fig. 6). The enzyme consists of 12 subunits each of which can have an AMP group added to it; the adenylylation state can therefore vary from 0 to 12. The mean adenylylation state is denoted by E_n. This reversible reaction is catalyzed by adenylyltransferase (ATase) an anzyme which itself can be activated by a protein called P_{II}. Like GS, P_{II} can itself exist in two states as a result of covalent nucleotide binding. Thus P_{IIA} can be converted to uridylyl-P_{II}D by the action of uridylyltransferase (UTase). The genes believed to be responsible for this cascade are *glnB, glnD* and *glnE* (Fig. 6). Adenylylated GS is unable to catalyze the synthesis of glutamine and the related Mg^{2+} stimulated biosynthetic reaction with hydroxylamine as substrate. However, adenylylated GS can still catalyze the Mn^{2+}-stimulated non-biosynthetic transferase reaction. Consequently it is possible to mea-

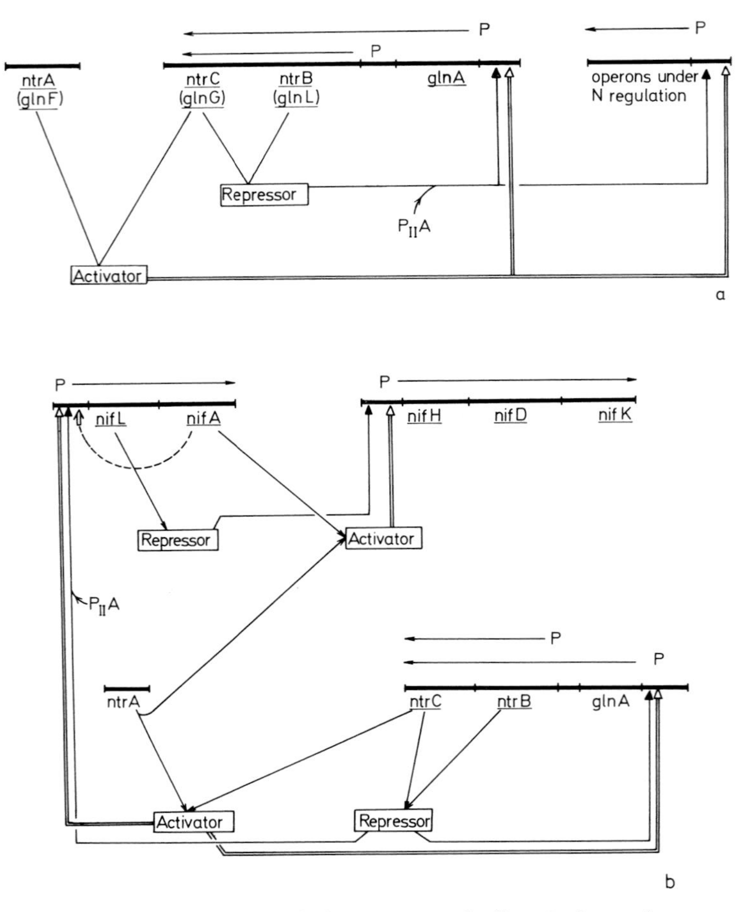

Fig. 5. Regulation of nitrogen metabolism in bacteria
(a) A model for the nitrogen regulatory control system in *E. coli* and *S. typhimurum* (after Merrick, 1982). The genes are given both their alternative (*ntr,* after Kustu *et al.,* 1979) and their original nomenclature *(gln)*. P indicates promoter regions.
(b) Regulatory control for the expression of *nif* genes in *K. pneumoniae*

sure the amount of GS present even when it is not functioning as a physiologically relevant catalyst. Deadenylylated GS is fully active in synthesizing glutamine. Large amounts of organic nitrogen or of ammonia lead to the adenylylation of GS and the repression of GS synthesis (and that of the other nitrogen regulatory operons). In contrast, nitrogen starvation causes deadenylylation and derepression of the *glnA* and other nitrogen operons. This control acts, at least in part, via a stimulatory effect of 2-oxoglutarate.

Recent work has shown that the adenylylation system is linked to the nitrogen regulatory system. Mutants of *K. pneumoniae* and *S. typhimurum* have been isolated (Foor *et al.,* 1980, Reuveny *et al.,* 1981, Bancroft *et al.,*

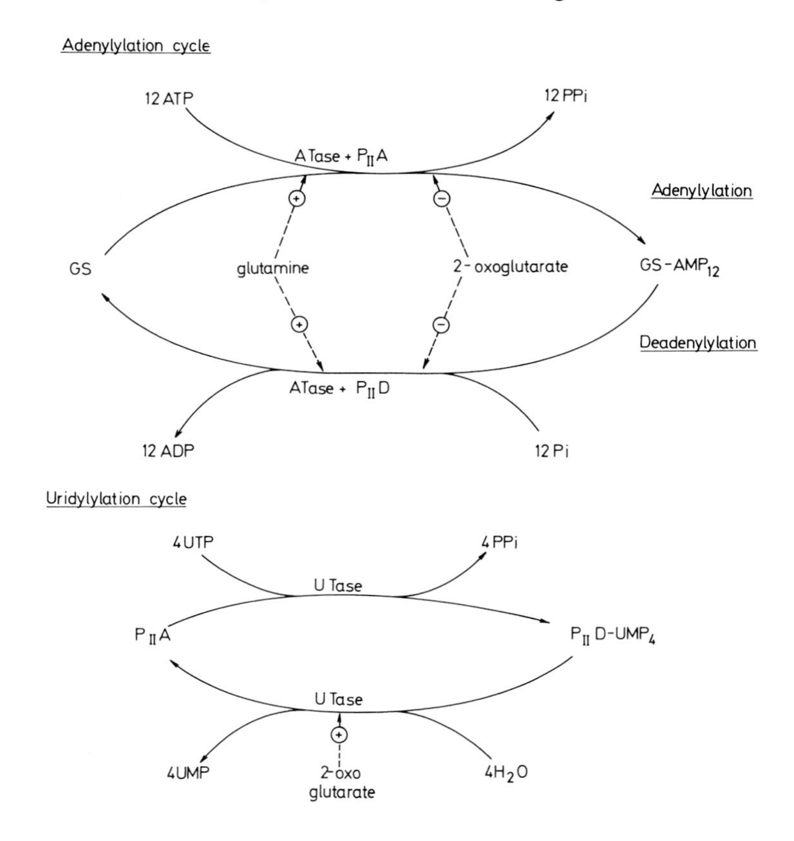

Fig. 6. The regulatory cascade operating on glutamine synthetase (after Stadtman *et al.*, 1980). The enzymes involved are ATase: adenylyltransferase and UTase: uridylyltransferase. The structural genes for the components of this cycle are *glnA:* GS, *glnB:* P_{II}, *glnD:* UTase and *glnE:* ATase. The positive and negative effects of certain metabolites are indicated by + and −. P_{II} is Shapiro's regulatory protein and consists of four subunits. Pi is inorganic phosphate, PPi inorganic pyrophosphate

1978) and have the following properties. The mutation *glnB3* (*glnB* is the putative structural gene for P_{II}) alters P_{II} so that it is no longer a substrate for UTase and the cells always have a high proportion of $P_{II}A$ (non-uridylylated P_{II}). The phenotype of such mutants is a low level of GS which is always adenylylated and not fully derepressed in the absence of ammonia. Mutations that are closely linked to *glnB3* abolish P_{II} production and result in a phenotype in which there is a lack of response to nitrogen excess and the cells have a large amount of GS even though the enzyme becomes highly adenylylated. In contrast, mutants lacking ATase have similar patterns of regulation to ATase containing strains (Bancroft *et al.*, 1978, Reuveney *et al.*, 1981). Thus the implication is that the adenylylation state of GS does not play a role in determining the amount of GS present in the cells. The results obtained with these mutants strongly suggests that P_{II} acts in relaying the information on the availability of fixed nitrogen not only to the adenylylation system but also to the nitrogen regulatory system, thereby modifying the expression of the various operons including that of *glnA*. It is likely that $P_{II}A$ in some way enhances the repression of operons under nitrogen regulatory control (Fig. 5a).

Recent results suggest that the nitrogen regulatory system extends to control the expression of the *nif* operons in *K. pneumoniae* (Fig. 5b). Within the nitrogen fixation gene cluster the genes *nifL* and *nifA* regulate the expression of the other operons. The *nifL* gene product represses *nif* transcription in response to fixed nitrogen and the presence of oxygen, whereas the *nifA* product activates *nif* transcription (Dixon *et al.,* 1977, Merrick *et al.,* 1982). Furthermore, it has been shown that *ntrC* but not *ntrB* (or *glnA*) is required for activation of *nif* expression (Espin *et al.,* 1982) probably by acting at the *nifLA* operon promoter (Mac Neil and Brill, 1980, Elmerich *et al.,* 1981). Recent work from two groups (Drummond *et al.,* 1983, Merrick, 1983, Ow and Ausubel, 1983) has shown, that the *nifA* and *ntrC* are evolutionarily closely related and can, to some extent, substitute for each others function although the *ntrC* product cannot act directly on the other operons of the nitrogen fixation cluster of *K. pneumoniae.* To some extent the *nifA* gene may be able to act as an autogenous regulator. It is also suggested that the *ntrA* gene product is required to obtain the maximum expression of the *nifLA* operon and that it is required together with the *nifA* product for activation of the *nifH* promoter (Sundaresan *et al.,* 1983a, Merrick, 1983). The proposed regulatory loops are shown in Fig. 5b. The parallels between the *ntrBC* and the *nifLA* operons thus become particularly striking and are further borne out by sequence analyses of the promoter regions of these and other nitrogen operons (Drummond *et al.,* 1983, Ow *et al.,* 1983).

Expression of *nif* genes also appears to be affected by *glnD* (the structural locus for UTase). Strains of *K. pneumoniae* in which *glnD* is missing are unable to transcribe from the *nifL* promoter under derepressing conditions although this effect can be overcome by the presence of multiple copies of *ntrC* (Ow and Ausubel, 1983). Presumably $P_{II}A$ can also have a repressive effect on the *nif* regulatory system that can be overcome by the *ntrC* product.

b) Control of Nitrogen Metabolism in *Rhizobium* by Nutrients

Brown and Dilworth (1975) showed that an excess of fixed nitrogen (e.g. nitrate or ammonium) added to *Rhizobium* cultures decreased total GS activity as measured in a biosynthetic assay. Glutamate synthase also decreased although the magnitude of the decrease differed between species. Other early studies of total GS activity in free living *Rhizobium* showed that excess fixed nitrogen caused an increased adenylylation of the enzyme and an apparent repression of activity (Bishop *et al.,* 1976, Upchurch and Elkan, 1978a). However, this work measured the mean effects on the two glutamine synthetases. Ludwig (1980a) has studied the effects on GS I and GS II separately. In the absence of ammonia, with 6.6 mM glutamate as a nitrogen source, both GS I and GS II of cowpea *Rhizobium* sp. 32HI were derepressed and GS I was unadenylylated. Addition of 10 mM ammonium caused a twofold decrease in GS I activity but a 50-fold repression of GS II. Besides the decrease in GS I transferase activity, GS I became highly adenylylated ($E_n = 7.5$) and thus rapidly lost biosynthetic activity. Growth on glutamine, a luxury nitrogen source, also caused adenylylation of GS I and repression of GS II. In general, ammonia affects expression of GS of 32H1 at much lower concentrations than in *K. aerogenes;* thus transient repression can be observed by adding 10 μM ammonium to cultures growing on glutamate (Ludwig, 1978).

These results suggest that *Rhizobium* has an adenylylation/deadenylylation system for GS I that in many ways resembles that in enteric bacteria. However the degree of repression of GS I synthesis is clearly much less in rhizobia whereas GS II (which does not exist in enteric bacteria) is much more responsive in terms of repression/derepression.

Much less is known about nitrogen regulatory genes in *Rhizobium* than in enteric bacteria although recent results suggest some parallels. Thus it has been shown that the products of *nifA* and *ntrC* can both activate the promoter of the *nifHDK* operon of *R. meliloti* (Sundaresan *et al.,* 1983 a, b). This effect of *ntrC* in *R. meliloti* is in contrast to its inability to affect the *nifH* promoter in *K. pneumoniae* (see Fig. 5b). Furthermore, the DNA sequences of the promoters of the *nifHDK* operons in *R. meliloti* and *K. pneumoniae* share similarities with each other and with other nitrogen regulatory promoters (Ow *et al.,* 1983, Drummond *et al.,* 1983). These results open up interesting hypotheses as to how regulatory mechanisms may have evolved and how certain subtle but distinct differences between free-living and symbiotic nitrogen fixers may be important in a physiological context. We might expect considerable advances in our understanding of the genetics of nitrogen control in *Rhizobium* to occur during the next few years.

c) Effect of O_2 Concentration on Nitrogen Metabolism of *Rhizobium*

For many years it was believed that rhizobia could only express nitrogenase and fix nitrogen in the symbiotic state. However, in 1975 three groups (Pagan *et al.,* 1975, Kurz and La Rue, 1975, McComb *et al.,* 1975) showed that dinitrogen fixation could be observed in cultures of slow growing rhizobia. One of the critical factors appears to be microaerobic conditions; when the O_2 tension is decreased to about 0.2%, dinitrogen fixation occurs (see Gibson *et al.,* 1977). Besides low O_2 tensions, a source of fixed nitrogen is also required, e.g. glutamate or glutamine.

Nitrogen fixation in culture has now been demonstrated for many representatives of all the 'slow-growing' rhizobia (see Bergersen and Gibson, 1978, Shanmugam *et al.,* 1978, for reviews) but not when similar conditions were used with members of the fast-growing group. But when cultures of *R. japonicum* were supplemented by dialysable material diffusing from plant cells they reduced acetylene (Reporter, 1976). This observation has now been extended to a wide range of free living rhizobia including *R. tri-*

folii, R. leguminosarum and *R. meliloti* (see Gresshoff *et al.*, 1981, for review). The exact nature of the active plant component is not yet defined but copper containing peptides have been implicated (Reporter, 1981).

These findings pose a number of questions: What happens to the assimilatory enzymes under these conditions? What mechanisms cause the derepression? Can the ammonia produced be used for growth? In almost every case it appears that the answer to the last question is no and that the ammonia produced is rapidly transported out of the rhizobia into the surrounding medium (O'Gara and Shanmugam, 1976, Bergersen and Turner, 1978). However a recent report (Dreyfus and Dommergues, 1981) showed that a strain of *Rhizobium* ORS571, isolated from the stem nodules of the tropical legume *Sesbania rostrata,* produced high activities of nitrogenase in culture and could grow at the expense of fixed nitrogen. Subsequent experiments in chemostat cultures (Gebhardt *et al.*, 1983) have shown that this strain is capable of prolonged growth on N_2 as the sole nitrogen source and that about 90% of the fixed nitrogen is retained within the cells. Interestingly this strain can fix nitrogen at a higher O_2 tension than other rhizobia, appears to have a higher respiratory efficiency (Gebhardt *et al.*, 1983) and can grow rapidly on ammonia (Dreyfus *et al.*, 1983). Another strain of interest is cowpea *Rhizobium* sp. R3205 which has constitutively high nitrogenase activity in O_2 limited cultures supplied with glutamate and succinate (Ludwig, 1980, Table 1) and apparently can also assimilate ammonia under nitrogen-fixing conditions. This suggests that Rc3205 might be able to grow on ammonia as glutamate becomes limiting, however, this is not the case as results indicate that the cells that are fixing in the culture are not actually growing (Ludwig and de Vries, 1982).

The effects of low O_2 tension on GS appear less clear cut. Bergerson and Turner (1978) have shown that O_2-limited cultures of cowpea *Rhizobium* sp. CB756, grown under steady state conditions, had high nitrogenase activities which were tolerant of excess ammonia but when the O_2 tension increased nitrogenase was repressed by ammonia. Observation of the changes in the adenylylation status of GS upon lowering of the O_2 tension suggested that the relative adenylylation decreased when O_2 was severely limited. In contrast Rao *et al.* (1978), using batch cultures of *R. japonicum* and measuring both GS I and GS II, found that GS II was absent and GS I was both repressed and relatively more adenylylated under nitrogen fixing conditions. Subsequently Darrow *et al.* (1981) confirmed and extended these results using cowpea *Rhizobium* sp. 32H1 and *R. japonicum.* They also compared the effect of replacing nitrogen in the gas phase by argon. With both argon and dinitrogen, GS II decreased in activity and GS I became adenylylated under O_2-limitation, however, the effects were more marked (and nitrogenase less active) with dinitrogen. The authors related these results to the amount of ammonia in the medium; but, even in argon some ammonia accumulated presumably from the breakdown of glutamate.

These apparently conflicting results have been partially explained by the results of Ludwig (1980b) on transient changes occurring during the change from air to 0.2% O_2. Initially GS I and GS II were about equal in activity and GS I was unadenylylated. After the O_2 tension was lowered GS I activity fell immediately whilst that of GS II went up before decreasing prior to the onset of nitrogenase activity at 24 hours. The adenylylation status of GS I increased at first to $E_n = 5$ but then dropped back to near zero at 20 hours before finally increasing to about 10 at 48 hours. Ludwig (1980b) points out that there are two correlations to note; one, GS I was unadenylylated prior to the appearance of nitrogenase and two, it became adenylylated after nitrogenase appeared, presumably in response to the build up of ammonia. Similar correlations were also noted when 10 mM ammonia was added to the cultures; 2 hours after the shift to low O_2 GS I became partially deadenylylated ($E_n = 3.5$) and nitrogenase appeared subsequently (23 hours later).

d) Studies with Mutants

A number of mutants of rhizobia have been isolated which have altered assimilatory enzymes (Table 1). The importance of the glutamate synthase cycle in ammonia assimilation is shown by mutants of *R. meliloti* (G41, 2620) and of cowpea *Rhizobium* sp. 32H1 (Rc 3202) which lack either glutamate synthase or total biosynthetic GS and are unable to assimilate ammonia despite the presence of NADPH- or NADH-dependent GDH. Only after the selection of revertants is there evidence for assimilation occurring by GDH. Thus Rm 2620–201 and Rm 2620–202 lack glutamate synthase, grow on ammonia at high concentrations (20 mM) and have about 10 times the amount of NADPH-GDH of the wild type (Osburne and Signer, 1980). Similarly glutamate synthase minus mutants (e.g. 2011A) of *R. meliloti* strain 2011 are unable to use 1 mM ammonia as a sole nitrogen source but can grow on 12 mM (Ali *et al.*, 1981). A mutant of particular interest with respect to control mechanisms is Rc 3201 which has a constitutively adenylylated GS I ($E_n = 11$) and a repressed GS II, and is a glutamine auxotroph. A revertant of this mutation exists (Rc 3205) which still lacks GS II but whose GS I is constitutively unadenylylated. Rc 3203 is not auxotrophic for glutamine and has a partially adenylylated GS I which will completely deadenylylate on a minimal medium. The site of these mutations is not known, nor whether they are all single mutations, however, it might be expected that some of them lie in steps of the nitrogen regulatory cycle (see Figs. 5 and 6).

Mutants lacking GS II (Rc 3201, 3203, 3205) have a further phenotype in that they appear to be partially auxotrophic for purines, since addition of purines decreases the doubling time even in the presence of glutamine (Ludwig, 1980a). Thus Ludwig has suggested that GS II functions in purine biosynthesis although whether or not the enzyme is important in ammonia assimilation remains open.

The various mutants also have different phenotypes with respect to their behaviour under microaerobic conditions. Whilst wild type *R. meliloti* and cowpea *Rhizobium* sp. 32H1 produce nitrogenase under such conditions, mutants (*R. meliloti* DB5, Rc 3201) lacking GS activity do not (Kon-

Table 1. Mutants of *Rhizobium* Altered in Their Nitrogen Metabolism

Strain	Phenotype	GS I	GS II	Glutamate synthase	NADPH GDH	NADH GDH	N_2 fixation free living	N_2 fixation symbiotic	Reference
R. meliloti									
2011	asm^+ wt	+	+	+	++				
2620	asm^-	+	−	−	+			+	Osburne, 1982
4001	glu^t	+	+	+	++			+	Osburne and
4002	glu^t	+	+	+	++			−	Signer, 1978
								−	
2620-101	asm^+ rev	+		+	+			−	
2620-201	asm^+ rev	+		−	++++			−	
41	asm^+ wt	+		+	−	low		+	Kondorosi
G41	asm^-	+		−	−	low		+	*et al.*,
Db5	asm^- gln^-	v. low		+	−	low		−	1977
2011A	asm^-	+		−	++			+	Ali *et al.*, 1981
R. Sp. 32H1									
Rc 3200	wt	+	+	+	−	+	+	+	
Rc 3201	gln^- asm^-	−	−	+	−	+	−	−	
= *gln5*									Ludwig, 1980a, b,
Rc 3202	gln^- asm^-	+	+	−	−	+			1981; Ludwig and
Rc 3203	asm^-	pAd	−	+	−	+			Signer, 1977
Rc 3204		Ad	+	+	−	+	−	−	
Rc 3205	gln^+	unAd	−	+	−	+	++const	++	
Rc 3213									

Ad — adenylylated all the time, pAd — partly adenylylated, unAd — unadenylylated all the time
gln^- — glutamine auxotroph, glu^t — require a trace amount of glutamate in order to grown on NH_4^+
asm^- — unable to grow on NH_4^+ (usually around 10 mM) as a nitrogen source
const — constitutive expression, rev — revertant

dorosi *et al.*, 1977, Ludwig and Signer, 1977). These results were originally interpreted in terms of the direct involvement of GS in nitrogenase expression, however, as this model no longer holds there is probably another mechanism operating. Subsequently, isolated revertants of Rc 3201 (e.g. Rc 3204) suggest that GS II does not appear to be involved in control and that the presence of a permanently adenylylated GS I is associated with lack of expression of *nif* in nodules.

Conversely, constitutive *nif* expression in culture and high nitrogenase activity in nodules found in a mutant (Rc 3205) in which GS I is permanently deadenylylated (Ludwig, 1980). Ludwig has suggested that the most likely explanation therefore is that the *gln5* (Rc 3201) mutation interferes with the adenylylation/deadenylylation system (see Fig. 6) and that the Rc 3205 type reversion occurs by some change in ATase. However, as discussed above in relation to mutants of enteric bacteria, the presence or absence of ATase does not affect regulation of the nitrogen operons (Bancroft *et al.*, 1978, Reuveney *et al.*, 1981) and thus, by analogy, a more likely site of action could be the putative homologue of the structural gene for P_{II} *(glnB)*. As discussed previously (Section B. III. a) mutations in *glnB* in *K. aerogenes* can cause the production of permanently adenylylated GS (cf. Rc 3204).

e) Ammonia Transport in *Rhizobium*

In considering the effect of externally applied compounds on gene expression it is necessary to take into account that, firstly, the added compound may not be the actual metabolite recognized by the regulatory sensor and, secondly, it is only likely to be effective if it is transported into the appropriate part of the cell. In terms of the actual regulatory metabolite we are unable to say which or how many combined nitrogen compounds are sensed by the nitrogen regulatory system, even in species in which that system is well defined (e.g. *K. aerogenes* or *K. pneumoniae*). All that we can conclude, based on the use of inhibitors of GS is that it is unlikely to be ammonia since addition of ammonia to cultures in which GS activity is inhibited does not repress nitrogenase (Gordon and Brill, 1974), nor does it cause GS to become adenylylated (Bishop *et al.*, 1975).

Thus, ammonia has to be taken up and metabolized before it is capable of repressing *nif* gene expression. Whilst unprotonated ammonia can diffuse readily through membranes, transport of NH_4^+ requires a carrier mechanism. Ammonium uptake has been studied in *Klebsiella* and *Rhizobium,* often by using methylammonium as an analog. The results from *Klebsiella* (see Kleiner *et al.*, 1981, for a review and Kleiner, 1982) have shown that NH_4^+ is transported across the bacterial membrane by a specific carrier capable of generating 100-fold gradients across the membrane and which is itself repressed by excess NH_4^+. Gober and Kashket (1983) have suggested that free-living cowpea *Rhizobium* sp. 32H1 contain an active methylammonium (or NH_4^+) transport system when grown under nitrogen-fixing conditions (0.2% O_2) and that this is capable of counter transport. However when 32H1 was grown under aerobic conditions the authors could not demonstrate the presence of a carrier. Dilworth and Glenn (1982) were also unable to find evidence for an active transport of methylammonium across the membranes of histidine-grown *R. leguminosarum*. Several authors have observed a loss of NH_4^+ into the medium when rhizobia are exposed to microaerobic conditions (O'Gara and Shanmugan, 1976, Bergersen and Turner, 1978, Darrow *et al.,* 1981). Ludwig (1980) has suggested that chemiosmotically coupled NH_4^+ export could occur after a shift to low O_2, presumably this would occur via the ammonium transport mechanism described by Gober and Kasket (1983). However transfer of ammonia (unprotonated) might also occur by diffusion across the membrane with accumulation occuring in the medium if its pH was markedly lower than inside the cells (see Kleiner *et al.,* 1981).

A suggestive link between ammonium transport and nitrogen fixation comes from mutants *R. meliloti* 4001 and 4002 (Table 1). Osburne (1982) considers that these mutants which are Fix^- may have a defective ammonium transport system.

f) Control by Energy Status

Nitrogen fixation has a high ATP requirement and it is presumably in the organism's interest to limit the process when its energy status is low. Appleby *et al.* (1975) found that there was a positive correlation between nitrogenase activity and the ATP/ADP ratio in bacteroids. Experiments with free-living nitrogen fixing cowpea *Rhizobium* sp. CB756 (Ching *et al.,* 1981) and *K. pneumoniae* (Upchurch and Mortenson, 1980) showed that there was an optimum energy charge of between 0.6 and 0.7 with little or no nitrogenase activity present in cells having energy charge values above 0.75 or below 0.53. Regulation of nitrogen metabolism has also been linked with the amounts of other nucleotides. Thus Upchurch and Elkan (1978a) found that 1 mM cyclic AMP decreased the activity of GS, glutamate synthase and GDH in free-living *R. japonicum.* Lim *et al.* (1979) found that cyclic GMP decreased nitrogenase synthesis when added exogenously and that, during a transition from a microaerobic to an aerobic environment, internal cyclic GMP increased 10 fold. Whether these effects are specific and direct upon nitrogenase is not clear. Certainly Van den Bos *et al.* (1983) found no specific effect of cyclic GMP on nitrogenase synthesis by bacteroids.

D. Nitrogen Metabolism in Rhizobia Under Symbiotic Conditions

Early work on the nitrogen metabolism of bacteroids suggested that ammonia was produced in the bacteroids and exported into the plant cytoplasm (Bergersen, 1965, Bergersen and Turner, 1967). Subsequent studies have shown that bacteroids do not appear to contain the necessary enzymes to assimilate the ammonia they produce (Brown and Dilworth, 1975, Robertson and Farnden, 1980). The two questions that are particularly relevant are: 'What are the changes that occur within the bacteroid that lead to the expression of nitrogenase and the export of ammonia?' and 'How does the bacteroid receive its nitrogen nutrition?'

Studies on the amount of GS in bacteroids show that in effective nodules there is considerably less activity than in free-living bacteria (Brown and Dilworth, 1975). Few studies have separated the bacteroid GS into its two forms although recently Cullimore *et al.* (1983) found that GS II was apparently not present in *R. phaseoli* bacteroids. Bishop *et al.* (1975) found that the bacteroid GS was partially adenylylated and did not change in response to a wide range of nitrogen nutrition. Planqué *et al.* (1978) also found that *R. leguminosarum* bacteroid GS was adenylylated both throughout the period of increasing nitrogenase activity and during the period when nitrogenase had constant activity.

Mutant evidence (see *R. meliloti* 2620, Rc 3205) (Table 1) indicates that GS II is not of importance in symbiotic nitrogen fixation. It is not clear why mutants with no GS I or with highly adenylylated GS I (e.g. Db 5, Rc 3201, Rc 3204) are ineffective whereas one (e.g. Rc 3205) with a constitutively deadenylylated GS I induces nodules with highly active nitrogenase. The metabolism of the nitrogen fixed in nodules formed from Rc 3205 infection might prove instructive. However, apart from this mutant the general trend appears to be that the rhizobial strains proving most effective in symbiotic nitrogen fixation are those with the lowest levels of GS activity (Upchurch and Elkan, 1978 a).

The amount of glutamate synthase in bacteroids differs between species (Brown and Dilworth, 1975) and the function of the enzyme may be important in ensuring the utilization of any glutamine supplied by the host rather than in ammonia assimilation. Since effective nodules can be formed by *Rhizobium* mutants lacking glutamate synthase (see Table 1) the enzyme is not crucial either to nitrogenase formation or to effective symbiotic nitrogen fixation. However, the reasons why some glutamate synthase mutants are non-fixing in nodules are unknown (Osburne, 1982).

A number of approaches have been made to measure the effect of combined nitrogen on nitrogenase. When ammonium nitrate was supplied to nodulated pea plants, acetylene reduction was diminished by up to 50%. This did not result from a reduced number of bacteroids or bacteroid proteins but was associated with a decreased amount of leghaemoglobin (Bisseling *et al.*, 1978). On the other hand, when isolated bacteroids were incubated with ammonium chloride then nitrogenase activity was unaffected (Bishop *et al.*, 1976, Laane *et al.*, 1980), results that generally parallel the lack of effect of ammonia on nitrogenase activity in free-living rhizobia (see above). In view of these results it is not surprising that externally supplied ammonia has no effect on the synthesis of nitrogenase protein either in nodules (Bisseling *et al.*, 1978) or in bacteroids (Scott *et al.*, 1979, Van den Bos *et al.*, 1983). In contrast low concentrations of O_2 inhibited nitrogenase activity in bacteroids (Laane *et al.*, 1978) and different results have been found on the effect of O_2 on nitrogenase synthesis. Van den Bos *et al.* (1983) using *R. leguminosarum* bacteroids found no effect of 100 μM

O_2 on nitrogenase synthesis even after 30 minutes exposure whereas Shaw (1983), using bacteroids of *Rhizobium* sp. NZP2257 isolated from *Lupinus angustifolius* nodules, found that the synthesis of nitrogenase declined under 20% O_2. Moreover, the rate of decline differed for the two nitrogenase components with the Fe-protein declining much more rapidly than the Mo-Fe protein. Why the different species of bacteroids should react differently to O_2 is not known.

Few studies have been done on flux of nitrogenous compounds across the bacteroid membrane, however it does appear that bacteroids like nitrogen-fixing, free-living rhizobia, rapidly lose ammonia to the surrounding medium (Laane *et al.*, 1980). Whether this is an active export mechanism mediated by a carrier system (Gober and Kashket, 1983) or a response to a Δ pH without a translocator (Laane *et al.*, 1980) is not clear. Such an outward flux of ammonia would be further enhanced by an efficient assimilatory system, such as the glutamate synthase cycle, in the plant cytoplasm (see section C. I.).

As regards the nitrogen nutrition of the bacteroids this must be provided by the plant and presumably be available in a number of forms since strains of *Rhizobium* auxotrophic for various amino acids are capable of forming effective symbioses (Table I, Pain, 1979, Ludwig, 1980 a, b). However, the effect of various sources of combined nitrogen (e.g. glutamine), or of other nutrients (e.g. 2-oxoglutarate or various sugars) on the expression of nitrogenase are as yet untested. Perhaps these might provide a way in which the nitrogen status of the plant could affect the rate of fixation by the bacteroids.

E. Summary

It is not easy to summarize the current situation neither with respect to the genes concerned with dinitrogen fixation and ammonia assimilation in *Rhizobum* nor as to how their expression may be regulated. In part this is because our direct knowledge of the genes involved is still slight and we are forced to argue by analogy from the enteric bacteria but also because we are faced with a number of paradoxes. These are:

(1) The central dogma on the regulation of biochemical pathways in prokaryotes is that they are turned on by substrates and off by endproducts. However, as emphasized by Ludwig (1978, 1980 a, b) non-fixing rhizobia, particularly *Rhizobium* sp. 32H1, manage to turn off their ammonia assimilatory pathway via GS in the presence of very small amounts of ammonia. Although GS in *K. pneumoniae* is turned off by greater concentrations of ammonia this is because there is a shift in assimilation to the GDH route; *Rhizobium* sp. 32H1 and some other rhizobia do not have this alternative pathway.

(2) Furthermore, in conditions in which rhizobia are fixing nitrogen and producing ammonia the expression and activity of nitrogenase is virtually insensitive to the end-product, ammonia.

(3) Whereas maximum fixation in free-living rhizobia and bacteroids can occur under conditions in which GS II is fully repressed, mutants with a constitutively adenylylated GS I are associated with a Nod^+ Fix^- symbiotic phenotype.

(4) At a time when rhizobia are fixing atmospheric nitrogen they are, with one exception, also auxotrophic for a source of organic nitrogen; in fact the provision of such a source is a prerequisite for fixation to occur.

The way in which these paradoxes may be resolved is not easy, however, focusing upon them may help to suggest the approaches necessary to develop an understanding of how nitrogen fixation is controlled in rhizobia and why this has evolved to the state where it flourishes most strongly in the symbiotic state. We suggest that the following aspects might be important.

(1) A full description of the links between adenylylation and repression/derepression of GS has not been achieved even for *E. coli,* however, the evidence points strongly towards there being one or more control loops linking these phenomena both in enteric bacteria and rhizobia. Mutations have been found, particularly in *glnB* and *glnD* which can disengage some of the links between the nitrogen regulatory system (Fig. 5) and the adenylylation system (Fig. 6). Similar effects could be relevant in rhizobia and a knowledge of the links seems vital for a full understanding of the control of nitrogenase expression and suppression of ammonia assimilation.

(2) The transient deadenylylation observed in free-living *Rhizobium* sp. 32H1 prior to the onset of nitrogenase synthesis (Ludwig, 1980) may also occur in bacteroids. However any linked control on nitrogenase expression would have to remain in the 'on' position even when GS became adenylylated since the half-life of nitrogenase mRNA in bacteroids appears to be only about 2 minutes (Van den Bos *et al.,* 1983). Such a situation could operate if the autogenous regulation of the *nifLA* operon by the *nifA* gene product occurs once transcription of the operon has been activated by the nitrogen regulatory system (i.e. the products of *ntrA* and *ntrC*) (see Ow and Ausubel, 1983).

(3) The ability of low concentrations of ammonia to provoke adenylyation of GS I and repression of GS I and GS II in aerobic rhizobia in comparison to the rapid loss of ammonia from microaerobic, nitrogen-fixing rhizobia and the inability of externally supplied ammonia to either inhibit fixation under such conditions or affect the transient deadenylylation in response to low O_2 tension suggests that the ability of rhizobia to take up or retain ammonia changes dramatically during the shift to a microaerobic environment. Such changes seem a more likely explanation for the inability of ammonia to repress nitrogenase than changes in the ability of the cells to monitor and respond to the changes in internal concentrations of ammonia (or a product of its assimilation).

(4) The information of *nif* operons, on their control and on the nature of the promoter sequences in rhizobia (particularly their differences from *K. pneumonia* e.g. see Ow *et al.,* 1983) should both provoke testable hypo-

theses for possible control circuits and eventually provide an explanation at the molecular level.
(5) The number of putative regulatory metabolites that have been tested in most studies has been limited (usually only ammonia and O_2). However, ammonia is probably not the actual regulatory molecule and the effect of (and internal concentrations of) glutamine and some of its products may be informative. Furthermore, little attention has been paid to non-nitrogen containing organic molecules such as 2-oxoglutarate. In *E. coli* it is known that nitrogen metabolism is controlled by the balance between 2-oxoglutarate and glutamine (see Stadtman *et al.,* 1980). The inherent complexity of the *E. coli* system is shown by the fact that over 40 metabolites have been shown to affect the cascade shown in Fig. 6. There is obviously no shortage of potential candidates for testing in rhizobia.

III. Plant Metabolism

The host contribution to dinitrogen fixation within the legume/*Rhizobium* symbiosis is considered in terms of: Ammonia assimilation; Synthesis of the nitrogenous transport compounds; Control of host genes involved in nodule nitrogen metabolism.

A. Ammonia Assimilation

It is now generally agreed that in most higher plant tissues assimilation of ammonia occurs via the glutamate synthase cycle (Fig. 4) in preference to assimilation by GDH (for recent reviews see Miflin and Lea, 1980, 1982). Evidence from studies using labelled substrates, inhibitors and measurements of enzyme activities are reviewed to support this view for the assimilation of ammonia derived from dinitrogen fixation in legume root nodules.

a) Labelling and Inhibitor Studies
Ammonia produced by dinitrogen fixation in the bacteroids is excreted into the plant nodule cytoplasm (see section II. D.). There, labelling studies using $^{15}N_2$ (Ohyama and Kumazawa, 1978, 1980a) and $^{13}N_2$ (Meeks *et al.,* 1978) have shown that glutamine amido-N followed by glutamate are the initial products of assimilation; the pattern of labelling follows a typical precursor: product relationship, indicative of the glutamate synthase cycle (see Wolk *et al.,* 1976, Meeks *et al.,* 1977). This pathway had already been suggested to be operating in nodules by Miflin and Lea (1976) following a reanalysis of data from $^{15}N_2$ labelling of serradella nodules (Kennedy, 1966). Moreover the inhibition of $^{15}N_2$ labelling into glutamate and glutamine by the inhibitor of GS, methionine sulphoximine, and the accumulation of label in glutamine when azaserine, an inhibitor of glutamate synthase, was used (Ohyama and Kumazawa, 1980b) further supports the sole

operation of this pathway in the assimilation of ammonia derived from dinitrogen fixation. However when ^{15}N is supplied to nodules as ^{15}N-labelled ammonia the results are less clear and the available data do not exclude the dual operation of GS and GDH (Fujihara and Yamaguchi, 1980, Schubert and Coker, 1981 a, b).

b) Enzyme Activites

Over 95% of the total nodule GS activity is present in the plant cytosol fraction (Dunn and Klucas, 1973, Brown and Dilworth, 1975, Planqué *et al.*, 1977, Streeter, 1977) and this enzyme may constitute up to 2% of the total soluble protein of nodules (Mc Parland *et al.*, 1976, Mc Cormack *et al.*, 1982, Cullimore *et al.*, 1983). Furthermore the activity of GS in nodules may be 10—500 fold higher than in non-nodulated roots (Robertson *et al.*, 1975 a, Sen and Schulman, 1980, Cullimore *et al.*, 1982); the discrepancy in values depending on whether results are expressed on a fresh weight or specific activity basis as nodules contain about 10 fold more extractable protein than roots (Cullimore *et al.*, 1982, Groat and Vance, 1982). The increase in GS activity during nodulation occurs over a time course similar to the synthesis of nitrogenase in the bacteroids and leghaemoglobin in the plant cytosol (Robertson *et al.*, 1975 a, Sen and Schulman, 1980, Werner

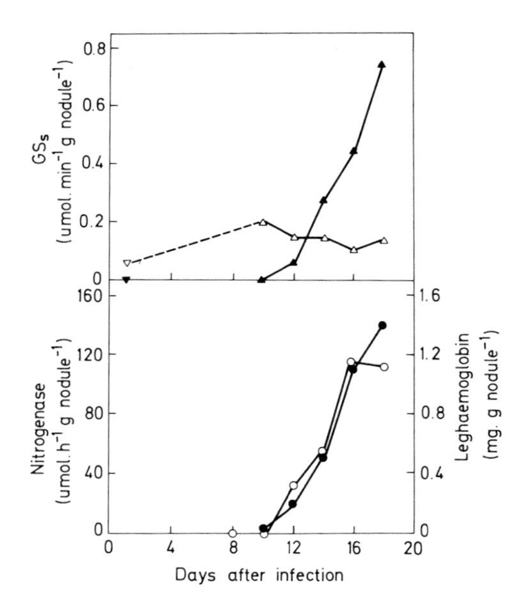

Fig. 7. Changes in the activity of different proteins during nodule formation. Nodules were formed on *P. vulgaris* in response to infection with *R. phaseoli* R3622 and the activity of (▲) GS_{n1}, (△) GS_{n2} and (●) nitrogenase and (○) the amount of leghaemoglobin were measured over 20 days. The activity of (▼) GS_{n1} and (▽) GS_{n2} in non-inoculated roots was also determined. GS_s : GS synthetase activity. (Reproduced with permission from Lara *et al.*, 1983)

et al., 1980, Atkins *et al.,* 1980, Schubert, 1981, Reynolds *et al.,* 1982 a, Lara *et al.,* 1982 a). In *Phaseolus* this increase is due to the production of a nodule specific form of the enzyme (GS_{n1}); the activity of the form (GS_{n2}), which has some similarities to root GS (GS_r) remains almost constant (Lara *et al.,* 1983) (Fig. 7). The activity of GS is much reduced in nodules which are unable to fix dinitrogen due to bacterial (Werner *et al.,* 1980, Sen and Schulman, 1980, Lare *et al.,* 1983) or host (Groat and Vance, 1982) genetic factors (see also section III. C.).

Problems in stabilizing glutamate synthase in cell free extracts (Robertson *et al.,* 1975 b, Boland and Benny, 1977) may have led to an underestimation of the activity of this enzyme, and most certainly to its lack of detection in earlier studies (Dunn and Klucas, 1973, Ryan and Fottrell, 1974, Brown and Dilworth, 1975, Planqué *et al.,* 1977). However, NADH dependent glutamate synthase (EC.1.4.1.14) has now been clearly demonstrated in the plant fraction of nodules (Robertson *et al.,* 1975, Boland and Benny, 1977) and its activity increases about 10 fold during nodulation of both lupin (Robertson *et al.,* 1975) and soybeans (Reynolds *et al.,* 1982 a). No such clear developmental changes however were observed by Sen and Schulman (1980). The activity of NADH-glutamate synthase has consistently been found to be about 10-fold lower than GS (for example see Boland *et al.,* 1978) but perhaps significantly the ratio of GS: glutamate synthase activity has been found to be higher in ureide transporting legumes which use glutamine amide-N directly for ureide synthesis than in those that transport asparagine (Boland *et al.,* 1978, Groat and Vance, 1981).

Ferredoxin-dependent-glutamate synthase (EC.1.4.7.1) although present in pea roots (Miflin and Lea, 1975) has not been demonstrated in root nodules of legumes.

Wide variations in the activity of GDH have been reported in the plant cytosol fraction of legume root nodules (Dunn and Klucas, 1973, Ryan and Fottrell, 1974, Brown and Dilworth, 1975, Streeter, 1977, Planqué *et al.,* 1977, Boland *et al.,* 1978, Groat and Vance, 1981). Some reports indicate that its activity is much lower than GS, and others that these two enzymes may attain almost equivalent activities. These discrepancies however are undoubtedly due to erroneous enzyme activity measurements, either by simultaneously measuring bacteroid enzymes to give erroneous high values or by failing in the assay conditions to attend to the Ca^{2+} dependence of higher plant GDH (see Stewart *et al.,* 1980) and hence underestimating its activity.

Changes in the activity of this enzyme do not appear to be correlated with nitrogen fixation. During nodule development in soybeans GDH activity declined markedly at a time when nitrogen fixation was increasing (Sen and Schulman, 1980) and in alfalfa, treatments which caused changes in the rate of nitrogen fixation were closely paralleled by changes in GS and glutamate synthase but no correlation was found with GDH (Groat and Vance, 1981). Moreover GDH has more than a 10-fold higher K_m for ammonia than GS and would therefore be inefficient at competing for this substrate in the presence of high activities of GS.

Taken together the labelling and enzyme studies provide convincing evidence that the glutamate synthase cycle provides the primary, and perhaps, in some legume nodules, the sole pathway of ammonia assimilation.

c) Enzyme Properties

Glutamine Synthetase

GS (EC.6.3.1.2) has been purified to apparent homogeneity from root nodules of soybean (McParland *et al.*, 1976), lupin (McCormack *et al.*, 1982) and *Phaseolus* (Cullimore *et al.*, 1983). The enzyme, like that from most higher plant tissues (see Stewart *et al.*, 1980), has a relative molecular mass (M_r) for the holoenzyme of about 380,000 with sub-units of 41,000—48,000. Electron microscopy studies have shown that there are eight sub-units arranged in two parallel sets of planar tetramers (McParland *et al.*, 1976). In alfalfa nodules GS has been shown to consist of two different sub-units with M_r of 40,000 and 45,000 (Groat and Schrader, 1983). Two different subunits of GS were also apparent in *Phaseolus* nodules; they had almost identical mobilities on denaturing polyacrylamide gels but could be separated by isoelectric focusing (Lara *et al.*, 1984).

Leaves of many legume and other species contain both chloroplastic and cytosolic forms of GS which can be separated by ion-exchange chromatography (for reviews see Miflin *et al.*, 1981, McNally *et al.*, 1983). In nodules of *Phaseolus* two forms of GS have also been shown to be present and these may differ from both leaf forms of the enzyme in this species (Cullimore *et al.*, 1983). The two nodule forms (GS_{n1} and GS_{n2}) could be separated by ion-exchange chromatography and by native polyacrylamide gel electrophoresis and had different transferase: synthetase activity ratios. GS_{n2} had many properties common to the root enzyme (GS_r) (Cullimore *et al.*, 1983, Cullimore and Miflin, 1983a) but made up only about 12% of the activity in nodules; GS_{n1} (the nodule specific form) accounted for over 85% of total nodule GS activity (Lara *et al.*, 1983). Recent evidence indicates that root GS is composed of two polypeptides (α and β) and that the production of GS_{n1} in nodules is associated with the synthesis of a nodule-specific polypeptide (γ and an increase in β (Lara *et al.*, 1984).

Antisera raised to GS_{n1} of *Phaseolus* could not distinguish between GS_{n2} and GS_r but showed that GS_{n1} was antigenically partially different (Cullimore and Miflin, 1983a). The antiserum also recognized the two leaf forms of *Phaseolus* GS and cross-reacted with the enzymes from other higher plant tissues but not with GS from lower plants, bacteria (including the two forms of *R. phaseoli* GS) or mammals. Antiserum raised to alfalfa nodule GS recognized both subunit types in this species (Groat and Schrader, 1983).

GS has a high affinity for ammonium with a K_m or $S_{0.5}$ value of 0.02—0.6 mM (Boland *et al.*, 1978, McCormack *et al.*, 1982, Cullimore *et al.*, 1983). Ammonia rather than ammonium ions may be the true substrate and the reaction mechanism appears to be sequential with Mg-ATP binding before glutamate and ammonia (McCormack *et al.*, 1982). The pH optimum of the synthetase activity is 7.5—8.0. *In vitro* the enzyme is inhibited by a number of amino acids, notably aspartate, at concentrations which may be physiological (McParland *et al.*, 1976, McCormack *et al.*, 1982). The enzyme activity is also affected by adenine nucleotides, possibly through energy charge control (McParland *et al.*, 1976).

GS activity is freely soluble following extraction of nodules and is generally thought to be cytosolic (Planqué *et al.*, 1977). Recently more stringent conditions for separating sub-cellular organelles from nodules have been applied and indeed in *Phaseolus* very little GS activity was associated with the organelles, almost all of it being in the plant cytosol supernatant (Awonaike *et al.*, 1981). This result has been supported by work on soybeans (Boland *et al.*, 1982, Boland and Schubert, 1983a) and cowpea (Shelp *et al.*, 1983) although in these studies a considerably higher proportion of the plastids and mitochondria were broken.

In all these studies the plant GS enzymes were clearly distinguishable by both their physical and kinetic properties (see also section B. II.) and antigenically (Cullimore and Miflin, 1983a) from the two forms of *Rhizobial* GS. A recent report however has suggested that GS II of *R. leguminosarum* and pea nodule GS show similar physical and antigenic properties (De Vries *et al.*, 1983). The antisera used in these experiments did not appear to be monospecific for GS.

Glutamate Synthase

NADH-specific glutamate synthase (EC.1.4.1.14) was detected in the plant fraction of lupin nodules by Robertson *et al.* (1975b) and was later purified from this tissue (Boland and Benny, 1977). The enzyme consists of a single polypeptide chain of M_r 235,000 and appears to contain two flavin prosthetic groups per molecule. The pH optimum is 8.5 and the enzyme has high affinities for its substrates (K_m values for 2-oxoglutarate, glutamine and NADH were 39 μM, 400 μM and 1.3 μM respectively). The reaction mechanism has been investigated (Boland, 1979) and NADH appears to be the compulsory first substrate to bind. Studies using analogues of 2-oxoglutarate (Boland and Court, 1981) and NAD^+ (Boland, 1981a) have shown that other 2-oxo compounds can bind to the active site but not be used as a substrate and that 3-acetylpyridine-adenine dinucleotide (reduced form) can act as an alternative, competitive, reductant to NADH.

Recent work has shown that glutamate synthase activity is probably entirely located within the plastids of nodules (Awonaike *et al.*, 1981, Boland *et al.*, 1982) and this contrasts with earlier studies, ignoring these organelles, which suggested either a mitochondrial or cytosolic localization for this enzyme (Ratajczak *et al.*, 1979).

Glutamate Dehydrogenase

GDH (EC.1.4.1.2) purified from lupin nodules has four charge isomers of M_r 270,000 with sub-units of M_r 45,000. A hexameric structure has thus been suggested (Stone *et al.*, 1979). A kinetic analysis of both the deaminating (Stone *et al.*, 1980a) and aminating (Stone *et al.*, 1980b) directions have shown that the reaction equilibrium favours reductive amination unless the

$NAD^+/NADH$ ratio is exceptionally high and the level of 2-oxoglutarate is low, in which situation deamination could become important. Some evidence however for a catabolic role for this enzyme was obtained in alfalfa. In most plant tissues NADH-GDH exceeds NAD-GDH activity by a factor of 10 or more (Givan, 1979) but in alfalfa nodules this ratio was found to be only 1.0 which suggested that, in this system, GDH could be functioning in a deaminating direction (Groat and Vance, 1981).

In the aminating direction the enzyme has a K_m for ammonia of 60—90 mM (Stone *et al.*, 1979) which is in agreement with estimates from impure extracts of a large number of legume species (Boland *et al.*, 1978). The enzyme has a pH optimum of 8.2 and a marked preference for NADH rather than NADPH as electron donor (Stone *et al.*, 1979). Antiserum was raised to purified GDH from lupin nodules (Tchan *et al.*, 1981). This antiserum recognized the enzyme but proved not to be monospecific and has not been used in further studies.

GDH appears to be localized within the mitochondria of the plant fraction of the nodule (Ratajczak *et al.*, 1979, Awonaike *et al.*, 1981) although in these studies the greater proportion of GDH activity was associated with the bacteroids (see also Streeter, 1977). Other reports have suggested however that the plant activity is higher (Dunn and Klucas, 1973, Brown and Dilworth, 1975).

B. Synthesis of Nitrogenous Transport Compounds

It is apparent that many 'temperate' legumes, including *Pisum, Lupinus* and *Medicago* export nitrogen from the nodule mainly as asparagine (Pate *et al.*, 1969, Streeter, 1972, Scott *et al.*, 1976, Atkins *et al.*, 1978). Others, mainly the 'tropical' legumes, for example *Glycine, Phaseolus* and *Vigna,* appear to change from transporting amino acids when grown on nitrate and during the early stages of nodulation to the export of ureides (mainly allantoin and allantoic acid) under nitrogen-fixing conditions (Atkins *et al.*, 1978, Matsumoto *et al.*, 1978, Herridge *et al.*, 1978, Schubert, 1981). The pathways of asparagine and ureide synthesis are described and the cost of their synthesis is discussed.

a) Asparagine Synthesis

i) *Labelling Studies*

Asparagine was identified as a major product of $^{15}N_2$ fixation in serradella nodules; the labelling pattern suggested that it was an end-product of the nodule nitrogen assimilatory pathway following primary incorporation into glutamate and glutamine (Kennedy *et al.*, 1966, Kennedy, 1966a, b). In these experiments aspartate was also quickly labelled but only after glutamate, and it was concluded that this compound was an intermediate in the nitrogen metabolic pathway (Kennedy *et al.*, 1966). In *Vicia* labelling studies using $^{14}CO_2$ have also identified glutamate and aspartate as primary products of assimilation, with label later appearing in asparagine (Lawrie and Wheeler, 1975). In soybeans, which transport mainly ureides, poor labelling of asparagine occured following feeding of $^{15}N_2$ or $^{13}N_2$ to nodules (Ohyama and Kumazawa, 1978, 1980b, Meeks *et al.*, 1978a). ^{15}N-labelled glutamine however was rapidly incorporated into asparagine, and label from $^{15}NH_4^+$ was initially detected in glutamine-amido N followed by glutamate and asparagine-amido N (Fujihara and Yamaguchi, 1980). These results indicated the presence of active glutamine-dependent asparagine synthesis in these nodules.

ii) *Enzyme Activities and Properties*

The pathway of asparagine synthesis in higher plants is considered to involve the two enzymes aspartate aminotransferase (EC.2.6.1.1) and asparagine synthetase (EC.6.3.5.4) (for a review see Lea and Miflin, 1980). This

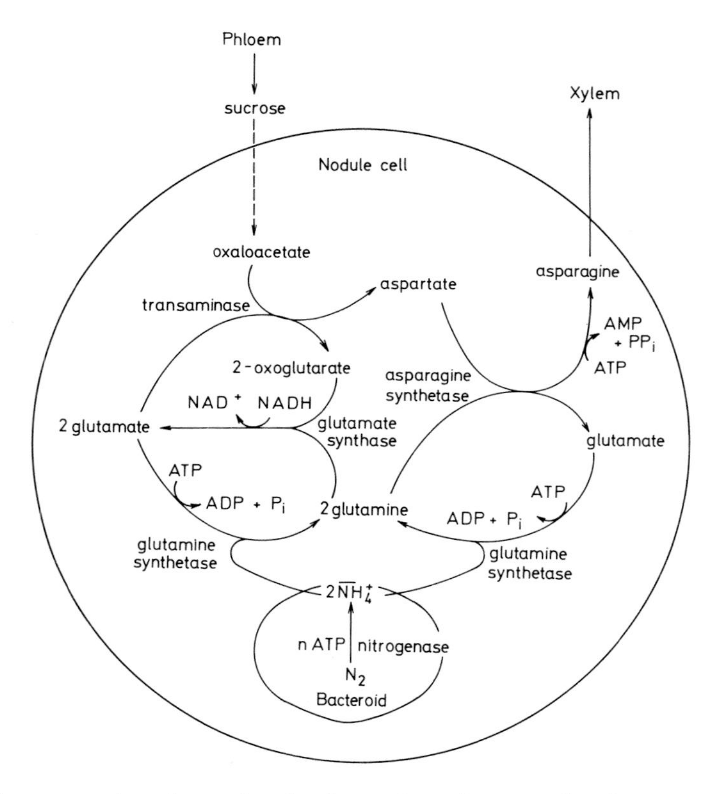

Fig. 8. A proposed pathway for the formation of asparagine from ammonia produced in dinitrogen fixation by lupin nodules. (Reproduced with permission from Scott *et al.*, 1976)

pathway is closely associated with the glutamate synthase cycle as glutamate and glutamine are directly involved in the synthesis of aspartate and asparagine respectively (Fig. 8). The carbon skeleton for this pathway is provided by oxaloacetate which may be produced in the TCA cycle or by direct carboxylation of phosphoenolpyruvate by PEP carboxylase (see section III. B. c).

In lupin, the activities of both aspartate aminotransferase (Reynolds and Farnden, 1979) and asparagine synthetase (Scott *et al.*, 1976) increased in parallel with the activities of GS, glutamate synthase and nitrogen fixation. Both enzyme activities also increased in soybeans during the initial stages of nodulation but as ureide biosynthesis proceeded their activities declined (Werner *et al.*, 1980, Reynolds *et al.*, 1982a).

In a comparison of several legume species a 7—10 fold higher activity of asparagine synthetase was found in amide than in ureide transporting legumes (Reynolds *et al.*, 1982b). The activity of this enzyme however, even in lupin, was lower than the nitrogen fixation rate but this was probably due to an underestimation of activity in view of its notorious instability (Scott *et al.*, 1976; see also Lea and Miflin, 1980).

Two forms of aspartate aminotransferase were found in the plant cytosol of nodules of lupin (Reynolds and Farnden, 1979) and soybean (Ryan *et al.*, 1972). In lupin the increase in activity during nodulation was associated with the increase in one form (AAT-P_2) whereas the activity of AAT-P_1 remained constant (Reynolds and Farnden, 1979). These two forms could be separated by their different electrophoretic properties and the different M_rs of both the holoenzymes (105,000 and 96,000) and the subunits (47,000 and 45,000) (Reynolds *et al.*, 1981). AAT-P_2 had lower K_m values for oxaloacetate and glutamate than AAT-P_1 but they had similar affinities for aspartate and 2-oxoglutarate. Both forms were highly specific for their substrates and the reaction appeared to involve a ping-pong-bi-bi mechanism (Reynolds *et al.*, 1981). Their pH optima were broad, centering around pH 8.

The two forms of aspartate aminotransferase in soybean nodules were separated by starch gel electrophoresis but they had identical M_rs of about 100,000 (Ryan *et al.*, 1972). One of these forms appeared to be localized in the proplastids and the other in the cytosol (Boland *et al.*, 1982). Recently Hanks *et al.* (1983) have shown that the cytosolic form is predominantly found in uninfected cells whereas the plastid enzyme is present in both uninfected and infected cells.

b) Ureide Synthesis

i) *Labelling and Inhibitor Studies*

Studies of Matsumoto and colleagues clearly established that nodulation was a prerequisite for the accumulation of ureides within soybean plants (Matsumoto *et al.*, 1977a, 1977b, 1978). Furthermore in labelling studies where the nodulated root system of soybean was fed with $^{15}N_2$ for 24 hours, label became incorporated into allantoin and allantoic acid in the nodules (Matsumoto *et al.*, 1977c). The amount of ureides however was greater in the roots in these experiments, but because of the higher ^{15}N atom % excess of the nodule ureides it was suggested that these compounds were synthesized in the nodules and then rapidly transported to the roots. In short-term labelling studies (15 minutes) with $^{13}N_2$, allantoin was identified as one of the major labelled products of fixation, but only in attached and not detached nodules of soybean (Meeks *et al.*, 1978a, Schubert and Coker, 1981b). The pattern of labelling however was entirely different when ^{15}N was supplied either as ammonium (Fujihara and Yamaguchi, 1978a; 1981) or nitrate (compare Ohyama and Kumazawa, 1978, 1979). Very little incorporation was observed into ureides, the label remaining predominantly in the amino acids.

These experiments have shown therefore that ureides are only synthesized in large quantities in nodules and then only from recently fixed dinitrogen and not from exogenously supplied reduced inorganic nitrogenous compounds.

The pathway of ureide synthesis in animals and microorganisms has been clearly shown to involve the degradation of purine bases although an alternative pathway via the direct condensation of glyoxylate and urea had also been considered (see Reinbothe and Mothes, 1962). Labelling studies,

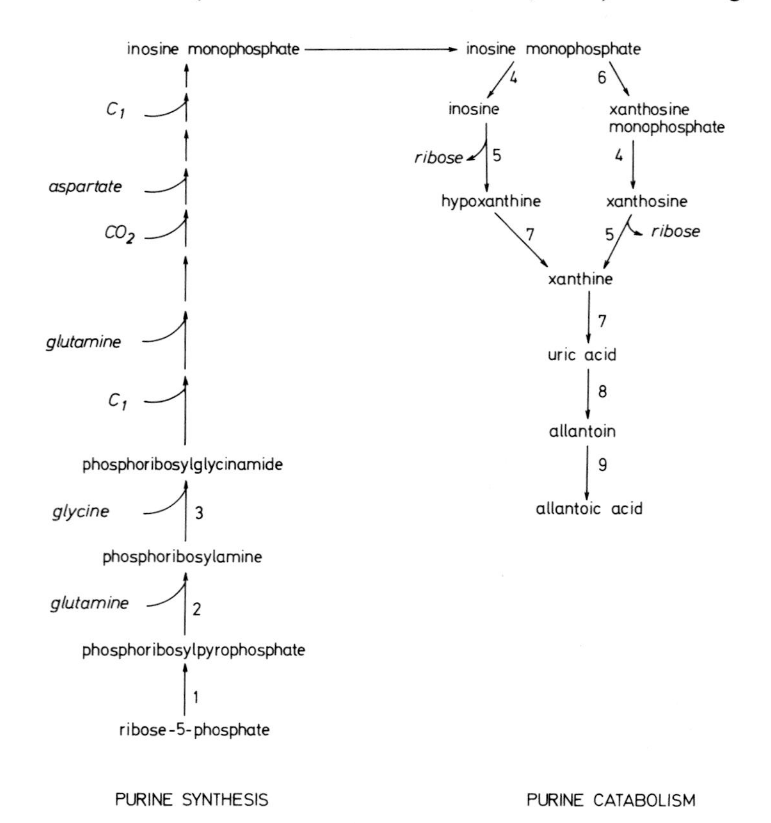

Fig. 9. A proposed pathway for ureide biosynthesis. The enzymes are (1) Phosphoribosylpyrophosphate synthetase, (2) Phosphoribosylamidotransferase, (3) Phosphoribosyglycinamide synthetase, (4) 5′ nucleotidase, (5) Purine nucleosidase, (6) Inosine monophosphate dehydrogenase, (7) Xanthine dehydrogenase, (8) Uricase, (9) Allantoinase

by themselves, have failed to provide convincing evidence of the pathway of allantoin synthesis in nodules although Matsumoto *et al.* (1977c) have suggested, from the observation that allantoin is more highly labelled than allantoic acid, that the pathway cannot involve the condensation reaction as allantoic acid should be labelled first.

However studies using inhibitors of the pathway and feeding unlabelled and ^{14}C labelled precursors have provided much positive evidence for the involvement in allantoin synthesis of the purine synthesis and degradation pathway shown in Fig. 9. This evidence is itemized below and is followed by a consideration of the enzymes involved.

1. $^{14}CO_2$ when fed to soybean nodules is incorporated into xanthine and uric acid (Boland and Schubert, 1982a).
2. ^{14}C-labelled glycine, a precursor of purine synthesis providing the N_7 of the purine ring, is incorporated preferentially into allantoin in nodule slices of cowpea (Atkins *et al.*, 1980).
3. Allantoic acid synthesis is strongly inhibited by azaserine, an inhibitor of glutamine transamidation reactions and hence *de novo* purine synthesis (Fujihara and Yamaguchi, 1978b, 1981).
4. Allopurinol, an inhibitor of xanthine dehydrogenase (XDH), caused an accumulation of xanthine in soybean nodules (Fujihara and Yamaguchi, 1978b, Boland and Schubert, 1982a) implicating the role of this enzyme and the purine degradative pathway in allantoin synthesis.
5. When allopurinol was simultaneously applied to nodules fed with $^{14}CO_2$ the xanthine that accumulated had the same ^{14}C specific activity as the xanthine in control plants (Boland and Schubert, 1982a). This indicated that the accumulated xanthine was the product of *de novo* purine biosynthesis and did not result from the breakdown of large pools of (unlabelled) purine containing compounds.
6. In $^{15}N_2$ labelling studies very little label was incorporated into RNA and DNA although substantial label was incorporated into allantoin (Matsumoto *et al.*, 1977c). This result suggested that the purine bases used for allantoin synthesis must form a separate pool to those required for normal nucleic acid synthesis.
7. Xanthine, hypoxanthine, inosine monophosphate (IMP) and xanthosine monophosphate (XMP) could all be used to synthesize allantoin and allantoic acid in cell free extracts of nodules (Woo *et al.*, 1980, 1981, Triplett *et al.*, 1980, Atkins, 1981) confirming the presence of the purine degradative pathway in nodules.

ii) *Enzyme Studies*

Purine synthesis may be considered to require the precursors, ribose, glutamine, glycine, aspartate and methylene-tetrahydrofolate and the end product is usually accepted to be IMP. The ureides are then produced by specific catabolism of IMP in a pathway involving as intermediates, xanthine and uric acid (Fig. 9). There are therefore a large number of enzymes involved; not only those directly involved in purine catabolism but also those responsible for purine synthesis and indeed for the synthesis of the precursors. We have already considered the enzymes concerned with glutamine and aspartate synthesis. Glycine can be synthesized by a number of alternative pathways in higher plants (see Keys, 1980) but in nodules it has been suggested to be produced by the phosphoglycerate rather than glycerate pathway and ends in the cleavage of serine. This reaction in addition to producing glycine also produces tetrahydrofolate which can then be oxidized to form another intermediate of purine biosynthesis, methylenetetrahydrofolate (Boland and Schubert, 1982b). Phosphoglycerate dehydrogenase (EC.1.1.1.95), serine hydroxymethyl transferase (EC.2.1.2.1) and methylenetetrahydrofolate dehydrogenase (EC.1.5.1.5), three enzymes involved in these reactions, have all been detected in soybean nodules at activities sufficient for their proposed function (Reynolds *et al.*, 1982a, Boland and Schubert, 1982b, 1983b). Their activities moreover appear to increase several fold during nodulation (Reynolds *et al.*, 1982a).

The initial two enzymes of the conventional purine biosynthetic pathway, phosphoribosylpyrophosphate (PRPP) synthetase (EC.2.7.6.1) and phosphoribosylamido transferase (PRAT) (EC.2.4.2.14) also increase in activity during nodulation of soybeans (Schubert, 1981, Reynolds *et al.*, 1982a). The third enzyme of the pathway, phosphoribosylglycinamide synthetase (EC.6.3.4.13), although not yet measured, is implicated to be present because ^{14}C-labelled glycine is rapidly incorporated into allantoin in nodule slices (Atkins *et al.*, 1980). Levels of PRPP synthetase, PRAT and phosphoglycerate dehydrogenase were much lower in lupin nodules which do not synthesize ureides (Reynolds *et al.*, 1982b). Other enzymes involved in IMP formation have not yet been measured.

The synthesis of xanthine from IMP is, in the conventional pathway, catalyzed by 5′ nucleotidase (EC.3.1.3.5) and purine nucleosidase (EC.3.2.2.1) to form inosine and hypoxanthine respectively and then oxidation of hypoxanthine by XDH. These three enzymes have been detected in soybean nodules; 5′ nucleotidase activity being 50 fold higher than in pea nodules (Christensen and Jochimsen, 1983). Purine nucleosidase results in the cleavage of the ribose moiety which may be recirculated for another round of *de novo* purine synthesis after being phosphorylated by ribokinase, also shown to be present in soybean nodules (Christensen and Jochimsen, 1983). In this pathway allopurinol would be expected to cause an accumulation of hypoxanthine but in experiments using this inhibitor of XDH only xanthine accumulated (Fujihara and Yamaguchi, 1978b, Atkins *et al.*, 1980, Boland and Schubert, 1982a). This result led Boland and Schubert (1982a) to suggest that hypoxanthine is not an intermediate in this pathway and that IMP is first converted to XMP by IMP dehydrogenase (EC.1.2.1.14) before xanthine formation. The presence of this enzyme has been implicated in soybean nodules by the observation that nodule plastids incubated with labelled IMP produce XMP in the presence of NAD^+ or $NADP^+$ (Boland and Schubert, 1983a). The exact pathway of IMP degradation to form xanthine has therefore not yet been elucidated and the two alternatives are shown in Fig. 9.

The final three enzymes in the pathway; XDH, uricase and allantoinase which catalyze the production of allantoic acid from xanthine via uric acid and allantoin, have been measured in a number of species (Tajima and Yamamoto, 1975, Herridge *et al.*, 1978, Atkins *et al.*, 1980, Triplett *et al.*, 1980, Schubert, 1981, Reynolds *et al.*, 1982a, Christensen and Jochimsen, 1983). Their activities increase substantially during nodule development in both soybean and cowpea (Schubert, 1981, Atkins *et al.*, 1980, Reynolds *et al.*, 1982a). XDH and uricase (see also Verma and Nadler, Chapter 3) are not generally found in root tissue and are considered as nodule specific.

Recently a scheme for the localization of ureide synthesis has been proposed (Boland *et al.*, 1982, Shelp *et al.*, 1983) based on enzyme localization studies (Hanks *et al.*, 1981, Boland *et al.*, 1982, Shelp *et al.*, 1983) and studies of feeding precursors and intermediates of the pathway to plastids (Boland and Schubert, 1983a). Ammonia assimilation by GS and the synthesis of PRPP by PRPP synthetase occur in the cell cytosol but then all the reactions of purine synthesis leading to the production of purine nucleotides are considered to take place in the plastids. It is speculated that the purines are catabolized in the cytosol so that the ribose moiety so released can be recycled and reused by PRPP synthetase. The next enzyme, XDH, has been shown to be cytosolic (Hanks *et al.*, 1981, Shelp *et al.*, 1983) and uricase, as in many organisms, is probably located in the peroxisomes where the hydrogen peroxide released by its activity can be degraded by catalase (Hanks *et al.*, 1981). There is some evidence to suggest that the final enzyme allantoinase is associated with the endoplasmic reticulum (Hanks *et al.*, 1981).

Ultrastructure studies of soybean nodules have shown that plastids are grouped at the periphery of infected cells and that these cells are surrounded by uninfected cells which contain many peroxisomes and a proliferation of smooth endoplasmic reticulum (Newcomb and Tandon, 1981). These authors have thus suggested that uninfected cells may carry out the final steps in ureide production. Recently Hanks *et al.* (1983) have separated protoplasts of infected and uninfected cells on sucrose density gradients and have shown that the peroxisomal enzymes uricase and catalase and also allantoinase and a number of enzymes involed in the synthesis of purine precursors are present at much higher specific activities in the uninfected cells. It is therefore possible that the whole of the purine synthesis and catabolism pathway occur predominately in the uninfected cells of the nodule using amino acids transported from the nitrogen fixing cells. This proposal has largely been supported by Shelp *et al.* (1983) although these authors envisage the involvement of both cell types in ureide production.

iii) *Enzyme Properties*

Xanthine dehydrogenase XDH (EC.1.2.1.37) has been purified from the plant cytosol of nodules of soybean (Triplett *et al.*, 1982) and navy bean (Boland, 1981b). The enzyme is a molybdoironflavoprotein of M_r 285,000, composed of two identical subunits of M_r about 141,000 (Triplett *et al.*, 1981). It is unusual for xanthine oxidizing enzymes in containing FMN rather than FAD as the flavin cofactor (Triplett *et al.*, 1982). The enzyme is able to oxidize both hypoxanthine and xanthine and both reactions are inhibited by allopurinol (Boland, 1981b). Studies on the kinetics of the enzyme have shown that the reaction involves a ping-pong mechanism and that a histidyl group on the protein may be essential in the catalysis (Boland *et al.,* 1983).

Uricase. Uricase (EC.1.7.3.3) was partially purified from soybean nodules and some kinetic properties were examined (Tajima and Yamamoto, 1975). The enzyme has a K_m for urate of about 8 μM and its activity is inhibited by xanthine and hypoxanthine. Uricase activity also requires molecular oxygen (K_m of about 29 μM) and its activity is severely restricted in oxygen limited conditions (Woo *et al.,* 1981). The enzyme from both soybean (Tajima and Yamamoto, 1975) and cowpea (Woo *et al.,* 1981) nodules has a pH optimum of about 9.5; activity increases about 7 fold as the pH is raised from 7.0—10.0. Recent evidence indicates that uricase is localized in the peroxisomes (Hanks *et al.,* 1981), of the uninfected cells of the nodules (Bergmann *et al.,* 1983, Hanks *et al.,* 1983).

Allantoinase. Allantoinase (EC.3.5.2.5) has been detected in leaves and fruits of cowpeas as well as in nodules of both cowpeas and soybeans (Tajima and Yamamoto, 1975, Herridge *et al.,* 1978, Atkins *et al.,* 1980, Woo *et al.,* 1981, Schubert, 1981). In crude extracts of soybean nodules, Tajima and Yamamoto (1975) found that the enzyme had a pH optimum of 5.0 but this has been questioned by the results of Woo *et al.* (1981) who found an optimum of 7—8. There are no reports as yet on the properties of the purified nodule enzyme. However the enzyme has been extensively studies in a number of plant and animal tissues including legume seeds (Lee and Roush, 1964, Vogels *et al.,* 1966, Vogels and Van der Drift, 1966) and more recently from shoot tissue of soybeans (Thomas *et al.,* 1983).

c) Costs and Efficiency of Transport Compound Synthesis

Dinitrogen fixation and subsequent synthesis of the nitrogenous transport compounds are energy consuming processes the cost of which is borne by the host plant (for a recent review see Schubert, 1982). Current estimates suggest that legumes deriving all their nitrogen from symbiotic nitrogen fixation spend 30% of their photosynthate on the nodule functions (Atkins *et al.,* 1978, Layzell *et al.,* 1979). Of this about 50% is lost through respiration as carbon dioxide.

By far the largest component in the fixation budget is the activity of nitrogenase (see Section B. I.). Ammonia assimilation within the plant requires additional energy, 1 ATP for GS activity and also NADH as cofactor for glutamate synthase (a total of 4 ATP equivalents). Many reports have compared the efficiency of asparagine and ureide synthesis (Atkins *et al.,* 1978, Herridge *et al.,* 1978, Layzell *et al.,* 1979, Pate *et al.,* 1980, Triplett *et al.,* 1980, Lea *et al.,* 1982). In terms of carbon conserved within the nodule, allantoin and allantoic acid are considered to be more efficient as transport compounds as their C : N ratio is 1 : 1 compared to 2 : 1 for asparagine. However in terms of energy required for synthesis, both compounds require approximately the same amount of ATP per molecule of nitrogen assimilated. Recent estimates based on a theoretical consideration of the two pathways suggest that asparagine synthesis requires a total of 7 ATP or 3.5 ATP/N whereas ureide synthesis requires 13 ATP or 3.25 ATP/N incorporated (Lea *et al.,* 1982). Triplett *et al.* (1980) have pointed out that the activity of XDH which produces 1 NADH for each xanthine molecule oxidized may be an important reaction in increasing the efficiency of ureide production. Moreover if IMP dehydrogenase is involved then these two exergonic reactions would reduce the cost of ureide synthesis to 2 ATP or 0.5 ATP/N assimilated (Schubert, 1981). In a comparison of the photosynthate use of cowpea and lupin Layzell *et al.* (1979) experimentally determined that the ureide producer was indeed more efficient, requiring 3.1 mg C/mg N fixed whereas the value for the asparagine producer was 6.6 mg C/mg N. The authors were unable to pinpoint the source of the difference in terms of the transport compound pathway (see Atkins *et al.,* 1978, for a theoretical consideration of these pathways) but suggested that differences in the hydrogen uptake system of the bacteroids or in the carbon dioxide fixation ability of the host nodule may have made an impact on the carbon economy of the two systems.

It is now clearly established that nodules have an active carbon dioxide fixation mechanism (for reviews see Wheeler, 1978, Minchin *et al.,* 1981, Rawsthorne *et al.,* 1980) but only recently has the magnitude and effect of this system on nodule efficiency been appreciated (Coker and Schubert, 1981, Vance *et al.,* 1983). Labelling studies using $^{14}CO_2$ and $^{11}CO_2$ have shown that this activity largely involves the enzyme phosphoenolpyruvate (PEP) carboxylase (EC.4.1.1.31).

$$\xrightarrow{\textit{PEP carboxylase}} \quad PEP + HCO_3^- \rightarrow \text{oxaloacetate} + P_i$$

Label was incorporated predominantly in both malate and also in the amino acids aspartate and glutamate (Lawrie and Wheeler, 1975, Laing *et al.,* 1979, Cookson *et al.,* 1980, de Vries *et al.,* 1980, Coker and Schubert, 1981) which is consistent with the activity of this enzyme as these organic and amino acids could both be synthesized via oxaloacetate.

In asparagine transporting legumes a direct correlation was found between PEP carboxylase activity and dinitrogen fixation (Christeller *et al.,* 1977, Laing *et al.,* 1979, Vance *et al.,* 1983) but whether the major effect of this enzyme is in producing oxaloacetate for aspartate and asparagine synthesis as suggested by Christeller *et al.* (1977) or in generally improving the efficiency of nodule metabolism has not yet been established. Work on ureide producers suggests the latter, as the activity of PEP carboxylase was found to be higher in soybean than in lupin nodules although no direct correlation could be observed between changes in the carbon dioxide fixation rate in soybeans and the rate of dinitrogen fixation (Coker and Schubert, 1981). In alfalfa it has been estimated that PEP carboxylase activity could provide 25% of the carbon required for the assimilation of symbiotically fixed dinitrogen (Vance *et al.,* 1983). The proportion of this used by the bacteroids has yet to be determined.

C. Plant Genes Involved in Nodule Nitrogen Metabolism

The legume-*Rhizobium* symbiosis provides an interesting system to study at the level of gene expression since alterations in either genome may result in an ineffective symbiosis, where the nodules produced are unable to fix dinitrogen (Bergersen, 1957, Horcher *et al.,* 1980, Ma *et al.,* 1982, Noel *et al.,* 1982, Vance and Johnston, 1983). Studies on the host genes responsible for an effective symbiosis however has lagged far behind work on the *Rhizobium* genes, partly because of the complexity of the plant genome (see chapter 3). In this section we wish to identify some of the plant genes involved in nitrogen assimilation within the nodule and describe two approaches used to study their identity and regulation.

a) Studies on Genetically Determined Non-fixing Nodules

Recently a number of genotypes of alfalfa *(Medicago sativum)* have been discovered which produce non-fixing nodules with strains of *R. meliloti* that form fixing nodules on normal alfalfa cultivars (Peterson and Barnes, 1981, Vance and Johnston, 1983). The nodules produced by these genotypes had different morphologies and degrees of establishment and differentiation of the bacteroids. Studies on the levels of nitrogen assimilatory enzymes showed that the activities of both GS and glutamate synthase were much lower in nodules of these ineffective genotypes (Groat and Vance, 1982, Vance and Johnson, 1983). Applying nitrate or ammonium did not increase the level of these enzymes which suggested that the increase in GS and glutamate synthase during normal nodulation is not due to the increased supply of nitrogen coming from dinitrogen fixation (Groat and Vance, 1982). However these experiments do not exclude the possibility that high concentrations of ammonia applied at a localized area in the specially differentiated nodule cells could promote the production of these enzymes. In roots, the levels of GS and glutamate synthase were about the same in normal and ineffective genotypes suggesting that the effects produced were specifically on nitrogen metabolism in nodules.

GDH activity was not substantially altered in either roots or nodules of the ineffective genotypes which indicated that this enzyme is not controlled by the genes causing ineffectiveness (Groat and Vance, 1982). The level of PEP carboxylase was reduced and its activity correlated positively with both glutamate synthase activity and nitrogen fixation (Vance *et al.,* 1983). It appears therefore that host-conditioned ineffectiveness produces alterations in the activities of a number of key enzymes involved in the assimilation of fixed dinitrogen.

Activities of plant ammonia assimilatory enzymes have also been determined in nodules formed with non-fixing strains of *R. japonicum* (Duke and Ham, 1976, Sen and Schulman, 1980, Werner *et al.,* 1980) and *R. phaseoli* (Lara *et al.,* 1983). In soybean nodules formed with the incompatible *R. japonicum* strain 61-A-24 glutamate synthase could not be detected (Sen and Schulman, 1980) and plant GS activity was very low and did not increase during nodule development (Werner *et al.,* 1980). GDH activity was not affected. Surprisingly plant aspartate aminotransferase activity was higher than in the fixing nodules but no attempt was made to measure separately the activities of the two isoforms, present in this species. In the studies with *Phaseolus,* specific mutants of *R. phaseoli* were used in which the transposon Tn5 had been inserted into the symbiotic plasmid of this strain (Lamb *et al.,* 1982). Nodules produced with two of these mutants had only about 25% of GS activity of normal fixing nodules and this was due to the almost complete absence of the nodule specific form of the enzyme, GS_{n1} (Lara *et al.,* 1983). Trace amounts of GS_{n1} and also leghaemoglobin were detectable suggesting that these plant proteins and nitrogenase activity were not directly correlated.

These studies, by following the response of the host proteins in bacterially conditioned nodules unable to fix dinitrogen, may help to identify those enzymes involved in a successful symbiosis. However, there are a great number of events that occur during nodulation at which defects in the bacterial genome could cause the inability to fix dinitrogen. Therefore it is difficult from these studies to identify the stimulus responsible for eliciting the production of the host ammonia assimilatory enzymes. Perhaps better information could be obtained by using well defined mutants in which the point and exact effect of the mutation is clearly known; this may also involve mutants able to fix dinitrogen but deficient in transport or in some other aspect of metabolism.

b) Studies on Expression of Nitrogen Assimilatory Enzymes

The studies described in sections C. I and C. II have identified a number of enzymes involved in nitrogen assimilation whose activities are modified during nodulation (see Verma and Nadler, Table 2). A number of these enzymes are present in other parts of the plant but their activities increase substantially during nodulation e.g. glutamate synthase, phosphoribosylpyrophosphate synthetase and serine hydroxymethyltransferase. Others appear to be nodule specific and may be classified as nodulins (see Verma and Nadler, Chapter 3) e.g. GS_{n1} of *Phaseolus,* aspartate aminotransferase AAT-P_2, uricase and xanthine dehydrogenase. Very little however is known of the regulation of these enzymes at the gene level. To date only GS (Cullimore and Miflin, 1983b) and uricase (Bergmann *et al.,* 1983) have been studied. In *Phaseolus* nodules, *in vitro* translation and immunoprecipitation of the products with anti-GS-antiserum has shown that, at a time when GS_{n1} activity is actively increasing, the levels of GS mRNA are substantially greater than in roots (Cullimore and Miflin, 1983b). The production of this form of the enzyme during nodulation therefore appears to be due to changes occurring at the transcriptional or, at least, pre-translational level. Purification of the mRNA for GS by immunopurification of nodule polysomes (Cullimore and Miflin, 1983b) has allowed the identification of a cDNA clone coding for GS (J. V. Cullimore, unpublished) which may be used to further study the expression of the GS genes during nodulation. Undoubtedly the isolation and study of expression of the genes coding for these and other nitrogen assimilatory enzymes will further our understanding of the regulatory mechanisms involved in host gene expression in the nodule.

IV. Conclusions

Nitrogen fixation is overwhelmingly a symbiotic endeavour. It effectively depends on linking together the photosynthetic capacity of plants to the bacterial ability to fix dinitrogen. Although both functions occur in cyanobacteria the symbiotic state particularly has successfully managed to separate the O_2 producing reaction of photosynthesis from the O_2-inhibited dinitrogen reduction. In addition the plant provides the correct architectural environment for the provision of energy and the correct amount of O_2 to the bacterium. In return the bacterium supplies an abundant supply of N to its host. Of the two partners rhizobia appear more committed to the symbiotic state in that generally they cannot utilize, neither in the free-living nor the symbiotic state, the ammonia that they fix. In contrast legumes can utilize N from any available fixed nitrogen source. In evolutionary terms it may be asked why the plant has not yet acquired the ability to fix N itself. The answer may be that the evolutionary pressure has either not been strong enough or, alternatively, it has not been exerted for long enough. It is an attractive idea (see Verma and Nadler, Chapter 3) that we are recording here a stage in the evolution of a new plant organelle to join the chloroplast and the mitochondrion.

For symbiotic nitrogen fixation to work continuously the end product ammonia should not accumulate at the site of fixation but should be continuously removed. This is probably the evolutionary pressure that has led to the division of labour between the bacterium and the plant in which the latter is responsible for ammonia assimilation.

To achieve this requires careful orchestration between the two organisms. The bacterium has had to evolve the ability to repress its assimilatory mechanisms at the same time as derepressing nitrogenase. The plant, for its part, has to be able to assimilate all the ammonia it needs for growth in a few concentrated specialized areas (the nodule cytosol) rather than over large diffuse areas (e.g. the whole root system or even the whole plant) as occurs when exogenous nitrate or ammonia is the source of supply. We do not yet know how all this is achieved. Our knowledge of rhizobial genes and their expression is advancing rapidly but still has a long way to go before we can define the nitrogen regulatory system with precision and also identify the factors within the nodule environment that lead to nitrogenase being expressed. Our knowledge of the plant system is even more fragmentary — all that we can say is that the plant has the ability to boost its ammonia assimilation capacity within the nodule cytosol by several-fold. There is suggestive evidence that it does this by turning on the expression of a nodule-specific GS gene in *Phaseolus* and uricase gene in soybean. Again there is no clue as to the stimulus that triggers the proposed expression of these genes. However, the advances that are being achieved in defining the genes in rhizobia that are important in symbiosis may lead to genes being identified that are crucial in specifically eliciting given host plant responses. Production of cloned cDNA probes for GS and other enzymes involved in nitrogen assimilation in the host should aid investigations on the expression of host genes during nodule development. They would also enable the corresponding structural genes to be identified and enable questions to be asked regarding the control of their expression. Thus the techniques and systems are now becoming available with which to probe in detail this joint endeavour of nitrogen fixation and hopefully these will provide knowledge that can be used to manipulate the association to further agricultural productivity.

Acknowledgements

We wish to express our thanks to all those many authors who generously sent us preprints of manuscripts in press. Our own research reported here has been supported by the Agricultural Research Council of the U. K. and J. V. C. gratefully acknowledges the receipt of a Pickering Research Fellowship from the Royal Society. We are also indebted to Susan Wilson for her skill in producing the finished manuscript.

V. References

Ali, H., Niel, C., Guillaume, J. B., 1981: The pathways of ammonium assimilation in *Rhizobium meliloti*. Arch. Microbiol. **129**, 391—394.

Appleby, C. A., Turner, G. L., MacNichol, P. K., 1975: Involvement of oxyleghaemoglobin and cytochromome P450 in an efficient oxidative phosphorylation pathway which supports nitrogen fixation in *Rhizobium*. Biochim. Biophys. Acta **387**, 461—474.

Atkins, C. A., 1981: Metabolism of purine-nucleotides to form ureides in nitrogen-fixing nodules of cowpea (*Vigna unguiculata* L. Walp). FEBS Letts. **125**, 89—93.

Atkins, C. A., Herridge, D. F., Pate, J. S., 1978: The economy of carbon and nitrogen in nitrogen-fixing annual legumes. Experimental observations and theoretical considerations. In: Isotopes in Biological Dinitrogen Fixation. Proc. FAO/IAEA Meeting Vienna, pp. 211—242.

Atkins, C. A., Rainbird, R. M., Pate, J. S., 1980: Evidence for a purine pathway of ureide synthesis in N_2-fixing nodules of cowpea (*Vigna unguiculata* L. Walp.). Z. Pflanzenphysiol. **97**, 249—260.

Auger, S., Verma, D. P. S., 1981: Induction and expression of nodule specific host genes in effective and ineffective root nodules of soybean *Glycine max* cultivar Prize. Biochemistry **20**, 1300—1306.

Awonaike, K. O., Lea, P. J., Miflin, B. J., 1981: The location of the enzymes of ammonia assimilation in root nodules of *Phaseolus vulgaris* L. Plant Sci. Lett. **23**, 189—195.

Bancroft, S., Rhee, S. G., Neumann, C., Kustu, S., 1978: Mutations that alter the covalent modification of glutamine synthetase in *Salmonella typhimurium*. J. Bacteriol. **134**, 1046—1055.

Banfalvi, Z., Randhawa, G. S., Kondorosi, E., Kiss, A., Kondorosi, A., 1983: Construction and characterization of R-prime plasmids carrying symbiotic genes of *R. meliloti*. Mol. Gen. Genet. **189,** 129—135.

Banfalvi, Z., Sakanyan, V., Koncz, C., Kiss, A., Dusha, I., Kondorosi, A., 1981: Location of nodulation and nitrogen fixation genes on a high molecular weight plasmid of *R. meliloti*. Mol. Gen. Genet. **184,** 318—325.

Bergersen, F. J., 1957: The structure of ineffective root nodules of legumes; an unusual new type of ineffectiveness, and an appraisal of present knowledge. Aust. J. Biol. Sci. **10,** 233—242.

Bergersen, F. J., 1965: Ammonia — an early stable product of nitrogen fixation by soybean root nodules. Aust. J. Biol. Sci. **18,** 1—9.

Bergersen, F. J., 1971: Biochemistry of symbiotic nitrogen fixation in legumes. Annu. Rev. Plant Physiol. **22,** 121—140.

Bergersen, F. J., Gibson, A. H., 1978: Nitrogen fixation by *Rhizobium* spp. in laboratory culture media. In: Dobreiner, J., Burris, R. H., Holander, A. (eds.): Limitations and potentials for biological nitrogen fixation in the tropics, pp. 263—274, New York: Plenum Press.

Bergersen, F. J., Turner, G. L., 1967: Nitrogen fixation by the bacteroid fraction of breis of soybean root nodules. Biochim. Biophys. Acta **141,** 507—515.

Bergersen, F. J., Turner, G. L., 1978: Activity of nitrogenase and glutamine synthetase in relation to availability of oxygen in continuous cultures of a strain of cowpea *Rhizobium* sp. supplied with excess ammonium. Biochim. Biophys. Acta **538,** 406—416.

Bergmann, H., Preddie, E., Verma, D. P. S., 1983: Nodulin-35: a subunit of specific uricase (uricase II) induced and localized in the uninfected cells of soybean nodules. The EMBO J. **2,** 2333—2339.

Beringer, J. E., Brewin, N. J., Johnston, A. W. B., 1980: The genetic analysis of *Rhizobium* in relation to symbiotic nitrogen fixation. Heredity **45,** 161—186.

Bishop, P. E., Guevara, J. G., Engelke, J. A., Evans, H. J., 1976: On the relation between glutamine synthetase and nitrogenase activities in the symbiotic association between *Rhizobium japonicum* and *Glycine max*. Plant Physiol. **57,** 542—546.

Bishop, P. E., McParland, R. H., Evans, H. J., 1975: Inhibition of the adenylylation of glutamine synthetase by methionine sulfone during nitrogenase derepression. Biochem. Biophys. Res. Commun. **67,** 774—781.

Bisseling, T., van den Bos, R. C., Weststrate, M. W., Hakkaart, M. J. J., van Kammen, A., 1978: The effect of ammonium nitrate on the synthesis of nitrogenase and the concentration of leghemoglobin in pea root nodules induced by *Rhizobium leguminosarum*. Biochim. Biophys. Acta **539,** 1—11.

Bisseling, T., Been, C., Klugkist, J., van Kammen, A., Nadler, K., 1983: Nodule specific host proteins in effective and ineffective root nodules of *Pisum sativum*. The EMBO J. **2,** 961—966.

Boland, M. J., 1979: Kinetic mechanism of NADH-dependent glutamate synthase from lupin nodules. Eur. J. Biochem. **99,** 531—532.

Boland, M. J., 1981a: NADH-dependent glutamate synthase from lupin nodules. Reactions with oxidised and reduced 3-acetyl-pyridine-adenine dinucleotide. Eur. J. Biochem. **115,** 485—489.

Boland, M. J., 1981b: NAD^+: xanthine dehydrogenase from nodules of navy beans: partial purification and properties. Biochem. Int. **2,** 567—574.

Boland, M. J., Benny, A. G., 1977: Enzymes of nitrogen metabolism in legume nodules. Purification and properties of NADH-dependent glutamate synthase from lupin nodules. Eur. J. Biochem. **79,** 344—362.

Boland, M. J., Court, C. B., 1981: Glutamate synthase (NADH) from lupin nodules. Specificity of the 2-oxoglutarate site. Biochim. Biophys. Acta **657,** 539—542.

Boland, M. J., Schubert, K. R., 1982a: Purine biosynthesis and catabolism in soybean root nodules: incorporation of ^{14}C from $^{14}CO_2$ into xanthine. Arch. Biochem. Biophys. **213,** 486—491.

Boland, M. J., Schubert, K. R., 1982b: The biosynthesis of glycine and methenyl tetrahydrofolate, precursors of ureide synthesis in soybean nodules. Plant Physiol. **69,** 618.

Boland, M. J., Schubert, K. R. 1983a: Biosynthesis of purines by a proplastid fraction from soybean nodules. Arch. Biochem. Biophys. **220,** 179—187.

Boland, M. J., Schubert, K. R., 1983b: Phosphoglycerate dehydrogenase from soybean nodules. Partial purification and some kinetic properties. Plant Physiol. **71,** 658—661.

Boland, M. J., Fordyce, A. M., Greenwood, R. M., 1978: Enzymes of nitrogen metabolism in legume nodules: a comparative study. Aust. J. Plant Physiol. **5,** 553—559.

Boland, M. J., Hanks, J. F., Reynolds, P. H. S., Blevins, D. G., Tolbert, N. E., Schubert, K. R., 1982: Subcellular organization of ureide biogenesis from glycolytic intermediates and ammonium in nitrogen-fixing soybean nodules. Planta **155,** 45—51.

Boland, M. J., Blevins, D. G., Randall, D. D., 1983: Soybean xanthine dehydrogenase: a kinetic study. Arch. Biochem. Biophys. **222,** 435—441.

Brewin, N. J., Beringer, J. E., Buchanan-Wollaston, A. V., Johnston, A. W. B., Hirsch, P. R., 1980: Transfer of symbiotic genes with bacteriocinogenic plasmids in *Rhizobium leguminosarum.* J. Gen. Microbiol. **116,** 261—270.

Brill, W. J., 1980: Biochemical genetics of nitrogen fixation. Microbiological Reviews **44,** 449—467.

Broughton, W. J. (ed.), 1982: Nitrogen fixation, Vol. 1. Oxford, U.K.: Oxford University Press.

Brown, C. M., MacDonald, Brown, D. S., Meers, J. L., 1974: Physiological aspects of microbial inorganic nitrogen assimilation. Adv. in Microb. Physiol. **11,** 1—52.

Brown, C. M., Dilworth, M. J., 1975: Ammonium assimilation by *Rhizobium* cultures and bacteroids. J. Gen. Microbiol. **86,** 39—48.

Buchanan-Wollaston, A. V., Beringer, J. E., Brewin, N. J., Hirsch, P. R., Johnston, A. W. B., 1980: Isolation of symbiotically defective mutants in *Rhizobium leguminosarum* by insertion of the transposon Tn5 into a transmissible plasmid. Mol. Gen. Genet. **178,** 185—190.

Ching, T. M., Bergersen, F. J., Turner, G. L., 1981: Energy status, growth and nitrogenase activity in continuous cultures of *Rhizobium* sp. strain CB 756 supplied with NH_4^+ and various rates of aeration. Biochim. Biophys. Acta **636,** 82—90.

Christeller, J. T., Laing, W. A., Sutton, W. D., 1977: Carbon dioxide fixation by lupin root nodules. 1. Characterization, association with phosphoenolpyruvate carboxylase, and correlation with nitrogen fixation during nodule development. Plant Physiol. **60,** 47—50.

Christensen, T. M. I. E., Jochimsen, B. U., 1983: Enzymes of ureide synthesis in pea and soybean. Plant Physiol., **72,** 56—59.

Coker, III, G. T., Schubert, K. R., 1981: Carbon dioxide fixation in soybean roots and nodules. 1. Characterization and comparison with N_2 fixation and composition of xylem exudate during early nodule development. Plant Physiol. **67,** 691—696.

Cookson, C., Hughes, H., Coombes, J., 1980: Effects of combined nitrogen on anapleurotic carbon assimilation and bleeding sap composition in *Phaseolus vulgaris* L. Planta **148,** 338—345.

Cullimore, J. V., Miflin, B. J., 1983a: Immunological studies on glutamine synthetase using antisera raised to the two plant forms of the enzyme from *Phaseolus* root nodules. J. Exp. Bot., in press.

Cullimore, J. V., Miflin, B. J., 1983b: Glutamine synthetase from the plant fraction of *Phaseolus* root nodules: purification of the mRNA and *in vitro* synthesis of the enzyme. FEBS Letts. **158,** 107—112.

Cullimore, J. V., Lara, M., Lea, P. J., Miflin, B. J., 1983: Purification and properties of the two forms of glutamine synthetase from the plant fraction of *Phaseolus* root nodules. Planta **157,** 245—253.

Cullimore, J. V., Lea, P. J., Miflin, B. J., 1982: Multiple forms of glutamine synthetase in the plant fraction of *Phaseolus* root nodules. Israel J. Bot. **31,** 151—162.

Darrow, R. A., Knotts, R. R., 1977: Two forms of glutamine synthetase in free-living root nodule bacteria. Biochem. Biophys. Res. Commun. **78,** 554—559.

Darrow, R. A., Crist, D., Evans, W. R., Jones, D. L., Keister, D. L., Knotts, R. R., 1981: Biochemical and physiological studies on two glutamine synthetases. In: Gibson, A.H., Newton, W. E. (eds.): Current perspectives in nitrogen fixation, pp. 182—185. Aust. Acad. Science, Canberra.

Denarie, J., Boistard, P., Casse-Delbart, F., Atherly, A. G., Berry, J. O., Russell, P., 1981: Indigenous plasmids of *Rhizobium.* Internat. Rev. Cytol. Supplement **13,** p. 225—246.

De Vries, G. E., In, T. P., Kijne, J. W., 1980: Production of organic acids in *Pisum sativum* root nodules as a result of oxygen stress. Plant Sci. Lett. **20,** 115—123.

De Vries, G. E., Oosterwijk, E., Kijne, J. W., 1983: Antigenic cross reactivity between *Rhizobium leguminosarum* glutamine synthetase II and *Pisum sativum* root nodule glutamine synthetases. Plant Sci. Lett., in press.

Dilworth, M. J., Glenn, A. R., 1982: Movements of ammonia in *Rhizobium leguminosarum.* J. Gen. Microbiol. **128,** 29—37.

Dixon, R. A., Kennedy, C., Kondorosi, A., Krishnapillai, V., Merrick, M., 1977: Complementation analysis of *Klebsiella pneumoniae* mutants defective in nitrogen fixation. Mol. Gen. Genet. **157,** 189—198.

Dougall, D. K., 1974: Evidence for the presence of glutamate synthase in carrot cell cultures. Biochem. Biophys. Res. Commun. **58,** 639—646.

Dreyfus, B. L., Elmerich, C., Dommergues, Y. R., 1983: Free-living *Rhizobium* strain able to grow under N_2 as the sole nitrogen source. Applied Environ. Microbiology **45,** in press.

Dreyfus, B. L., Domergues, Y. R., 1981: Skin nodules on the tropical legume *Sesbania rostrata.* In: Gibson, A.H., Newton, W. E. (ed.): Current perspectives in nitrogen fixation. Australian Academy of Science, Canberra, p. 471.

Drummond, M., Clements, J., Merrick, M., Dixon, R., 1983: Positive control and autogenous regulation of the nif LA promoter in *Klebsiella pneumoniae.* Nature **301,** 302—306.

Duke, S. H., Ham, G. E., 1976: The effect of nitrogen addition on N_2-fixation and on glutamate dehydrogenase and glutamate synthase activities in nodules and roots of soybeans inoculated with various strains of *Rhizobium japonicum.* Plant and Cell Physiol. **17,** 1037—1044.

Dunn, S. D., Klucas, R. V., 1973: Studies on possible routes of ammonia assimilation in soybean root nodule bacteroids. Can. J. Microbiol. **19,** 1493—1499.

Elmerich, C., Sibold, L., Guerineau, M., Tanndeau de Marsac, N., Chocat, P., Gerbaud, C., Aubert, J.-P., 1981: The *nif* genes of *Klebsiella pneumoniae:* Characterization of a *nif* specific constitutive mutant and cloning in yeast. In: Gibson, A. H., Newton, W. E.: Current perspectives in nitrogen fixation, pp. 157—160, Australian Academy of Science, Canberra.

Espin. G., Alvarez-Morales, A., Cannon, F., Dixon, R., Merrick, M., 1982: Cloning of the glnA, ntrB and ntrC genes of *Klebsiella pneumoniae* and studies on their role in regulation of the nitrogen fixation (nif) gene cluster. Mol. Gen. Genet. **186,** 518—524.

Foor, F., Reuveny, Z., Magasanik, B., 1980: Regulation of the synthesis of glutamine synthetase by the P II protein in *Klebsiella aerogenes.* Proc. Natl. Acad. Sci. (U. S. A.) **77,** 2636—2640.

Fuchs, R. L., Keister, D. L., 1980a: Identification of two glutamine synthetases in *Agrobacterium.* J. Bacteriol. **141,** 996—998.

Fuchs, R. L., Keister, D. L., 1980b: Comparative properties of glutamine synthetases I and II in *Rhizobium* and *Agrobacterium* spp. J. Bacteriol. **144,** 641—648.

Fujihara, S., Yamaguchi, M., 1978a: Probable site of allantoin formation in nodulating soybean plants. Phytochemistry **17,** 1239—1243.

Fujihara, S., Yamaguchi, M., 1978b: Effects of allopurinol on the metabolism of allantoin in soybean plants. Plant Physiol. **62,** 134—138.

Fujihara, S., Yamaguchi, M., 1980: Asparagine formation in soybean nodules. Plant Physiol. **66,** 139—141.

Fujihara, S., Yamaguchi, M., 1981: Assimilation of $^{15}NH_3$ by root nodules detached from soybean plants. Plant and Cell Physiol. **22,** 797—806.

Gebhardt, C., Turner, G. L., Dreyfus, B. L., Gibson, A. H., Bergersen, F. J., 1983: Nitrogen-fixing growth of a strain of *Rhizobium* sp. in continuous culture J. Gen. Microbiol. (submitted).

Gibson, A. H., Scowcroft, W. R., Pagan, J. D., 1977: Nitrogen fixation in plants: an expanding horizon? In: Newman, W., Postgate, J. R., Rodriguez-Barrueco, C. (eds.): Recent developments in nitrogen fixation, pp. 387—417. London: Academic Press.

Givan, C. V., 1979: Metabolic detoxification of ammonia in tissues of higher plants. Phytochemistry **18,** 375—382.

Gober, J. W., Kashket, E. R., 1983: Methylammonium uptake by *Rhizobium* sp. *strain* 32H1. J. Bacteriol. **153,** 1196—1201.

Gordon, J. K., Brill, W. J., 1974: Derepression of nitrogenase synthesis in the presence of excess NH_4^+. Biochem. Biophys. Res. Commun. **59,** 967—971.

Gresshoff, P. M., Carroll, B., Mohaptra, S. S., Reporter, M., Shine, J., Rolfe, B. G., 1981: Host factor control of nitrogenase function. In: Gibson, A. H., Newton, W. E. (eds.): Current perspectives in nitrogen fixation, pp. 209—212, Australian Academy of Science, Canberra.

Groat, R. G., Vance, C. P., 1981: Root nodule enzymes of ammonia assimilation in alfalfa (*Medicago sativa* L.). Developmental patterns and response to applied nitrogen. Plant Physiol. **67,** 1198—1203.

Groat, R. G., Vance, C. P., 1982: Root and nodule enzymes of ammonia assimilation in two plant-conditioned symbiotically ineffective genotypes of alfalfa (*Medicago sativa* L.). Plant Physiol. **69,** 614—618.

Groat, R. G., Schrader, L. E., 1983: Isolation and immunochemical characterization of plant glutamine synthetase in alfalfa (*Medicago sativa* L.) nodules. Plant Physiol. **70,** 1759—1761.

Hanks, J. F., Tolbert, N. E., Schubert, K. R., 1981: Localization of enzymes of ureide biosynthesis in peroxisomes and microsomes of nodules. Plant Physiol. **68,** 65—69.

Hanks, J. F., Schubert, K., Tolbert, N. E., 1983: Isolation and characterization of infected and uninfected cells from soybean nodules. Plant Physiol. **71,** 869—873.

Herridge, D. F., Atkins, C. A., Pate, J. S., Rainbird, R. M., 1978: Allantoin and allantoic acid in the nitrogen economy of the cowpea *(Vigna unguiculata).* Plant Physiol. **62,** 495—498.

Horcher, R., Wilcockson, J., Werner, D., 1980: Screening for mutants of *Rhizobium japonicum* with defects in nitrogen fixing ability. Z. Naturforsch. **35c,** 729—732.

Kennedy, C., Cannon, F., Cannon, M., Dixon, R., Hill, S., Jensen, J., Kumar, S., McLean, P., Merrick, M., Robson, R., Postgate, J., 1981: Recent advances in the genetics and regulation of nitrogen fixation. In: Gibson, A. H., Newton, W. E. (eds.): Current Perspectives in Nitrogen Fixation, pp. 46—156, Australian Academy of Science, Canberra.

Kennedy, I. R., 1966a: Primary products of symbiotic nitrogen fixation I. Short-term exposures of serradella nodules to $^{15}N_2$. Biochim. Biophys. Acta **130,** 285—294.

Kennedy, I. R., 1966b: Primary products of symbiotic nitrogen fixation II. Pulse-labelling of serradella nodules with $^{15}N_2$. Biochim. Biophys. Acta **130,** 295—303.

Kennedy, I. R., Parker, C. A., Kidby, D. K., 1966: The probable site of nitrogen fixation in root nodules of *Ornithopus sativum.* Biochim. Biophys. Acta **130,** 517—519.

Keys, A. J., 1980: Synthesis and interconversion of glycine and serine. In: Miflin, B. J. (ed.): The biochemistry of plants, Vol. 5, pp. 359—374, New York: Academic Press.

Kleiner, D., 1982: Ammonium (methylammonium) transport by *Klebsiella pneumoniae.* Biochim. Biophys. Acta **688,** 702—708.

Kleiner, D., Phillips, S., Fitzke, E., 1981: Pathways and regulatory aspects of N_2 and NH_4^+ assimilation in N_2-fixing bacteria. In: Bothe, H., Trebst, A.: Biology of inorganic nitrogen and sulphur, pp. 131—140, Berlin - Heidelberg - New York: Springer.

Kondorosi, A., Svab, Z., Kiss, G. B., Dixon, R. A., 1977: Ammonium assimilation and nitrogen fixation in *Rhizobium meliloti.* Mol. Gen. Genet. **151,** 221—226.

Krol, A. J. M., Hontelez, J. G., van den Bos, J. G., van Kammen, A., 1980: Expression of large plasmids in the endo-symbiotic form of *Rhizobium leguminosarum.* Nucl. Acid Res. **8,** 4337—4347.

Kurz, W. G. W., LaRue, T. A., 1975: Nitrogenase activity in rhizobia in absence of plant host. Nature **256,** 407—409.

Kustu, S., Burton, D., Garcia, E., McCarter, L., McFarland, N., 1979: Nitrogen control in *Salmonella:* regulation by the *glnR* and *glnF* gene products. Proc. Natl. Acad. Sci (U. S. A.) **76,** 4576—4580.

Laane, C., Haaker, H., Veeger, C., 1978: Involvement of the cytoplasmic membrane in nitrogen fixation by *Rhizobium leguminosarum* bacteroids. Eur. J. Biochem. **87,** 147—153.

Laane, C., Krone, W., Konings, W., Haaker, H., Veeger, C., 1980: Short-term effect of ammonium chloride on nitrogen fixation by *Azotobacter vinelandii* and by bacteroids of *Rhizobium leguminosarum*. Eur. J. Biochem. **103,** 39—46.

Laing, W. T., Christeller, J. T., Sutton, W. D., 1979: Carbon dioxide fixation by lupin nodules II. Studies with ^{14}C-labelled glucose, the pathway of glucose metabolism and the effect of some treatments that inhibit nitrogen fixation. Plant Physiol. **63,** 450—454.

Lamb, J. W., Hombrecher, G., Johnston, A. W. B., 1982: Plasmid-determined nodulation and nitrogen-fixing abilities in *Rhizobium phaseoli*. Molec. Gen. Genet. **186,** 449—452.

Lara, M., Cullimore, J. V., Lea, P. J., Miflin, B. J., Johnston, A. W. B., Lamb, J. W., 1983: Appearance of a novel form of plant glutamine synthetase during nodule development in *Phaseolus vulgaris* L. Planta **157,** 254—258.

Lara, M., Porta, H., Padilla, J., Folch, J., Sanchez, F., 1984: Heterogeneity of glutamine Synthase polypeptides in *Phaseolus vulgaris*. In: Advances in Nitrogen Fixation Research eds. c. Veeger W. E. Newton p. 601. The Hague, The Netherlands: Martinus Nijhoff.

Lawrie, A. C., Wheeler, C. T., 1975: Nitrogen fixation in the root nodules of *Vicia faba* L. in relation to the assimilation of carbon II. The dark fixation of carbondioxide. New Phytol. **74,** 437—445.

Layzell, D. B., Rainbird, R. M., Atkins, C. A., Pate, J. S., 1979: Economy of photosynthate use in nitrogen-fixing legume nodules. Observations on two contrasting systems. Plant Physiol. **64,** 888—891.

Lea, P. J., Miflin, B. J., 1974: An alternative route for nitrogen assimilation in higher plants. Nature (Lond.) **251,** 614—616.

Lea, P. J., Miflin, B. J., 1980: Transport and metabolism of asparagine and other nitrogen compounds within the plant. In: Miflin, B. J. (ed.): The biochemistry of plants, pp. 569—607, Vol. 5, New York: Academic Press.

Lea, P. J., Awonaike, K. O., Cullimore, J. V., Miflin, B. J., 1982: The role of ammonium assimilatory enzymes during nitrogen fixation in root nodules. Israel J. Bot. **31,** 140—154.

Lee, K. W., Roush, A. H., 1964: Allantoinase assays and their application to yeast and soybean allantoinases. Arch. Biochem. Biophys. **108,** 460—467.

Legocki, R. P., Verma, D. P. S., 1980: Identification of nodule specific host proteins (nodulins) involved in the development of *Rhizobium*-legume symbiosis. Cell **20,** 153—164.

Lim, S. T., Hennecke, H., Scott, D. B., 1979: Effect of cyclic guanosine 3′, 5′-monophosphate on nitrogen fixation in *Rhizobium japonicum*. J. Bacteriol. **139,** 256—263.

Ludwig R. A., 1978: Control of ammonium assimilation in *Rhizobium* 32H1. J. Bacteriol. **135,** 114—123.

Ludwig, R. A., 1980a: Physiological roles of glutamine synthetases I and II in ammonium assimilation in *Rhizobium* sp. 32H1. J. Bacteriol. **141,** 1209—1216.

Ludwig, R. A., 1980b: Regulation of *Rhizobium* nitrogen fixation by the unadenylylated glutamine synthetase I system. Proc. Natl. Acad. Sci. U. S. A. **77,** 5817—5821.

Ludwig, R. A., Signer, E. R., 1977: Glutamine synthetase and control of nitrogen fixation in *Rhizobium*. Nature **267,** 245—248.

Ludwig, R. A., de Vries, G. E., 1982: Free living N_2-fixation by *Rhizobium* sp. occurs in a cell-state divorced from that of growth. Abstracts 1st International Symposium on Molecular-Genetics of Bacteria-Plant Interaction, Bielefeld, FRG, p. 22.

Ma, Q.-S., Johnston, A. W. B., Hombrecher, G., Downie, J. A., 1982: Molecular genetics of mutants of *Rhizobium leguminosarum* which fail to fix nitrogen. Mol. Gen. Genet. **187,** 166—171.

McComb, J. A., Elliott, J., Dilworth, M. J., 1975: Acetylene reduction by *Rhizobium* in pure culture. Nature **256,** 409—410.

McCormack, D. K., Farnden, K. J. F., Boland, M. J., 1982: Purification and properties of glutamine synthetase from the plant cytosol fraction of lupin nodules. Arch. Biochem. Biophys. **218,** 561—571.

McFarland, N., McCarter, L., Artz, S., Kustu, S., 1981: Nitrogen regulatory locus *glnR* of entric bacteria is composed of cistrons *ntrB* and *ntrC:* Identification of their protein products. Proc. Natl. Acad. Sci. U. S. A **78,** 2135—2139.

McNally, S., Hirel, B., Gadal, P., Mann, A., F., Stewart, G. R., 1983: Glutamine synthetase of higher plants. Plant Physiol. **72,** 22—25.

MacNeil, D., Brill, W. J., 1980: Mutations in *nif* genes that cause *Klebsiella pneumoniae* to be derepressed for nitrogenase synthesis in the presence of ammonium. J. Bacteriol. **144,** 744—775.

McParland, R. H., Guevara, J. G., Becker, R. R., Evans, H. J., 1976: The purification of the glutamine synthetase from the cytosol of soya-bean root nodules. Biochem. J. **153,** 597—606.

Magasanik, B., 1982: Genetic control of nitrogen assimilation in bacteria. Annu. Rev. Genet. **16,** 135—168.

Matsumoto, T., Yatazawa, M., Yamamoto, Y., 1977a: Effects of exogenous nitrogen-compounds on the concentrations of allantoin and various constituents in several organs of soybean plants. Plant and Cell Physiol. **18,** 613—624.

Matsumoto, T., Yatazawa, M., Yamamoto, Y., 1977b: Distribution and changes in the contents of allantoin and allantoic acid in developing nodulating and non-nodulating soybean plants. Plant and Cell Physiol. **18,** 353—359.

Matsumoto, T., Yatazawa, M., Yamamoto, Y., 1977c: Incorporation of ^{15}N into allantoin in nodulated soybean plants supplied with $^{15}N_2$. Plant and Cell Physiol. **18,** 459—462.

Matsumoto, T., Yatazawa, M., Yamamoto, Y., 1978: Allantoin metabolism in soybean plants as influenced by grafts, a delayed inoculation with *Rhizobium,* and a late supply of nitrogen-compounds. Plant and Cell Physiol. **19,** 1161—1168.

Meeks, J. C., Wolk, C. P., Schilling, N., Shaffer, P. W., Avisar, Y., Chien, W.-S., 1978a: Initial organic products of fixation of [^{13}N] dinitrogen by root nodules of soybean *(Glycine max.).* Plant Physiol. **61,** 980—983.

Meeks, J. C., Wolk, C. P., Thomas, J., Lockau, W., Shaffer, P. W., Austin, S. M., Chien, W.-S., Galonsky, A., 1977b: Pathways of assimilation of $^{13}NH_4^+$ by the cyanobacterium *Anabaena cylindrica.* J. Biol. Chem. **252,** 7894—7900.

Merrick, M., 1982: A new model for nitrogen control. Nature **297,** 362—363.

Merrick, M. J., 1983: Nitrogen control of the *nif* regulon in *Klebsiella pneumonia:* involvement of the *ntrA* gene and analogies between *ntrC* and *nifA.* EMBO Journal **2,** 39—44.

Merrick, M., Hill, S., Hennecke, H., Hahn, M., Dixon, R., Kennedy, C., 1982: Repressor properties of the *nifL* gene product in *Klebsiella pneumoniae.* Mol. Gen. Genet. **185,** 75—81.

Miflin, B. J., Lea, P. J., 1975: Glutamine and asparagine as nitrogen donors for reductant-dependent glutamate synthesis in pea roots. Biochem. J. **149,** 403—407.

Miflin, B. J., Lea, P. J., 1976: The pathway of nitrogen assimilation in plants. Phytochemistry **15,** 873—885.

Miflin, B. J., Lea, P. J., 1980: Ammonia assimilation. In: Miflin, B. J. (ed.): The Biochemistry of Plants, pp. 169—202, New York: Academic Press.

Miflin, B. J., Wallsgrove, R. M., Lea, P. J., 1981: Glutamine metabolism in higher plants. Curr. Topics in Cell Regulat. **20,** 1—43.

Miflin, B. J., Lea, P. J., 1982: Ammonia assimilation and amino acid metabolism. In: Boulter, D., Parthier, B. (eds.): Encyclopedia of Plant Physiology Vol. 14A Nucleic Acids and Proteins in Plants I. pp. 5—64, Berlin - Heidelberg - New York: Springer.

Minchin, F. R., Summerfield, R. J., Hadley, P., Roberts, G. H., Rawsthorne, S., 1981: Carbon and nitrogen nutrition of nodulated roots of grain legumes. Plant Cell and Envir. **4,** 5—26.

Mortenson, L. E., 1962: Inorganic nitrogen assimilation and ammonia incorporation. In: Gunsalus, I. C., Stanier, R. Y. (eds.): The Bacteria, Vol. 3, pp. 119—166, New York: Academic Press.

Newcomb, E. H., Tandon, S. R., 1981: Uninfected cells of soybean root nodules: ultrastructure suggest key role in ureide production. Science **21,** 1394—1396.

Noel, K. D., Stacey, G., Tandon, S. R., Silver, L. E., Brill, W. J., 1982: *Rhizobium japonicum* mutants defective in symbiotic nitrogen fixation. J. Bacteriol. **152,** 485—494.

Nuti, M. P., Lepidi, A. A., Prakash, R. K., Schilperoort, R. A., Cannon, F. C., 1979: Evidence for nitrogen fixation genes on indigenous *Rhizobium* plasmids. Nature **282,** 533—535.

O'Gara, F., Shanmugan, K. T., 1976: Regulation of nitrogen fixation by *Rhizobia,* export of fixed N_2 as NH_4^+. Biochim. Biophys. Acta **437,** 313—321.

Ohyama, T., Kumazawa, K., 1978: Incorporation of ^{15}N into various nitrogenous compounds in intact soybean nodules after exposure to $^{15}N_2$ gas. Soil Sci. Plant Nutr. (Tokyo) **24,** 525—533.

Ohyama, T., Kumazawa, K., 1979: Assimilation and transport of nitrogenous compounds originated from $^{15}N_2$ fixation and $^{15}NO_3$ absorption. Soil Sci. Plant Nutr. **25,** 9—19.

Ohyama, T., Kumazawa, K., 1980a: Nitrogen assimilation in soybean nodules II. $^{15}N_2$ assimilation in bacteroid and cytosol fractions of soybean nodules. Soil Sci. Plant Nutr. **26,** 205—213.

Ohyama, T., Kumazawa, K., 1980b: Nitrogen assimilation in soybean nodules I. The role of GS/GOGAT system in the assimilation of ammonia produced by N_2 fixation. Soil Sci. Plant Nutr. **26,** 109—115.

Osburne, M. S., 1982: *Rhizobium meliloti* mutants altered in ammonium utilization. J. Bacteriol. **151,** 1633—1636.

Osburne, M. S., Signer, E. R., 1980: Ammonium assimilation in *Rhizobium meliloti.* J. Bacteriol. **143,** 1234—1240.

Ow, D. W., Ausubel, F. M., 1983: Regulation of nitrogen metabolism genes by *nifA* gene product in *Klebsiella pneumoniae.* Nature **301,** 397—313.

Ow, D. W., Sundaresan, V., Rothstein, D., Brown, S. E., Ausubel, F. M., 1983: Promoters regulated by the *glnG(ntrC)* and *nifA* gene products share a heptameric consensus sequence in the –15 region. Proc. Natl. Acad. Sci. U. S. A **80,** 2524—2528.

Pagan, J. D., Child, J. J. Scowcroft, W. R., Gibson, A. H., 1975: Nitrogen fixation by *Rhizobium* cultured on a defined medium. Nature **256,** 406—407.

Pahel, G., Rothstein, D. M., Magasanik, B., 1982: Complex glnA-glnL-glnG operon of *Escherichia coli.* J. Bacteriol. **150,** 202—213.

Pain, A. N., 1979: Symbiotic properties of antibiotic-resistant and auxotrophic mutants of *Rhizobium leguminosarum*. J. Appl. Bacteriol. **47,** 53—64.

Pate, J. S., Gunning, B. E. S., Briarty, L., 1969: Ultrastructure and functioning of the transport system of the leguminous root nodule. Planta **85,** 11—34.

Pate, J. S., Atkins, C. A., White, S. J., Rainbird, R. M., Woo, K. C., 1980: Nitrogen nutrition and xylem transport of nitrogen in ureide producing grain legumes. Plant Physiol. **65,** 961—965.

Peterson, M. A., Barnes, D. K., 1981: Inheritance of ineffective nodulation and non-nodulation traits in alfalfa. Crop Sci. **21,** 611—616.

Planqué, K., Kennedy, I. R., de Vries, G. E., Quispel, A., van Brussel, A. A. N., 1977: Location of nitrogenase and ammonia-assimilatory enzymes in bacteroids of *Rhizobium leguminosarum* and *Rhizobium lupini*. J. Gen. Microbiol. **102,** 95—104.

Planqué, K., de Vries, G. E., Kijne, J. W., 1978: The relationship between nitrogenase and glutamine synthetase in bacteroids of *Rhizobium leguminosarum* of various ages. J. Gen. Microbiol. **106,** 173—178.

Postgate, J. R., 1982: The Fundamentals of Nitrogen Fixation. Cambridge, U. K.: Cambridge University Press.

Prakash, R. K., Schilperoort, R. A., and Nuti, M. P., 1981: Large plasmids of fast growing Rhizobia: homology studies and location of structural nitrogen fixation *(nif)* genes. J. Bacteriol. **145,** 1129—1136.

Quinto, C., de la Vega, H., Flores, M., Fernandez, L., Ballado, T., Soberon, G., Palacios, R., 1982: Reiteration of nitrogen fixation gene sequences in *Rhizobium phaseoli*. Nature **299,** 724—726.

Rao, V. R., Darrow, R. A., Keister, D. L., 1978: Effect of oxygen tension on nitrogenase and on glutamine synthetases I and II in *Rhizobium japonicum* 61A76. Biophys. Res. Commun. **81,** 224—231.

Ratajczak, L., Ratajczak, W., Mazurowa, H., Wozny, A., 1979: Localization of glutamate dehydrogenase and glutamate synthase in roots and nodules of *Lupinus* seedlings. Biochem. Physiol. Pflanzen **174,** 289—295.

Rawsthorne, S., Minchin, F. R., Summerfield, R. J., Cookson, C., Coombes, J., 1980: Carbon and nitrogen metabolism in legume root nodules. Phytochemistry **19,** 341—355.

Reinbothe, H., Mothes, K., 1962: Urea, ureides and guanidines in plants. Annu. Rev. Plant Physiol. **13,** 129—150.

Reporter, M., 1976: Synergetic cultures of *Glycine max.* root cells and rhizobia separated by membrane filters. Plant Physiol. **57,** 651—655.

Reporter, M., 1981: Do small metallo-peptides affect nitrogen fixation in legumes? In: Gibson, A. H., Newton, W. E. (eds.): Current Perspectives in Nitrogen Fixation, pp. 214—215, Australian Academy of Science, Canberra.

Reuveny, Z., Foor, F., Magasanik, B., 1981: Regulation of glutamine synthetase by regulatory protein PII in *Klebsiella aerogenes* mutants lacking adenylyltransferase. J. Bacteriol. **146,** 740—745.

Reynolds, P. H. S., Farnden, K. J. F., 1979: The involvement of aspartate aminotransferases in ammonium assimilation in lupin nodules. Phytochemistry **18,** 1625—1630.

Reynolds, P. H. S., Boland, M. J., Farnden, K. J. F., 1981: Enzymes of nitrogen metabolism in legume nodules: partial purification and properties of the aspartate aminotransferases from lupin nodules. Arch. Biochem. Biophys. **209,** 524—533.

Reynolds, P. H. S., Boland, M. J., Blevins, D. G., Schubert, K. R., Randall, D. D.,

1982a: Enzymes of amide and ureide biogenesis in developing soybean nodules. Plant Physiol. **69,** 1334—1338.

Reynolds, P. H. S., Blevins, D. G., Boland, M. J., Schubert, K. R., Randall, D. D., 1982b: Enzymes of ammonia assimilation in legume nodules: a comparison between ureide- and amide-transporting plants. Physiol. Plant. **55,** 255—260.

Rhodes, D., Sims, A. P., Folkes, B. F., 1980: Pathway of ammonia assimilation in illuminated *Lemna*. Phytochemistry **19,** 357—365.

Robertson, J. G., Farnden, K. J. F., 1980: Ultrastructure and metabolism of the developing legume root nodule. In: Miflin, B. J. (ed.): The Biochemistry of Plants, Vol. 5, pp. 65—113, New York: Academic Press.

Robertson, J. G., Farnden, K. J. F., Warburton, M. P., Banks, J. M., 1975a: Induction of glutamine synthetase during nodule development in lupin. Aust. J. Plant Physiol. **2,** 265—272.

Robertson, J. G., Warburton, M. P., Farnden, K. J. F., 1975b: Induction of glutamate synthase during nodule development in lupin. FEBS Letts. **55,** 33—37.

Rosenberg, C., Boistard, P., Denarie, J., Casse-Delbart, F., 1981: Genes controlling early and late functions in symbiosis are located on a megaplasmid in *Rhizobium meliloti*. Mol. Gen. Genet. **184,** 326—333.

Ruvkun, G. B., Ausubel, F. M., 1980: Interspecies homology of nitrogenase genes. Proc. Natl. Acad. Sci. U. S. A. **77,** 191—195.

Ryan, E., Fottrell, P. F., 1974: Subcellular localization of enzymes involved in the assimilation of ammonium by soybean root nodules. Phytochemistry **13,** 2647—2652.

Ryan, E., Bodley, F., Fottrell, P. F., 1972: Purification and characterization of aspartate aminotransferases from soybean root nodules and *Rhizobium japonicum*. Phytochemistry **11,** 957—963.

Schubert, K. R., 1981: Enzymes of purine biosynthesis and catabolism in *Glycine max*. I. Comparison of activities with N_2 fixation and composition of xylem exudate during nodule development. Plant Physiol. **68,** 1115—1122.

Schubert, K. R., 1982: The energetics of biological nitrogen fixation. Plant Physiol. Workshop Supplement 1.

Schubert, K. R., Ryle, G. J. A., 1980: The energy requirements for nitrogen fixation in nodulated legumes. In: Summerfield, R. H., Bunting, H. (eds.): Advances in Legume Science, pp. 85—96, Royal Botanic Gardens, Kew, England.

Schubert, K. R., Coker III, G. T., 1981a: Ammonium assimilation in *Alnus glutinosa* and *Glycine max*. Short term studies using [^{13}N] ammonium. Plant Physiol. **67,** 662—665.

Schubert, K. R., Coker III, G. T., 1981b: Nitrogen and carbon assimilation in N_2-fixing plants. In: Root, J. W., Krohn, K. A. (eds.): Advances in Chemistry Series No. 197. Short-lived radionucleotides in chemistry and biology, pp. 317—339, American Chemical Society.

Schubert, K. R., Evans, H. J., 1976: Hydrogen evolution: a major factor affecting the efficiency of nitrogen fixation in nodulated symbionts. Proc. Natl. Acad. Sci. U. S. A. **73,** 1207—1211.

Scott, D. B., Hennecke, H., Lim, S. T., 1979: The biosynthesis of nitrogenase MoFe protein polypeptides in free-living cultures of *Rhizobium japonicum*. Biochim. Biophys. Acta **565,** 365—378.

Scott, D. B., Robertson, J. G., Farnden, K. J. F., 1976: Ammonia assimilation in lupin nodules. Nature **263,** 703—708.

Scott, K. F., Rolfe, B. G., Shine, J., Sundaresan, V., Ausubel, F. M., 1981: Nucleotide sequence of the gene coding for *Klebsiella pneumoniae* nitrogenase iron protein. In: Gibson, A. H., Newton, W. E. (eds.): Current Perspectives in Nitrogen Fixation, pp. 393—395, Australian Academic of Science, Canberra.

Sen, D., Schulman, H. M., 1980: Enzymes of ammonia assimilation in the cytosol of developing soybean root nodules. New Phytol. **85,** 243—250.

Shanmugam, K. T., O'Gara, F., Andersen, K., Valentine, R. C., 1978: Biological nitrogen fixation. Ann. Rev. Plant Physiol. **29,** 263—276.

Shaw, B. D., 1983: Non-coordinate regulation of *Rhizobium* nitrogenase synthesis by oxygen: studies with bacteroids from nodulated *Lupinus angustifolius.* J. Gen. Microbiol. **129,** 849—857.

Shelp, B. J., Atkins, C. A., Storer, P. J., Canvin, D. T., 1983: Cellular and subcellular organization of pathways of ammonia assimilation and ureide synthesis in nodules of cowpea (*Vigna unguiculata* L. Walp.). Arch. Biochem. Biophys. **224,** 429—441.

Stadtman, E. R., Mura, U., Chock, P. B., Rhee, S. G., 1980: The interconvertible enzyme cascade that regulates glutamine synthetase activity. In: Mora, J., Palacios, R., Glutamine: Metabolism, Enzymology and Regulation, pp. 41—59, Academic Press, N. Y.

Stewart, G. R., Mann, A. F., Fentem, P. A., 1980: Enzymes of glutamate formation: glutamate dehydrogenase, glutamine synthetase and glutamate synthase. In: Miflin, B. J. (ed.): The Biochemistry of Plants, Vol. 5, pp. 272—327, New York: Academic Press.

Stone, S. R., Copeland, L., Kennedy, I. R., 1979: Glutamate dehydrogenase of lupin nodules: purification and properties. Phytochemistry **18,** 1273—1278.

Stone, S. R., Copeland, L., Heyde, E., 1980a: Glutamate dehydrogenase of lupin nodules: kinetics of the deamination reaction. Arch. Biochem. Biophys. **199,** 550—559.

Stone, S. R., Heyde, E., Copeland, L., 1980b: Glutamate dehydrogenase of lupin nodules: kinetics of the aminating reaction. Arch. Biochem. Biophys. **199,** 560—571.

Streeter, J. G., 1972: Nitrogen nutrition of field-grown soybean plants 1. Seasonal variations in soil nitrogen and nitrogen composition of stem exudates. Agron. J. **64,** 311—314.

Streeter, J. G., 1977: Asparaginase and asparagine transaminase in soybean leaves and root nodules. Plant Physiol. **60,** 235—239.

Sundaresan, V., Ow, D. W., Ausubel, F. M., 1983a: Activation of *Klebsiella pneumoniae* and *Rhizobium meliloti* nitrogenase promoters by *gln* regulatory proteins. Proc. Natl. Acad. Sci. U. S. A., **80,** 4030—4034.

Sundaresan, V., Jones, J. D. G., Ow, D. W., Ausubel, F. M., 1983b: *Klebsiella pneumoniae nifA* product activates the *Rhizobium meliloti* nitrogenase promoter. Nature **301,** 728—732.

Tajima, S., Yamamoto, Y., 1975: Enzymes of purine catabolism in soybean plants. Plant and Cell Physiol. **16,** 271—282.

Tchan, Y. T., Wyszomirska-Dreher, Z., Kennedy, I. R., 1981: Preparation of monospecific antiserum to lupin nodule glutamate dehydrogenase. Aust. J. Biol. Sci. **34,** 161—169.

Tempest, D. W., Meers, J. L., Brown, C. M., 1970a: Synthesis of glutamate in *Aerobacter aerogenes* by a hitherto unknown route. Biochem. J. **114,** 405—407.

Tempest, D. W., Meers, J. L., Brown, C. M., 1973: Glutamate synthetase (GOGAT): a key enzyme in the assimilation of ammonia by prokaryotic organisms. In: Prusiner, S., Stadtman, E. R. (eds.): The enzymes of Glutamine Metabolism, pp. 167—182, New York: Academic Press.

Thomas, R. J., Schrader, L. E., 1981: Ureide metabolism in higher plants. Phytochemistry **20,** 361—371.

Thomas, R. J., Meyers, S. P., Schrader, L. E., 1983: Allantoinase from shoot tissues of soybeans, Phytochemistry **22,** 1117—1120.

Triplett, E. W., Blevins, D. G., Randall, D. D., 1980: Allantoic acid synthesis in soybean root nodule cytosol via xanthine dehydrogenase. Plant Physiol. **65,** 1203—1206.

Triplett, E. W., Blevins, D. G., Randall, D. D., 1982: Purification and properties of soybean xanthine dehydrogenase. Arch. Biochem. Biophys. **219,** 39—46.

Tyler, B. M., 1978: Regulation of the assimilation of nitrogen compounds. Annu. Rev. Biochem. **47,** 1127—1162.

Ueno-Nishio, S., Backman, K. C., Magasanik, B., 1983: Regulation of the gln L-Operator-Promoter of the complex gln ALG operon of *Escherichia coli.* J. Bacteriol. **153,** 1247—1251.

Upchurch, R. G., Elkan, G. H., 1978 a: Ammonium assimilation in *Rhizobium japonicum* colonial derivatives differing in nitrogen-fixing efficiency. J. Gen. Microbiol. **204,** 219—225.

Upchurch, R: G., Elkan, G. H., 1978 b: The role of ammonia, L-glutamate, and cyclic adenosine 3′–5′-monophosphate in the regulation of ammonia assimilation in *Rhizobium japonicum.* Biochem. Biophys. **538,** 244—248.

Upchurch, R. G., Mortenson, L. E., 1980: *In vivo* energetics and control of nitrogen fixation. J. Bacteriol. **143,** 274—284.

Vance, C. P., Johnson, L. E. B., 1983: Plant induced ineffective nodules in alfalfa (*Medicago sativa* L.): structural and biochemical comparisons. Can. J. Bot. (In press.)

Vance, C. P., Stade, S., Maxwell, C. A., 1983: Alfalfa root nodule carbon dioxide fixation: 1. Association with nitrogen fixation and incorporation into amino acids. Plant Physiol. (In press.)

Van den Bos, R. C., Schetgens, Th. M. B., Hontelez, J. G. J., Bakkeren, G., van Dun, C., Bisseling, T., van Kammen, A., 1983: Expression of nodule-specific genes in both partners in the *Rhizobium* legume symbiosis. In: Proceedings of the 2nd International Colloquium on Endocytobiology, Tübingen, F. R. G. – Berlin: de Gruyter. (In press.)

Van den Bos, R. C., Schots, A., Hontelez, J., van Kammen, A., 1983: Nitrogenase synthesis in isolated *Rhizobium leguminosarum* bacteroids: constitutive synthesis from *de novo* transcribed mRNA. Biochim. Biophys. Acta **740,** 313—322.

Van den Bos, R. C., Schetgens, T. M. P., Bisseling, T., Hontelez, J. G. J., van Kammen, A., 1983: Analysis of nodule-specific plant and bacteroid proteins in pea plants inoculated by transposon mutagenized *Rhizobium leguminosarum.* In: Molecular Genetics of Plant Bacterial Interactions ed A. Pühler, p. 121—129. Berlin – Heidelberg – New York: Springer.

Vogels, G. D., van der Drift, C., 1966: Allantoinases from bacterial, plant and animal sources. II. Effects of bivalent cations and reducing substances on the enzymic activity. Biochim. Biophys. Acta **122,** 497—509.

Vogels, G. D., Trijbels, F., Uffink, A., 1966: Allantoinases from bacterial, plant and animal sources. 1. Purification and enzymic properties. Biochim. Biophys. Acta **122,** 482—496.

Werner, D., Morschel, E., 1978: Differentiation of nodules of *Glycine max.* Planta **141,** 169—177.

Werner, D., Morschel, E., Stripf, R., Winchenbach, B., 1980: Development of nodules of *Glycine max.* infected with an ineffective strain of *Rhizobium japonicum.* Planta **147,** 320—329.

Wheeler, C. T., 1978: Carbon dioxide fixation in the legume root nodule. Ann. Appl. Biol. **88,** 481—484.

Wolk, C. P., Thomas, J., Shaffer, P. W., Austin, S. M., Galonsky, A., 1976: Pathway of nitrogen metabolism after fixation of ^{13}N-labelled nitrogen gas by the cyanobacterium, *Anabaena cylindrica.* J. Biol. Chem. **251,** 5027—5034.

Woo, K. C., Atkins, C. A., Pate, J. S., 1980: Biosynthesis of ureides from purines in a cell-free system from the nodule extracts of cowpea *(Vigna unguiculata).* Plant Physiol. **66,** 735—739.

Woo, K. C., Atkins, C. A., Pate, J. S., 1981: Ureide synthesis in a cell free system from cowpea *(Vigna unguiculata)* nodules. Plant Physiol. **67,** 1156—1160.

Yates, M. G., 1980: Biochemistry of nitrogen fixation. In: Miflin, B. J. (ed.): The Biochemistry of Plants, pp. 1—64, New York: Academic Press.

Chapter 6

Hydrogenase and Energy Efficiency in Nitrogen Fixing Symbionts

N. J. Brewin

John Innes Institute, Colney Lane, Norwich NR4 7UH, U.K.

With 2 Figures

Contents

I. Hydrogenase — A Suitable Candidate for Genetic Manipulation?

Biological nitrogen fixation is confined to prokaryotes and catalysed by an enzyme complex, nitrogenase, that is biochemically similar throughout the diverse groups of bacteria that harbour it. In all cases, the enzyme system is very sensitive to oxygen damage, has a low turnover number, and a large requirement for chemical energy in the form of ATP and reducing potential. In addition to reducing nitrogen (N_2) to ammonia (NH_3), nitrogenase also evolves hydrogen (H_2) as a by-product of the nitrogen fixation reaction. For this reason, nitrogen-fixing legume root nodules are often found to evolve significant quantities of H_2, which must add considerably to the energy costs for nitrogen fixation without any known benefit. However, some, but not all, rhizobia possess an oxygen-dependent enzyme system, termed "uptake hydrogenase", that is capable of recycling the H_2 released during N_2 fixation. The possession of such a hydrogen oxidase system is thought to confer a number of benefits to the nitrogen fixation process: chemical energy is regenerated in the form of ATP or reducing power, and in addition, H_2 and oxygen (O_2) are both removed from the active site of nitrogenase, where they might act respectively as reversible and irreversible inhibitors of nitrogen fixation.

In view of these *prima facie* arguments concerning the importance of the uptake hydrogenase *(hup)* system to the energy efficiency of nitrogen fixation, it is not surprising that a major objective in the field of *Rhizobium* molecular genetics is to isolate the hydrogenase *(hup)* genes in a form that would allow them to be introduced into rhizobial strains that lack this trait. At the present state of knowledge on the biochemistry and genetics of nitrogen fixation, it is probably fair to say that hydrogenase is the only system from *Rhizobium* that could conceivably be introduced into a commercial inoculant strain with any prospect of achieving a 10—20% stimulation in plant growth dependent on symbiotic nitrogen fixation.

The object of this chapter is to examine the hydrogen recycling system of *Rhizobium* from the point of view of bioenergetics, biochemistry and genetics in order to evaluate the prospects for manipulating the *hup* genes. Several recent reviews have been devoted to H_2 metabolism and H_2 recycling, in legume root nodules in particular (Dixon, 1978; Evans *et al.*, 1980, 1981; Eisbrenner and Evans, 1983) and nitrogen-fixing microorganisms in general (Robson and Postgate, 1980; Mortensen, 1978). In addition, the relevance of uptake hydrogenase to the bioenergetics of nitrogen fixation by legumes has been considered by Schubert (1982), Phillips (1980) and Pate *et al.* (1981).

II. Hydrogen Evolution by Nitrogenase

Hydrogen evolution from soybean root nodules was first observed by Hoch *et al.* (1957). Nitrogenase was subsequently purified from the nitrogen-fixing bacteria (bacteroids) contained within these nodules (Koch *et al.*, 1967) and this enzyme system was shown to evolve hydrogen in the presence of ATP and sodium dithionite. Similarly, nitrogenase preparations from many other diazotrophs were shown to liberate hydrogen *in vitro* in the presence of ATP and reductant; for example *Azotobacter vinelandii* (Bulen and LeComte, 1966), *Rhodospirillum rubrum* (Burns and Bulen, 1966), *Anabaena cylindrica* (Haystead *et al.*, 1970), and actinorhizal root nodules (Benson *et al.*, 1979).

It has been estimated that, at a minimum, 25—30% of the flux of ATP and electrons through nitrogenase is consumed in the reduction of protons, and the remainder is used in the reduction of nitrogen. However, several factors were shown to increase the level of the reaction leading to hydrogen evolution. When the normal substrate (N_2) was replaced by an inert gas, argon, the flux of ATP and electrons through nitrogenase remained unchanged but the only product was H_2 gas (Bulen *et al.,* 1965). Similarly, the proportion of H_2 evolved increased when the supply of reductant to nitrogenase was sub-optimal (Hageman and Burris, 1980). On the other hand, acetylene, (C_2H_2), which is also a substrate for nitrogenase (being reduced to ethylene, C_2H_4), almost entirely inhibits the reduction of protons to hydrogen gas by nitrogenase (Hageman and Burris, 1980).

The available evidence suggests that hydrogen evolution by nitrogenase is an inescapable by-product of the nitrogen fixation reaction. Because the exact mechanism for nitrogen reduction by nitrogenase is not understood, there is no generally accepted explanation for this unavoidable release of hydrogen (Robson and Postgate, 1980; Mortenson, 1978). Chatt (1981) has recently proposed a simple model which is consistent with the available data. It is based on a series of possible redox states for molybdenum which,

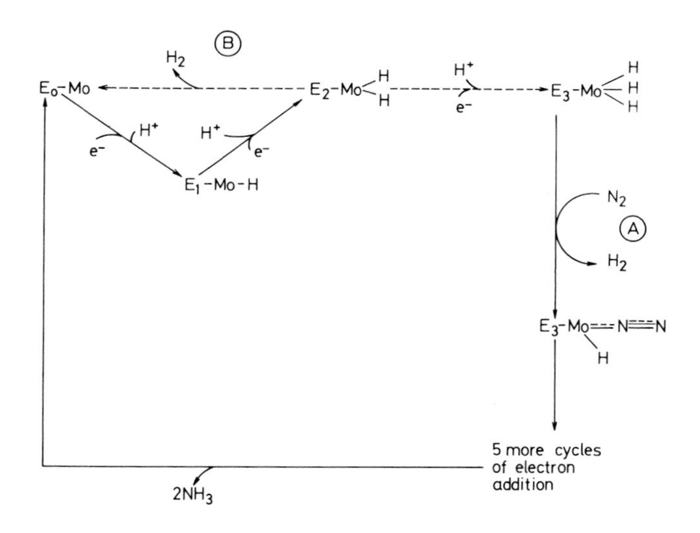

Fig. 1. Model for nitrogenase action, illustrating the two main sources of H_2 evolution. (A) Stoichiometric evolution of 1 molecule of H_2 for each molecule of N_2 bound to the active site of nitrogenase FeMo protein. (B) Proton-catalysed discharge of H_2, following dissociation of the Fe protein from the reduced FeMo protein

This model for nitrogenase activity is based on a cycle of eight successive electron transfers from the Fe protein to the active centre of the FeMo protein (Thorneley & Lowe, 1981, 1982). The solid arrows $E_1 \rightarrow E_2$ etc., represent a rapid electron transfer, followed in each case by a slower rate-limiting dissociation of the oxidised Fe protein from the reduced FeMo protein. Progressive reduction of the FeMo protein probably leads to a series of redox states for molybdenum (Chatt, 1981), of which only the tri-hydride state (MoH_3) is capable of binding substrateN_2, with concomitant release of one molecule of H_2. Acetylene (C_2H_2) inhibits H_2 evolution from nitrogenase because it is probably able to associate with the active centre at an early stage in the catalytic cycle, thereby avoiding the two major sources of hydrogen loss with iron, is thought to be present at the active site of nitrogenase. Depending on the supply of electrons through the nitrogenase system, it is postulated that Mo could exist as either the mono-, di- or tri-hydride state after having complexed with the appropriate number of protons (Fig. 1). It is further suggested that substrate nitrogen (N_2) can only bind to the trihydride state, with concomitant release of one molecule of H_2 for every molecule of N_2 bound to nitrogenase: this N_2 is subsequently reduced to ammonia. The proposed mechanism neatly explains the observed minimum stoichiometry for hydrogen evolution during nitrogen reduction (Table 1). Moreover, it is suggested that with a reduced electron flux through nitrogenase an increased proportion of the active centres would carry molybdenum dihydride or monohydride instead of the trihydride form. These forms are still able to release hydrogen after interacting with additional protons but they are unable to bind nitrogen, and hence the relative proportion of hydrogen evolved is increased under conditions of reduced electron flux through nitrogenase (Hageman and Burris, 1980).

A revised version of this model has also been suggested (Thorneley and Lowe, 1981, 1982). It is based on a cycle of eight successive electron transfers to the MoFe protein (component 1 of nitrogenase) from the Fe protein (component II of the nitrogenase complex). After each electron transfer, the oxidized Fe protein dissociates from the MoFe protein (Hageman and Burris, 1978). The model of Thorneley and Lowe emphasises that this dissociation step is of critical importance to the overall level of H_2 evolution, because, at each stage in the cycle, the oxidised Fe protein protects the active centre of the MoFe protein from reacting with protons and releasing H_2. In particular, hydrogen may be evolved during the protein dissociation that follows the second electron addition in the cycle (starting from Mo at the resting state of zero): after this second electron addition, Mo at the active site is in the dihydride state. Interaction with protons at this stage releases H_2 and prevents the third electron addition which would have generated the species MoH_3, to which substrate N_2 can bind. As with the model of Chatt, binding of N_2 to the trihydride form of Mo is associated with the stoichiometric release of one molecule of H_2 (Fig. 1).

Six electrons are needed to reduce a molecule of N_2 to ammonia ($2NH_3$) at the active site of nitrogenase. Each electron transferred to the active site involves the consumption of 2 molecules of ATP by nitrogenase component II (Hageman and Burris, 1978; Eady *et al.*, 1980). This energy cost is the same regardless of whether the electrons are ultimately used to reduce N_2 to NH_3 or protons to H_2. The proximal electron donor for the nitrogenase complex (e.g. in soybean nodule bacteroids) is probably ferredoxin (Carter *et al.*, 1980), which has a mid-point redox potential of -485 mV. Instead of being used in nitrogen fixation, the electrons from ferredoxin could otherwise have been used to generate ATP by oxidative phosphorylation through the respiratory chain. Thus, the diversion of reducing power towards nitrogen fixation and away from oxidative phosphorylation represents a further loss of chemical energy. Altough P : O ratios have not been determined for *Rhizobium* or any other diazotroph, a value of 2 seems probable in the light of studies with other bacteria (Hinkle and McCarty, 1978).

From the above considerations, it is possible to draw up an energy balance sheet for nitrogen fixation with concomitant hydrogen evolution. This is illustrated in Table 1, and it should be emphasised that the values given for hydrogen losses represent the observed minimum, i.e. 25% of the energy flux through nitrogenase; under many conditions hydrogen losses are far higher than this. The ratio of nitrogenase protein components, ATP levels, and temperature have been reported to affect the electron allocation to N_2 in purified nitrogenase preparations (Thorneley and Eady, 1977). Hageman and Burris (1980) have presented evidence that the electron flux through the nitrogenase complex may be the controlling factor, more electrons being allocated to N_2 at higher electron flux.

Table 1. Minimum Energy Utilisation for N_2 Fixation and Concomitant H_2 Evolution

Portion of Reaction$^+$	Reducing equivalents needed (e^-)	ATP needed (moles)	Total energy (ATP-equivalents)*
$2H^+ \rightarrow H_2$	2	4	6
$N_2 \rightarrow 2NH_3$	6	12	18
Total requirement	8	16	24

$^+$Assuming optimum stoichiometry, as indicated by the reaction

$$N_2 + 8e^- + 10H^+ + 16ATP \rightarrow 2NH_4^+ + H_2 + 16ADP + 16Pi$$

*Assuming that each pair of electrons used for N_2 fixation could otherwise have generated 2 molecules of ATP by oxidative phosphorylation.

III. Hydrogen Loss by Root Nodules

It has been estimated that more than a million tonnes of H_2 are released annually into the atmosphere from N_2-fixing legume root nodules (Conrad and Seiler, 1980). The rate of hydrogen evolution by detached nodules can conveniently be monitored using a hydrogen electrode (Wang, 1980) or on a gas chromatograph fitted with a thermal conductivity detector (Nelson and Child, 1981). Some indication of the extent of energy losses associated with hydrogen evolution during nitrogen fixation can be obtained if the total energy flux through the nitrogenase system is also measured at the same time. There are two ways to obtain these measurements. Either the rate of conversion of acetylene to ethylene ($C_2H_2 \rightarrow C_2H_4$) can be measured, because under these conditions hydrogen losses from nitrogenase are extremely small: alternatively, all substrates for nitrogenase can be withheld by putting the nodules under argon, and under these conditions the entire energy flux through nitrogenase is channeled into the reduction of protons and hence into hydrogen evolution.

Hydrogen evolution during nitrogen fixation can thus be expressed as a function of total energy flux through nitrogenase. The term "electron allocation coefficient" (EAC) has been used to describe this relationship (Burns and Hardy, 1975).

$$\mathrm{EAC} = \left(\frac{\text{Electrons to exogenous reducible substrate}}{\text{Total electron flux to exogenous substrate and protons}}\right)$$

As mentioned previously, when acetylene (C_2H_2) is used as substrate the electron allocation coefficient approaches 1.00, because hydrogen evolution is virtually undetectable. With nitrogen (N_2) as substrate at physiological levels the electron allocation coefficient for nitrogenase is always less than 0.75. Another term, relative efficiency, has also been used to reflect the extent of hydrogen losses during N_2 fixation (Schubert and Evans, 1976). Operationally, relative efficiency is equivalent to the electron allocation coefficient (in the absence of other interacting enzyme systems which will be discussed below).

$$\mathrm{RE} = 1 - \left(\frac{\text{Rate of } H_2 \text{ evolution in air}}{\text{Rate of } C_2H_2 \text{ reduction}}\right)$$

Using this concept, Schubert and Evans (1976) surveyed a number of nodulated N_2 fixing plants. Many legumes evolved H_2 at high rates, corresponding to relative efficiencies of 0.4 to 0.7. However, a few legume and many of the actinorhizal species (i. e., alder, *Purshia*, *Ceanothus* and *Myrica*) did not evolve H_2 in air and had apparent relative efficiencies approaching 1.00. High relative efficiency could in most cases be correlated with the particular bacterial strain used as inoculant (Carter *et al.* 1978). These apparently more efficient symbionts were subsequently found to possess a hydrogenase system which functioned to reoxidize the H_2 formed as part of the nitrogen fixation reaction. Such hydrogenases were first reported in N_2 fixing microorganisms by Phelps and Wilson (1941) and were studied subsequently by Dixon (1967, 1968).

Clearly, the presence of a hydrogen-recycling system within N_2 fixing nodules distorts the measurement of "relative efficiency" and makes it impossible to estimate what proportion of the electron flux through nitrogenase is expended in the reduction of protons. If the analysis is confined to those symbionts that lack an uptake hydrogenase, it is seen that relative efficiency values vary widely, depending on the particular bacterial and plant combinations, and the environmental conditions (Schubert and Evans, 1976; Nelson and Child, 1981). There has been, however, no systematic analysis of the physiological factors that influence the proportion of the energy flux allocated to hydrogen evolution during nitrogen fixation for legumes under normal growth conditions. If the level of H_2 production by nitrogenase could be controlled to the minimum value

Table 2. Surveys of *Rhizobium* Strains for Their Ability to Recycle Hydrogen Produced During N_2 Fixation

Rhizobium species	Host legume	Number of strains examined	Number of Hup^+ strains*	Reference
A. *Fast-growing strains*$^+$				
R. leguminosarum	Pea	15	2	Ruiz-Argüeso *et al.*, 1978
		108	12	Nelson and Child, 1981
		48	2	N. J. Brewin, unpublished
R. meliloti	Alfalfa	19	4	Ruiz-Argüeso *et al.*, 1979
R. trifolii	White clover	7	0	Ruiz-Argüeso *et al.*, 1979
R. phaseoli	*Phaseolus vulgaris*	10	0	N. J. Brewin, unpublished
B. *Slow-growing strains*				
R. japonicum	Soybean	1400	300	Lim *et al.*, 1981
		32	7	Carter *et al.*, 1978
Cowpea *Rhizobium*	Cowpea	13	11	Schubert *et al.*, 1977, 1978
Mungbean *Rhizobium*	Mungbean	30	28	Pahwa and Dogra, 1981

* Criteria for a Hup^+ strain vary slightly in different surveys, e. g. relative efficiency >0.80; significant levels of uptake of hydrogen or tritated hydrogen by nodules or isolated bacteroids.

$^+$ Note that the specific activity for hydrogenase from bacteroids of any of the fast-growing rhizobia was less than 1% of the specific activity typical for slow-growing strains.

($1\,H_2 : 1\,N_2$), this would represent a very significant energy saving for the N_2 fixation process.

IV. Occurrence of Uptake Hydrogenase in *Rhizobium*

Many *Rhizobium* strains have been screened for the presence or absence of hydrogenase (Table 2). Several interesting points can be made from these surveys. What is perhaps most remarkable is the very sporadic occurrence of hydrogenase in *Rhizobium*, particularly among the fast-growing strains *R. meliloti, R. trifolii, R. leguminosarum* and *R. phaseoli;* amongst these, only 5—10% of the *R. leguminosarum* and *R. meliloti* isolates have significant levels of hydrogenase uptake activity (Hup^+). This level of activity, however, is often insufficient to recycle all the H_2 generated within N_2-fixing nodules. Similarly in *R. japonicum*, a slow-growing species, only about 25% of field isolates possess hydrogenase activity, although in *Rhizobium* isolates from cowpea and mungbean hydrogenase activity is more commonly found. In all these slow-growing strains, hydrogenase activity, when present, is normally more than adequate to prevent any H_2 evolution form N_2 fixing nodules. The specific activity of hydrogenase isolated from bacteroids of the best hydrogenase positive strain of *R. meliloti* was less than. 0.1% of the specific activity of the hydrogenase system in bacteroids formed by Hup^+ *R. japonicum* USDA 110 (Ruiz-Argüeso *et al.,* 1979c).

Table 3. Assays for Hydrogenase Activity

Method	Reference
Hydrogen uptake —	
hydrogen electrode	Wang, 1980
gas chromatography	Nelson and Child, 1981
Tritium (H-T) uptake	Lim, 1978 Bethlenfalvay and Phillips, 1979 Brewin *et al.,* 1982
H_2-dependent chemoautotrophic growth	Lepo *et al.,* 1980 Maier, 1981
H_2-dependent dye reduction —	
methylene blue	Haugland *et al.,* 1983
phenazine methosulphate	Arp and Burris, 1979
triphenyl tetrazolium	Maier *et al.,* 1978b
H_2-evolution from reduced methyl-viologen	Arp and Burris, 1979
Deuterium exchange	Arp and Burris, 1982

* Although the capacity to reduce triphenyltetrazolium chloride was used to distinguish wild type Hup^+ and mutant Hup^- mutants of *R. japonicum* (Maier *et al.,* 1978), this method was not reliable for screening Hup^- point mutants (Eisbrenner and Evans, 1983).

A further difference between the hydrogenases of slow-growing and fast-growing rhizobial strains is that the former can easily be induced *in vitro* along with nitrogenase, when bacterial cultures are grown under conditions of low oxygen tension and low carbon availability (Maier *et al.,* 1978a). Indeed, these bacteria can even be induced to grow chemoautotrophically using H_2 as their energy source and CO_2, via ribulosebisphosphate carboxylase, as their sole carbon source (Lepo *et al.,* 1980).

It should also be noted that in several of the surveys described in Table 2 the sole criterion for a Hup$^+$ strain was a high relative efficiency (>0.80). Undoubtedly some Hup$^+$ strains might have low relative efficiencies for some unknown reasons (e.g. Nelson and Child, 1981): hence, a more direct measurement of hydrogenase activity is needed. The hydrogen uptake system can be monitored qualitatively and quantitatively by a variety of assays using whole plants, detached nodules, free-living cultures induced for hydrogenase activity, and in cell-free extracts. A summary of the methods available for studying hydrogenase activity is presented in Table 3. Many of these techniques have been used in the further study of the physiology, biochemistry and genetics of hydrogenases from *Rhizobium.*

V. Potential Benefits Associated with Uptake Hydrogenase

Assuming that the activity of root nodules is limited by the supply of photosynthate from the plant (Hardy and Havelka, 1975; Bethlenfalvay and Phillips, 1977; Williams *et al.,* 1981), any process that reduces the energy cost for nitrogen fixation might be expected to result in increased nitrogen fixation and consequently in increased plant growth. The loss of more thant 25% of the energy flux through nitrogenase as hydrogen evolution (Table 1) might be compensated, at least in part, by the activity of the hydrogen recycling system. H_2-dependent ATP synthesis has been demonstrated in *R. leguminosarum* bacteroids (Dixon, 1968), although in some strains it has been reported that H_2 oxidation is not coupled to ATP synthesis (Nelson and Salminen, 1982). In *R. japonicum* bacteroids, Emerich *et al.* (1979) have shown that the ATP concentration was increased by 20—40% during H_2 oxidation. It has been suggested that 2 molecules of ATP would be regenerated for each molecule of H_2 recycled by the oxyhydrogenase system (Evans *et al.,* 1981). It is theoretically possible that electrons from hydrogenase could also be used as a source of reducing power within the cell, instead of being used to generate ATP by oxidative phosphorylation, although there is no direct evidence for this. Several studies have shown that the rates of CO_2 evolution during N_2 fixation were substantially reduced in *Rhizobium* bacteroids or nodules formed by Hup$^+$ strains (Dixon, 1968; Emerich *et al.,* 1980a; Arima, 1981; Drevon *et al.,* 1982). These results suggest that H_2 can be used as an energy source, thereby conserving carbohydrate during nitrogen fixation.

It is also possible that an uptake hydrogenase system within nodules could be advantageous in several other ways. For example, hydrogen (evolved by nitrogenase) is a competitive inhibitor of N_2 fixation by nitrogenase, and Dixon *et al.* (1981) have suggested that hydrogen could accumulate to inhibitory levels (c. 10%, v/v) within the nodule: thus, hydrogenase might serve to relieve this inhibition.

The hydrogen oxidase system might also be regarded as a system for protecting nitrogenase form oxygen damage. Nelson and Salminen (1982) observed that the optimal O_2 concentration for acetylene reduction by bacteroids from Hup^+ strains of *R. leguminosarum* was increased by 0.01 atm. They suggest that oxygen protection may be the primary role for hydrogenase within pea nodules and an inverse correlation between hydrogenase activity and leghaemoglobin content was noted (Nelson and Child, 1981). Similarly in Hup^+ *R. japonicum* bacteroids, the addition of H_2 increased by 0.01 atm the optimum concentration of O_2 for acetylene reduction (Ruiz-Argüeso *et al.*, 1979b): in addition, the maximum rate of C_2H_2 reduction was increased by two or three-fold.

The discovery that several slow-growing *Rhizobium* strains can grow as hydrogen autotrophs (Lepo *et al.*, 1980) raises another interesting possibility, namely that hydrogenase is of importance to the bacterium in some ecological niche in the soil. For example, hydrogen and CO_2 evolved from a legume root nodule occupied by a Hup^- *Rhizobium* strain could provide the necessary energy and carbon sources for hydrogen-dependent autotrophic growth of a Hup^+ *Rhizobium* strain on the surface of that nodule.

In view of the many potential benefits thought to be associated with the hydrogen recycling system in *Rhizobium*, it might be expected that legumes inoculated with Hup^+ strains of *Rhizobium* should grow better and fix more nitrogen than their counterparts inoculated with Hup^- strains. These expectations have been considered theoretically by Evans *et al.* (1981) for soybean seedlings growing exponentially under carbon-limited conditions in the absence of combined nitrogen. The calculations were based on a number of assumptions, namely that 32% of the electron flux through nitrogenase was lost as H_2 evolution by a Hup^- strain; that 9.1% of the total energy flux through nitrogenase was recoverable in a Hup^+ strain by H_2-dependent ATP regeneration; and that this energy saving was converted into a 9.1% stimulation in exponential growth rate for the soybean plants, beginning 15 days after seed planting at the onset of nitrogen fixation by the nodules. Taking all these parameters into consideration, and excluding any other potential benefits associated with the hydrogenase system (e.g. oxygen protection, relief of H_2 inhibition, etc.) it was predicted that soybean seedlings inoculated with a Hup^+ strain and grown for a total of 40 days would show a 21% increase in dry weight compared to plants in which hydrogen-recycling did not occur.

A series of experiments have been conducted in which Hup^+ and Hup^- strains of *Rhizobium* have been compared as inoculants for legumes grown under controlled environmental conditions and in the field. The results of these studies have been summarised by Eisbrenner and Evans (1983). In many cases significant increases in plant yield and total N content were observed, and in a few cases the increases were very dramatic indeed. However, the general criticism of most of these experiments is that the comparisons may not reflect differences that are specifically attributable to the presence or absence of the hydrogenase gene product. In other words, the Hup^+ and Hup^- strains being compared were not isogenic, except perhaps in the case of revertible Hup^- mutants (Lepo *et al.*, 1981; Drevon *et al.*, 1982).

Most of these studies have involved comparisons of groups of Hup^+ and Hup^- field isolates: Schubert *et al.* (1978), Hanus *et al.* (1981) and Pahwa and Dogra (1981) were able to demonstrate significant increases ($>10\%$) associated with the Hup^+ character, whereas Gibson *et al.* (1981), Nelson and Child (1981) and Rainbird *et al.*, (1983) found no improvement in plant growth. DeJong *et al.*, (1982) compared two Hup^- field isolates of *R. leguminosarum* with derivatives that had acquired an additional symbiotic plasmid carrying the Hup^+ determinants: although siginifant increases in plant dry weight and N content were obtained, there is no direct evidence that this was due to the introduction of the *hup* gene system, rather than to any of the other determinants for nodule formation or nitrogen fixation that were also introduced on this new symbiotic plasmid.

Mutants defective in some aspect of the hydrogenase system have also been examined for their effects on plant growth. Many of these mutants are not revertible (Albrecht *et al.*, 1979), and so the observed depression in plant growth dependent on symbiotic nitrogen fixation could not be attributed directly to a mutation in hydrogenase. Furthermore, these Hup^- mutant strains were found to have different plasmid patterns from the Hup^+ parental strain (Cantrell *et al.*, 1982).

Revertible mutations that affect hydrogenase expression have been isolated in *R. japonicum* by Lepo *et al.*, (1981). Significant depression of soybean plant growth was observed when two Hup^- mutants, PJ 17 or PJ 18, were compared with the original Hup^+ parent strain SR (Lepo *et al.*, 1981). However, when Hup^- mutant, PJ 17, and Hup^+ revertant, PJ 17-1, were compared in growth experiments (Eisbrenner and Evans, 1983), there was no significant difference in plant growth, although nodules formed by the Hup^- mutant strain, PJ 17, evolved CO_2 at a rate about 10% higher than that of nodules formed by the Hup^+ revertant strain, PJ 17-1 (Drevon *et al.*, 1982). For another revertible mutant, PJ 18, the Hup^+ revertant gave a 10% improvement in plant dry weight and nitrogen content, compared to PJ 18 itself.

Taken together, all these experiments provide circumstantial evidence in favour of a beneficial role for hydrogenase. Failure to detect any improvement associated with hydrogenase activity may perhaps result from the use of growth conditions in which carbon supply to nodules was not limiting nitrogen fixation activity. Phillips and co-workers have suggested that in short-term experiments during early nodule development an increased rate of photosynthesis may not lead to increased N_2 fixation (DeJong and Phillips, 1981; Williams, DeJong and Phillips, 1981). At this stage in plant development, the simultaneous development of root nodules and leaf photosynthetic area could place a great demand on the reserves of organic nitrogen on the plant. Thus, plant growth would be nitrogen-limited, not carbon limited. An initial application of 2 mM nitrate may alleviate this transitory period of nitrogen stress in young seedlings and allow the plant to develop maximum rates of N_2 fixation (DeJong *et al.,* 1982).

VI. Relationships Between Hydrogenase Determinants, Plasmids and Other Symbiotic Genes

Because the hydrogenase determinants of both fast and slow-growing *Rhizobium* strains are expressed within the root nodule, it would be interesting to know whether they resemble other genes concerned with symbiotic nitrogen fixation in terms of their organisation and the control of their expression.

In fast-growing *Rhizobium* strains the structural genes for nitrogenase (*nif*) are located on large plasmids (see chapter 4 for details). On these same large plasmids are also found genetic determinants essential for the induction of root nodules (*nod*) and N_2 fixation (*fix*). In a Hup^+ *R. leguminosarum* strain, 128 C 53, the symbiotic plasmid pRL6JI (MW 190 megadaltons) could be identified by hybridisation to cloned *nif* DNA (Hombrecher *et al.,* 1981). This plasmid, pRL6JI, was eliminated from strain 128 C 53 and replaced by another symbiotic plasmid (pRL1JI) that was derived from a Hup^- field isolate of *R. leguminosarum* (Brewin *et al.,* 1982). The resulting strain was Nod^+, Fix^+, but Hup^-, indicating that the *hup* determinants had probably been lost from strain 128 C 53 when plasmid pRL6JI was eliminated.

Additional evidence that the hydrogenase determinants of 128 C 53 were linked to plasmid-borne *nod* and *fix* determinants was obtained by cotransfer of all these markers to a Hup^- recipient strain of *R. leguminosarum* (Brewin *et al.,* 1980, 1982). Because pRL6JI was not self-transmissible, its *sym* determinants could only be mobilised following recombination with a transmissible plasmid derived from another strain of *Rhizobium leguminosarum.* These data argue strongly that all the determinants necessary to convert a Hup^- field isolate of *R. leguminosarum* into a Hup^+ derivative are to be found linked to other symbiotic genes on the plasmid pRL6JI from 128 C 53. Similarly in another Hup^+ field isolate of *R. leguminosarum,* strain 18 A from Denmark, the *hup* determinants are located on the symbiotic plasmid (N. Brewin and G. Sorensen, unpublished results).

Not only are the *hup* determinants of *R. leguminosarum* (128 C 53) genetically linked to *nod* and *fix* determinants, but the expression of hydrogenase within the pea root nodule is closely coupled to the expression of *fix* determinants. The hydrogenase determinants derived from 128 C 53 gave less than 1% of the wild-type level of hydrogenase activity when present in nodules formed by Fix^- mutant strains of *Rhizobium* (Brewin *et al.,* 1983) even when exogenous hydrogen was supplied. Thus the *hup* determinants of *R. leguminosarum* strains such as 128 C 53 and 18 A have all the characteristics that might be expected for classical symbiotic genes, except that they are inessential for symbiotic N_2 fixation and sporadic in occurrence, being present in only a minority of *R. leguminosarum* field isolates.

A possible relationship between hydrogenase genes and plasmids has also been sought in *R. japonicum* (Cantrell *et al.,* 1982). The plasmid profiles of a series of Hup^+ and Hup^- field isolates were compared: whereas most of the Hup^- field isolates harboured plasmids with molecular weights ranging from 50—300 megadaltons, the 7 field isolates that were strongly Hup^+ in phenotype yielded no plasmids visible after agarose gel electrophoresis. No satisfactory explanation is offered for this surprising observation. One is tempted to speculate that Hup^+ and Hup^- *R. japonicum* strains may represent two taxonomic or ecological races that perhaps occupy slightly different niches in the soil environment. On this argument, the plasmids seen in Hup^- strains may have no direct relevance to symbiotic N_2 fixation or H_2 metabolism. It should be noted that in slow-growing *Rhizobium* strains such as *R. japonicum* the nitrogen fixation (*nif*) genes are also probably not on plasmids (Haugland and Verma, 1981; Masterson *et al.,* 1982).

Also unexpected was the discovery by Cantrell *et al.* (1982) of two plasmids in each of three non-reverting Hup^- mutants derived from the $Hup^{R.}$ *japonicum* strain SR (Maier *et al.,* 1978). No plasmids could be found in the parent strain, SR, nor in any of three revertible Hup^- mutants derived from SR in a subsequent screening programme (Lepo *et al.,* 1981). Cantrell *et al.* (1981) postulate that hydrogenase determinants in the Hup^+ *R. japonicum* strains SR are either present on the chromosome or alternatively they may be carried on a plasmid that was too large to be recovered by standard physical techniques (perhaps greater than 300 megadaltons): the non-reverting Hup^- mutants might have arisen from a deletion and recombination event giving rise to two smaller plasmids that could be visualised after agarose gel electrophoresis. It remains unclear what relationship these plasmids have to the plasmids seen in Hup^- field isolates of *R. japonicum.*

VII. Biochemical Components of the Hydrogenase System

If the hydrogenase system of *Rhizobium* is to be regarded as a suitable candidate for genetic manipulation, it is essential to know how many components are involved specifically in the oxidation of hydrogen and also how many of these components are missing from the natural Hup^- field isolates of *Rhizobium.* This problem can be approached in two ways, by biochemical and by genetical analysis.

The hydrogenase enzyme is defined biochemically as catalysing a reaction in which H_2 yields two protons and two electrons. A highly purified preparation of hydrogenase has been isolated from bacteroids of *R. japonicum* after solubilisation of the membranes with Triton X 100 (Arp and Burris, 1979). It was found to be an oxygen-sensitive iron-sulphur protein, with a single polypeptide component, molecular weight 63.3 Kdal. This enzyme readily transferred electrons to methylene blue, ferricyanide and 2—6 dichlorophenol-indophenol, but not to NADP, FAD, FMN or O_2; in addition, H_2 evolution could be driven by reduced methyl viologen and a deuterium exchange reaction between H_2 and water could be demonstrated.

The properties of hydrogenase from *R. japonicum* bacteroids (McCrae *et al.,* 1978, Arp and Burris, 1981, 1982) are basically similar to those of membrane hydrogenases from other microorganisms, for example *Azotobacter* (Van der Werf and Yates, 1978) *Alcaligenes eutrophus* (Schink and Schlegel, 1979) and the actinorhizal endophyte (Benson *et al.,* 1980). The general properties of hydrogenases have been reviewed recently (Adams *et al.,* 1981, Bowien and Schlegel, 1981) and a specific requirement for nickel has been emphasised (Eisbrenner and Evans, 1983; Klucas *et al.,* 1983).

In addition to hydrogenase itself, the hydrogen oxidation reaction may well require other components and the possible involvement of electron transport factors has been examined. On the basis of spectral evidence, Eisbrenner and Evans (1982b) identified in Hup^+ strains of *R. japonicum* a b-type cytochrome, component 559-H_2 which became reduced only when H_2 was supplied. The cellular concentration of component 559-H_2 was correlated with the level of hydrogenase activity for a range of strains and mutants grown under a range of different physiological conditions. Component 559-H_2 was not detected in the revertible Hup^- mutants PJ 17 and PJ 18 (Lepo *et al.,* 1981) under conditions which allowed detection in the Hup^+ parent, strain SR. These correlations indicate that component 559-H_2 may be an electron carrier that couples hydrogenase to the main electron pathways of *Rhizobium* (Eisbrenner and Evans, 1982b), as indicated in Fig. 2. Although the difference spectrum of component 559-H_2 resembles

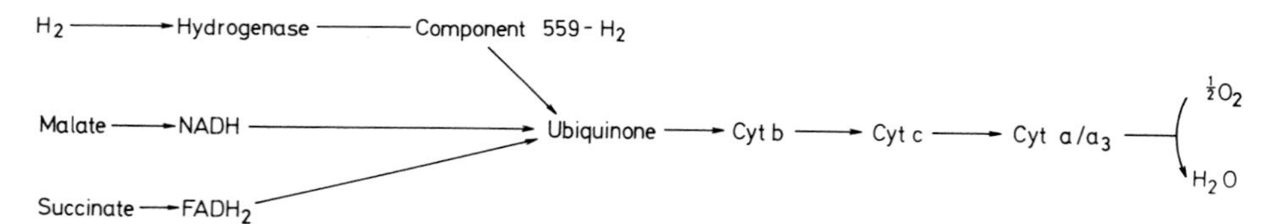

Fig. 2. Possible relationship between hydrogenase and the main components of the electron transport pathway of *Rhizobium japonicum* (Eisbrenner and Evans, 1982b)

that of a b-type cytochrome, Eisbrenner *et al.* (1982) presented evidence that its oxidation-reduction potential was lower than that of ubiquinone, perhaps as a result of its close association with the membrane-bound hydrogenase.

Subsequent to component 559-H_2, there is evidence for the involvement of ubiquinone as a carrier in electron transfer from H_2 to O_2 (Eisbrenner and Evans, 1982a), because dibromothymoquinone (DBMIB), an ubiquinone antagonist, inhibited oxygen-dependent H_2 uptake by bacteroids. The involvement of b- and c-type cytochromes has also been implicated because, when endogenous respiration was inhibited by iodoacetate or malonate, the addition of H_2 resulted in a marked increase in the rate of cytochrome reduction (Eisbrenner and Evans, 1982a; Eisbrenner *et al.*, 1982). Finally, a cyanide-sensitive terminal oxidase (cytochrome a-a_3) is involved in the transfer of electrons from H_2 to O_2 (Ruiz-Argüeso *et al.* 1979). It is probable that all these components from ubiquinone to cytochrome a-a_3 are common to the electron transport pathways from hydrogen and from other substrates.

Another component that seems to be associated with the hydrogenase system under some situations is ribulosebisphosphate (RuBP) carboxylase (Purohit *et al.*, 1982; Manian and O'Gara, 1982). When hydrogenase activity is expressed by *R. japonicum* cultures under free-living conditions, RuBP carboxylase is co-ordinately induced (Simpson *et al.*, 1979), and this is the basis for H_2-dependent chemoautotrophic growth in *R. japonicum* (Lepo *et al.*, 1980; Hanus *et al.*, 1979). However, when hydrogenase activity is expressed in the bacteroid forms of Hup^+ *R. japonicum*, there was no evidence for bacterial RuBP carboxylase activity (Simpson *et al.*, 1979). It is therefore unlikely that this enzyme is in any way necessary for the proper recycling of hydrogen within the nodule. Moreover, mutants in RuBP carboxylase activity still expressed hydrogenase activity within the nodule (Maier, 1981).

Thus, biochemical studies provide evidence that at least two specific components are needed in order for the hydrogenase system to function effectively within nodule bacteroids, namely hydrogenase and component 559-H_2.

VIII. Genetic Components of the Hydrogenase System

Mutants of *R. japonicum* that are defective in hydrogenase activity have been isolated: the *in vitro* screening systems used depended either on detecting the impairment of H_2-dependent reduction of a dye, methylene blue (Maier *et al.*, 1978b; Haugland *et al.*, 1983), or the impairment of H_2-dependent chemoautotrophic growth (Maier, 1981, Lepo *et al.*, 1981). Various classes of Hup^- mutant have been isolated which support the view that the operation of several genes is required for a functional hydrogenase system.

Maier and Merberg (1982) isolated some *R. japonicum* mutants that were hypersensitive to repression of H_2 uptake by oxygen: however, these mutants showed normal hydrogenase activity as bacteroids within root nodules. Other mutants (Maier, 1981) took up hydrogen only when methylene blue was present as the electron acceptor instead of oxygen, and another class did not interact with hydrogen under any conditions (Lepo *et al.*, 1981). By mixing *in vitro* extracts from bacteroids of different Hup^- mutants, Maier and Mutaftschiev (1982) were able in one case to demonstrate biochemical complementation for methylene blue-dependent hydrogen uptake. Both the range of mutant phenotypes and the observation of *in vitro* complementation argue that several different genes involved in the Hup system may have been indentified by mutation. So far, none of these Hup⁻ mutations has been completely characterised biochemically. Bacteroids isolated from the revertible Hup⁻ mutants PJ 17 and PJ 18 (Lepo *et al.*, 1981) have both been examined by two dimensional gel electrophoresis and shown to lack a protein which is probably hydrogenase (Drevon *et al.*, 1982). In addition, bacteroids of both mutants lacked the spectral signals normally associated with component 559-H_2 after exposure to hydrogen (Eisbrenner and Evans, 1982b), a result that would be expected if hydrogenase was required for the H_2-dependent reduction of component 559-H_2.

IX. Cloning the Hydrogenase Genes

In order to isolate genetic components of the hydrogenase system from *R. japonicum* 122 DES, a gene bank was constructed from an EcoR 1 partial digest of DNA. This was cloned into the broad host range cosmid vector pLAFR-1, which confers tetracycline resistance (Cantrell *et al.*, 1983). The gene bank in *E. coli* was mated with a revertible Hup⁻ point mutant, PJ 17 (which is itself a derivative of strain 122 DES). Tetracycline resistant derivatives of PJ 17 were screened for the hydrogen-dependent ability to grow chemolithotrophically and to reduce methylene blue (Haugland *et al.*, 1981). Hup⁺ transconjugants were isolated from the gene bank at a frequency of 6×10^{-3}. These Hup⁺ transconjugants were found to carry cloned DNA that restored the Hup⁻ mutant PJ 17 to a Hup⁺ phenotype, both for free-living bacterial cultures and for bacteroids isolated from soybean root nodules. Variability in hydrogenase expression after plant tests was probably related to the significant degree of instability of the cosmid clones during growth of the *Rhizobium* strain within the nodules. Bacteroids showed specific activities for H_2 uptake that were 5—25% of those for the wild-type control strains. DNA from eleven of the cosmid clones that suppressed PJ 17 was compared after digestion with *Eco*R 1. All the clones had three DNA restriction fragments in common; 13, 2.9 and 2.3 kilobases in length. Presumably this common region contains the cloned wild-type copy of the gene that is mutated in PJ 17. Each clone also carried a variety of additional EcoR 1 fragments which probably represented the sequences flanking the central region that was common to all plasmids.

One of the cosmid clones, termed pHU 1, that suppressed the Hup⁻ mutant PJ 17 was then introduced into a related Hup⁻ revertible strain, PJ 18. In this case, when tetracycline-resistant colonies were screened for hydrogenase activity, only one colony per thousand showed activity in the methylene blue assay. The spontaneous reversion rate for PJ 18 from Hup⁻ to Hup⁺ has previously been reported as being 10^{-9} (Lepo *et al.*, 1981). There are several possible explanations for this low frequency suppression of PJ 18 by pHU 1. Perhaps the most likely is that only part of the wild-type gene has been cloned (Cantrell *et al.*, 1983). Thus a functional gene (or transcriptional unit) could only arise as a result of recombination between the "wild-type" gene fragment on cosmid pHU 1 and the mutant genomic sequence in PJ 18. This model implies that mutants PJ 17 and PJ 18 carry mutations in different but adjacent genes. It might have been expected that, at least occassionally, both genes would have been cloned together. However, none of the eight cosmid clones tested which suppressed PJ 17 at high frequency also suppressed PJ 18 at high frequency: in each case approximately 10^{-3} Hup^{+} derivatives of PJ 18 were obtained. This observation is itself interesting and relevant to the model of a truncated *hup* gene clone, because similar frequencies of suppression of PJ 18 would only be likely if all eight cosmids terminated in the same *Eco*R 1 fragment, unless a very long operon is involved.

An alternative explanation for the low frequency suppression of PJ 18 might be that the entire wild-type allele is carried by pHU 1, but that it is recessive to a trans-dominant mutation carried in PJ 18: hence the wild-type allele could only find expression following recombination with the genome and subsequent elimination of the dominant mutation. (It is also remotely conceivable that the relevant wild-type gene may not even be present on pHU 1 and related cosmids: perhaps the presence of pHU 1 in some way stimulates a high frequency $Hup^{-} \rightarrow Hup^{+}$ reversion rate for PJ 18).

Thus, from the available data, it can be seen that pHU 1 and related cosmids probably carry the entire *hup* gene that is mutant in PJ 17, and some, all or possibly none of the *hup* gene that is mutant in PJ 18. Obviously, subcloning and localised mutagenesis will tell us much more about the gene organisation on these cosmids. It will also be interesting to learn how many of the Hup^{-} mutants of *R. japonicum* are suppressible by these clones and whether they carry sufficient information to confer hydrogenase activity to a Hup^{-} field isolate.

X. Problems of Gene Stability and Gene Expression for a Cloned *hup* System

The potential value of a hydrogenase (*hup*) clone is of course to convert a Hup^{-} field isolate of *Rhizobium* into a Hup^{+} derivative. Even if it proves possible to clone all the essential components of the *hup* system into a single plasmid vector, there might still be a number of obstacles to overcome before the system would be suitable for the genetic improvement of commercial inoculants.

One difficulty already noted by Cantrell *et al.* (1983) is that pLAFR cosmids are very unstable in *R. japonicum* unless selection for tetracycline resistance is maintained. Thus, it will be necessary to develop a more stable delivery system as a cloning vehicle for the *hup* genes, or alternatively a system will have to be devised whereby the introduced *hup* genes can be stably integrated by recombination into the bacterial genome.

Even after stable establishment of the *hup* gene system, there are still potential problems associated with the expression of these genes. The induction of hydrogenase within nodules appears to be controlled in a very complex manner: parameters that might be involved include the availability of hydrogen (Simpson *et al.*, 1979), nickel (Klucas *et al.*, 1983), a suitable carbon source (Maier *et al.*, 1979), a low level of oxygen (Maier and Merberg, 1982), the level of expression of other nitrogen fixation genes (Ruiz-Argüeso *et al.*, 1981, Brewin *et al.*, 1983), and the physiological state of the host plant (Schubert *et al.*, 1977, Bethlenfalvay and Phillips, 1979). Some of these parameters may vary with the genotype of the host plant: although Carter *et al.* (1978) examined a series of soybean cultivars and found no host legume effect on hydrogenase expression, Gibson *et al.* (1981) reported that "cowpea" *Rhizobium* strains CB756 and 32H1 were phenotypically Hup^- within nodules of *Vigna radiata* but expressed hydrogenase activity within nodules of *Vigna mungo* and *Vigna unguiculata*. A controlling effect of the host legume on hydrogenase expression has also been observed for *R. leguminosarum* (Dixon, 1972; Lopez *et al.*, 1983; Bedmar *et al.*, 1983).

An interesting parallel with the cloned *hup* gene system from *R. japonicum* is to be found in the plasmid, pIJ1008, which carries all the genetic information necessary to convert a Hup^- field isolate of *R. leguminosarum* into a Hup^+ strain (Brewin *et al.*, 1980, 1982). This plasmid of molecular weight 195 megadaltons is, in effect, a symbiotic plasmid from a Hup^+ strain of *R. leguminosarum*. It is transmissible at high frequency (10^{-3} per recipient) between strains and species of *Rhizobium*, and carries a selectable drug resistance marker (kanamycin resistance). It was formed by *in vivo* recombination between the symbiotic plasmid from *R. leguminosarum* strain 128C53 (which carries hydrogenase determinants linked to other determinants for symbiotic functions), and pVW5JI, a transposon Tn5 marked derivative of a transmissible plasmid derived from another field isolate of *R. leguminosarum* (Brewin *et al.*, 1982).

When pIJ1008 was introduced into a strain of *R. meliloti* (B287) it did not affect the level of nodulation or nitrogen fixation on alfalfa, when compared to the original strain (Bedmar, Phillips and Brewin, unpublished results). All bacteria reisolated from the alfalfa nodules still contained the introduced plasmid, and hence the criterion for the stable introduction of *hup* genes from *R. leguminosarum* into *R. meliloti* seems to have been fulfilled. However, almost no hydrogenase activity could be detected within alfalfa nodules by the tritium incorporation technique. This illustration highlights the problems that may be encountered when cloned hydrogenase genes are introduced into different species of *Rhizobium*, and suggests that further studies in the organisation, regulation and genetic manipulation of the *hup* system may be necessary in order to achieve satisfactory expression of the cloned hydrogenase genes in the environment of the new host.

XI. Conclusions

Altough in theory the potential benefits associated with hydrogen recycling seem obvious, the most appropriate experimental systems for testing this hypothesis have yet to be established. What is needed is a series of clearly defined isogenic lines that differ only in the expression of the hydrogenase gene. Point mutations in the hydrogenase system of *R. japonicum* do exist, but none of these has been completely characterised biochemically. Also, although some genetic components of the hydrogenase system have been cloned, these clones are unstable, biochemically undefined, and probably genetically incomplete: it is therefore very unlikely that they could be used to convert a Hup^- *Rhizobium* field isolate into a Hup^+ derivative. Despite all these reservations, the growth experiments that have been conducted so far with soybeans have done nothing to dispel the belief in the importance of the hydrogen recycling system (Eisbrenner and Evans, 1983), although under field conditions the advantages of the hydrogenase system will be modulated by other factors affecting whole-plant physiology, for example carbon availability and moisture stress (Gibson *et al.,* 1981; Rainbird *et al.,* 1983). Among the fast growing *Rhizobium* strains, advantages associated with the hydrogen recycling system have not been convincingly established (Nelson and Child, 1981; Nelson and Salminem, 1982), but perhaps this is because hydrogenase expression by *Rhizobium* isolates within nodules of peas or alfalfa is orders of magnitude lower than for nodules of cowpea or soybean.

Acknowledgements

I am indebted to H. J. Evans and associates for helpful discussions and for making available material prior to publication. I would also like to thank J. A. Downie and S. A. Kagan for their criticisms of the manuscript.

XII. References

Adams, M. W. W., Mortenson, L. E., Chen, J.-S., 1981: Hydrogenase. Biochim. Biophys. Acta **594**, 105—176.

Albrecht, S. L., Maier, R. J., Hanus, F. J., Russell, S. A., Emerich, D. W., Evans, H. J., 1979: Hydrogenase in *R. japonicum* increases nitrogen fixation by nodulated soybeans. Science **203,** 1255—1257.

Arima, Y., 1981: Respiration and efficiency of nitrogen fixation by nodules formed with a hydrogen uptake positive strain of *Rhizobium japonicum.* Soil Sci. Plant Nutr. **27,** 115—120.

Arp, D. J., Burris, R. H., 1979: Purification and properties of the particulate hydrogenase from the bacteroids of soybean root nodules. Biochim. Biophys. Acta **570,** 221—230.

Arp, D. J., Burris, R. H., 1981: Kinetic mechanism of the hydrogen oxidizing hydrogenase from soybean nodule bacteroids. Biochemistry **20,** 2234—2240.

Arp, D. J., Burris, R. H., 1982: Isotope exchange and discrimination by the H_2-oxidizing hydrogenase from soybean root nodules. Biochim. Biophys. Acta **700,** 7—15.

Bedmar, E. J., Edie, S. A., Phillips, D. A., 1983: Host plant cultivar effects on hydrogen evolution by *Rhizobium leguminosarum.* Plant Physiol. **72,** 1011—1015.

Benson, D. R., Arp, D. J., Burris, R. H., 1979: Cell-free nitrogenase and hydrogenase from actinorhizal root nodules. Science **205,** 688—689.

Benson, D. R., Arp, D. J., Burris, R. H., 1980: Hydrogenase in actinorhizal root nodules and root nodule homogenates. J. Bacteriol. **142,** 138—144.

Bethlenfalvay, G. J., Phillips, D. A., 1977: Effect of light intensity on efficiency of carbon dioxide and nitrogen reduction in *Pisum sativum* L. Plant Physiol. **60,** 868—871.

Bethlenfalvay, G. J., Phillips, D. A., 1979: Variation in nitrogenase and hydrogenase activity of Alaska pea root nodules. Plant Physiol. **63,** 816—820.

Bowien, B., Schlegel, H. G., 1981: Physiology and biochemistry of aerobic hydrogen oxidising bacteria. Ann. Rev. Microbiol. **35,** 405—452.

Brewin, N. J., DeJong, T. M., Phillips, D. A., Johnston, A. W. B., 1980: Co-transfer of determinants for hydrogenase activity and nodulation ability in *Rhizobium leguminosarum.* Nature **288,** 77—79.

Brewin, N. J., Wood, E. A., Johnston, A. W. B., Dibb, N. J., Hombrecher, G., 1982: Recombinant nodulation plasmids in *Rhizobium leguminosarum.* J. Gen. Microbiol. **128,** 1817—1827.

Brewin, N. J., Dibb, N. J., Hombrecher, G., 1983: Natural variation in *Rhizobium* plasmids. In: Proceedings of the 1st Symposium on Bacteria-Plant Interactions. (A. Puhler, ed.) In the press.

Bulen, W. A., Burns, R. C., LeComte, J. R., 1965: Nitrogen fixation: hydrosulfite as electron donor with cell-free preparations of *Azotobacter vinelandii* and *Rhodospirillum rubrum.* Proc. Natl. Acad. Sci. U. S. A., **53,** 532—539.

Bulen, W. A., LeComte, J. R., 1966: The nitrogenase system from *Azotobacter:* two-enzyme requirement for N_2 reduction, ATP-dependent H_2 evolution and ATP hydrolysis. Proc. Natl. Acad. Sci. U.S.A. **56,** 979—986.

Burns, R. C., Bulen, W. A., 1966: A procedure for the preparation of extracts from *Rhodospirillum rubrum* catalyzing N_2 reduction and ATP-dependent H_2 evolution. Arch. Biochim. Biophys. **113,** 461—463.

Burns, R. C., Hardy, R. W. F. (eds.), 1975: Nitrogen Fixation in Bacteria and Higher Plants. (Molecular Biology, Biochemistry and Biophysics, Vol. 21.) Berlin - Heidelberg - New York: Springer.

Cantrell, M. A., Haugland, R. A., Evand, H. J., 1983: Construction of a *Rhizobium japonicum* gene bank and use in the isolation of a hydrogen uptake gene. Proc. Natl. Acad. Sci. U. S. A. **80,** 181—185.

Cantrell, M. A., Hickok, R. E., Evans, H. J., 1982: Identification and characterization of plasmids in hydrogen uptake positive and hydrogen uptake negative strains of *Rhizobium japonicum.* Arch. Microbiol. **131,** 102—106.

Carter, K. R., Jennings, N. T., Hanus, J., Evans, H. J., 1978: Hydrogen evolution and uptake by nodules of soybeans inoculated with different strains of *Rhizobium japonicum.* Can. J. Microbiol. **24,** 307—311.

Carter, K. R., Rawlings, J., Orme-Johnson, W. H., Becker, R. R., Evans, H. J., 1980: Purification and characterization of a ferredoxin from *Rhizobium japonicum* bacteroids. J. Biol. Chem. **255,** 4213—4223.

Chatt, J., 1981: Towards new catalysts for nitrogen fixation. In: Gibson, A. H.,

Newton, W. E. (eds.), Current perspectives in nitrogen fixation, pp. 15—21. New York: Elsevier/North Holland.

Conrad, R., Seiler, W., 1980: Contribution of hydrogen production by biological nitrogen fixation to the global hydrogen budget. J. Geophys. Res. **85,** 5493—5498.

DeJong, T. M., Phillips, D. A., 1981: Nitrogen stress and apparent photosynthesis in symbiotically grown *Pisum sativum.* Plant Physiol. **68,** 309—313.

DeJong, T. M., Brewin, N. J., Johnston, A. W. B., Phillips, D. A., 1982: Improvement of symbiotic properties in *Rhizobium leguminosarum* by plasmid transfer. J. Gen. Microbiol. **128,** 1829—1838.

Dixon, R. O. D., 1967: Hydrogen uptake and exchange by pea root nodules. Ann. Bot. **31,** 179—188.

Dixon, R. O. D., 1968: Hydrogenase in pea root nodule bacteroids. Arch. Microbiol. **62,** 272—283.

Dixon, R. O. D., 1972: Hydrogenase in legume root nodule bacteroids: occurrence and properties. Arch. Microbiol. **85,** 193—201.

Dixon, R. O. D., 1978: Nitrogenase-hydrogenase interrelationships in rhizobia. Biochimie **60,** 233—236.

Dixon, R. O. D., Blunden, E. A. G., Searl, J. W., 1981: Inter-cellular space and hydrogen diffusion in root nodules of pea *Pisum sativum* and lupine *Lupinus alba.* Plant Sci. Lett. **23,** 109—116.

Drevon, J. J., Frazier, L., Russell, S. A., Evans, H. J., 1982: Respiratory and nitrogenase activities of soybean nodules formed by hydrogen uptake negative (Hup^-) mutant and revertant strains of *Rhizobium japonicum* characterized by protein patterns. Plant Physiol. **70,** 1341—1346.

Eady, R. R., Imam, S., Lowe, D. J., Miller, R. W., Smith, B. E., Thorneley, R. N. F., 1980: The molecular enzymology of nitrogenase. In: Stewart, W. D. P., Gallon, J. R. (eds.), Nitrogen Fixation, pp. 19—35. London: Academic Press.

Eisbrenner, G., Evans, H. J., 1982a: Carriers in the electron transport from molecular hydrogen to oxygen in *Rhizobium japonicum* bacteroids. J. Bacteriol. **149,** 1005—1012.

Eisbrenner, G., Evans, H. J., 1982b: Spectral evidence for a component involved in hydrogen metabolism of soybean nodule bacteroids. Plant Physiol. **70,** 1667—1672.

Eisbrenner, G., Hickok, R. E., Evans, H., 1982: Cytochrome patterns in *Rhizobium japonicum* cells grown under chemolithotrophic conditions. Arch. Microbiol. **132,** 230—235.

Eisbrenner, G., Evans, H. J., 1983: Aspects of hydrogen metabolism in nitrogen-fixing legumes and other plant microbe interactions. Ann. Rev. Plant Physiol. **34,** 105—136.

Emerich, D. W., Ruiz-Argüeso, T., Ching, T. M., Evans, H. J., 1979: Hydrogen-dependent nitrogenase activity and ATP formation in *R. japonicum* bacteroids. J. Bacteriol. **137,** 153—160.

Emerich, D. W., Albrecht, S. L., Russell, S. A., Ching, T. M., Evans, H. J., 1980a: Oxyleghaemoglobin-mediated hydrogen oxidation by *Rhizobium japonicum* USDA 122 DES bacteroids. Plant Physiol. **65,** 605—609.

Emerich, D. W., Ruiz-Argüeso, T., Russell, S. A., Evans, H. J., 1980b: Investigation of the H_2 oxidation system in *Rhizobium japonicum* 122 DES nodule bacteroids. Plant Physiol. **66,** 1061—1066.

Evans, H. J., Emerich, D. W., Ruiz-Argüeso, T., Maier, R. J., Albrecht, S. L., 1980: Hydrogen metabolism in the legume-*Rhizobium* symbiosis. In: Newton, W. E.,

Orme-Johnson, W. E. (eds.), Nitrogen fixation II, pp. 69—86. Baltimore: University Park Press.

Evans, H. J., Purohit, K., Cantrell, M. A., Eisbrenner, G., Russell, S. A., Hanus, F. J., Lepo, J. E., 1981: Hydrogen lossess and hydrogenases in nitrogen-fixing organisms. In: Gibson, A. H., Newton, W. E. (eds.), Current perspectives in nitrogen fixation, pp. 84—96. New York: Elsevier/North Holland.

Gibson, A. H., Dreyfus, B. L., Lawn, R. J., Sprent, J. I., Turner, G. L., 1981: Host and environmental factors affecting hydrogen evolution and uptake. In: Gibson, A. H., Newton, W. E. (eds.), Current perspective in nitrogen fixation, p. 373. New York: Elsevier/North Holland.

Hageman, R. V., Burris, R. H., 1978: Nitrogenase and nitrogenase reductase associate and dissociate with each catalytic cycle. Proc. Natl. Acad. Sci. U.S.A. **75,** 2699—2702.

Hageman, R. V., Burris, R. H., 1980: Electron allocation to alternative substrates of *Azotobacter* nitrogenase is controlled by the electron flux through dinitrogenase. Biochim. Biophys. Acta **591,** 63—75.

Hanus, F. J., Maier, R. J., Evans, H. J., 1979: Autotrophic growth of H_2-uptake positive strains of *R. japonicum* in an atmosphere supplied with hydrogen gas. Proc. Natl. Acad. Sci. U. S. A. **76,** 1788—1792.

Hanus, F. J., Albrecht, S. L., Zablotowicz, R. M., Emerich, D. W., Russell, S. A., Evans, H. J., 1981: The effect of the hydrogenase system in *Rhizobium japonicum* inocula on the nitrogen content and yield of soybean seed in field experiments. Agron. J. **73**, 368—372.

Haugland, R., Verma, D. P. S., 1981: Interspecific plasmid and genomic DNA sequence homologies and localization of *nif* genes in effective and ineffective strains of *Rhizobium japonicum.* J. Mol. Appl. Genet. **1,** 205—217.

Haugland, R. A., Hanus, F. J., Cantrell, M. A., Evans, H. J., 1983: A rapid screening method for identifying hydrogenase activity in *Rhizobium japonicum.* Appl. and Environ. Microbiol. **45,** 892—897.

Hardy, R. W. F., Havelka, U. D., 1975: Nitrogen fixation research: a key to world food. Science **188,** 633—643.

Haystead, A., Robinson, R., Stewart, W. D. P., 1970: Nitrogenase activity in extracts of heterocystous and non-heterocystous blue-green algae. Arch. Microbiol. **72,** 235—243.

Hinkle, P. C., McCarty, R. E., 1978: How cells make ATP. Sci. Amer. **238,** 104—123.

Hoch, G. E., Little, H. N., Burris, R. H., 1957: H_2 evolution from soybean root nodules. Nature **179,** 430—431.

Hombrecher, G., Brewin, N. J., Johnston, A. W. B., 1981: Linkage of genes for nitrogenase and nodulation ability on plasmids in *Rhizobium leguminosarum* and *R. phaseoli.* M. G. G. **182,** 133—136.

Klucas, R. V., Hanus, F. J., Russell, S. J., Evans, H. J., 1983: Nickel: A micronutrient element for hydrogen-dependent growth of *Rhizobium japonicum* and for expression of urease activity in soybean leaves. Proc. Natl. Acad. Sci. U. S. A. **80,** 2253—2257.

Koch, B., Evans, H. J., Russell, S. A., 1967: Properties of the nitrogenase system in cell-free extracts of bacteroids from soybean root nodules. Proc. Natl. Acad. Sci. U.S.A. **58,** 1343—1350.

Lepo, J. E., Hanus, F. J., Evans, H. J., 1980: Chemoautotrophic growth of hydrogen uptake positive strains of *R. japonicum.* J. Bacteriol. **141,** 664—670.

Lepo, J. E., Hickok, R. E., Cantrell, M. A., Russell, S. A., Evans, H. J., 1981: Revertible hydrogen uptake-deficient mutants of *Rhizobium japonicum.* J. Bacteriol. **146,** 614—620.

Lim, S. T., 1978: Determination of hydrogenase in free-living cultures of *Rhizobium japonicum* and energy efficiency of soybean nodules. Plant Physiol. **62,** 609—611.

Lim, S. T., Uratsu, S. L., Weber, D. F., Keyser, H. H., 1981: Hydrogen uptake (hydrogenase) activity of *Rhizobium japonicum* strains forming nodules in soybean production areas of the U. S. A. In: Lyons, J. M., Valentine, R. C., Philips, A. A., Rains, A. W., Huffaker, R. C. (eds.), Genetic engineering of symbiotic nitrogen fixation and conservation of fixed nitrogen, pp. 159—171. New York: Plenum Press.

Lopez, M., Carbonero, V., Cubrera, E. Ruiz-Argüeso, T., 1983: Effects of host on the expression of the H_2 uptake hydrogenase of *Rhizobium* in legume nodules. Plant Sci. Lett. **29,** 191—199.

Maier, R. J., 1981: *Rhizobium japonicum* mutant strains unable to grow chemoautotrophically. J. Bacteriol. **145,** 533—540.

Maier, R. J., Campbell, N. E. R., Hanus, F. J., Simpson, F. B., Russel S. A., Evans, H. J., 1978a: Expression of hydrogenase activity in free-living *Rhizobium japonicum.* Proc. Natl. Acad. Sci. U. S. A. **75,** 3258—3262.

Maier, R. J., Postgate, J. R., Evans, H. J., 1978: Mutants of *R. japonicum* unable to utilize hydrogen. Nature **276,** 494—495.

Maier, R. J., Hanus, F. J., Evans, H. J., 1979: Regulation of hydrogenase in *R. japonicum.* J. Bacteriol. **137,** 824—829.

Maier, R. J., Merberg, D. M., 1982: *Rhizobium japonicum* mutants that are hypersensitive to repression of H_2 uptake by oxygen. J. Bacteriol. **150,** 161—167.

Maier, R. J., Mutaftschiev, S., 1982: Revonstitution of H_2 oxidation activity from H_2 uptake negative mutants of *Rhizobium japonicum* bacteroids. J. Biol. Chem. **257,** 2092—2096.

Manian, S. S., O'Gara, F., 1982: Derepression of ribulose bisphosphate carboxylase activity in *Rhizobium meliloti.* FEMS Microbiology Lett. **14,** 95—99.

Masterson, R. V., Russell, P. R., Atherly, A. G., 1982: Nitrogen fixation genes and large plasmids of *Rhizobium japonicum.* J. Bacteriol. **152,** 928—931.

McCrae, R. E., Hanus, J., Evans, H. J., 1978: Properties of the hydrogenase system in *Rhizobium japonicum* bacteroids. Biochem. Biophys. Res. Comm. **80,** 384—390.

Mortenson, L. E., 1978: The role of dihydrogen and hydrogenase in nitrogen fixation. Biochimie **60,** 219—223.

Nelson, L. M., Child, J. J., 1981: Nitrogen fixation and hydrogen metabolism in *Rhizobium leguminosarum* isolates in pea *Pisum sativum* root nodules. Can. J. Microbiol. **27,** 1028—1034.

Nelson, L. M., Salminen, S. O., 1982: Uptake hydrogenase activity and ATP formation in *Rhizobium leguminosarum* bacteroids. J. Bacteriol. **151,** 989—995.

Pahwa, K., Dogra, R. C., 1981: Hydrogen recycling system in mung bean *Vigna radiata Rhizobium* in relation to nitrogen fixation. Arch. Microbiol. **129,** 380—383.

Pate, J. S., Atkins, C. A., Rainbird, R. M., 1981: Theoretical and experimental costing of nitrogen fixation and related processes in nodules of legumes. In: Gibson, A. H., Newton, W. E. (eds.), Current Perspectives in Nitrogen Fixation, pp. 105—116. New York: Elsevier/North Holland.

Phelps, A. S., Wilson, P. M., 1941: Occurrence of hydrogenase in nitrogen fixing organisms. Proc. Soc. Exp. Biol. Med. **47,** 473—476.

Phillips, D. A., 1980: Efficiency of symbiotic nitrogen fixation in legumes. Ann. Rev. Plant Physiol. **31,** 29—49.

Purohit, K., Becker, R. R., Evans, H. J., 1982: D-Ribulose-1,5-bisophosphate carboxylase/oxygenase from chemolithotrophically-grown *Rhizobium japonicum* and inhibition by D-4-phosphoerythronate. Biochim. Biophys. Acta **715,** 320—329.

Rainbird, R. M., Atkins, C. A., Pate, J. S., Sandford, P., 1983: Significance of hydrogen evolution in the carbon and nitrogen economy of nodulated cowpea. Plant Physiol. **71,** 122—127.

Robson, R. L., Postgate, J. R., 1980: Oxygen and hydrogen in biological nitrogen fixation. Ann. Rev. Microbiol. **34,** 183—207.

Ruiz-Argüeso, T., Hanus, J., Evans, H. J., 1978: Hydrogen production and uptake by pea nodules as affected by strains of *Rhizobium japonicum.* Arch. Microbiol. **116,** 113—118.

Ruiz-Argüeso, T., Emerich, D. W., Evans, H. J., 1979a: Characteristics of the hydrogen oxidizing system in soybean nodule bacteroids. Arch. Microbiol. **121,** 199—206.

Ruiz-Argüeso, T., Emerich, D. W., Evans, H. J., 1979b: Hydrogenase system in legume nodules: a mechanism for providing nitrogenase with energy and protection from oxygen damage. Biochim. Biophys. Res. Commun. **86,** 259—264.

Ruiz-Argüeso, T., Maier, R. J., Evans, H. J., 1979c: Hydrogen evolution from alfalfa and clover nodules and hydrogen uptake by free-living *R. meliloti.* Appl. and Environ. Microbiol. **37,** 582—587.

Ruiz-Argüeso, T., Cabrera, E., Bertalmeo, M. B., 1981: Induction of hydrogenase in non-nitrogen fixing soybean nodules produced by a Hup^+ strain of *Rhizobium japonicum.* In: Gibson, A. H., Newton, W. E. (eds.), Current perspectives of nitrogen fixation, p. 337. New York: Elsevier/North Holland.

Schink, B., Schlegel, H. G., 1979: The membrane-bound hydrogenase of *Alcaligenes eutrophus.* Biochim. Biophys. Acta **567,** 315—324.

Schubert, K. R., 1982: The energetics of biological nitrogen fixation. Workshop Summary 1, American Society of Plant Physiologists, pp. 1—30.

Schubert, K. R., Evans, H. J., 1976: Hydrogen evolution: a major factor affecting the efficiency of nitrogen fixation in nodulated symbionts. Proc. Natl. Acad. Sci. U. S. A. **73,** 1207—1211.

Schubert, K. R., Engelke, J. A., Russell, S. A., Evans, H. J., 1977: Hydrogen reactions of nodulated leguminous plants I. Effects of rhizobial strain and plant age. Plant Physiol. **60,** 651—654.

Schubert, K. R., Jennings, N. T., Evans, H. J., 1978: Hydrogen reactions of nodulated leguminous plants II. Effects on dry matter accumulation and nitrogen fixation. Plant Physiol. **61,** 398—401.

Simpson, F. B., Maier, R. J., Evans, H. J., 1979: Hydrogen-stimulated CO_2 fixation and coordinate induction of hydrogenase and ribulose-bisophosphate carboxylase in a H_2-uptake positive strain of *Rhizobium japonicum.* Arch. Microbiol. **123,** 1—8.

Thorneley, R. N. F., Eady, R. R., 1977: Nitrogenase of *Klebsiella pneumoniae.* Distinction between proton-reducing and acetylene-reducing forms of the enzyme: effect of temperature and component protein ratio on substrate reduction kinetics. Biochem. J. **167,** 457—461.

Thorneley, R. N. F., Lowe, D. J., 1981: Pre-steady state kinetic studies with nitrogenase from *Klebsiella pneumoniae.* In: Gibson, A. H., Newton, W. E. (eds.), Current perspectives in nitrogen fixation, p. 360. New York: Elsevier/North Holland.

Thorneley, R. N. F., Lowe, D. J., 1982: Mechanistic studies on nitrogenase from *Klebsiella pneumoniae* using the rapid quench technique. Israeli Journ. Bot. **31,** 1—11.

Van der Werf, A. N., Yates, M. G., 1978: Hydrogenase from nitrogen-fixing *Azotobacter chroococcum.* In: Schlegel, H. G., Schneider, K. (eds.), Hydrogenases: their catalytic activity, structure and function, pp. 307—326. Göttingen: Erich Goltze KG.

Wang, R. T., 1980: Amperometric hydrogen electrode. Methods in Enzymol. **69,** 409—412.

Williams, L. E., DeJong, T. M., Phillips, D. A., 1981: Carbon and nitrogen limitations on soybean seedling development. Plant Physiol. **68,** 1206—1209.

Chapter 7

Symbiotic Relationships in Actinorhizae

A. Moiroud

Ecologie Microbienne ERA CNRS 848, Laboratoire de Microbiologie Physiologique et Appliquée, Université Claude Bernard Lyon I, F-69622 Villeurbanne, Cedex, France

V. Gianinazzi-Pearson

Station d'Amélioration des Plantes, INRA, BV 1540, 21034 Dijon, France

With 1 Figure

Contents

I. Introduction

The importance of nitrogen-fixing non-legumes for the nitrogen economy of certain ecosystems has been known for a long time (see Silvester, 1977), and some of these plants probably played an essential part in soil reconstruction during and after the formation of glaciers in the Pleistocene age when species of *Dryas, Hippophaë, Elaeagnus, Alnus* and *Shepherdia* covered large areas of northern Europe and Canada (Lawrence *et al.*, 1967, Silvester, 1974). Although the nodules of non-legumes were first described in 1829 and nodule formation by an endophyte was demonstrated in 1866 (see Goodchild, 1977), these plants were considered of little agricultural

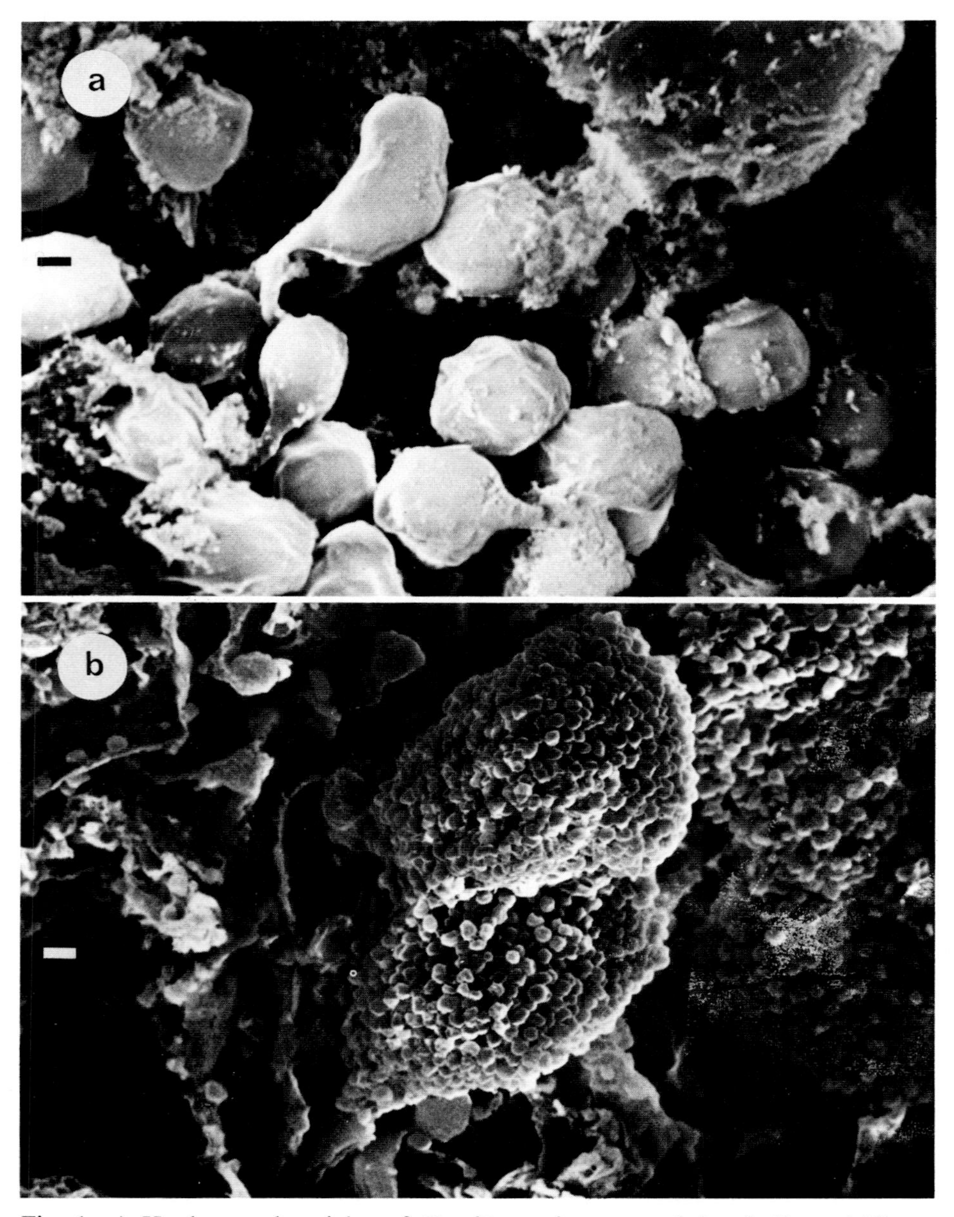

Fig. 1. a) Hyphae and vesicles of *Frankia* sp. in root-nodule of *Alnus viridis* sp. *viridis* .S.E.M. bar = 1μm (Photograph by A. Capellano) (× 5000)
b) Sporangia of *Frankia* sp. in root-nodule of *Alnus viridis* sp. *viridis* .S.E.M. bar = 2,5μm (Photograph by A. Capellano) (× 2000)

interest and received virtually no attention for nearly a century. It was only in the 1950's, when their potentiality for forestry was recognized, that studies really began on the symbiotic nitrogen-fixing associations formed by non-legumes. However, fundamental knowledge concerning this type of symbiosis is still very fragmentary, as compared to that known about the symbiotic associations between *Rhizobia* and legumes, and there is a particular lack of information concerning the microbial partner, the infection processes and the physiological relationships existing within the symbiosis.

II. The Symbiotic Association

The nodules formed by nitrogen-fixing non-legumes are grouped together into two main types, the *Alnus* type and the *Myrica* type, according to their morphology and the plants where they were first observed (Becking, 1977). The *Alnus* type, a coralloid root nodule, consists of short, fat, ramified lobes formed by dichotomous branching and originating from lateral roots with inhibited or very slowly growing apical meristems. These develop on plants from numerous genera including *Dryas, Purshia, Elaeagnus, Coriaria* and *Discaria.* Nodules of the *Myrica* type have thin lobes and the apex of each nodule lobe produces a negatively geotropic root so that the nodule becomes clothed with upward growing rootlets. This type of nodule is found in such genera as *Casuarina, Comptonia* or *Rubus.* The endophyte is localized within the cortical parenchyma tissue of the nodules and it never invades the nodule vascular tissue nor the meristem. In the distal part of the nodule, the endophyte forms hyphae which fill the host cell and which spread from one cell to another by perforating the host cell walls (Lalonde and Knowles, 1975, Newcomb *et al.*, 1978, Capellano and Moiroud, 1979, Newcomb and Pankhurst, 1982a). Below this region the hyphae swell and give rise to vesicles (Fig. 1a), the shape and size of which can vary from one host plant to another. The endophyte also develops sporangia filled with spores (Fig. 1b) and these are mainly localized in the middle and basal parts of the nodule. Contrary to the hyphae and vesicles, sporangia are not always present and up to now they have only been found in the genera *Alnus, Myrica, Dryas* and *Purshia* (Akkermans *et al.,* 1979).

A. The Host Plant

Nitrogen-fixing non-legumes, generally termed *actinorhizal* plants, are all angiosperms and, apart from two species of *Datisca* (Chaudary, 1978), they all belong to woody species. About 178 species are recognized at present (Table 1) and these belong to 20 different genera in eight families and seven orders (Akkermans and Houwers, 1979, Akkermans and Roelofsen, 1980). However, there has been no systematic study of *actinorhizal* plants, especially amongst tropical species, and it is highly probable that the number of plant species forming actinorhizae is much higher. Within a given genus, the number of species that form nodules can vary greatly; for example, nearly all the species in the genera *Alnus, Dryas, Purshia, Shepherdia* or *Coriaria* nodulate whilst about halft of those belonging to the genera *Casuarina, Elaeagnus* and *Ceanothus* are non-nodulating. Furthermore, certain species are not consistently nodulated in their native habitats and this is particularly true of some *Casuarina* species in Australia (Lawrie, 1982) and of *D. octopetela* in the mountain ranges of Europe (Bond, 1976a).

Table 1. Genera of *Dicotyledonous* Nitrogen-Fixing Plants with *Frankia* Symbioses

Genus	Number of nodulated species
Alnus	34
Casuarina	25
Ceanothus	31
Cercocarpus	4
Chamaebatia	1
Colletia	3
Coriaria	14
Cowania	1
Datisca	2
Discaria	6
Dryas	3
Elaeagnus	17
Hippophaë	1
Kentrothamnus	1
Myrica	26
Purshia	2
Rubus	1
Shepherdia	3
Talguena	1
Trevoa	2

Apart from the tropical *Casuarina*, actinorhiza-forming species are mainly found in the cold or temperate zones of the northern hemisphere (Becking, 1977, Silvester, 1977) and certain, like *D. integrifolia*, are limited to the circumpolar regions. Those species of *Cercocarpus, Coriaria* or *Alnus* present in the tropics only occur, in fact, at high altitude (Rodriguez-Barrueco, 1868; Hoeppel and Wollum, 1971). Actinorhizal plants are able to colonize very different ecosystems ranging from those with organic acid soils to dry sand dunes. For example, certain *Alnus, Coriaria* and *Shepherdia* species are common in wet habitats along rivers or as pioneer colonizers of glacier moraines, whilst *H. rhamnoides* and species of *Casuarina* or *Elaeagnus* are able to grow in the particularly dry sands of coastal dunes.

B. The Endophyte

Electron microscope studies of *Alnus* and *Myrica* nodules first revealed, in 1964, that the actinorhizal endophyte is a procaryote, closely related to the Actinomycetes (Becking *et al.*, 1964, Silver, 1964, Gardner, 1965) and in

1970 Becking introduced into this class a new genus, *Frankia*, which groups all the endophytes of actinorhizal plants. Later, Callaham *et al.* (1978) successfully isolated and cultured the endophyte from actinorhizal nodules and there have since been many reports of endophyte isolation from different nitrogen-fixing non-legumes (Baker and Torrey, 1979, Baker *et al.,* 1979, Berry and Torrey, 1979, Burggraaf *et al.,* 1981, Lalonde *et al.,* 1981, Benson, 1982, Diem *et al.,* 1982). In pure culture, actinorhizal endophytes behave as microaerophilic and mesophilic microorganisms (Burggraaf and Shipton, 1982, Moiroud and Faure-Raynaud (in press). Isolates usually have much-divided, septate hyphae which develop vesicles and sporangia with immobile spores; this morphology resembles that described within the actinorhizal nodules (Callaham *et al.,* 1978, Baker *et al.,* 1979, Lalonde and Calvert, 1979, Faure-Raynaud *et al.,* 1982, Normand and Lalonde, 1982). The number of vesicles formed in pure culture depends both on the isolate and the culture conditions. Endophytes from *Alnus* nodules form few vesicles but their production can be stimulated by certain culture conditions (Tjepkema *et al.,* 1980, 1981); isolates from *Casuarina, Elaeagnus,* and *Hippophaë,* on the contrary, produce many vesicles spontaneously (Diem *et al.,*1982, Lalonde, personal communication). The number of sporangia produced in pure culture can also be modified by the addition of various compounds to the culture medium (Lalonde and Calvert, 1979, Perradin 1982, Simonet, unpublished data). Certain endophyte cultures have a characteristic pigmentation; those obtained from *Casuarina, Elaeagnus* and *Hippophaë* are rose-coloured due to the formation of a crystallised pigment in the culture (Moiroud and Arpin, unpublished data) and Horrière *et al.* (1982) have reported an endophyte with black spores.

The genus *Frankia* is now generally recognized and it is characterized by isolates which, although differing in their nutrient requirements (Lalonde and Calvert, 1979, Blom, 1981, 1982, Blom *et al.,* 1980, Burggraaf *et al.,* 1981, Lechevalier *et al.,* 1982, Faure-Raynaud and Moiroud, 1983), all form nitrogen-fixing nodules with a suitable host plant, develop sporangia with immobile spores *in vitro* (Lechevalier and Lechevalier, 1979) and appear to contain specific sugars (Lalonde, personal communication). There have been several attempts to define natural species within the genus *Frankia* based on 1) host-endophyte specificity (Becking, 1970) — now considered invalid (Lalonde and Calvert, 1979, Rodriguez-Barrueco and Miguel, 1979, Baker and Torrey, 1980, Baker *et al.,* 1980, Gauthier *et al.,* 1981 b, 2) serological characterization of *Frankia* isolates vis-a-vis host compatibility (Baker *et al.,* 1981, Baker 1982), 3) analysis of endophyte cell walls and detection of particular compounds within them (Lechevalier and Lechevalier, 1979, Akkermans and Roelofsen, 1980) and 4) the nutrient requirements of the endophytes in pure culture (Lechevalier *et al.,* 1982, Shipton and Burggraaf, 1982, Faure-Raynaud and Moiroud, 1983), but none of these approaches have given satisfactory results. Certain techniques used in molecular biology, such as DNA-DNA hybridization, have been successfully applied to studies on the taxonomy of bacteria (Deley, 1979) and other Actinomycetes (Bradley *et al.,* 1973) and such an approach may prove useful for taxonomic studies of *Frankia*. With improved lysis (Normand *et al.,* 1983) and DNA extraction techniques, it has in fact recently been possible to apply the DNA-DNA hybridization method to pure cultures of *Frankia* (Simonet *et al.,* in press). By applying these studies to a large number of *Frankia* isolates, it should be possible to advance towards a new classification within this group of microorganisms.

Two distinct types of *Frankia* can however be recognized amongst isolates, one of which forms sporangia within the actinorhizal nodules (Sp^+ nodules) and another which does not form sporangia or sometimes forms very small sporangia with few spores in the host tissues (Sp^- nodules) (van Dijk, 1978). The behaviour of the two types also differs in other respects: the Sp^+ endophyte is generally less efficient, grows much more slowly, has a lower biomass and is more difficult to isolate than the Sp^- type (Hall *et al.,* 1979, Burggraaf *et al.,* 1981, Normand and Lalonde, 1982). Both types can infect the same host plant and the fact that in pure culture they both develop sporangia suggests that the expression of this character within the host tissues may be genetically determined (van Dijk, 1978). Successive subculturing of *Frankia* isolates on synthetic media can cause the loss or modification of certain properties of the endophytes, including the loss of the sporulating ability of Sp^+ isolates with reinfection of the host plant (Bruggraaf *et al.,* 1981, Lechevalier *et al.,* 1982, Moiroud, unpublished data). The reasons for these modifications remain obscure.

Very little is known about the genetics of *Frankia* endophytes and absolutely nothing is known of the plant or microbial genes involved in the processes of nodule formation or nitrogen fixation. However, Ruvkun and Ausubel (1980) have shown that the DNA of *Frankia* can hybridize with a *nif* cluster of *Klebsiella pneumoniae.* Furthermore, the recent discovery of low molecular weight plasmids in certain isolates of *Frankia* (Normand *et al.,* 1983), together with the obtention of protoplasts from pure *Frankia* culture (Faure-Raynaud *et al.,* in press), opens interesting possibilities for future genetic studies of these microorganisms.

C. Host Plant-Endophyte Specificity

Although there have been several studies on host plant-endophyte specificity in actinorhizal symbioses, it is difficult to draw clearcut conclusions from most of them since crushed nodules were used as inoculum and the possibility of contaminations cannot be excluded. Some results have been obtained using isolated *Frankia* cultures (see Quispel and Burggraaf, 1981) and these indicate that cross inoculations are possible between endophytes and plants form the genera *Alnus, Myrica* and *Comptonia* whilst those from genus *Elaeagnus* seem to show a greater degree of host-endophyte specificity. However, more extensive investigations are necessary in order to clearly define the level of specificity that may exist between actinorhizal endophytes and their host plants.

D. Host Plant-Endophyte Interactions

The formation of a symbiotic association supposes that a whole series of events must occur which lead to the establishment of a complex, balanced relationship between the participating organisms. These events must involve a certain degree of recognition between the two symbionts and although nothing is known of the recognition phenomena that exist in actinorhizal symbioses, it is reasonable to think that they may bear some similarity to those reported in the *Rhizobium*-legume associations (Bauer, 1981, see also chapter 3 and 4) where lectin-like molecules are believed to play a role in the cellular recognition between host and endophyte. Chaboud and Lalonde (1982) have localized certain specific sugar residues on the outer walls of hyphae, vesicles and sporangia of *in vitro Frankia* cultures but whether these compounds are somehow involved in interactions with the host plant is not known. Most of the information concerning host plant-endophyte interactions in actinorhizae have in fact been obtained from cytological studies using light and electron microscopy.

In presence of the host plant, the endophyte seems to proliferate in the rhizosphere surrounding the elongating region of the root (Lalonde, 1977) and, as in *Rhizobium*-legume interactions, this is accompanied by extensive curling of a large number of developing root hairs in this zone (Becking, 1977, Callaham *et al.,* 1979). This root hair deformation may be due to hormonal substances excreted by the endophyte, although these have yet to be identified. It is not, however, a specific response as it can also be induced by other bacteria (Berry and Torrey, 1982) or by cell-free nodule extracts (Pizelle, 1972). Penetration of the endophyte into a root hair proceeds by hyphal growth (Angulo Carmona, 1974, Lalonde, 1977, Callaham and Torrey, 1977) and occurs at the root hair tip where it is most strongly curved. In *Alnus* associations the penetrating hypha is always surrounded by a layer of material which is continuous with the external layer covering the root hair cell and root epidermis, and it has been suggested that this layer may have a role during the symbiotic host-endophyte interactions leading to formation of the nodule (Lalonde, 1977).

With penetration of the root hair, the host cell wall and plasmalemma invaginate to surround the endophyte. The infected root hair is the site of intense metabolic activity whilst the adjacent uninfected root hairs rapidly degenerate. As the endophyte enters the root hair, it becomes separated from the host plasmalemma by an encapsulating material which probably consists of cell wall material produced by the host. The hypha or hyphae of the endophyte migrate towards the base of the root hair cell, accompanied by the host nucleus, and before the endophyte reaches the inner cell layers, numerous divisions occur in the cortical cells of the root situated below the root hair (Angulo Carmona, 1974, Callaham and Torrey, 1977, Lalonde, 1977). This results in a slight thickening of one side of the root giving rise to what is called the pre-nodule. The formation of a pre-nodule appears to be a developmental stage which is essential for nodule formation in actinorhizal plants. Only a few cells of the pre-nodule are invaded by the endophyte; these are much larger than uninfected cells and they have a very dense, granular cytoplasm, a swollen, often lobed nucleus and numerous small vacuoles. It is not known whether these cells, or those of the true nodule, are polyploide as in legume nodules. The uninfected cells of the pre-nodule contain large amounts of starch and their vacuoles have polyphenolic or tannin contents (Angulo Carmona, 1974, Callaham and Torrey, 1977, Prin, unpublished results). In the pre-nodule stage, the endophyte is already often present in its vesicular form but these pre-nodules do not fix nitrogen. The pre-nodule remains in a disorganized state and never gives rise directly to the true nodule. The subsequent events which lead to the formation of a true nodule differ depending on whether the nodule is an *Alnus* type or a *Myrica* type.

In the *Alnus* type, root primordia originate from the pericycle and develop avoiding the infected root cells, thus giving rise to the nodule *sensu-stricto.* When a primordium is sufficiently developed, the cortical parenchym cells become infected by hyphae of the endophyte coming from the pre-nodule. The nodule lobe grows through the root epidermis, elongates and the meristem divides repeatedly to give a dichotomously branched noduled which has been called a *"rhizothamnium".* The nodule therefore corresponds to an adventitious root whose initiation and development is induced by the presence of the actinorhizal endophyte within the root tissues. The mechanisms responsible for this induction and those which inhibit subsequent elongation of the nodule are completely unknown. In the *Myrica* type of nodules, the pre-nodule stage is succeeded by three different stages of root nodule development: nodule lobe formation, a transitional phase of arrested growth and nodule root development. Several root primordia originate almost simultaneously with pre-nodule formation, probably in response to stimuli coming from the endophyte (Callaham and Torrey, 1977, Torrey and Callaham, 1978). These primordia are equivalent to lateral roots since their development involves pericycle, endodermis and cortical cell derivates. The primordia develop slowly to form a cluster of nodule lobes, the cortical cells of which are invaded by the actinorhizal endophyte. This phase of development is followed by a period of arrested growth which can extend from a few days to several weeks, and after which the activity of the nodule lobe meristem changes to give rise to a nodule root. The latter, which is negatively geotropic is thinner than the nodule lobe and is completely devoid of the endophyte. It has been suggested that these nodule roots may facilitate gas diffusion to the endophyte situated in the nodule lobe. Thus the *Myrica* type of nodule also corresponds to an adventitious root induced by the endophyte. Certain authors have suggested that hormone modifications could be responsible for the morphological changes occurring in actinorhizal roots (Silver *et al.,* 1966, Dullaart, 1970, Bermudez de Castro *et al.,* 1977) but whether actinorhizal endophytes produce plant hormones or whether they can induce changes in the plant hormone equilibrium is not known. It is evident that no single mechanism will determine the different stages in the processes of nodule initiation and development. The fact that the presence of pre-nodules is not always followed by nodule formation and that the endophyte is strictly limited to a well-defined zone in the host tissues indicates that the host plant somehow controls endophyte development. There is some evidence that phenolic compounds may contribute to host control of the actinorhizal infection. These are abundant in the roots and nodules of actinorhizal plants, and the infected zone of the pre-nodule is delimited by a layer of tannin-rich cells (Angulo Carmona, 1974). Furthermore, certain cinnamic acid derivatives which exist in *Alnus* roots inhibit the growth of *Frankia* in pure culture and benzoic phenolic acids strongly modify its morphology (Perradin, 1982, Perradin *et al.*, in press).

Numerous ultracytological studies have provided some insight into the complex interactions occurring between actinorhizal host and endophyte at the cellular level. In the nodule, the infection spreads by hyphae passing from one host cell to another; when the *Frankia* endophyte infects new host cells it penetrates the cell wall, probably by a mechanism involving enzymic digestion rather than by physical force (Lalonde and Knowles, 1975, Newcomb *et al.*, 1978), and then enters the cell by an endocytic process similar to that described for *Rhizobium* infections in legumes (Lalonde and Devoe, 1976). The host plasmalemma invaginates to surround the hyphae of the endophyte as they proliferate within the host cell and, in the inner regions of the nodule, the hyphal tips swell to form the vesicles. The morphology of these vesicles varies from one host plant to another: they are round and septate in *Alnus* (Becking *et al.*, 1964), *Elaeagnus* (Henry, 1978) or *Discaria* (Newcomb and Pankhurst, 1982b), whilst in *Comptonia* or *Myrica* they form ill-defined masses (Mian *et al.*, 1976, Newcomb *et al.*, 1978) and in *Casuarina* they are difficult to distinguish from the hyphae (Torrey, 1976, Tyson and Silver, 1979, Dommergues *et al.*, 1983). Septa have not been observed in vesicles formed in actinorhizal nodules of *Ceanothus* and nitrogen-fixing members of the Rosaceae (Bond, 1976b, Strand and Laetsch, 1977, Newcomb, 1981). These observations have led certain authors to suggest that the morphology of *Frankia* vesicles is under the control of the host plant (Lalonde, 1978, Lalonde and Calvert, 1979) but as only natural infections have been compared, it is difficult to exclude the possibility that different endophytes are involved. The vesicles are generally grouped together in a dense layer along the cell periphery and orientated towards the cell exterior but occasionally they are dispersed throughout the host cytoplasm, as in *Discaria toumatou* (Newcomb and Pankhurst, 1982b), or orientated towards the middle of the cell, as in *Datisca* (Calvert *et al.*, 1979) or *Coriaria* (Newcomb and Pankhurst, 1982a).

Hyphae, vesicles and sporangia are always separated from the host plasmalemma by a layer of material which is slightly electron-opaque: the capsule (Lalonde and Knowles, 1975, Lalonde and Devoe, 1975, 1976, van Dijk and Merkus, 1976). The capsule around the hyphae is usually fairly thick (0.5 to 3 µm) and can surround two or more hyphae together, whilst around the vesicles it is thinner, more electron-dense and fibrillar in nature. Although the capsule sometimes appears to be continuous with the wall of the host cell (Lalonde and Knowles, 1975, van Dijk and Merkus,

1976, Capellano and Moiroud 1979), cytochemical analyses have shown that it has a different composition. The capsule has a polysaccharidic nature and is made up of galacturonic acid monomers but, unlike the host cell wall, it does not contain cellulose (Lalonde and Knowles, 1975, Newcomb *et al.,* 1978, Newcomb and Pankhurst, 1982a). Electron-dense granules are also localized in the capsule and Lalonde and Devoe (1976) have interpreted these as being enzymes. An electron-translucent zone can often be observed between the vesicle wall and the capsule but this seems to be a fixation artefact which could be due to the elimination of several lipid layers from the vesicle envelope (Lalonde and Knowles, 1975, Lalonde and Devoe, 1976, Torrey and Callaham, 1982). The origin of the capsule is still not perfectly clear but it is thought that the host cell participates in its formation and it is, in fact, absent from the surface of hyphae and vesicles developing in pure cultures of *Frankia.* According to Lalonde and Knowles (1975) the capsule material in *Alnus* is produced within secretory vesicles in the host cytoplasm which migrate towards the host plasmalemma and deposit their contents on the endophyte wall, whilst Newcomb and Pankhurst (1982a) have recently reported that the endoplasmic reticulum of infected cells of *Coriaria* is swollen and has a fibrillar matrix where precursors or constituents of the capsule could be localized.

The actinorhizal infection induces several ultrastructural modifications in the host cell which illustrate the host's recognition of and reaction to the presence of the endophyte. Starch grains diminish whilst host mitochondria, ribosomes and dictyosomes increase in number. There is proliferation of the endoplasmic reticulum in the host cytoplasm, the nucleus becomes hypertrophied, frequently lobed with a large nucleolus, and numerous small vacuoles often appear near the cell wall. The changes occurring in cells that are only filled with hyphae and those containing mainly vesicles are slightly different (Newcomb and Pankhurst, 1982a). Another, rather surprising, modification that has been observed in infected cells in the actinorhizal genera *Coriaria* and *Datisca* is an important increase in the number of host nuclei (up to six) with the development of the endophyte (Calvert *et al.,* 1979, Newcomb and Pankhurst, 1982a). Newcomb and Pankhurst (1982a) also recently reported the appearance of structures, interpreted as being peroxysomes, uniquely in the infected cells of actinorhizal nodules; peroxysomes have already been observed in legume nodules but only in uninfected cells (Newcomb and Tandon, 1981, Newcomb, 1982, see also chapter 3). Changes can occur in the walls of infected cells in some actinorhizal associations; they accumulate suberin-like substances, as in *Casuarina* (Berg, 1982), or lignin, as in *Myrica* (Newcomb, 1982). Much more research is clearly necessary in order to understand the significance of these ultrastructural modifications and their implication in the host-endophyte interactions in actinorhizae.

In the older zones of the nodule, the host cells degenerate and their cytoplasmic contents become disorganized. The endophyte present in these cells also shows signs of an important disorganization, especially the vesicles (Becking *et al.,* 1964, Capellano and Moiroud, 1979, Schwintzer *et al.,*

1982), and the sporangia in the lysed host cells of Sp^+ nodules have generally released most of their spores (Suetin *et al.*, 1979).

III. Nitrogen Fixation

The efficiency of the nitrogen-fixing activity of an actinorhizal association depends on the ability of the *Frankia* endophytes both to form nodules and to fix nitrogen (Dillon and Baker, 1982). Sp^+ type endophytes are less efficient than the Sp^- type and certain *Frankia* isolates can even form nodules that are totally inefficient. Such variations underline the importance of host-endophyte compatibility in these symbiotic associations.

Only true nodules fix nitrogen in actinorhizae and there is now convincing evidence that this nitrogen fixing activity is localized within the vesicles of the *Frankia* endophyte. A strong correlation exists between the presence of these structures and the ability of nodules to reduce acetylene to ethylene (Mian *et al.*, 1976, Mian and Bond, 1978, Baker *et al.*, 1980, Lechevalier *et al.*, 1982, Schwintzer *et al.*, 1982), and vesicles, or cell-free extracts of vesicles, isolated from nodules show a strong reducing activity (Akkermans *et al.*, 1977, van Straten *et al.*, 1977, Benson *et al.*, 1979). Furthermore, pure cultures of *Frankia* can reduce acetylene and N^{15}-labelled nitrogen and although the intensity of this reducing activity is always very weak, it is directly related to the number of vesicles present in the cultures (Tjepkema *et al.*, 1980, Gauthier *et al.*, 1981a, Tjepkema *et al.*, 1981, Torrey *et al.*, 1981). The exact site of the nitrogenase within the vesicles is not known but cytological studies also show that they have a strong reducing activity (Akkermans, 1971) which appears to be localized close to the internal septa (Capellano and Moiroud, unpublished data).

The process of nitrogen reduction to ammonium by the nitrogenase is irreversibly inactivated by oxygen (Benson *et al.*, 1979, Akkermans and Roelofsen, 1980). Unlike nodules formed by *Rhizobium* in legumes, the partial pressure of oxygen in actinorhizal nodules is generally high and close to atmospheric pressure, although this can be much lower in certain zones where the intercellular spaces are absent and which probably correspond to the infected tissue (Tjepkema, 1979, 1983). There is evidence that actinorhizal nodules contain a pigment analogous to the leghaemoglobin which protects the nitrogenase in legume nodules. However, a globin-like protein has been recently observed in nodules of a non-legume, *Parasponia* (Appleby *et al.*, 1983). Certain isolates of *Frankia* can fix nitrogen in pure culture in the presence of partial pressures of oxygen that are nearly atmospheric (Tjepkema *et al.*, 1980) which indicates that the vesicles themselves must be able to protect the nitrogenase from oxygen. The vesicles have a complex, multi-layered envelope, certain layers of which contain lipids (Torrey and Callaham, 1982). In this aspect, the envelope closely resembles the internal lamellar layer in heterocysts of nitrogen-fixing cyanobacteria and which is thought to play an important role in maintaining a low oxygen pressure within the heterocyst. Torrey and Callaham (1982) have suggested that the lipid layers of the vesicle envelope could similarly create a barrier for gas exchange and thus protect the nitrogenase from the external oxygen. The protection of the nitrogenase in actinorhizae would therefore depend mainly on the endophyte with little participation from the host plant, which is a system that differs greatly from that occurring in nitrogen-fixing legume nodules.

The ammonium formed by nitrogen fixation is assimilated by a glutamine synthetase/glutamate synthase enzyme system which is localized uniquely within the host cell (Akkermans and Roelofsen, 1980, Blom *et al.*, 1981, Schubert and Coker, 1981). Little is known of the form in which the fixed nitrogen is subsequently transported to other parts of the actinorhizal plant. In *Alnus* it appears to be transported in the form of citrulline (Leaf *et al.*, 1958) whose synthesis involves an ornithine carbamyl transferase present mainly in the host cytosol (Martin *et al.*, 1982).

Nitrogen fixation is a process which has a high energy requirement and, actinorhizae being symbiotic associations, the endophyte obtains the necessary carbon compounds from the host plant (Gordon and Wheeler, 1978, Johnsrud, 1978). The exact nature of the compounds furnished by the host to the endophyte is still unknown. They do not appear to be uniquely simple sugars as many, but not all, *Frankia* isolates are unable to use these as a carbon source in pure culture (Baker and Torrey, 1980, Blom and Hardink, 1981, Blom, 1982, Faure-Raynaud and Moiroud, 1983). These microorganisms do not possess glycolytic enzymes and obtain their carbon preferentially from lipids (Blom and Hardink, 1981). Maudinas *et al.* (1982) have shown that *Alnus* nodules are, in fact, rich in lipids and saturated and unsaturated fatty acids which could provide a source of carbon to the endophyte. In pure culture, *Frankia* breaks down lipids into acetyl-CoA which can be oxidized in the tricarboxylic acid cycle with the production of energy. Endophyte vesicles from *A. glutinosa* possess the enzymes belonging to the tricarboxylic acid cycle (Akkermans *et al.*, 1981) and could therefore, as in pure *Frankia* cultures, metabolize carbon compounds by this pathway.

During the reduction of atmospheric nitrogen in the nitrogen-fixing process, protons are also reduced to give hydrogen. The production of hydrogen greatly reduces the yield from nitrogen fixation but this loss of energy by hydrogen formation is recuperated in most actinorhizal species. The actinorhizal endophyte has very efficient hydrogenase by which it can recycle all the hydrogen that is formed (Evans *et al.*, 1979, Roelofsen and Akkermans, 1979, Akkermans and Roelofsen, 1980, Benson *et al.*, 1980) and this enzyme can even oxidize atmospheric hydrogen in the soil, thus increasing the energy resources of the endophyte (Roelofsen and Akkermans, 1979).

IV. Conclusions

Although large gaps still exist in our knowledge concerning actinorhizal symbioses, considerable progress has been made over recent years, especially with respect to the isolation and culture of *Frankia*. Research into the molecular biology and genetics of these microorganisms is now beginning. With its development it should be possible not only to have a better understanding of how the actinorhizal association functions, but also to make considerable advances towards resolving such problems as endophyte taxonomy (and therefore identification), efficiency and instability, and therefore envisage the possibility of a better utilization of the symbiosis to improve the production of actinorhizal species in forestry.

V. References

Akkermans, A. D. L., 1971: Nitrogen fixation and nodulation of *Alnus* and *Hippophaë* under natural conditions. Thesis, 85 p. University of Leyden, Holland.

Akkermans, A. D. L., Houwers, A., 1979: Symbiotic nitrogen fixers available for use in temperate forestry. In: Gordon, J., Wheeler, C., Perry, D. (eds.): Symbiotic nitrogen fixation in the management of temperate forests, pp. 23—35. Oregon State University Press.

Akkermans, A. D. L., Roelofsen, W., 1980: Symbiotic nitrogen fixation by Actinomycetes in *Alnus*-type root nodules. In: Stewart, W., Gallon, J. (eds.): Nitrogen fixation (Proc. of the Phytochemical Society of Europe Symposium, Sussex), pp. 279—299. London - New York: Academic Press.

Akkermans, A. D. L., Huss-Danell, K., Roelofsen, W., 1981: Enzymes of the tricarboxylic acid cycle and the malate aspartate shuttle in the N_2-fixing endophyte of *Alnus glutinosa*. Physiol. Plant. **53,** 289—294.

Akkermans, A. D. L., Roelofsen, W., Blom, J., 1979: Dinitrogen fixation and ammonia assimilation in actinomycetous root nodules of *Alnus glutinosa*. In: Gordon, J., Wheeler, C., Perry, D. (eds.): Symbiotic nitrogen fixation in the management of temperate forests, pp. 160—174. Oregon State University Press.

Akkermans, A. D. L., van Straten, J., Roelofsen, W., 1977: Nitrogenase activity of nodules homogenates of *Alnus glutinosa:* a comparison with the pea-system. In: Newton, W., Postgate, J., Rodriguez-Barrueco, C. (eds.): Recent development in nitrogen fixation, pp. 591—603. London - New York: Academic Press.

Appleby, C. A., Tjepkema, J. D., Trinick, M. J., 1983: Hemoglobin in a nonleguminous plant, *Parasponia*. Possible genetic origin and functions in nitrogen fixation. Science **220,** 951—953.

Angulo Carmona, A., 1974: La formation des nodules fixateurs d'azote chez *Alnus glutinosa* (L) Vill. Acta Bot. Neerl. **23,** 257—303.

Baker, D., 1982: Serological and host compatibility relationship among the isolated *Frankiae*. Communication at the Conference on *Frankia*, Madison, U. S. A.

Baker, D., Torrey, J., 1979: The isolation and cultivation of actinomycetous root nodule endophytes. In: Gordon, J., Wheeler, C., Perry, D. (eds.): Symbiotic nitrogen fixation in the management of temperate forests, pp. 38—56. Oregon State University Press.

Baker, D., Torrey, J., 1980: Characterization of an effective actinorhizal microsymbiont, *Frankia* sp. ArcI 1 (Actinomycetales). Can. J. Bot. **26,** 1066—1071.

Baker, D., Newcomb, W., Torrey, J., 1980: Characterization studies of an ineffective actinorhizal microsymbiont, *Frankia* sp. EuI 1 (Actinomycetales). Can. J. Bot. **26,** 1072—1089.

Baker, D., Pengelly, W., Torrey, J., 1981: Immunochemical analyses of relationship among isolated *Frankiae* (Actinomycetales). Int. J. Syst. Bacteriol. **31,** 148—151.

Baker, D., Torrey, J., Kidd, G., 1979: Isolation by sucrose-density fractionation and cultivation *in vitro* of actinomycetes from nitrogen-fixing root nodule. Nature **281,** 76—78.

Bauer, W., 1981: Infection of legumes by *Rhizobia.* Ann. Rev. Plant Physiol. **32,** 407—449.

Becking, J., 1970: *Frankiaceae* fam. nov. (Actinomycetales) with one new combination and six new species of the genus *Frankia* Brunchorst 1886. Int. J. Syst. Bacteriol. **20,** 201—220.

Becking, J., 1977: Nitrogen fixation in higher plants other than legumes. In: Hardy, R., Silver, W. (eds.): Dinitrogen fixation, vol. 2, pp. 185—275. New York: Wiley and Sons Inc. Publ.

Becking, J., de Boer, W., Houwink, A., 1964: Electron microscopy of the endophyte of *Alnus glutinosa.* Anton. Leeuwenhoek **30,** 342—376.

Benson, D., 1982: Isolation of *Frankia* strains from Alder actinorhizal root nodules. Appl. Environ. Microbiol. **44,** 461—465.

Benson, D., Arp, D., Burris, R., 1979: Cell-free nitrogenase and hydrogenase from actinorhizal root nodules. Science **205,** 688—689.

Benson, D., Arp, J., Burris, R., 1980: Hydrogenase in Actinorhizal root nodule and homogenates. J. Bacteriol. **142,** 138—144.

Berg, R., 1982: Preliminary evidence for the involvement of suberization in infection of *Casuarina.* Communication at the Conference on *Frankia,* Madison, U. S. A.

Bermudez de Castro, F., Canizo, A., Costa, A., Miguel, C., Rodriguez-Barrueco, C., 1977: Cytokinins and nodulation of the non legumes *Alnus* and *Myrica.* In: Newton, W., Postgate, J., Rodriguez-Barrueco, C. (eds.): Recent development in nitrogen fixation, pp. 539—550. London - New York: Academic Press.

Berry, A., Torrey, J., 1979: Isolation and characterization *in vivo* and *in vitro* of an actinomycetous endophyte from *Alnus rubra* Bong. In: Gordon, J., Wheeler, C., Perry, D. (eds.): Symbiotic nitrogen fixation in management of temperate forests, pp. 69—83. Oregon State University Press.

Berry, A., Torrey, J., 1982: Root hair deformation in the infection process of *Alnus rubra* Bong. Communication at the Conference on *Frankia,* Madison, U. S. A.

Blom, J., 1981: Utilization of fatty acids and NH_4^+ by *Frankia* ArcI 1. FEMS Microbiol. Lett. **10,** 143—145.

Blom, J., 1982: Carbon and nitrogen source requirements of *Frankia* strains. FEMS Microbiol. Lett. **13,** 51—55.

Blom, J., Harkink, R., 1981: Metabolic pathways for gluconeogenesis and energy generation in *Frankia* AvcI 1. FEMS Microbiol. Lett. **11,** 221—224.

Blom, J., Roelofsen, W., Akkermans, A. D. L., 1980: Growth of *Frankia* AvcI 1 on media containing Tween 80 as C-source. FEMS Microbiol. Lett. **9,** 131—135.

Blom, J., Roelofsen, W., Akkermans, A. D. L., 1981: Assimilation of nitrogen in root nodules of Alder *(Alnus glutinosa).* New Phytol. **89,** 321—326.

Bond, G., 1976a: The results of the IBP survey of root nodule formation in nonleguminous angiosperms. In: Nutman, P. (ed.): Symbiotic nitrogen fixation in plants, IBP 7, pp. 443—474. Cambridge University Press.

Bond, G., 1976b: Observation on the root nodules of *Purshia tridentata*. Proc. R. Soc. Lond. B. **193,** 127—135.

Bradley, S., Brownell, G., Clark, J., 1973: Genetic homologies among *Nocardiae* and other Actinomycetes. Can. J. Microbiol. **19,** 1007—1014.

Burggraaf, A., Shipton, W., 1982: Estimates of *Frankia* growth under various pH and temperature regimes. Plant and Soil **69,** 135—147.

Burggraaf, A., Quispel, A., Tak, T., Valstar, J., 1981: Methods of isolation and cultivation of *Frankia* species from actinorhizas. Plant and Soil **61,** 157—168.

Callaham, D., Torrey, J., 1977: Prenodule formation and primary nodule development in roots of *Comptonia (Myricaceae)*. Can. J. Bot **55,** 2306—2318.

Callaham, D., Del Tredici, P., Torrey, J., 1978: Isolation and cultivation *in vitro* of the actinomycete causing root nodulation in *Comptonia*. Science **199,** 899—902.

Callaham, D., Newcomb, W., Torrey, J., Peterson, R., 1979: Root hair infection in actinomycete-induced root nodule initiation in *Casuarina, Myrica* and *Comptonia*. Bot. Gaz. **140,** Suppl., 1—9.

Calvert, H., Chaudary, A., Lalonde, M., 1979: Structure of an unusual root nodule symbiosis in a non leguminous herbaceous dicotyledon. In: Gordon, J., Wheeler, C., Perry, D. (eds.): Symbiotic nitrogen fixation in management of temperate forests, pp. 474—475. Oregon State University Press.

Capellano, A., Moiroud, A., 1979: Etude de la dynamique de l'azote à haute altitude. II. Etude ultrastructurale de l'endophyte des nodules d'*Alnus viridis* Chaix. Bull. Soc. Linn. **7,** 383—441.

Chaboud, A., Lalonde, M., 1982: Lectin binding on surfaces of *Frankia* strains. Communication at the Conference on *Frankia,* Madison, U. S. A.

Chaudary, A. 1978: The discovery of root nodules in the new species of non leguminous angiosperms from Pakistan and their significance. In: Döbereiner, J., Burris, R., Hollaender, A. (eds.): Limitations and potentials for biological nitrogen fixation in the tropics, p. 356. New York - London: Plenum Press.

Deley, J., 1971: Hybridization of DNA. In: Norris, J., Ribbons, D. (eds.): Methods in Microbiology, pp. 311—329. London - New York: Academic Press.

Diem, H., Gauthier, D., Dommergues, Y., 1982: Isolation of *Frankia* from nodules of *Casuarina equisetifolia*. Can. J. Microbiol. **28,** 526—530.

van Dijk, C., 1978: Spore formation and endophyte diversity in root nodules of *Alnus glutinosa* (L.) Vill. New Phytol. **81,** 601—615.

van Dijk, C., Merkus, E., 1976: A microscopical study of the development of a spore-like stage in the life cycle of the root nodule endophyte of *Alnus glutinosa* (L.) Gaertn. New Phytol. **77,** 73—91.

Dillon, J., Baker, D., 1982: Variations in nitrogenase activity among pure-cultured *Frankia* strains tested on an actinorhizal plants as an indication of symbiotic compatibility. New Phytol. **92,** 215—219.

Domergues, Y., Diem, H., Gauthier, D., 1983: Les symbioses actinorhiziennes tropicales. Communication au colloque Fixation Biologique de l'azote. SMF, Paris, pp. 1—2.

Dullaart, J., 1970: The auxin content of root nodules and roots of *Alnus glutinosa* (L.) Vill. J. Exp. Bot. **21,** 975—984.

Evans, H., Emerich, D., Maier, R., Hanus, F., Russell, S., 1979: Hydrogen cycling within the nodule of legumes and non legumes and its role in nitrogen fixation. In: Gordon, J., Wheeler, C., Perry, D. (eds.), Symbiotic nitrogen fixation in the management of temperate forests, pp. 196—207. Oregon State University Press.

Faure-Raynaud, M., Moiroud, A., 1983: Symbiose *Frankia*-Aulne: culture *in vitro* de *Frankia* sp. sur différentes sources de carbone et d'azote. C. R. Acad. Sc. Paris **296,** 757—760.

Faure-Raynaud, M., Horriere, F., Simonet, P., Moiroud, A., 1982: Caractéristiques principales de quelques souches pures de *Frankia* isolées de nodules d'Aulnes de la flore française. In: Energies renouvelables en milieu rural. Colloque Limoges, France, pp. 121—125.

Faure-Raynaud, M., Bonnefoy, M. A., Perradin, Y., Simonet, P., Moiroud, A., 1983: Protoplasms formation from *Frankia* strains. Microbios (in press).

Gardner, I., 1965: Observations on the fine structure of the endophyte of the root nodule of *Alnus glutinosa* (L.) Gaertn. Archiv. für Mikrobiol. **51,** 365—383.

Gauthier, D., Diem, H., Dommergues, Y., 1981 a: *In vitro* nitrogen fixation by two actinomycete strains isolated from *Casuarina* nodules. Appl. Environ. Microbiol. **41,** 306—308.

Gauthier, D., Diem, H., Dommergues, Y., 1981 b: Infectivité et effectivité de souches de *Frankia* isolées de nodules de *Casuarina equisetifolia* et d'*Hippophaë rhamnoides*. C. R. Acad. SC. Paris **293,** 489—491.

Goodchild, D., 1977: The ultrastructure of root nodules in relation to nitrogen fixation. In: Int'l Review of Cytology, Supplement 6, pp. 235—288. London and New York: Academic Press.

Gordon, J., Wheeler, C., 1978: Whole plant studies on photosynthesis and acetylene reduction in *Alnus glutinosa*. New Phytol. **80,** 179—186.

Hall, R., Mc Nabb, H., Maynard, C., Green, T., 1979: Toward development of optimal *Alnus glutinosa* symbioses. Bot. Gaz. **140,** Suppl., 120—126.

Henry, M., 1978: Sur la localisation et la morphologie des microorganismes présents dans les nodules racinaires d'*Elaeagnus angustifolia*. 103° Congrès Nat. Soc. Sav. Nancy, Fasc. I, pp. 359—370.

Hoeppel, R., Wollum II, A., 1971: Histological studies of ectomycorrhizae and root nodules from *Cercocarpus montanus* and *Cercocarpus paucidentalus*. Can. J. Bot. **49,** 1315—1318.

Horriere, F., Lechevalier, M., Lechevalier, H., 1982: *In vitro* morphogenesis and ultrastructure of a *Frankia* sp ArI3 (Actinomycetales) from *Alnus rubra* Bong and a morphologically similar isolate AirI2 from *Alnus incana* (L.) Moench. subsp rugosa (Duroi) Clausen. Communication at the Conference on *Frankia*, Madison, U. S. A.

Johnsrud, S., 1978: Nitrogen fixation by root nodules of *Alnus incana* in a Norwegian forest ecosystem. Oikos **30,** 475—479.

Lalonde, M., 1977: The infection process of the *Alnus* root nodule symbiosis. In: Newton, W., Postgate, J., Rodriguez-Barrueco, C. (eds.): Recent development in nitrogen fixation, pp. 569—589. London – New York: Academic Press.

Lalonde, M., 1978: Confirmation of the infectivity of a free-living actinomycete isolated from *Comptonia peregrina* (L.) Coult. root nodules by immunological and ultrastructural studies. Can. J. Bot. **56,** 2621—2635.

Lalonde, M., Calvert, H., 1979: Production of *Frankia* hyphae as an infective inoculant for *Alnus* species. In: Gordon, J., Wheeler, C., Perry, D. (eds.): Symbiotic nitrogen fixation in the management of temperate forests, pp. 95—110. Oregon State University Press.

Lalonde, M., Devoe, I., 1975: Scanning electron microscopy of the *Alnus crispa* var. mollis Fern. root nodule endophyte. Archiv für Mikrobiol. **105,** 87—94.

Lalonde, M., Devoe, I., 1976: Orgin of the membrane enveloppe enclosing *Alnus crispa* var. mollis Fern. root nodule endophyte as revealed by freeze-etching microscopy. Physiol. Plant Pathol. **8,** 123—129.

Lalonde, M., Knowles, R., 1975: Ultrastructure of the *Alnus crsipa* var. mollis Fern. root nodule endophyte. Can. J. Microbiol. *21,* 1058—1080.

Lalonde, M., Calvert, H., Pine, S., 1981: Isolation and use of *Frankia* strains in actinorhizae formation. In: Gibson, A., Newton, W. (eds.): Current perspectives in nitrogen fixation, pp. 296—299. Camberra: Australian Academy of Science.

Lawrence, D., Schoenike, R., Quispel, A., Bond, G., 1967: The role of *Dryas drummondii* in vegetation development following ice recession at Glacier Bay, Alaska, with special reference to its nitrogen fixation by root nodules. J. Ecology **55,** 793—813.

Lawrie, A., 1982: Field nodulation in nine species of *Casuarina* in Victoria. Aust. J. Bot. **30,** 447—460.

Leaf, G., Gardner, I., Bond, G., 1958: Observations on the composition and metabolism of nitrogen-fixing root nodules of *Alnus.* J. Exp. Bot. **9,** 320—331.

Lechevalier, M., Lechevalier, H., 1979: The taxonomic position of the actinomycetic endophytes. In: Gordon, J., Wheeler, C., Perry, D., (eds.): Symbiotic nitrogen fixation in the management of temperate forests, pp. 111—122. Oregon State University Press.

Lechevalier, M., Baker, D., Horriere, F., 1982: Physiology, chemistry, serology and infectivity of two *Frankia* isolates from *Alnus incana* (L.) Moench subsp. rugosa (Duroi) Clausen. Communication at the Conference on *Frankia,* Madison, U. S. A.

Martin, F., Hirel, B., Gadal, P., 1982: Sur l'activité enzymatique ornithine carbamyl transferase des actinorhizes d'*Alnus glutinosa* (L.) Gaertn. C. R. Acad. Sci. Paris **295,** 557—559.

Maudinas, B., Chemardin, M., Gadal, P., 1982: Fatty acid composition of roots and root nodules of *Alnus* species. Phytochemistry **21,** 1271—1273.

Mian, S., Bond, G., 1978: The onset of nitrogen fixation in young alder plants and its relation to differentiation in the nodular endophyte. New Phytol. **80,** 187—192.

Mian, S., Bond, G., Rodriguez-Barrueco, C., 1976: Effective and ineffective root nodules in *Myrica faya.* Proc. R. Soc. Lond. B. **194,** 285—293.

Moiroud, A., Faure-Raynaud, M., 1983: Influence de basses températures sur la croissance et la survie de souches pures de *Frankia* isolées de nodules d'Aulne. Plant and Soil (in press).

Newcomb, W., 1981: Fine structure of the root nodules of *Dryas drummondii* Richards (Rosaceae). Can. J. Bot. **59,** 2500—2514.

Newcomb, W., 1982: Nodule morphogenesis and differentiation. In: Atherly, A., Giles, K. (eds.): The biology of the *Rhizobiaceae.* International review of cytology, Suppl. 13, pp. 247—296. London - New York: Academic Press.

Newcomb, W., Pankhurst, C., 1982a: Fine structure of actinorhizal root nodules of *Coriaria arborea.* New Zealand J. Bot. **20,** 93—103.

Newcomb, W., Pankhurst, C., 1982b: Ultrastructure of actinorhizal root nodules of *Discaria toumatou* Raoul *(Rhamaceae).* New Zealand J. Bot **20,** 105—113.

Newcomb, W., Tandon, S. 1981: Uninfected cells of soybean root nodules: ultrastructure suggests key role in ureide production. Science **212,** 1394—1396.

Newcomb, W., Peterson, R., Callaham, D., Torrey, J., 1978: Structure and host-actinomycete interactions in developing root nodules of *Comptonia peregrina.* Can. J. Bot. **56,** 502—531.

Normand, P., Lalonde, M., 1982: Evaluation of *Frankia* strains isolated from provenances of two *Alnus* species. Can. J. Microbiol. **28,** 1133—1142.

Normand, P., Simonet, P., Butour, J., Rosenberg, C., Moiroud, A., Lalonde, M., 1983: Plasmids in *Frankia* sp. J. Bacteriol. **155,** 32—35.

Perradin, Y., 1982: Etude des acides phénoliques de *Populus balsamifera* et d'*Alnus crispa* var. mollis Fern. et de leur influence sur *Frankia* AcN1ag, Thèse 3e cycle, 145 p. University of Lyon, France.

Perradin, Y., Mottet, M., Lalonde, M., 1983: Influence of phenolics on the *in vitro* growth of *Frankia* strains. Can. J. Bot. (in press.)

Pizelle, G., 1972: Observations sur les racines de plantules d'Aulne glutineux (*Alnus glutinosa* Gaertn.) en voie de nodulation. Bull. Soc. bot. Fr. **119,** 571—580.

Quispel, A., Burggraaf, A., 1981: *Frankia,* the diazotrophic endophyte from actinorhizas. In: Gibson, A., Newton, W. (eds.): Current perspectives in nitrogen fixation, pp. 229—236. Canberra: Australian Academy of Science.

Rodriguez-Barrueco, C., 1968: The occurrence of nitrogen fixing root nodules on non-legume plants. Bot. Linn. Soc. **62,** 77—84.

Rodriguez-Barrueco, C., Miguel, C., 1979: Host plant endophyte specificity in actinomycete-nodulated plants. In: Gordon, J., Wheeler, C., Perry, D. (eds.): Symbiotic nitrogen fixation in the management of temperate forests, pp. 143—159. Oregon State University Press.

Roelofsen, W. Akkermans, A. D. L., 1979: Uptake and evolution of H_2 and reduction of C_2H_2 by root nodules and nodule homogenates of *Alnus glutinosa.* Plant and Soil **52,** 571—578.

Ruvkun, G., Ausubel, F., 1980: Interspecies homology of nitrogenase genes. Proc. Nat. Acad. Sci. U. S. A **77,** 191—195.

Schubert, K., Coker III, G., 1981: Ammonia assimilation in *Alnus glutinosa* and *Glycine max.* Plant Physiol. **67,** 662—665.

Schwintzer, C., Berry, A., Disney, L., 1982: Seasonal pattern of root nodule growth endophyte morphology, nitrogenase activity and shoot development in *Myrica gale.* Can. J. Bot. **60,** 746—757.

Shipton, W., Burggraaf, A., 1982: A comparaison of the requirements for various carbon and nitrogen sources and vitamins in some *Frankia* isolates. Plant and Soil **69,** 149—161.

Silver, W., 1964: Root nodule symbiosis. I. Endophyte of *Myrica cerifera* L. J. Bacteriol. **87,** 416—421.

Silver, W., Bendana, F., Powell, R., 1966: Root nodule symbiosis. II. The relation of auxin to root geotropism in root and root nodules of non legumes. Physiol. Plant. **19,** 207—218.

Silvester, W., 1974: Ecological and economic significance of the non legume symbioses. In: Newton, W., Nyman, C. (eds.): Proc. of the 1st International Symposium on Nitrogen Fixation, vol. 2, pp. 489—506. Washington State University Press.

Silvester, W., 1977: Dinitrogen fixation by plant associations excluding legumes. In: Hardy, R., Gibson, A. (eds.): A treatise on Dinitrogen fixation. IV. Agronomy and Ecology, pp. 141—190. New York: Wiley and Sons Inc. Publ.

Simonet, P., Moiroud, A., Heizmann, P.: Use of molecular methods in *Frankia* sp. strains taxonomy. Plant and Soil (submitted).

Strand, R., Laetsch, W., 1977: Cell and endophyte structure of the nitrogen-fixing root nodules of *Ceanothus integerrimus* H. and A. 1. Fine stucture of the nodule and its endosymbiont. Protoplasma **93,** 165—178.

van Straten, J., Akkermans, A. D. L., Roelofsen, W., 1977: Nitrogenase activity of endophyte suspensions derived from root nodules of *Alnus, Hippophaë, Shepherdia* and *Myrica* spp. Nature **266,** 257—258.

Suetin, S., Parijskaja, A., Kalakoutskii, L., 1979: Structural aspects of endosymbiont development and interaction with host cytoplasm in *Alnus glutinosa* L. Gaertn root nodules. Microbios. Letters **12,** 83—92.

Tjepkema, J., 1979: Oxygen relations in leguminous and actinorhizal nodules. In: Gordon, J., Wheeler, C., Perry, D. (eds.): Symbiotic nitrogen fixation in the management of temperate forests, pp. 175—186. Oregon State University Press.

Tjepkema, J., 1983: Oxygen concentration within the nitrogen-fixing root nodules of *Myrica gale* L. Amer. J. Bot. **70,** 59—63.

Tjekema, J., Ormerod, W., Torrey, J., 1980: Vesicle formation and acetylene reduction activity in *Frankia* sp. CpI 1 cultured in defined nutrient media. Nature **287,** 633—635.

Tjepkema, J., Ormerod, W., Torrey, J., 1981: Factors affecting vesicle formation and acetylene reduction (nitrogen activity) in *Frankia* sp. CpI 1. Can. J. Microbiol. **27,** 815—823.

Torrey, J., 1976: Initiation and development of root nodules of *Casuarina (Casuarinaceae).* Amer. J. Bot. **63,** 335—344.

Torrey, J., Callaham, D., 1978: Determinate development of nodule roots in actinomycete-induced root nodules of *Myrica gale.* Can. J. Bot. **56,** 1357—1364.

Torrey, J., Callaham, D., 1982: Structural features of the vesicles of *Frankia* sp. CpI 1 in culture. Can. J. Microbiol. **28,** 749—757.

Torrey, J., Tjepkema, J., Bergersen, J., Gibson, A., 1981: Dinitrogen fixation by culture of *Frankia* sp. CpI 1 demonstrated by $^{15}N_2$ incorporation. Plant Physiol. **68,** 983—984.

Tyson, J., Silver, W., 1979: Relationship of ultrastructure to acetylene reduction (N_2 fixation) in root nodules of *Casuarina.* Bot. Gaz. **140,** Suppl., 44—48.

Chapter 8

Host-Fungus Specificity, Recognition and Compatibility in Mycorrhizae

V. Gianinazzi-Pearson

Station d'Amélioration des Plantes, INRA, BV 1540, F-21034 Dijon, France

With 3 Figures

Contents

I. Introduction

Susceptibility in plants to parasites (for definition see Vanderplank, 1978) is considered to be a relatively rare phenomenon in nature; a striking exception to this rule is found in mycorrhizal associations. Mycorrhizae, a term describing a range of mutualistic associations between soil fungi and plant roots, are no doubt the most frequent examples of compatibility between plants and microbes. The mycorrhizal habit has a long evolutionary history (Nicolson, 1975, Pirozynski and Malloch, 1975, Boullard, 1979) and today more than 90% of all plant taxa, ranging from thallophytes to angiosperms, form associations of one type or another with mycorrhizal fungi. Only a small number of plant species belonging mainly to the Cuperaceae, Chenopodiaceae, Cruciferae, Juncaceae and Proteaceae are nonmycorrhizal and can be considered as incompatible towards mycorrhizal fungi. A characteristic feature of the fungi forming mycorrhizae is that, although they are generally widespread in soils, they exhibit a strong biotrophic dependence on their host plants and are rarely free-living saprophytes.

Mycorrhizae are usually divided into three morphologically distinct groups depending on whether or not there is fungal penetration of the root cells: ectomycorrhizae, endomycorrhizae and ectendomycorrhizae. These groups encompass widely differing forms whose structural and functional diversity is determined by the plants and fungi involved.

Ectomycorrhizae are formed nearly exclusively be tree species on short roots that do not have secondary thickening. They are characterized by an important development of fungal mycelium on the surface of the host root which becomes completely enclosed by a fungal sheath or mantle. This often results in marked changes in root morphology and colour which can be used to distinguish the different kinds of ectomycorrhizae (Dominik, 1969). The mycorrhizal roots are short and depending on the host plant can be undivided, as in oak, or intensely branched, as in pine. The structure of the fungal sheath is determined by the fungus involved; it can have a smooth surface with only a few hyphae radiating out into the soil, be covered by stiff bristlelike hyphae projecting from the surface or develop an external network of mycelium made up of either individual hyphae or of aggregates forming strands and rhizomorphs. On the inside of the fungal sheath, hyphae develop in the intercellular spaces of the root to form a network or "Hartig net" which extends between the outer cell layers of the root and sometimes up to the endodermis in roots of small diameter. The fungal hyphae remain intercellular and never penetrate the living cells of the host root (Fig. 1 a).

In endomycorrhizae, on the contrary, fungal development close to the root surface is much less than in ectomycorrhizae but there is an important colonization of the inner root tissues with penetration and proliferation of fungi within the cortical cells. Endomycorrhizae do not usually cause observable changes in root morphology and their presence cannot be detected without appropriate staining. Three distinct types of endomycorrhiza with different infection patterns are formed by various plant and fungal taxa: ericoid, orchid and vesicular-arbuscular (VA) mycorrhizae.

▷

Fig. 1. Morphological features of different types of mycorrhiza. a) ectomycorrhizal association *Tuber brumale* × *Cistus* species; b) ectendomycorrhiza of wild *Arbutus unedo;* c) VA endomycorrhizal infection in field-grown *Triticum aestivum;* d) orchid endomycorrhiza of wild *Dactylorchis maculata;* e) ericoid endomycorrhizal association *Pezizella ericae* × *Vaccinium myrtillus*

ap, appressorium; ar, arbuscule; eh, external hypha; h, Hartig net region; ic, intracellular hyphal coils; is, intercellular spaces; n, host nucleus; fs, fungal sheath; vc, vascular cylinder. Photographs by courtesy of A. Fusconi (a), P. Bonfante-Fasolo (b, d, e) and A. Trouvelot (c)

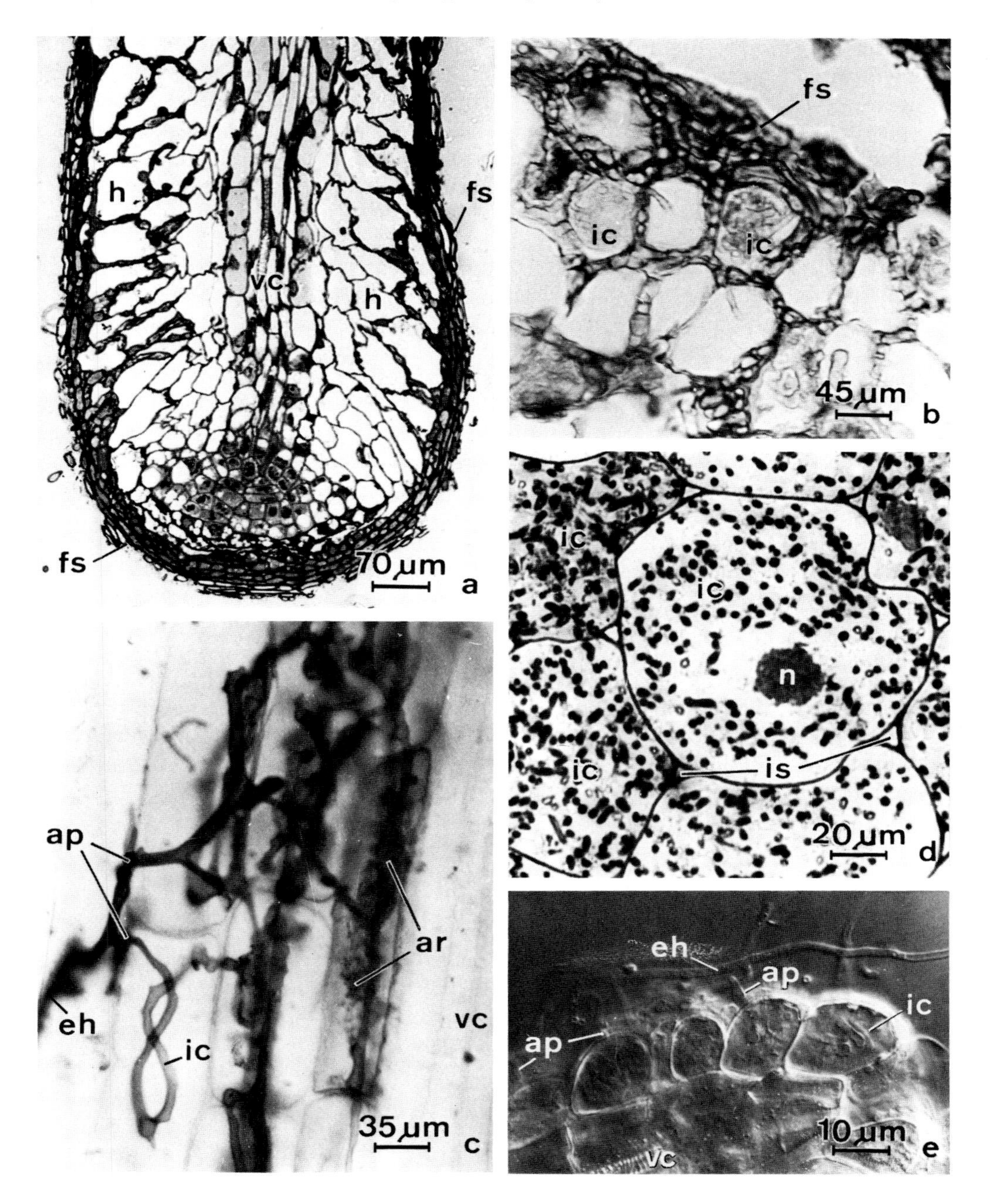
h
fs
vc
h
fs
70 µm
a
fs
ic
ic
45 µm
b
ap
ar
eh
ic
vc
35 µm
c
ic
ic
n
ic
is
20 µm
d
eh
ap
ic
ap
vc
10 µm
e

Ericoid mycorrhizae are found exclusively in the Ericaceae where the mycorrhizal fungi colonize the fine terminal roots of host plants and form loops or coils within the host cells (Fig. 1 e). In orchidaceous species the term mycorrhizae is used in a much broader sense to describe mutualistic plant-fungus associations that can be established at different moments in the life-cycle of the plant. The mycorrhizal fungi colonize either the embryo of the minute orchid seeds or the roots of chlorophyllous and achlorophyllous orchid species; they penetrate the host cells and form intracellular hyphal coils (Fig. 1 d). Vesicular-arbuscular mycorrhizae are not restricted to any one plant family and, considering the extremely large variation in host plants, the pattern of fungal development within the host tissues is remarkably constant. The mycorrhizal fungi spread through the primary cortex of unsuberised roots mainly in the intercellular spaces and after penetration of the cortical cells they form a characteristic much divided haustorium, the arbuscule (Fig. 1 c). The hyphae of most VA fungi also develop inter- or intracellular spherical swellings, vesicles, which have lipid-rich contents. The presence of these two characteristic structures is responsible for the name given to this particular type of mycorrhiza.

Ectendomycorrhizae, as the term suggests, represent an intermediate form between the previously described mycorrhizal associations. There is an organized fungal sheath as well as inter- and intracellular penetration in the root cortex (Fig. 1 b). They are formed by a limited number of plants belonging to the Arbutae, Pyrolaceae and Monotropaceae of the Ericales and by seedlings of certain conifers. Little is known of the fungi involved except that they seem to be Basidiomycetes. In the Arbutae (arbutoid mycorrhizae), Pyrolaceae and conifers the fungi form coils within host cells whilst in the Monotropaceae (monotropoid mycorrhizae) the intracellular penetration is limited to a peg-like structure.

II. Plant-Fungus Specificity

Although the fungi involved in the different types of mycorrhizae are commonly present together in soils, they each have a clear-cut host range and very rarely trespass into the plant taxa belonging to another mycorrhizal type. As for other plant-microbe interactions, the nature and degree of specificity between the host plant and mycorrhizal fungus are determined by either of the symbionts and they are quite variable from one type of mycorrhizal association to another.

Of all mycorrhizae, the greatest degree of specificity can be found in the ericoid endomycorrhizae which are formed by a very limited number of plant species and by fungi which are taxonomically extremely close. The ericoid mycorrhizal fungi that have been isolated in pure culture are not yet all taxonomically defined but different studies agree that they must belong to or be closely related to the same ascomycetous species, *Pezizella ericae* Read (Pearson and Read, 1973, Read, 1974, Webster, 1976, Vegh *et al.,* 1979). Infection tests have shown that these fungi only form mycorrhizal associations with a limited number of species in the Ericaceae. They do not establish mycorrhizae with other plant families (Gianinazzi-Pearson, Jacquelinet-Jeanmougin, Bonfante-Fasolo, unpublished data) and Freisleben (1936) obtained ericoid mycorrhiza with only 15 out of the 23 ericaceous species he tested. Little host-fungus specificity seems to exist, however, within this potential host range of ericoid mycorrhizae. Mycorrhizal fungi isolated from *Calluna vulgaris, Vaccinium myrtillus, V. oxycoccus, Erica cinerea, E. carnea* and *Rhododendron ponticum,* for example, can form mycorrhiza with each of these different host plants (Pearson and Read, 1973, Duclos 1981, Gianinazzi-Pearson, unpublished data). There have been very occasional reports of basidiomycetous fungi being associated with ericaceous plants (Bonfante-Fasolo, 1980, Peterson *et al.,* 1980) but their significance in terms of mycorrhizae is far from clear and they must therefore be considered exceptional.

Orchid endomycorrhizae are also limited to one plant family, the Orchidaceae, but the mycorrhizal fungi that have been isolated and grown in pure culture come from several genera in the Basidiomycetes, the most common of which are *Ceratobasidium, Tulasnella, Sebacina, Armillaria, Thanetephorus* and *Marasmius.* All known orchid mycorrhizal fungi are also either normal soil saprophytes or parasites on other plants; *Thanetephorus cucumeris* (Syn. *Rhizoctonia solani*) for example, has received considerable attention as a root pathogen (see Harley and Smith, 1983). There appears to be a range of specificity in orchid mycorrhizae. Some groups of orchids like *Caladenia* are commonly associated with one or a few fungal species (Warcup, 1971) whilst others like *Dactylorhiza purpurella* or *Goodyera repens* can establish mycorrhiza with a wide variety of different fungi (Hadley, 1970). The same fungi can colonise the seed embryo or root tissues.

Ectomycorrhizae are, apart from a few exceptions, restricted to a certain number of temperate forest tree species whilst the fungi involved are not limited to any one taxonomic group. These associations mainly concern about 40 plant genera in the Pinaceae, Fagaceae, Betulaceae, Ulmaceae, Tiliaceae, Cupressaceae and Myrtaceae. In contrast to this relatively specialized host range, some 2000 species of fungi are recognized as being potentially ectomycorrhizal with different host plants (Trappe, 1977). Trappe in 1962 already listed ectomycorrhizal fungi belonging to as many as 99 genera, 30 families and 11 orders in the Basidiomycetes and Ascomycetes. Other fungi have since been added to this impressive list (Mosse *et al.,* 1981) and one genus in the Zygomycetes, *Endogone,* has also been recognized as forming ectomycorrhizae (Fassi and Palenzona, 1969, Warcup, 1975, Bonfante-Fasolo and Scannerini, 1977). A certain number, but not all, ectomycorrhizal fungi can be grown in pure culture on synthetic media. Morphologically distinct kinds of ectomycorrhizae can be formed by a single plant species, each kind usually depending on the type of fungus involved (Trappe, 1962, Trappe, 1967, Zak, 1973). In addition, a given species of fungus can form morphologically similar ectomycorrhiza with many hosts (Trappe, 1962, Trappe, 1964, Molina and Trappe, 1982a). This lack of specificity between ectomycorrhizal fungi and their host plants is underlined by the fact that fungi known to form ectomycorrhizae can also be responsible for the formation of ectendomycorrhizae in the Arbutae and Monotropaceae. Molina and Trappe (1982b) reported that out of 28 species of known ectomycorrhizal fungi isolated from diverse hosts and habitants, all but three formed arbutoid mycorrhizae with *Arbutus menziesii* and *Arctostaphylos uva-ursi.* Björkman (1960) showed that a fungus forming mycorrhizae with *Monotropa hypopytis* was simultaneously ectomycorrhizal with *Pinus.* Some exceptions do seem to exist, however, to this general lack of host-fungus specificity in ectomycorrhizal associations. Certain fungi like *Suillius* and *Rhizopogon* species probably associate exclusively with conifers belonging to the Pinaceae (Malajzuck *et al.,* 1982) and plant species belonging to the genus *Alnus* appear to form ectomycorrhiza with only a small number of fungi (Molina, 1981). Furthermore, a few fungi are apparently genus-specific vis-à-vis their host plant, for example, *Alpova diplophloeus* with *Alnus* (Molina and Trappe, 1982a) and *Hymenogaster albellus* with *Eucalyptus* (Malajzuck *et al.,* 1982).

VA mycorrhizae are characterised by a general lack of host specificity; the fungi involved infect an extremely wide range of plants coming from very different taxonomic groups. It is estimated that VA mycorrhizae are formed by about 80% of species in the plant kingdom, ranging from Thallophytes and Pteridophytes to Gymnosperms and Angiosperms. This type of mycorrhiza concerns most agricultural and horticultural plants, fruit trees and a large number of forest tree species. It is so ubiquitous that it is easier to list the plants not forming VA mycorrhizae than to try to present an exhaustive list of those where they have been observed. Most species in the Cruciferae, Chenopodiaceae, Proteaceae and the Juncaceae are immune to VA infection and apart from the rare examples of plants forming both ecto- and VA mycorrhizae (some species of *Populus, Alnus* and *Eucalyptus*), plants forming other types of mycorrhizae are not infected by VA fungi. There are many examples illustrating the complete lack of specificity of VA fungi vis-à-vis their host plants. For instance, Koch (1961) demonstrated that the VA mycorrhizal infection in roots of *Atropa belladonna* could be transmitted to 45 plant species belonging to 22 different families. However, if specificity is considered from a fungal point of view, VA mycorrhizae are formed by fungi from a fairly limited taxonomic range. Of the 88 fungal species that have been described and are presumed to be mycorrhizal (Trappe, 1982), only about 30 have been confirmed as being VA mycorrhiza-forming fungi. These are limited to four genera *(Glomus, Gigaspora, Acaulospora, Sclerocystis)* belonging to one family (Endogonaceae) in the Zygomycetes. Thus, although VA mycorrhizae are aspecific as far as the host plant is concerned, from a fungal view point they can be considered to show a fair degree of specificity as compared, for example, to ectomycorrhizae. VA fungi have not yet been grown in pure culture separated from a host plant.

There has been very little investigation into the genetics of host-fungus interactions in mycorrhizae. Studies on VA mycorrhizae in maize and wheat cultivars (Hall, 1978, Bertheau *et al.*, 1980, Azcon and Ocampo, 1981) and on ectomycorrhizae in control-pollinated half-sib progenies of slash pine (Marx and Bryan, 1971) have shown that although mycorrhizae develop on all host plants, the genotype of the latter can influence the intensity of infection by a given fungal strain or species. Similarly, different strains or species of VA or ectomycorrhizal fungi can produce varying extents of mycorrhizal development on an appropriate host species (Mason, 1975, Sanders *et al.*, 1977, McGraw and Schenck, 1980, Molina, 1979, 1981, Plenchette *et al.*, 1982). No species-to-species interaction has been found between fungi and host plants in mycorrhizae and this makes the existence of a strain-genotype interaction highly improbable. There is no evidence for the moment, therefore, that specific genes affecting virulence (or avirulence) in mycorrhizal fungi or compatibility (or incompatibility) in hosts are involved in mycorrhiza formation. However, some interesting caryological observations have been made on ectomycorrhizal fungi. The spores of mycorrhizal Ascomycetes and Basidiomycetes are mononucleate (Fasolo-Bonfante and Brunel, 1972, Mims, 1980) and give rise to monocaryotic hyphae but the mycelia of the fungal sheath formed in ectomycorrhizal associations are composed of binucleate hyphae (Fasolo-Bonfante and Brunel, 1972, Bonfante and Giovanetti, 1982). Detailed studies on the Ascomycete *Tuber melanosporum* indicate that ectomycorrhizae are only formed by the heterocaryotic mycelium of this fungus which develops after anastomosis between monocaryotic hyphae with genetic compatibility (Fasolo-Bonfante and Brunel, 1972, Grente *et al.*, 1972, Rouquerol and Payre, 1974), whilst *Hebeloma cylindrosporum*, a Basidiomycete, seems to be able to form mycorrhiza also in a monocaryotic state under laboratory conditions although it is usually dicaryotic in nature (Athanassiou, 1979). Both these systems offer interesting possibilities for studying the genetic determinants which control the mycorrhiza-forming ability of ectomycorrhizal fungi.

In mycorrhizae the specificity between plant and fungus has not developed to the same extent as in biotrophic plant-pathogen associations. A single species or strain of mycorrhizal fungus is not restricted to a single species, genotype or cultivar of host plant. Nonetheless the different types of mycorrhizae involve well-defined host plants and fungi, and the infection intensities developing in roots can be affected by the genotype of either symbiont. Thus although specificity in mycorrhizae is not very close, it must in fact exist to a certain degree in the host-fungus interactions which determine mycorrhizal establishment and development.

III. Host-Fungus Interactions

As in other biotrophic associations, a basic requirement for the establishment of mycorrhizae (whether they are highy specific or not) is the association of appropriate fungus and plant in a compatible pair. Unfortunately very little research has been focussed on the mechanisms and factors determining either host-fungus specificity or compatibility in mycorrhizal associations. However, the numerous studies on specificity in other biotrophic associations like pathogenic infections or *Rhizobium* symbiosis provide a starting point in a search for clues to the mechanisms responsible for the establishment of functional mycorrhizal associations. Briefly, the success or failure of a microbial infection can be determined by one or several factors acting at the various stages of colonization of the plant tissues, for example, the rhizosphere environment (antagonistic micro-organisms, pH, presence or absence of specific growth compounds or toxins), recognition between fungus and plant, presence within plant tissues of metabolites essential for the growth of the microorganism, production of antimicrobial compounds by the plant. There is now some evidence that one or a combination of these mechanisms probably contributes to host-fungus specificity in mycorrhizae.

A. Rhizosphere Environment

In all mycorrhizae, establishment of the infection is preceded by some growth of the fungi on the root surface of the host-plant. VA and ectomycorrhizal fungi can also apparently spread along roots of non-hosts (Theodorou and Bowen, 1971, Bevege and Bowen, 1975, Ocampo *et al.*, 1980). Hyphal growth is however generally stimulated by the presence of roots of a host plant and a number of observations indicate that the rhizosphere environment may be a factor contributing to mycorrhizal specificity in some associations. For instance, spore germination by the ectomycorrhizal fungus *Thelephora terrestris* is only stimulated by the roots of host plants (Birraux and Fries, 1981) and the growth of some ectomycorrhizal fungi is inhibited by substances in the root exudates of ericaceous species which form ericoid mycorrhizae (Robinson, 1972). Recent *in vitro* studies on specificity in ericoid mycorrhizae have also shown that the rhizosphere of a non-host plant can adversely affect the development of the ericoid fungus *P. ericae;* hyphae segment on the root surface of *Trifolium pratense* and hardly develop along roots of *Gentiana lutea* (V. Gianinazzi-Pearson, S. Jacquelinet-Jeanmougin, P. Bonfante-Fasolo, unpublished data). There has been no experimentation so far to explain these different observations.

B. Recognition Phenomena

When the hyphae and root tissues of appropriate mycorrhizal associates come into close contact, they react to each others' presence and develop a structurally complex and functionally compatible relationship. As Heslop-Harrison (1978) pointed out *"a cell that reacts in a special way in consequence of association with another must do so because it acquires 'information' from that other, information that must be conveyed through chemical and/or physical signals"*. In other terms, some kind of recognition system or systems must be active in developing mycorrhizal associations. Furthermore, the fact that a certain degree of specificity does exist in the choice of partners in mycorrhizae means that some discrimination or recognition between potential associates must occur. What, where and when are recognition phenomena likely to occur in mycorrhizae is not a simple question and the answer is even less simple since no single recognition process could determine the formation of such complex associations. Some insight into the recognition processes that might exist can be obtained by closely examining the processes involved in the establishment and development of the different mycorrhizae.

The hyphae of mycorrhizal fungi grow close to the surface of outer root cells and some recent observations suggest that the fungal hyphae might "recognize" and become attached to the wall of the host cells prior to their penetration of the root tissue. Ultracytological studies of ericoid mycorrhiza synthesized in axenic culture have shown that hyphae of the mycorrhizal fungus *P. ericae* growing around the roots of *C. vulgaris* or *V. myrtillus* develop an abundant sheath of fibrils which is never present in mycelium in pure culture (Bonfante-Fasolo and Gianinazzi-Pearson, 1982). The fibrils extend from the outer fungal wall and closely adher to the outer surface of host cells as hyphae come into contact with the root (Fig. 2a). Similar observations have been made on *Terfezia leptoderma,* the desert truffle, which is ectomycorrhizal with *Helianthemum salicifolium* (Cistaceae). Hyphae close to the root surface produce fibrils on their outer wall (Dexheimer, unpublished data) and fibrillar material seems to connect the fungal wall to the host cell wall (Fig. 2b). These are the first observations indicating any binding activity between mycorrhizal fungi and host plants prior to the infection process. They recall similar phenomena observed in other plant-microbe systems like *Rhizobium* /legume and plant/ pathogen interactions (Callow, 1975, Murray and Maxwell, 1975, Mengden, 1978, Dazzo, 1980, see also chapter 1 of this volume) where the involvement of lectins in cell-to-cell contact and recognition has been proposed. Lectins are proteins or glycoproteins which can bind to carbohydrate-containing molecules; they can show considerable variation in their binding specificities and may play some role in the greater or lesser specificity of certain relationships (Heslop-Harrison, 1978). Although lectins have yet to be demonstrated in mycorrhizae, the fibrils produced by the mycorrhizal fungus *P. ericae* are known to have a polysaccharide-protein composition (Bonfante-Fasolo and Gianinazzi-Pearson, 1982). It is indeed possible therefore that, as Harley and Smith (1983) envisaged, a relatively nonspecific binding by fungal lectins to the plant cell wall could be the first recognition step between mycorrhizal associates.

The cell to cell contact established between mycorrhizal fungus and host root leads to changes in fungal activity which differ according to the type of mycorrhiza to be formed.

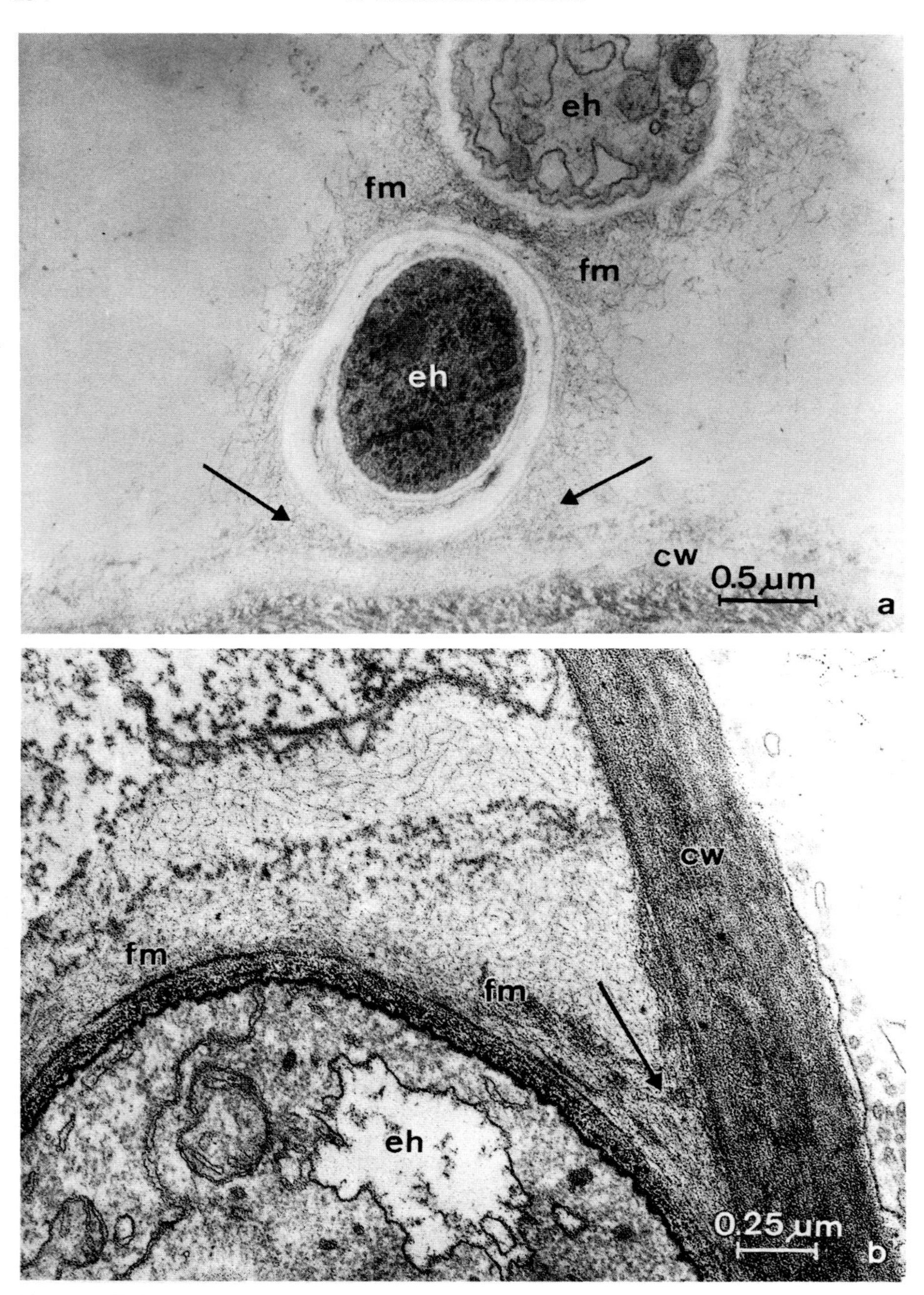

Fig. 2. Fibrillar material (fm) produced on the walls of external hyphae (eh) of mycorrhizal fungi in presence of the host root. a) *Pezizella ericae* with *Vaccinium myrtillus;* b) *Terfezia leptoderma* with *Helianthemum salicifolium.* Arrows indicate where fibrils adhere to the outer surface of the host cell walls (cw.) Photographs by courtesy of P. Bonfante-Fasolo (a) and J. Dexheimer (b)

i) Ectomycorrhizae

In ectomycorrhizal associations the change in fungal activity on contact with the host cell results in the formation of a differentiated pseudoparenchymatous tissue, the fungal sheath or mantle, and hyphal colonization of the intercellular spaces of the epidermis and cortex (Hartig net) (Fig. 1 a). These two processes of fungal differentiation can occur together or one can precede the other. There is little evidence that specific recognition factors are necessary for sheath formation (see Harley and Smith, 1983). It can occur on both long and short roots and has even been simulated using silicone tubes (Read and Armstrong, 1972). Hartig net development, on the contrary, seems to depend on special host-fungus interactions and morphological changes in the fungus appear to be host-induced since they only occur in response to the close contact between hyphae and living host cells (Debaud *et al.,* 1981, Nylund and Unestam, 1982). With penetration of the root tissue hyphae branch profusely and septa formation becomes irregular and often incomplete. This growth form has been likened to that induced in pure cultures of nonmycorrhizal fungi by growth inhibitors (Valla, 1979), L-sorbose (Tatum *et al.,* 1949, Galpin *et al.,* 1977) or cysteine (Bianchi and Turian, 1967) and research for paramorphogenic compounds in ectomycorrhiza-forming roots may provide some clues to the fungal morphogenesis of the Hartig net. In adult root systems Hartig net formation only takes place on the shortest host roots, probably because these are most compatible with the fairly slow process of infection (Harley, 1969, Bonfante and Giovanetti, 1982) and it is limited to a well-defined zone lying behind the dividing root meristem. More precisely, hyphae only penetrate tissues that are in a physiological state of cell-wall building and cell maturation. The mycorrhizal fungus "recognizes", therefore, host-tissue particularly favourable for hyphal penetration and proliferation. The basis for this recognition is not known but it could be partly due do a nutritional attraction, leakage of nutrients being greater in the subapical root region, and partly because here tissues offer less physical resistance to penetration than in regions where cell wall formation is complete. The Hartig net can be formed outside the zone of cell elongation in axenic cultures if the root system is in a sufficiently juvenile state (Nylund and Unestam, 1982).

Host tissues, on the contrary, show little if no specific reaction to fungal infection in ectomycorrhizae. The typical development of short roots where ectomycorrhizal associations are established is not necessarily brought about by a fungal stimulus as once thought; this pattern of rhizomorphogenesis appears to be genetically determined by the host plant since it can occur in uninfected roots (Harley, 1937, Clowes, 1951, Faye *et al.,* 1981). However, the fungus affects root longevity since abortion of rootlets is prevented and root branching is more pronounced with mycorrhizae formation, probably due to fungus-produced hormones (Slankis, 1958, Harley, 1969). In mycorrhizal roots, host cells in contact with the Hartig net mycelium do not undergo any marked cytological changes and their plasmodesmata remain intact during development of the fungus (Scannerini, 1975,

Strullu and Gourret, 1980, Nylund and Unestam, 1982, Scannerini and Bonfante-Fasolo, 1983). Nevertheless, there appears to be some control of fungal activity by the host cells which ensures that fungal development remains intercellular. Although few ectomycorrhizal fungi produce cell-wall degrading enzymes in pure culture (see Harley and Smith, 1983), they must possess the necessary enzyme equipment for cell-wall breakdown since some intracellular penetration of dead or senescent root cells occurs in most mycorrhizal roots (Harley, 1969, Strullu and Gourret, 1980, Nylund and Unestam, 1982, Scannerini and Bonfante-Fasolo, 1983) and several ectomycorrhiza-forming fungi develop ectendomycorrhizal associations with other host-plants (Molina and Trappe, 1982b). Cortical cells of mycorrhizal roots frequently contain polyphenolic compounds or tannins within their vacuoles and whilst some workers have claimed that these can inhibit cellulases of ectomycorrhizal fungi (Scannerini, 1975, Marks and Foster, 1973), this has not been confirmed as a general phenomenon. Furthermore, it does not represent a specific host response to mycorrhiza infection since phenolic compounds are also found in nonmycorrhizal roots (Piché *et al.,* 1981). Another possibility is that the fungal symbiont somehow interferes with cell-wall polymerization processes so that host cells continually release soluble cell-wall precursor carbohydrates which in turn inhibit any repressible cell-wall degrading fungal enzymes. Harley and Smith (1983) have proposed, for example, that the host enzymes which polymerize cell wall precursors could be inhibited or complexed by proteins or enzymes at the hyphal apex or on the hyphal walls. Alternatively, cell-wall degrading enzymes of the fungi may be latent and only activated or induced under certain specific conditions. None of these mechanisms for the control of fungal activity require a specific host reaction to fungal penetration.

ii) Ectendomycorrhizae

Studies on ectendomycorrhizae have been minimal and consequently very little is known of the host-fungus interactions in these associations. One detailed investigation of an arbutoid mycorrhiza in *A. uva-ursi* reports that the Hartig net is poorly developed and hyphae penetrate living cortical cells with vacuoles rich in tannins. The intracellular hyphae form coils which are surrounded by a dense layer of material deposited between the fungal wall and the intact host plasmalemma (Scannerini and Bonfante-Fasolo, 1983). Whether this material is of fungal origin or is produced as a host reaction to infection has yet to be determined. Monotropoid mycorrhiza have received little more attention (Lutz and Sjolund, 1973, Duddridge and Read, 1982, Robertson and Robertson, 1982). Here Hartig net and hyphal penetration into the host cells are strictly limited to the outermost layer of root cells. As individual hyphae enter each cell, the host wall is not ruptured but invaginates around the advancing fungus and produces numerous wall ingrowths into the cell cytoplasm. Once inside the cell fungal growth quickly ceases and a peg-like structure is formed, the tip of which appears to open and a membranous sac to extend from it into the host cytoplasm. All this structure is surrounded by continuous host plasmalemma which proliferates and extensively invaginates, illustrating an evident reaction by the host cell to the fungal intrusion. Although the pattern of fungal peg formation suggests that the processes involved are regulated by both fungus and plant, nothing is known of the controlling mechanisms; cytochemical and enzyme studies could provide a useful approach to shed some light on these.

iii) Endomycorrhizae

Endomycorrhizae only form in unsuberised tissues. With cell to cell contact between the two symbionts, the external hyphae frequently swell to form a more or less well-defined appressorium from which the infecting hypha develops and penetrates the cell wall (Fig. 1 c, e). Appressoria formation is another illustration that a certain degree of recognition has occurred between fungus and root. Although endomycorrhizae can develop anywhere on young roots without secondary thickening, certain root tissues may be more susceptible to infection than others. Mathematical models of infection processes in VA mycorrhizae indicate that the root tissues behind the meristematic zone may be considerably more infectible than other parts of the root (Smith and Walker, 1981) and in ericoid mycorrhiza a truly biotrophic association is only established in the root region immediately behind the apex (Bonfante-Fasolo and Gianinazzi-Pearson, 1979, Bonfante-Fasolo *et al.*, 1981 a, Bonfante-Fasolo and Gianinazzi-Pearson, 1982). This is reminiscent of the localization of Hartig net formation in ectomycorrhiza and deserves further study.

Infection occurs mainly be direct penetration of the outermost host cells or root hairs, although in VA mycorrhizae some hyphae enter via the intercellular spaces. The morphogenesis of the fungi within host tissues and the way in which they spread varies from one type of endomycorrhiza to another. Orchid and ericoid mycorrhizal fungi show little or no development within intercellular spaces (Fig. 1 d) and whilst hyphal spread in the former is directly from one cell to another, in the latter this is very rare and host cells are usually infected by hyphae originating directly from the external mycelium (Fig. 1 e). In both types of mycorrhiza the intracellular hyphae proliferate to form haustoria as dense hyphal coils or peletons (Fig. 1 d, e).

The morphogenesis of VA fungi within host tissues is more complex. In most cases colonization of the epidermal or outermost cortical cells is sparse and here the fungi form simple intracellular coils. Intercellular hyphae develop with increasing frequency as the fungi spread into the inner cortex and infection development is most intense in the deeper layers of the root cortex where fungi form complex much-branched intracellular haustoria or arbuscules (Fig. 1 c). Not only the morphology of VA fungi changes with development in the host tissues but also the architectural structure and composition of their hyphal walls (Bonfante-Fasolo, 1982, Bonfante-Fasolo and Grippioli, 1982). Whilst the fungus typically forms chitin in its hyphal walls outside the root, polymerization of the N-acetyl glycosamine units into chitin seems to be somehow blocked during the intracellular development of arbuscules. This modification in fungal wall structure with infection of root tissue could be an indication that fungal development is determined by recognition and/or interaction processes within the plant. Morphological patterns and distribution of the hyphal system in VA mycorrhizae vary depending on the plant or fungus involved (see Bonfante-Fasolo, 1983). This may be partly due to a nutritional attraction; the greater the distance the root cortical cells are from the vascular tissue, the lower may be their content in substances favourable to fungal growth. This could perhaps explain why arbuscule formation in certain plants, for example clover or soybean, is localized in cells close to the central cylinder whilst in others like onion it can occur in a much greater region of the cortex.

Electron microscope studies indicate that at the ultrastructural level host reaction to penetration and intracellular proliferation of the fungi is quite similar in the different types of endomycorrhizae (see for example Strullu and Gourret, 1974, Hadley, 1975, Scannerini, 1975, Bonfante-Fasolo and Gianinazzi-Pearson, 1979 and 1982, Dexheimer *et al.,* 1979, Strullu and Gourret, 1980, Bonfante-Fasolo *et al.,* 1981 a and b, Gianinazzi-Pearson *et al.,* 1981, Carling and Brown, 1982, Dexheimer *et al.,* 1982, Serrigny, 1982, Bonfante-Fasolo, 1983). No changes in the contents of host cells are evident before an infecting hypha breaches the cell wall. With hyphal penetration into a cell, which probably involves a combination of enzymic and mechanical processes, the host plasmalemma invaginates and elongates to extend around the invading fungus. In all endomycorrhiza an interface is formed between host cell and fungus which is limited by the fungal wall and newly-formed host plasmalemma. As in some infections involving pathogenic fungi (Bracker and Littlefield, 1973), the infecting hypha is surrounded by an osmiophilic fibrillar material which is deposited by the host in the interface and which is continuous with and morphologically similar to the host primary wall. In the case of the poorly developed intracellular coils of VA fungi developing in outermost root cells, this wall material remains intact and seems to separate the two partners by acting as a physical barrier. This is the only marked host reaction to fungal invasion in these cells; cytoplasmic content of the host increases only slightly and its composition resembles that of uninfected cells. This is completely different to what happens in host cells where the fungus proliferates forming haustoria, whether these be dense coils as in orchid and ericoid mycorrhizae or arbuscules as in VA mycorrhizae. Here fungal development is accompanied by a dramatic increase in the volume of host cytoplasm, proliferation of host organelles and hypertrophy of the host nuclei, all of which are signs of a metabolically active condition which recall the situation in juvenile plant cells. Furthermore, endoreplication of nuclear DNA has been shown in infected cells of orchid embryos (Williamson and Hadley, 1969) and it may well occur in the hypertrophied nuclei in other types of mycorrhiza. The quantity of host wall material in the host-fungus interface decreases with development of the intracellular hyphae and it can become reduced to scattered polysaccharide and protein fibrils in ericoid and VA endomycorrhizae. Newly-formed host plasmalemma extends around and closely surrounds the developing hyphae, forming membranous configurations in the interface. A large surface of contact is thus formed between the fungal symbiont and the host cell. No significant morphological or cytochemical differences have been observed between the host plasmalemma which is formed around coils or arbuscules and the normal peripherally situated cell membrane in any type of endomycorrhiza. Furthermore, it possesses neutral phosphatase (IDPase) and ATPase activities characteristic of this plant membrane in uninfected cells (Marx *et al.*, 1982, Serrigny, 1982, Gianinazzi *et al.*, 1983, Gianinazzi-Pearson, Dexheimer and Bonfante-Fasolo, unpublished data). This is contrary to what has been observed in some fungal pathogen infections where the properties of the host plasmalemma can be more or less modified as it forms around the fungus (Bracker and Littlefield, 1973, Gil and Gay, 1977, Spencer-Phillips and Gay, 1981). Moreover, whilst peroxidase acitivity usually increases in plants after fungal pathogen infections (Stahmann and Demorest, 1973), this decreases in host tissues in the presence of a mycorrhizal fungus (Bonfante-Fasolo and Scannerini, 1980, Grippiolo, 1982). Endomycorrhizae are characterized therefore by the creation of an intracellular relationship which is compatible with the life of both the fungus and plant cell.

In cells of VA mycorrhizae where arbuscules develop, the ATPase activity bound to the host plasmalemma becomes specifically associated with the membrane which closely surrounds the fine hyphal branches and the enzyme activity along the uninvaginated membrane diminishes (Marx *et al.*, 1982, Gianinazzi *et al.*, 1983). This host reaction is linked to the presence of living fungus since the ATPase activity strongly diminishes around senescent or dead hyphae. It is evident that some sort of recognition process must be involved to explain such a redistribution of enzyme activity on the host plasmalemma.

The presence of neutral phosphatase, thought to participate in cell wall building processes in plants (Goff, 1973, Maruyama, 1974, Zerban and Werz, 1975), along the host plasmalemma extending around the intracellular hyphae in all types of endomycorrhiza, together with the deposition of host wall material in the host-fungus interface, suggests that the host plasmalemma extending around a growing endomycorrhizal fungus has a cell wall synthesizing activity, similar to that found in elongating cells. In orchids, cell wall material appears to be persistently deposited around the actively developing fungus in both seed and root tissues (Strullu and Gourret, 1974, Hadley, 1975, Serrigny, 1982). This is quite surprising since orchid mycorrhizal fungi can produce pectolytic and cellulolytic enzymes in appreciable amounts (see Smith, 1974). Plant recognition and repression of such destructive enzyme activity must therefore exist so that pathogenicity of the fungal partners is controlled in orchid associations. Studies mainly on orchid tubers have shown that fungal infection elicits the production of antifungal compounds like orchinol, hircinol and loroglyssol (dihydroxyphenanthrenes) by host cells (see Neusch, 1963). Fisch *et al.*

(1973) have proposed that the ability of mycorrhizal fungi to both induce and slowly degrade such compounds might keep the infection within acceptable limits and thus contribute to maintaining a balanced relationship between plant and fungus. In photosynthetic orchids high levels of soluble carbohydrates in the host cells might also be expected to repress the production or activity of cellulolytic enzymes of invading mycorrhizal fungi. In ericoid mycorrhizae and in the arbuscular phase of VA mycorrhizae, however, very little fibrillar material can be observed around the actively growing intracellular hyphae which seems to suggest that the fungi degrade the cell-wall components or precursors of their hosts. Ericoid mycorrhizal fungi do indeed show pectolytic and weak cellulolytic activities in pure culture (Pearson and Read, 1975) but nothing is known for VA fungi, except that they can spread intercellularly and cross cell walls. However, once inside the host cell these enzyme activites, if they exist, risk being repressed by the high level of soluble carbohydrates. An alternative explanation proposed by Harley and Smith (1983) is that in these endomycorrhizae polymerization of host-produced cell-wall precursors is somehow prevented in the host-fungus interface. As in ectomycorrhizae, host enzymes involved in polymerization processes might be inhibited or complexed by molecules on the walls of actively growing hyphae. In relation to this it is interesting to observe that the arbuscule walls of a VA fungus have been shown to possess a binding activity with the wheat germ lectin agglutinin (Bonfante-Fasolo, 1982).

With decreasing fungal activity and deterioration of the intracellular hyphae in VA and orchid mycorrhizae, host plasmalemma-bound neutral phosphatase activity persists and host material indeed accumulates around the fungal remains. Fungal senescence begins as soon as morphological development is complete but it is not clear whether this process is somehow provoked by the host cell. In orchids, for example, this could be caused by an imbalance in the host-fungus relationship due to the accumulation of antifungal compounds in the cells as fungal activity decreases. With hyphal senescence the fungal cytoplasm breaks down by autolysis, hyphae collapse and their walls aggregate in clumps. The host plasmalemma around the individual hyphae fuses to enclose the whole structure. Host membrane integrity is maintained during this process and the host cytoplasmic content diminishes to re-establish cytological characteristics typical of an uninfected cell. In ericoid mycorrhizae, fungal hyphae usually outlive the host cytoplasm but it is not known whether the presence of the fungus actually enhances senescence of the host cell.

These different morphological and cytological studies of the various types of mycorrhizal associations illustrate that mycorrhiza development depends on a series of recognition processes. Some of these are probably related to the detection of a food source by the fungal symbiont whilst others are clearly surface phenomena which involve molecules located in or on the walls or membranes of the plant or fungus.

C. Interactions of Mycorrhizal Fungi with Non-Host Plants

This aspect has been given very little attention in studies of specificity or compatibility in mycorrhizae. The best known example is that of *Thanetephorus cucumeris* (syn. *Rhizoctonia solani*) which is a common fungal symbiont in orchid mycorrhiza. The same fungal strains of this species can also be destructive parasites of a number of other plants where they rapidly colonise all the host tissues (Williamson and Hadley, 1970). Preliminary investigations with the ericoid mycorrhizal fungus. *P. ericae* have recently shown that it is able to penetrate the roots of a non-host plant, red clover. The infection which develops in red clover is atypical of ericoid mycorrhizae. The fungus grows straight across the root tissues without forming hyphal loops in the epidermal or cortical cells and frequently penetrates the vascular cylinder (Gianinazzi-Pearson and Bonfante-Fasolo, unpublished data). Interestingly, similar observations have been made on non-host plants with VA fungi. Species of the Chenopodiaceae, Cruciferae and Amantharaceae which do not form mycorrhizae in nature can become slightly infected if they grow close to a strongly infected mycorrhizal host-plant (Hirrel *et al.*, 1978, Ocampo *et al.*, 1980). The infection is restricted to intercellular hyphae and vesicles, no intracellular development into arbuscules occurs and the VA fungi can penetrate the vascular cylinder of the non-host root.

An important characteristic of mycorrhizal associations is that fungal colonization is restricted to epidermal and/or cortical host tissue and the fungal symbionts never enter the meristematic nor the vascular regions of the root. The fact that in non-host roots, on the contrary, fungal development in the cortex is very poor and the vascular tissue is penetrated reinforces the idea that fungal development in compatible mycorrhizal associations is under host control and must involve recognition phenomena. As has already been seen, in orchids this may be due to resistance phenomena, host cells producing antifungal compounds or being able to repress fungal production of destructive enzymes. In VA mycorrhizae it seems to be host susceptibility that is determinate and this may be related to the supply of root exudates that can be used by the fungal symbiont. For instance, foliar application of the herbicide simazine to a non-mycorrhiza forming plant species increases root exudation of carbohydrates, sugars and amino acids and stimulates a weak but typical VA infection with arbuscule formation and no hyphal penetration into the stele (Schwab *et al.*, 1982).

Further research on both compatible and incompatible associations are clearly necessary for a better understanding of the mechanisms underlying specificity and compatibility in mycorrhizae. It is evident that the influence of these mechanisms is determined by the genetical make-up of the fungus and plant but since they have not been sufficiently studied in mycorrhizae it is not yet possible to offer any genetical interpretation for them in mycorrhizal associations. The only presumption that can be made is that, in the words of Vanderplank (1978), "(in contrast to pathogen associations) *mutations to resistance in mycorrhizal plants are eliminated by selection because they are disadvantageous; and the elimination also eliminates a major source of specificity*".

IV. Functional Compatibility

Being mutualistic or symbiotic in their associations, the notion of compatibility in mycorrhizae is complicated by the fact that fungus and plant may show complete compatibility as far as establishment for infection is concerned but the resulting association may not provide optimal benefit to either partner. This introduces another notion, that of functional compatibility, which must be determined by the physiological activity of the two partners. This functional compatibility will be a prerequisite for the phenotypic expression of the mycorrhizal association (mycorrhizal effectiveness) and will depend not only on the genetic make-up of fungus and plant but also on factors external to the association which influence the expression of their genomes (Fig. 3).

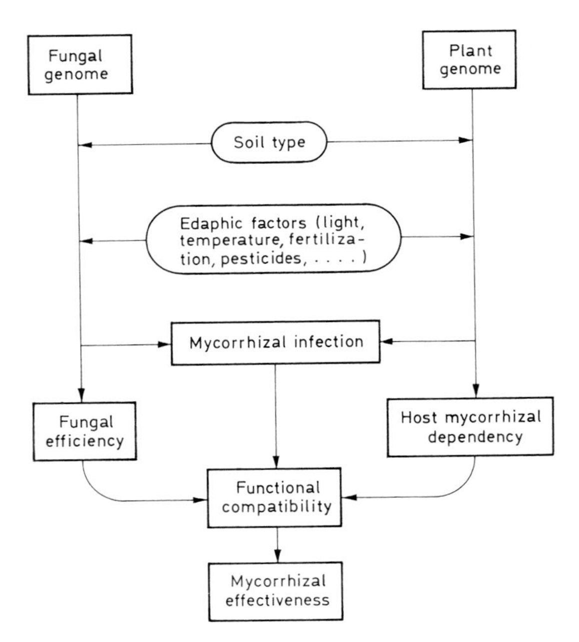

Fig. 3. Interactions between different components governing mycorrhizal effectiveness

The principal role of mycorrhizae is to improve plant nutrition, whether it be the carbon nutrition of minute orchid seeds or achlorophyllous seedlings (Smith, 1967, Purves and Hadley, 1975), the phosphate uptake by VA mycorrhizal plants or both the nitrogen and phosphate assimilation by ectomycorrhizal and ericoid mycorrhizal plant species (Gianinazzi-Pearson, 1982). There is much convincing evidence that the better nutrition of mycorrhizal plants is directly due to the activity of the associated fungi (see Harley and Smith, 1983). These can absorb nutrients from the soil at a distance from the root, accumulate and transport them in relatively important amounts within their hyphae and release them to the host tissues. The effect of mycorrhizae on plant growth and nutrition will therefore depend on the activity of the fungal symbiont (fungal efficiency), the capacity of the host plant to satisfy its nutritional needs (mycorrhizal dependency) and the functional or physiological compatibility between the two partners. Any factors influencing these different elements will affect the overall effectiveness of the mycorrhizal association (Fig. 3).

Mycorrhizal dependency, measured as plant growth improvement following infection by a defined fungal inoculum, can vary greatly from one plant species to another (see for example Lamb and Richards, 1971, Marx, 1975, Trappe, 1977, McGraw and Schenck, 1980, Gianinazzi, 1982, Plenchette, 1982) and between cultivars, clones or individuals of different provenance within a single plant species (Hall, 1978, Lambert *et al.,* 1980, Bertheau *et al.,* 1980, Azcon and Ocampo, 1981, Powell and Bates, 1981, Cline and Reid, 1982, Powell *et al.,* 1982, Granger *et al.,* 1983). In the case of VA mycorrhizae, this genetic variability in mycorrhizal dependency has been attributed to the nature of the root system (Baylis, 1972, St. John, 1980), plants with coarse root systems and limited root hair development being theoretically less efficient in soil exploration and nutrient absorption and therefore more mycorrhiza dependent than those having fibrous root systems with abundant root hairs. In the few studies that have tested this hypothesis, results are contradictory (Baylis, 1975, Azcon and Ocampo, 1981, McGraw and Schenck, 1980) and it is evident that additional factors controlled by the plant genome such as growth rate (and hence nutrient demand), root absorption efficiency, root longevity must also determine the magnitude of mycorrhizal dependency.

Inter- and intraspecific differences also exist in the growth-promoting abilities of mycorrhizal fungi (see for example Young, 1940, Levisohn, 1959, Theodorou and Bowen, 1970, Mosse, 1972, Marx, 1975, Sanders *et al.,* 1977, Trappe, 1977, Carling and Brown, 1980, McGraw and Schenck, 1980, Gianinazzi, 1982, Jensen, 1982, Plenchette *et al.,* 1982). This variability in fungal efficiency is not necessarily directly related to the rate or degree of infection development within roots but depends also on the intrinsic physiological characteristics of each fungus. In particular, the rôle of the external mycelium has frequently been emphasized since the production of hyphae which extensively spread in the soil and have high nutrient uptake and translocating capacities will be a vital factor in mycorrhizal effectivenness (Bowen, 1968, Sanders *et al.,* 1977, Gianinazzi-Pearson and Gianinazzi, 1981, 1983, Bethlenfalvay *et al.,* 1982a, Harley and Smith, 1983). Although knowledge concerning the physiology of mycorrhizal fungi has made considerable progress during recent years (see Gianinazzi *et al.,* 1982, Harley and Smith, 1983), there is still need for much more investigation since research has up to now concerned only a few fungi and there has been very little comparative work.

The functional relationship between fungus and plant is far from simple and a plant showing a relatively important mycorrhizal dependency may not necessarily benefit from an association with an efficient fungus. This is particularly well demonstrated in McGraw and Schenck's (1980) experiments: *G. macrocarpus* was as beneficial to plant growth as other VA fungi tested when in association with peach but it reduced growth and fruit production of tomato plants under the same conditions in which the other VA fungi were beneficial. Similar observations have also been made in ectomycorrhizal associations (Trappe, 1977). Functional compatibility between mycorrhizal associates not only depends on the plant and fungus involved but it is also affected by external factors which can cause an imbalance in the symbiotic relationship so that one of the partners dominates the other. As Harley and Smith (1983) pointed out, growth responses to most mycorrhizae will be determined by two opposing processes: provision of soil-derived nutrients by the fungus to the plant and consumption of carbon compounds derived from host photosynthate by the fungus. Maximal mycorrhizal effectiveness occurs when nutritional conditions for the host are suboptimal but if these become severely deficient the fungus can become aggressive and have no positive, or even negative, effects on plant growth, probably because any nutritional improvement of the host does not sufficiently counteract the carbon drain imposed by the fungus (Slankis, 1967, Dumbroff, 1968, Stribley and Read, 1976, Bethenfalvay *et al.*, 1982b). With improving nutritional conditions a reciprocally beneficial host-fungus relationship is established but as nutrient supply reaches an optimum for the host plant, mycorrhizal effectiveness is reduced to zero and the host plant begins to dominate the fungal infection with subsequent rejection of the fungal associate from the host roots (Sanders and Tinker, 1973, Slankis, 1974, Stribley and Read, 1976, Asimi *et al.*, 1980, Graham *et al.*, 1981). However, studies on VA mycorrhizae have shown that mycorrhizal fungi are not affected to the same extent by an improved external nutrient supply to the host. A differential tolerance to phosphate supply is observed not only in relation to infection development (Jasper *et al.*, 1979) but also in the ability of fungal species or isolates to improve plant growth at different levels of fertilisation (Carling and Brown, 1980, Gianinazzi, 1982).

There is some evidence that soil nutrients affect mycorrhizal acitivity at the level of the host-fungus interaction and it has been suggested that the internal nutrient status of the host plant is an important determining factor (Hatch, 1937, Sanders, 1975, Menge *et al.*, 1978, Graham *et al.*, 1981, Plenchette, 1982). In VA mycorrhizae, changes in plant membrane integrity have been proposed as a possible physiological explanation for the effects of high phosphate supply (Ratnayake *et al.*, 1978, Graham *et al.*, 1981) and observations on mycorrhiza-specific alkaline phosphatases (Gianinazzi-Pearson and Gianinazzi, 1976, Gianinazzi *et al.*, 1979) suggest that increasing phosphate fertilisation can also disturb physiological processes within the fungi (Gianinazzi-Pearson and Gianinazzi, 1983). VA and ectomycorrhizal fungi apparently accumulate large quantities of phosphate by sequestering it in the osmotically inactive form of polyphosphate within their vacuoles (Ashford *et al.*, 1975, Ling-Lee *et al.*, 1975, Callow *et al.*, 1978, Cox *et al.*, 1980, Strullu *et al.*, 1981) but it is not known what triggers the release of this phosphate from the fungus to the plant and the mechanisms by which it is controlled. It is evident that the host-fungus interface must be an important site for determining the rate of nutrient transfer between the two symbionts. The fact that in VA mycorrhizae the plasmalemma-bound ATPase activity of the host is specifically localized at the intracellular interface of the arbuscule branches, where phosphate transfer to the host plant is thought to mainly occur, underlines this point (Marx *et al.*, 1982, Gianinazzi *et al.*, 1983). Any factors affecting the structure and the physiology of the interfacial zone in mycorrhizae, whether originating in fungus or host, will therefore have direct repercussions on the functional compatibility of these symbiotic associations. Improved knowledge of the processes occurring in the host-fungus interface, the mechanisms controlling them and the factors affecting them would greatly advance our understanding of functional compatibility in mycorrhizal associations, so that better use could be made of their potential for improving plant yields.

Acknowledgements

The author is very grateful to S. Gianinazzi and S. E. Smith for critically reading the manuscript.

V. References

Ashford, A. E., Ling Lee, M., Chilvers, G. A., 1975: Polyphosphate in eucalypt mycorrhizas: A cytochemical demonstration. New Phytol. **74,** 447—453.

Asimi, S., Gianinazzi-Pearson, V. Gianinazzi, S., 1980: Influence of increasing soil phosphorus levels on interactions between vesicular-arbuscular mycorrhizae and *Rhizobium* in soybeans. Can. J. Bot. **58,** 2200—2205.

Athanassiou, Z., 1979: Contribution à l'étude des mycorrhizes du pin d'Alep. Doctorate Thesis, p. 89, University of Paris-Sud, France.

Azcon, R., Ocampo, J. A., 1981: Factors affecting the vesicular-arbuscular infection and mycorrhizal dependency of thirteen wheat cultivars. New Phytol. **87,** 677—685.

Baylis, G. T. S., 1972: Minimum levels of phosphorus for non-mycorrhizal plants. Plant and Soil **36,** 233—234.

Baylis, G. T. S., 1975: The magnoloid mycorrhiza and mycotrophy in root systems derived from it. In: Sanders, F. E., Mosse, B., Tinker, P. B. (eds.): Endomycorrhizas, pp. 373—389. London – New York: Academic Press.

Bertheau, Y., Gianinazzi-Pearson, V., Gianinazzi, S., 1980: Développement et expression de l'association endomycorhizienne chez le Blé. I. Mise en évidence d'un effet variétal. Ann. Amél. Plantes **30,** 67—78.

Bethlenfalvay, G. J., Brown, M. S., Pacovsky, R. S., 1982a: Relationships between host and endophyte development in mycorrhizal soybeans. New Phytol. **90,** 537—543.

Bethlenfalvay, G. J., Brown, M. S., Pacovsky, R. S., 1982b: Parasitic and mutualistic associations between a mycorrhizal fungus and soybean: development of the host plant. Phytopathology **72,** 889—893.

Bevege, D. I., Bowen, G. D., 1975: *Endogone* strain and host plant differences in development of vesicular-arbuscular mycorrhizas. In: Sanders, F. E., Mosse, B., Tinker, P. B. (eds.): Endomycorrhizas, pp. 77—86. London - New York: Academic Press.

Bianchi, D. E., Turian, G., 1967: The effect of nitrogen source and cysteine on the morphology conidiation and a cell wall fraction of conidial and aconidial *Neurospora crassa*. Z. Allg. Mikrobiol. **7,** 257—262.

Birraux, D., Fries, N., 1981: Germination of *Thelephora terrestris* basidiospores Can. J. Bot. **59,** 2062—2064.

Björkman, E., 1960: *Monotropa hypopitys* L. an epiparasite on tree roots. Physiol. Plant **13,** 308—327.

Bonfante, P., Giovannetti, M., 1982: Le micorrize. In: Quaderni di Biologia **8,** p. 46. Padova - Roma - Milano: Piccin.

Bonfante-Fasolo, P., 1980: Occurrence of a basidiomycete in living cells of mycorrhizal hair roots of *Calluna vulgaris*. Trans. Brit. mycol. Soc. **75,** 320—325.

Bonfante-Fasolo, P., 1982: Cell wall architectures in a mycorrhizal association as revealed by cryoultramicrotomy. Protoplasma **111,** 113—120.

Bonfante-Fasolo, P., 1983: Anatomy and morphology. In: Powell, C. Ll., Bagyaraj, D. G. (eds.): VA Mycorrhizas. CRC Reviews (in press).

Bonfante-Fasolo, P., Berta, G., Gianinazzi-Pearson, V., 1981a: Ultrastructural aspects of endomycorrhiza in the Ericaceae. II. Host-endophyte relationships in *Vaccinium myrtillus*. New Phytol. **89,** 219—224.

Bonfante-Fasolo, P., Dexheimer, J., Gianinazzi, S. Gianinazzi-Pearson, V., Scannerini, S., 1981b: Cytochemical modifications in the host-fungus interface during intracellular interactions in vesicular-arbuscular mycorrhizae. Plant Science Letters **22,** 13—21.

Bonfante-Fasolo, P., Gianinazzi-Pearson, V., 1979: Ultrastructural aspects of endomycorrhiza in the Ericaceae. I. Naturally infected hair roots of *Calluna vulgaris* L. Hull. New Phytol. **83,** 739—744.

Bonfante-Fasolo, P., Gianinazzi-Pearson, V., 1982: Ultrastructural aspects of endomycorrhiza in the Ericaceae. III. Morphology of the dissociated symbionts and modifications occurring during their reassociation in axenic culture. New Phytol. **91,** 691—704.

Bonfante-Fasolo, P., Grippioli, R., 1981: Ultrastructural and cytochemical changes in the wall of a vesicular-arbuscular mycorrhizal fungus during symbiosis. Can. J. Bot. **60,** 2303—2312.

Bonfante-Fasolo, P., Scannerini, S., 1977: Cytological observations on the mycorrhiza *Endogone flammicorona — Pinus strobus*. Allionia **22,** 23—34.

Bonfante-Fasolo, P., Scannerini, S., 1980: Ultrastructural localization of peroxidase activity in endomycorrhizal roots. Caryologia **33,** 126—127.

Boullard, B., 1979: Considérations sur la symbiose fongique chez les ptéridophytes. Syllogeus **19,** 1—58.

Bowen, G. D., 1968: Phosphate uptake by mycorrhizas and uninfected roots of *Pinus radiata* in relation to root distribution. Trans. 9th Int. Congr. Soil. Sc. **2,** 219.

Bracker, C. E., Littlefield, L. J., 1973: Structural concepts of host-pathogen interfaces. In: Byrde, R. J. W., Cutting, C. V. (eds.): Fungal pathogenicity and the plant's response, pp. 159—317. London - New York: Academic Press.

Callow, J. A., 1975: Plant lectins. Curr. Adv. Pl. Sci. **7,** 181—193.

Callow, J. A., Capaccio, L. C. M., Parish, G., Tinker, P. B., 1978: Detection and estimation of polyphosphate in vesicular-arbuscular mycorrhizas. New Phytol. **80,** 125—134.

Carling, D. E., Brown, M. F., 1980: Relative effect of vesicular-arbuscular mycorrhizal fungi on the growth and yield of soybeans. Soil. Sci. Soc. Am. J. **44,** 528—532.

Carling, D. E., Brown, M. F., 1982: Anatomy and physiology of vesicular-arbuscular and nonmycorrhizal roots. Phytopathology **72,** 1108—1114.

Cline, M. L., Reid, C. P. P., 1982: Seed source and mycorrhizal fungus effects on growth of containerized *Pinus contorta* and *Pinus ponderosa* seedlings. Forest Sci. **28,** 237—250.

Clowes, F. A. L., 1951: The structure of mycorrhizal roots of *Fagus sylvatica.* New Phytol. **50,** 1—16.

Cox, G., Moran, K. J., Sanders, F., Nockolds, C., Tinker, P. B., 1980: Translocation and transfer of nutrients in vesicular-arbuscular mycorrhizas. III. Polyphosphate granules and phosphorus translocation. New Phytol. **84,** 649—659.

Dazzo, F. B., 1980: Absorption of micro-organisms to roots and other plant surfaces. In: Bitton, G., Marshall, K. C. (eds.): Absorption of microorganisms to surfaces, pp. 253—316. New York: John Wiley & Sons.

Debaud, J. C., Pepin, R., Bruchet, G., 1981: Ultrastructure des ectomycorhizes synthétiques à *Hebeloma alpinum* et *Hebeloma marginatulum* de *Dryas octopetela.* Can. J. Bot. **59,** 2160—2166.

Dexheimer, J., Gianinazzi, S., Gianinazzi-Pearson, V., 1979: Ultrastructural cytochemistry of the host-fungus interfaces in the endomycorrhizal association *Glomus mosseae / Allium cepa.* Z. Pflanzenphysiol. **92,** 191—206.

Dexheimer, J., Gianinazzi-Pearson, V., Gianinazzi, S., 1982: Acquisitions récentes sur la physiologie des mycorhizes VA au niveau cellulaire. In: Gianinazzi, S., Gianinazzi-Pearson, V., Trouvelot, A. (eds.): Les Mycorhizes, Partie Intégrante de la Plante: Biologie et Perspectives d'Utilisation (Les Colloques de l'INRA, vol. 13), pp. 61—73. Paris: INRA.

Dominik, T., 1969: Key to ectotrophic mycorrhizae. Folia For. Poln. Lesn. **15,** 309—328.

Duclos, J. L., 1981: Contribution à l'étude expérimentale des endomycorhizes des Ericacées. Thèse 3è cycle, p. 133, University of Nancy, France.

Duddridge, J. A., Read, D. J., 1982: An ultrastructural analysis of the development of mycorrhizas in *Monotropa hypopitys* L. New Phytol. **92,** 203—214.

Dumbroff, E. B., 1968: Some observations on the effects of nutrient supply on mycorrhizal development in pine. Plant and soil **28,** 463—466.

Fasolo-Bonfante, P., Brunel, A., 1972: Caryological features in a mycorrhizal fungus: *Tuber melanosporum* Vitt. Allionia **18,** 5—11.

Fassi, B., Palenzona, M., 1969: Sintesi micorrizica tra *Pinus strobus, Pseudotsuga douglasii* ed *Endogone lactiflua.* Allionia **15,** 105—114.

Faye, M., Rancillac, M., David, A., 1981: Determinism of the mycorrhizogenic root formation in *Pinus pinaster* sol. New Phytol. **87,** 557—565.

Fisch, M. H., Flick, B. H., Arditti, J., 1973: Structure and antifungal activity of hircinol, loroglossol and orchinol. Phytochemistry **12,** 437—441.

Freisleben, R., 1936: Weitere Untersuchungen über die Mykotrophie der Ericaceen. Jb. Wiss. Bot. **82,** 413—459.

Galpin, M. F. J., Jennings, D. H., Thornton, J. D., 1977: Hyphal branching of *Dendryphiella saline:* effect of various compounds and the further elucidation of the effect of sorbose and the role of cyclic AMP. Trans. Brit. mycol. Soc. **69,** 175—182.

Gianinazzi, S., 1981: L'endomycorhization contrôlée en agriculture, en horticulture et en arboriculture: problèmes et progrès. In: Gianinazzi, S., Gianinazzi-Pearson, V., Trouvelot, A. (eds.): Les Mycorhizes, Partie Intégrante de la Plante: Biologie et Perspectives d'Utilisation (les Colloques de l'INRA, Vol. 13), pp. 231—241. Paris: INRA.

Gianinazzi, S., Gianinazzi-Pearson, V., Dexheimer, J., 1979: Enzymatic studies on the metabolism of vesicular-arbuscular mycorrhiza. III. Ultrastructural localization of acid and alkaline phosphatase in onion roots infected by *Glomus mosseae* (Nicol & Gerd). New Phytol. **82,** 127—132.

Gianinazzi, S., Gianinazzi-Pearson, V., Trouvelot, A. (eds.): 1982: Les Mycorhizes, Partie Intégrante de la Plante: Biologie et Perspectives d'Utilisation (les Colloques de l'INRA, vol. 13), p. 397. Paris: INRA.

Gianinazzi, S., Dexheimer, J., Gianinazzi-Pearson, V., Marx, C., 1983: Role of the host-arbuscule interface in the symbiotic nature of VA mycorrhizal associations. Plant and Soil **71,** 211—215.

Gianinazzi-Pearson, V., 1982: Importance des mycorhizes dans la nutrition et la physiologie des plantes. In: Gianinazzi, S., Gianinazzi-Pearson, V., Trouvelot, A. (eds.): Les Mycorhizes, Partie Intégrante de la Plante: Biologie et Perspective d'Utilisation (les Colloques de l'INRA, vol. 13), pp. 51—59. Paris: INRA.

Gianinazzi-Pearson, V., Gianinazzi, S., 1976: Enzymatic studies on the metabolism of vesicular-arbuscular mycorrhiza. I. Effect of mycorrhiza formation and phosphorus nutrition on soluble phosphatase activities in onion roots. Physiol. Vég. **14,** 833—841.

Gianinazzi-Pearson, V., Gianinazzi, S., 1981: The role of endomycorrhizal fungi in phosphorus cycling in the soil. In: Wicklow, D. T., Carrol, G. C. (eds.): The Fungal Community, Its Organisation and Role in the Ecosystem, pp. 637—652. New York: Marcel Dekker.

Gianinazzi-Pearson, V., Gianinazzi, S., 1983: The physiology of vesicular-arbuscular mycorrhizal roots. Plant and Soil **71,** 197—209.

Gianinazzi-Pearson, V., Morandi, D., Dexheimer, J., Gianinazzi, S., 1981: Ultrastructural and ultracytochemical features of a *Glomus tenuis* mycorrhiza. New Phytol. **88,** 633—639.

Gil, F., Gay, J. L., 1977: Ultrastructural and physiological properties of the host interfacial components of haustoria of *Erysiphe pisi in vivo* and *in vitro.* Physiol. Plant. Path. **10,** 1—12.

Goff, C. W., 1973: Localization of nucleoside diphosphatase in the onion root tip. Protoplasma **78,** 397—416.

Graham, J. J., Leonard, R. T., Menge, J. A., 1981: Membrane-mediated decrease in root exudation responsible for phosphorus inhibition of vesicular-arbuscular mycorrhiza formation. Plant Physiol. **68,** 548—552.

Granger, R. L., Plenchette, C., Fortin, J. A., 1983: Effect of a vesicular-arbuscular (VA) endomycorrhizal fungus *(Glomus epigaeum)* on the growth and mineral content of two apple clones propagated *in vitro.* Can. J. Plant Sci. **63,** 551—555.

Grente, J., Chevalier, G., Pollacsek, A., 1972: La germination de l'ascospore de *Tuber melanosporum* et la synthèse sporale des mycorhizes. C. R. Acad. Sc. **275,** Série D, 743—746.

Grippiolo, R., 1982: Peroxidase activity in the wall of Star of Bethlehem *(Ornithogallum umbellatum)* VA mycorrhizae. Caryologia **35,** 384—385.

Hadley, G., 1970: Non-specificity of symbiotic infection in orchid mycorrhiza. New Phytol. **69,** 1015—1023.

Hadley, G., 1975: Organisation and fine structure of orchid mycorrhiza. In: Sanders, F. E., Mosse, B., Tinker, P. B. (eds.): Endomycorrhizas, pp. 335—351. London - New York: Academic Press.

Hall, I. R., 1978: Effect of vesicular-arbuscular mycorrhizas on two varieties of maize and one of sweetcorn. N. Z. J. Agric. Res. **21,** 517—519.

Harley, J. L., 1937: Ecological observations on the mycorrhiza of beech. J. Ecol. **25,** 421—423.

Harley, J. L., 1969: The biology of mycorrhiza, 2nd. Ed., p. 334. Oxford: Leonard Hill.

Harley, J. L., Smith, S. E., 1983: Mycorrhizal symbiosis, p. 483. London - New York: Academic Press.

Hatch, A. B., 1937: The physical basis of mycotrophy in *Pinus.* Black Rock Forest. Bull. **6,** 1—168.

Heslop-Harrison, J., 1978: Cellular Recognition Systems in Plants (Institute of Biology, Studies in Biology n° 100), p. 60. London: Edward Arnold.

Hirrel, M. C., Mehravan, H., Gerdemann, J. W., 1978: Vesicular-arbuscular mycorrhizae in the Chenopodiaceae and Cruciferae: do they occur? Can. J. Bot. **56,** 2813—2817.

Jasper, D. A., Robson, A. D., Abbott, L. K., 1979: Phosphorus and the formation of vesicular-arbuscular mycorrhizas. Soil Biol. Biochem. **11,** 501—505.

Jensen, A., 1982: Influence of four vesicular-arbuscular mycorrhizal fungi on nutrient uptake and growth in barley *(Hordeum vulgare).* New Phytol. **90,** 45—50.

Koch, H., 1961: Untersuchungen über die Mykorrhiza der Kulturpflanzen unter besonderer Berücksichtigung von *Althaea officinalis* L., *Atropa belladonna* L., *Helianthus annua* L. und *Solanum lycopersicum* L. Gartenbauwiss. **26,** 5.

Lamb, R. J., Richards, B. N., 1971: Effect of mycorrhizal fungi on the growth and nutrient status of slash and radiata pine seedlings. Aust. For. **35,** 1—7.

Lambert, D. H., Cole, H., Baker, D. E., 1980: Variation in the response of alfalfa clones and cultivars to mycorrhizae and phosphorus. Crop Sci. **20,** 615—618.

Levisohn, I., 1959: Strain differentiation in a root-infecting fungus. Nature **183,** 1065—1066.

Ling Lee, M., Chilvers, G. A., Ashford, A. E., 1975: Polyphosphate granules in three kinds of tree mycorrhiza. New Phytol. **75,** 551—554.

Lutz, R. W., Sjolund, R. D., 1973: *Monotropa uniflora:* ultrastructural details of its mycorrhizal habit. Amer. J. Bot. **60,** 339—345.

Marks, G. C., Foster, R. C., 1973: Structure, morphogenesis and ultrastructure of ectomycorrhizae. In: Marks, G. C., Kozlowski, T. T. (eds.): Ectomycorrhizae, pp. 1—41. New York - London: Academic Press.

Malajczuk, N., Molina, R., Trappe, J. M., 1982: Ectomycorrhiza formation in *Eucalyptus.* I. Pure culture synthesis, host specificity and mycorrhizal compatibility with *Pinus radiata.* New Phytol. **91,** 467—482.

Maruyama, K., 1974: Localization of polysaccharides and phosphatases in the golgi apparatus of *Tradescantia* pollen. Cytologia **39,** 767—776.

Marx, D. H., 1975: Mycorrhizae of exotic trees in the Peruvian Andes and synthesis of ectomycorrhizae on Mexican pines. Forest Sci. **21,** 353—358.

Marx, D. H., Bryan, W. C., 1971: Formation of ectomycorrhizae on halfsib progenies of slash-pine in aseptic culture. Forest Sci. **17,** 488—492.

Marx, C., Dexheimer, J., Gianinazzi-Pearson, V., Gianinazzi, S., 1982: Enzymatic studies on the metabolism of vesicular-arbuscular mycorrhizas. IV. Ultrastructural evidence (ATPase) for active transfer processes in the host-arbuscule interface. New Phytol. **90,** 37—43.

Mason, P., 1975: The genetics of mycorrhizal associations between *Amanita muscaria* and *Betula verrucosa.* In: Torrey, J. G., Clarkson, D. T. (eds.): The development and function of roots, pp. 367—374. London - New York: Academic Press.

McGraw, A. C., Schenck, N. C., 1980: Growth stimulation of citrus, ornamental and vegetable crops by select mycorrhizal fungi. Proc. Fla. State Hort. Sci. **93,** 201—205.

Mengden, K., 1978: Attachment of bean rust cell wall material to host and non-host plant tissue. Arch. Microbiol. **119,** 113—117.

Menge, J. A., Stierle, D., Bagyaraj, D. J., Johnson, E. L. V., Leonard, R. T., 1978: Phosphorus concentrations in plants responsible for inhibition of mycorrhizal infection. New Phytol. **80,** 575—578.

Mims, C. W., 1980: Ultrastructure of basidiospores of the mycorrhizal fungus *Pisolithus tinctorius.* Can. J. Bot. **58,** 1525—1533.

Molina, R., 1979: Ectomycorrhizal inoculation of containerized Douglas-fir and lodgepole pine seedlings with six isolates of *Pisolithus tinctorius.* Forest Sci. **25,** 585—590.

Molina, R., 1981: Ectomycorrhizal specificity in the genus *Alnus.* Can. J. Bot. **59,** 325—334.

Molina, R., Trappe, J. M., 1982a: Patterns of ectomycorrhizal host specificity and potential among pacific Northwest conifers and fungi. Forest Sci. **28,** 423—458.

Molina, R., Trappe, J. M., 1982b: Lack of mycorrhizal specificity by ericaceous hosts *Arbutus meriziesii and Arctostaphylos uva-ursi.* New Phytol. **90,** 495—509.

Mosse, B., 1972: Influence of soil type and *Endogone* strain on the growth of mycorrhizal plants in phosphate deficient soils. Rev. Ecol. Biol. Sol. **9,** 529—537.

Mosse, B., Stribley, D. P., Le Tacon, F., 1981: Ecology of mycorrhizae and mycorrhizal fungi. Adv. Microb. Ecol. **5,** 137—210.

Murray, G. M., Maxwell, D. P., 1975: Penetration of *Zea mays* by *Helminthosporium carbonum.* Can. J. Bot. **53,** 2872—2883.

Neusch, J., 1963: Defense reactions in orchid bulbs. Symp. Soc. Gen. Microbiol. **13,** 335—343.

Nicolson, T. H., 1975: Evolution of vesicular-arbuscular mycorrhizas. In: Sanders, F. E., Mosse, B., Tinker, P. B. (eds.): Endomycorrhizas, pp. 25—34. London - New York: Academic Press.

Nylund, J. E., Unestam, T., 1982: Structure and physiology of ectomycorrhizae. I. The process of mycorrhiza formation in Norway spruce *in vitro.* New Phytol. **91,** 63—79.

Ocampo, J. A., Martin, J., Hayman, D. S., 1980: Influence of plant interactions on vesicular-arbuscular mycorrhizal infections. I. Host and non-host plants grown together. New Phytol. **84,** 27—35.

Pearson, V., Read, D. J., 1973: The biology of mycorrhiza in the Ericaceae. I. The isolation of the endophyte and the synthesis of mycorrhizas in aseptic culture. New Phytol. **72,** 371—379.

Pearson, V., Read, D. J., 1975: The physiology of the mycorrhizal endophyte of *Calluna vulgaris*. Trans. Brit. mycol. Soc. **64,** 1—7.

Peterson, T. A., Mueller, W. C., Englander, L., 1980: Anatomy and structure of a *Rhododendron* root-fungus association. Can. J. Bot. **58,** 2421—2433.

Piché, Y., Fortin, J. A., Lafontaine, J. G., 1981: Cytoplasmic phenols and polysaccharides in ectomycorrhizal and non-mycorrhizal short roots of pine. New Phytol. **88,** 695—703.

Pirozynski, K. A., Malloch, D. W., 1975: The origin of land plants: a matter of mycotrophism. Biosystems **6,** 153—164.

Plenchette, C., 1982: Recherches sur les endomycorhizes à vésicules et arbuscules. Influence de la plante-hôte, du champignon et du phosphore sur l'expression de la symbiose endomycorhizienne. Ph. D. Thesis, p. 170, University of Laval, Canada.

Plenchette, C., Furlan, V., Fortin, J. A., 1982: Effects of different endomycorrhizal fungi on five host plants grown in a calcined montmorillonite clay. J. Amer. Hort. Sci. **107,** 535—538.

Powell, C. L., Bates, P. M., 1981: Ericoid mycorrhizas stimulate fruit yield of blueberry. Hort Science **16,** 655—656.

Powell, C. L., Clark, G. E., Verberne, N. J., 1982: Growth response of four onion cultivars to several isolates of VA mycorrhizal fungi. N. Z. J. Agric. Res. **25,** 465—470.

Purves, S., Hadley, G., 1975: Movement of carbon compounds between the partners in orchid mycorrhiza. In: Sanders, F. E., Mosse, B., Tinker, P. B. (eds.): Endomycorrhizas, pp. 175—194. London - New York: Academic Press.

Ranayake, M., Leonard, R. T., Menge, J. A., 1978: Root exudation in relation to supply of phosphorus and its possible relevance to mycorrhizal formation. New Phytol. **81,** 543—552.

Read, D. J., 1974: *Pezizella ericae* sp. nov. the perfect state of a typical mycorrhizal endophyte of Ericaceae. Trans. Mycol. Soc. **63,** 381—383.

Read, D. J., Armstrong, W., 1972: A relationship between oxygen transport and the formation of ectotrophic mycorrhizal sheath in conifer seedlings. New Phytol. **71,** 49—53.

Robertson, D. C., Robertson, J. A., 1982: Ultrastructure of *Pterospora andromedea* Nuttall and *Sarcodes sanguinea* Torrey mycorrhizas. New Phytol. **92,** 539—551.

Robinson, R. K., 1972: The production by roots of *Calluna vulgaris* of a factor inhibitory to growth of some mycorrhizal fungi. J. Ecol. **60,** 219—224.

Rouquerol, T., Payre, H., 1974: Observations sur le comportement de *Tuber melanosporum* dans un site naturel. Revue Mycol. **39,** 107—117.

Sanders, F. E., 1975: The effect of foliar-applied phosphate on the mycorrhizal infections of onion roots. In: Sanders, F. E., Mosse, B., Tinker, P. B. (eds.): Endomycorrhizas, pp. 261—276. London - New York: Academic Press.

Sanders, F. E., Tinker, P. B., 1973: Phosphate flow into mycorrhizal roots. Pestic. Sci. **4,** 385—395.

Sanders, F. E., Tinker, P. B., Black, R. L. B., Palmerley, S. M., 1977: The development of endomycorrhizal root systems. I. Spread of infection and growth-promoting effects with four species of vesicular-arbuscular endophyte. New Phytol. **78,** 257—268.

Scannerini, S., 1975: Le ultrastrutture delle micorrize. Giorn. Bot. Ital. **109,** 109—144.

Scannerini, S., Bonfante-Fasolo, P., 1983: Comparative ultrastructural analysis of mycorrhizal associations. Can. J. Bot. **61,** 917—943.

Schwab, S. M., Johnson, E. L. V., Menge, J. A., 1982: Influence of simazine on formation of vesicular-arbuscular mycorrhizae in *Chenopodium quinona* Willd. Plant and Soil **64,** 283—287.

Serrigny, J., 1982: Etudes cytologique, cytochimique, et cytoenzymologique d'une orchidée tropicale: *Epidendrum* sp. Diplôme d'Étude Approfondie, p. 32, University of Nancy, France.

Slankis, V., 1958: The role of auxin and other exudates in mycorrhizal symbiosis of forest trees. In: Thimann, K. V. (ed.): Physiology of forest trees, pp. 427—443. New York: Ronald Press.

Slankis, V., 1967: Renewed growth of ectotrophic mycorrhizae as an indication of an unstable symbiotic relationship. Proc. 14th Congr. I. U. F. R. O. Pt. V, Sect. **24,** 84—99.

Slankis, V., 1974: Soil factors influencing formation of mycorrhizae. Ann. Rev. Phytopath. **12,** 437—456.

Smith, S. E., 1967: Carbohydrate translocation in orchid mycorrhizas. New Phytol. **66,** 371—378.

Smith, S. E., 1974: Mycorrhizal fungi. In: CRC Critical Reviews in Microbiology **3,** 275—315.

Smith, S. E., Walker, N. A., 1981: A quantitative study of mycorrhizal infection in *Trifolium:* separate determination of the rates of infection and of mycelial growth. New Phytol. **89,** 225—240.

Spencer-Phillips, P. T. N., Gay, J. L., 1981: Domains of ATPase in plasma membranes and transport through infected cells. New Phytol. **89,** 393—400.

Stahmann, M. A., Demorest, D. M., 1973: Changes in enzymes of host and pathogen with special reference to peroxidase interaction. In: Byrde, R. J. W., Cutting, C. V. (eds.): Fungal pathogenicity and the Plant's Response, pp. 405—422. London – New York: Academic Press.

St. John, T. V., 1980: Root size, root hairs and mycorrhizal infection: a re-examination of Baylis's hypothesis with tropical trees. New Phytol. **84,** 483—487.

Stribley, D. P., Read, D. J., 1976: The biology of mycorrhiza in the Ericaceae. VI. The effects of mycorrhizal infection and concentration of ammonium nitrogen on growth of cranberry (*Vaccinium macrocarpon* Ait.) in sand culture. New Phytol. **77,** 63—72.

Strullu, D. G., Gourret, J. P., 1974: Ultrastructure et évolution du champignon symbiotique des racines de *Dactylorchis maculata* (L.) Verm. J. Microscopie **20,** 285—294.

Strullu, D. G., Gourret, J. P., 1980: Données ultrastructurales sur l'intégration cellulaire de quelques parasites ou symbiotes de plantes. II. Champignons mycorrhiziens. Bull. Soc. Bot. Fr. **127,** Actual. Bot., 97—106.

Strullu, D. G., Gourret, J. P., Garrec, J. P., 1981: Microanalyse des granules vacuolaires des ectomycorrhizes, endomycorrhizes et endomycothalles. Physiol. Vég. **19,** 367—378.

Tatum, E. L., Barratt, R. W., Cutter, V. M., 1949: Chemical induction of colonial paramorphs in *Neurospora* and *Syncephalastrum.* Science **109,** 509—511.

Theodoru, C., Bowen, G. D., 1970: Mycorrhizal responses of radiata pine in experiment with different fungi. Aust. For. **34,** 183—191.

Theodorou, C., Bowen, G. D., 1971: Effects of non-host plants on growth of mycorrhizal fungi of radiata pine. Aust. For. **35,** 17—22.

Trappe, J. M., 1962: Fungus associates of ectrotrophic mycorrhizae. Bot. Rev. **38,** 538—606.

Trappe, J. M., 1964: Mycorrhizal hosts and distribution of *Cenococcum graniforme*. Lloydia **27,** 100—106.
Trappe, J. M., 1967: Pure culture synthesis of Douglas-fir mycorrhizae with species of *Hebeloma, Suillius, Rhizopogon* and *Astraeus*. Forest Sci. **13,** 121—130.
Trappe, J. M., 1977: Selection of fungi for ectomycorrhizal inoculation in nurseries. Ann. Rev. Phytopath. **15,** 203—222.
Trappe, J. M., 1982: Synoptic keys to the genera and species of zygomycetous mycorrhizal fungi. Phytopathology **72,** 1102—1108.
Valla, G., 1979: Effect of griseofulvin on cytology, growth, mitosis and branching of *Polyporus arcularius*. Trans. Br. mycol. Soc. **73,** 135—139.
Vanderplank, J. E., 1978: Genetic and molecular basis of plant pathogenesis. In: Bommer, D. F. R., Salvey, B. R., Thomas, G. W., Vaadia, Y., Van Vleck, L. D. (eds.): Advanced series in agricultural sciences **6,** p. 167. Berlin - Heidelberg - New York: Springer.
Vegh, I., Fabre, E., Gianinazzi-Pearson, V., 1979: Présence en France de *Pezizella ericae* Read, Champignon endomycorhizogène des Ericacées horticoles. Phytopath. Z. **96,** 231—243.
Warcup, J. H., 1971: Specificity of mycorrhizal association in some Australian terrestrial orchids. New Phytol. **70,** 41—46.
Warcup, J. H., 1975: A culturable *Endogone* associated with Eucalypts. In: Sanders, F. E., Mosse, B., Tinker, P. B. (eds.): Endomycorrhizas, pp. 53—63. London - New York: Academic Press.
Webster, J., 1976: *Pezizella ericae* is homothallic. Trans. Brit. mycol. Soc. **66,** 173.
Williamson, B., Hadley, G., 1969: DNA content of nuclei in orchid protocorms symbiotically infected with *Rhizoctonia*. Nature **222,** 582—583.
Williamson, B., Hadley, G., 1970: Penetration and infection of orchid protocorms by *Thanetephorus cucumeris* and other *Rhizoctonia* isolates. Phytopathology **60,** 1092—1096.
Young, H. E., 1940: Mycorrhizae and the growth of *Pinus* and *Araucaria*. The influence of different species of mycorrhiza-forming fungi on seedling growth. J. Aust. Inst. Agric. Sci. **6,** 21—25.
Zak, B., 1973: Classification of ectomycorrhizae. In: Marks, G. C., Kozlowski, T. T. (eds.): Ectomycorrhizae, pp. 43—78. New York: Academic Press.
Zerban, H., Werz, G., 1975: Localization of nucleoside diphosphatases and thiamine pyrophosphatase in various stages of the life cycle of the green algae *Acetabularia ciftonii* and *Acetabularia mediterranea*. Cytobiologie **12,** 13—27.

Chapter 9

Molecular Biology of Stem Nodulation

R. P. Legocki and A. A. Szalay

Boyce Thompson Institute for Plant Research, Cornell University, Ithaca, N. Y., U. S. A.

With 3 Figures

Contents

I. Introduction

Due to the agricultural importance of legumes and their symbiosis with nitrogen-fixing rhizobia, the term "*Rhizobium*-legume symbiosis" has become almost synonymous with the term "root nodule symbiosis". Recent studies, however, have shown that nodulation and symbiotic nitrogen fixation also occur on the stem of some legumes, including *Aeschynomene* and *Sesbania.* Upon association with specific *Rhizobium* strains, generally referred to as "stem rhizobia", the *Aeschynomene* and *Sesbania* plants can form nodules on either roots or stems, or both (Yatazawa and Yoshida, 1979; Dreyfus and Ommergues, 1981; Eaglesham and Szalay, 1983). Unlike the formation of root nodules, stem nodulation in *Aeschynomene* is not significantly inhibited at increased concentrations of applied nitrogen, whereas the efficiency of nitrogen fixation in stem nodules is comparable to that of soybean root nodules (Legocki *et al.*, 1983 a). These observations, along with the fact that repeated inoculations of *Aeschynomene* plants with stem *Rhizobium* result in an increased number of nodules (up to 450 per plant, Legocki *et al.*, 1983 a), have recently led to an intensive research of the stem nodule symbiosis.

Due to a substantial structural resemblance, and due to the obligatory and not associative (Burns and Hardy, 1972) character of the stem nodule symbiosis, the interaction of stem *Rhizobium* with *Aeschynomene* or *Sesbania* plants strongly resembles the root nodule symbiosis of common legumes. Although the presence of stem nodules on *Aeschynomene* plants was reported as early as in 1928 by Hagerup, our knowledge of the physiology and genetics of stem *Rhizobium* and the host plant remains very incomplete. Since a single *Rhizobium* strain has the ability to form effective nodules on the root and stem of a host plant, yet within a very narrow host range (see below), this *Rhizobium*-legume symbiosis may represent a unique system for studying the expression of plant and *Rhizobium* genes involved in the symbiotic process.

II. Occurrence of Stem-Nodulating Legumes

To this date, only three genera of the family *Leguminosae* have been reported to form stem nodules: *Aeschynomene* and *Sesbania*, both of the subfamily *Papilionoideae*, and *Neptunia*, of the *Mimosoideae* subfamily.

The genus *Aeschynomene* assembles 150—250 species, of which approximately one-half are hydrophytes, occurring in rice paddies, flooded areas, and along lake and river margins. Species are mostly tropical, numerous in the Americas, and less common in Africa, Asia, Australia and Pacific areas (Allen and Allen, 1981). *A. aspera* L. in tropical Asia and *A. elaphroxylon* in Africa are important sources of ambatch wood used in the manufacture of art paper products. *A. americana* L. and *A. indica* L. are considered good forage and green-manuring plants. The former species, now undergoing extensive trials in the southern United States, is suitable for cattle browse (Allen and Allen, 1981). *A. cristata* Vatke, *A. fluitans* Peter, *A. indica*, and *A. nilotica* Taub. are an efficient fodder on the sandy soils of the Western Province of Zambia (Verboom, 1966). *A. fluitans* is a particularly valuable graze crop. Its young shoots contain about 15—27% of crude protein, and are readily browsed by cattle.

▷

Fig. 1. Stem nodules of *Aeschynomene evenia* (A), *A. scabra* (B) and *Sesbania rostrata* (C) grown under greenhouse conditions. (D) Electron micrograph of the boundary between the peripheral and central cortex of stem nodule of *A. scabra* (magnification 10,000 ×), showing the presence of bacteroids (BD), peribacterial membrane (PBM), and chloroplasts (CHL). (E) Scanning electron micrograph showing host cells of stem nodules of *A. scabra* containing bacteroids (magnification 16,000 ×). Photographs in panels A and B were kindly provided by Dr. A.R.J. Eaglesham from Boyce Thompson Institute, Cornell University; photograph shown in panel C was a gift from Dr. B. L. Dreyfus of Dakar, Senegal; electron micrograph in panel D was kindly provided by Dr. F. Sack of Boyce Thompson Institute, Cornell University

A
B
C
D
CHL
PBM
BD
E

The genus *Sesbania* is represented by about 70 species in the warmer latitudes of both hemispheres. *Sesbania* species are common along stream and swamp banks and in moist and inundated bottomlands (Allen and Allen, 1981). Several species are good sources of fiber suitable for rope-making and fish nets, while stems of *S. aegyptiaca, S. cannabina* Roxb. and *S. aculeata* Poir. yield fiber used as a substitute for hemp and jute in Japan and India. *S. grandiflora* is a source of ascorbic acid and gum, and its leaf juices have diuretic and laxative properties (Allen and Allen, 1981). The seeds of *S. sesban* contain saponins, a group of glucosides widely used in medicine. This species, along with *S. speciosa* Taub., are good forage and fodder plants in India and Taiwan. The former appears to be a suitable cultivar for saline areas subjected to flooding. However, the consumption of leaves and seeds of *S. drummondii* (Rydb.) Cory, *S. exaltata, S. punicea* (Cav.) DC., and *S. vesicarium* (Jacq.) Ell. in North America, has been reported to cause symptoms of poisoning among cattle, sheep, goats and poultry (Marsh and Clawson, 1920; Muenscher, 1939; Kingsbury, 1964). The toxic substances are believed to be saponins.

The biology and agricultural significance of *Neptunia* have not been studied in much detail to this date. The 10—15 species of this genus are indigenous to tropical and subtropical regions of the Old and New Worlds. The slender-stemmed terrestrial species, growing mostly in moist and swampy areas, are excellent soil-binders. *N. lutea* (Leavenw.) Benth., for example, forms dense mats on sites that are partially submerged during wet seasons in the southern United States (Allen and Allen, 1981). The sprouts and leaves of *N. oleracea* Lour., the best known member of this genus, are consumed in Vietnam as a pot herb. No reports are available, however, describing successful application of *Neptunia* in agricultural practice.

III. Structure of Stem Nodules

Stem nodules of *Aeschynomene* are usually ellipsoidal, with an average size of 4—5 mm (Figs. 1 A and 1 B), whereas the nodules of *Sesbania* are spherical, and their average size is about 6 mm (Fig. 1 C). Both types of nodules are visible approximately 5—7 days after inoculation. They are green on the surface and contain chloroplasts in the peripheral cortex (Fig. 1 D). Light and electron microscopy shows the presence of a large number of bacteroids in the central portion of stem nodules (Figs. 1 D and 1 E), however, none of the infected plant cells appear to contain chloroplasts. The percentage of infected cells in stem nodules of *A. scabra* was found to range from 36—42% of total cells. The *Aeschynomene* and *Sesbania* stem and root nodules contain an abundant pink pigment, leghemoglobin, and their ultrastructure shows the presence of a peribacterial membrane surrounding the bacteroids, as well as numerous deposits of poly-β-hydroxybutyrate within the bacterial cells (Fig. 1 D). As shown in Fig. 1 D, stem nodules of *Aeschynomene* contain a single bacteroid cell per peribacterial membrane. Thus, with the exception of chloroplasts, the ultrastructure of stem nodules resembles that of root nodules (e.g. peanut or soybean). The initial mode of rhizobial infection of the stem remains unknown. Whether or not the development of nodule structure initiates from the stem lenticles and/or loci of adventitious roots (B. L. Dreyfus and A. R. J. Eaglesham, personal communications), and whether or not the infection process involves the formation of infection threads, are subjects for further investigation. A recent study by Tsien *et al.* (1983) showed the presence of branched intercellular infection threads in the development of stem nodules of *Sesbania rostrata,* however, no infection threads have been observed in *Aeschynomene* plants to this date.

Our knowledge of the structure and physiology of *Neptunia* stem nodules is very incomplete. Nodules formed on adventitious water roots of *N. oleracea* were described as being as large as hazelnuts (Schaede, 1940), whereas those borne on earth roots were smaller and fewer in number (Allen and Allen, 1981).

IV. Host Specificity and Physiology of Stem Nodulation in *Aeschynomene*

Stem *Rhizobium* strain BTAil, isolated by Eaglesham and Szalay (1983), was found to have a very restricted host range. Of 16 different species of *Aeschynomene,* only 7 developed nodules upon inoculation with this *Rhizobium* strain: *A. denticulata, A. evenia, A. indica, A. pratensis, A. rudis, A. scabra.,* and *A. sesitiva* (Legocki *et al.*, 1983 a). Interestingly, all 7 hosts were able to form nodules on both stems and roots, and no plant has been found that nodulates on either stems or roots only. Although indirectly, this observation may further confirm the genetic involvement of the host plant in the nodulation process (see Verma and Nadler, Chapter 3). Moreover, comparative studies among the Nod^+ and Nod^- species of *Aeschynomene,* through development of isogenic lines, may contribute to our understanding of the genetic involvement of these legumes in stem and root nodulation.

Stem *Rhizobium* BTAil was also tested for nodulation on several other legumes, including soybean, pea, bean, alfalfa, cowpea, lupin, clover, peanut and pigeon pea. No nodules were observed on any of these hosts (Legocki *et al.*, 1983 a). On the other hand, none of numerous *Rhizobium* species other than stem *Rhizobium* BTAil, was able to develop stem or root nodules on *Aeschynomene* (A. R. J. Eaglesham, personal communication). Of 39 *Rhizobium* strains isolated from 6 *Sesbania* species, all strains were able to nodulate all of the *Sesbania* species, although with different degrees of efficiency (Johnson and Allen, 1952). Some of those *Rhizobium* strains ineffectively nodulated beans and cowpea plants, but nodulation was not reciprocal. Similar to stem *Rhizobium* BTAil of *Aeschynomene,* rhizobia isolated from *Sesbania* did not nodulate alfalfa, clover, pea, soybean, or lupin.

The formation of stem nodules on *Aeschynomene* is enhanced in flooded soil, although waterlogging is not a prerequisite for stem nodulation (Eaglesham and Szalay, 1983). Hence, the ability to control or alternate the stem and root nodulation of *Aeschynomene* in the field, e.g. depending on the environmental conditions and availability of inoculum, may represent an important goal in view of agricultural applicability of this system. Furthermore, in contrast to root-nodulating legumes, certain spe-

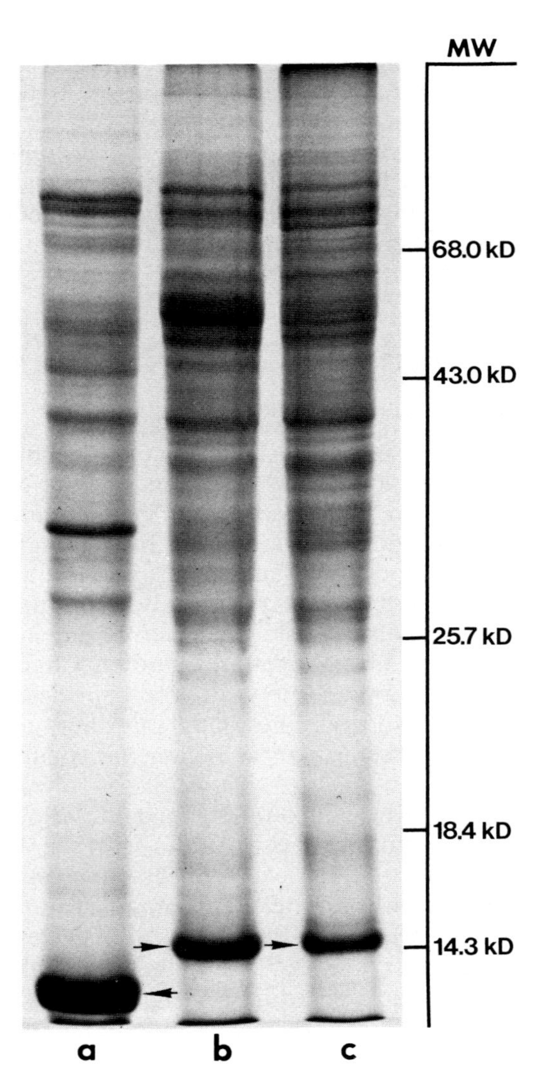

Fig. 2. Polyacrylamide gel electrophoresis in the presence of sodium dodecyl sulphate (SDS-PAGE) of plant cytoplasmic proteins from soybean root nodules (a), stem (b) and root (c) nodules of *A. scabra* (Coomassie blue stain). Gels were prepared and processed as described by Laemmli (1970). Arrows indicate leghemoglobins

cies of *Aeschynomene*, including *A. scabra*, can form stem nodules in the presence of increased (up to 17 mM) levels of applied nitrogen (Legocki *et al.*, 1983 a). It is important to note, however, that while the observed number of stem nodules is not substantially affected by high nitrogen concentrations, the nodule dry weight is significantly reduced (Eaglesham and Szalay, 1983). On the other hand, similar to root-nodulating legumes, formation of root nodules on *Aeschynomene* is strongly inhibited by increased concentrations of applied nitrogen.

V. Identification of Stem and Root Leghemoglobins of *Aeschynomene*

Both stem and root nodules of *Aeschynomene* and *Sesbania* contain a red pigment, leghemoglobin, characteristic of root nodules of all legumes examined to date. One-dimensional polyacrylamide gel electrophoresis (PAGE) of soluble cytoplasmic proteins from mature stem and root nodules of *Aeschynomene scabra* indicated that the stem and root leghemoglobins are identical in size, approximately 14,500 MW (Fig. 2). In mature (5 week-old) nodules, leghemoglobins are more abundant in stem (about 11%) than in root nodules (about 7% w/w of total cytoplasmic protein) (Legocki *et al.*, 1983 a; see also Fig. 2). Whether or not the higher level of leghemoglobin in stem nodules is due to the presence of chloroplasts in this tissue, and hence elevated oxygen concentrations, remains unknown.

To further identify the stem and root leghemoglobins of *Aeschynomene*, they were subjected to a variety of electrophoretic techniques, including non-denaturing polyacrylamide gel electrophoresis (ND-PAGE), non-denaturing and denaturing isoelectric focusing (ND-IEF and D-IEF), as well as two dimensional electrophoresis (2 D-PAGE). Extracts of soluble cytoplasmic proteins (post-ribosomal supernatant) were prepared from mature stem and root nodules as described by Legocki and Verma (1979), and leghemoglobins were enriched by ammonium sulphate fractionation at 50—80% saturation (Appleby *et al.*, 1975). Analysis of leghemoglobins from stem and root nodules by ND-PAGE (Davis, 1964) showed the presence of a single electrophoretically-distinguishable component in both tissues, as opposed to the slow (LbS) and fast-moving (LbF) components of soybean leghemoglobins (Ellfolk, 1972) examined in parallel as a control (Fig. 3 A). Under these conditions, R_f values of the *Aeschynomene* leghemoglobin, relative to the soybean LbS and LbF components, were found to be approximately 0.8 and 0.6, respectively (Legocki *et al.*, 1983 a).

ND-IEF (Fuchsman and Appleby, 1979) of purified stem and root leghemoglobins indicated an apparent polypeptide heterogeneity in the *Aeschynomene* leghemoglobins. Fig. 3 B shows that the stem and root leghemoglobins resolve into 4 components, designated Lbα, Lbβ_1, Lbβ_2, and Lbβ_3. Whereas the 3 Lbβ components occur at about the same level in stem and root nodules, Lbα appears to be substantially more abundant in the stem tissue (Fig. 3 B). Leghemoglobins Lbα, Lbβ_1, Lbβ_2 and Lbβ_3 were each isolated using preparative ND-IEF and, following removal of heme by acid

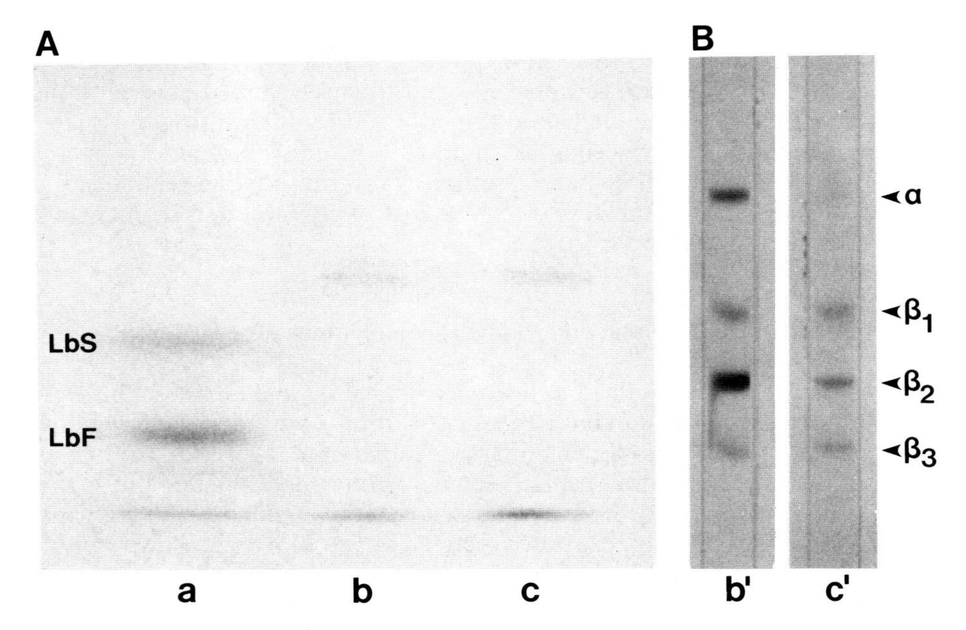

Fig. 3. Electrophoretic identification of *Aeschynomene* leghemoglobins by ND-PAGE (A) and ND-IEF (B). (a) the slow (LbS) and fast (LbF)-moving components of soybean leghemoglobin, used as a control; (b, b′) stem and (c, c′) root leghemoglobins of *A. scabra*. Both types of gels were run at 4° C, and leghemoglobins were photographed without staining

methylethylketone (Teale, 1959), the resulting apoprotein was subjected to D-IEF (O'Farrell, 1975). While Lbα remained a distinctly different component, components $Lb\beta_1$, $Lb\beta_2$ and $Lb\beta_3$ were found to represent a single leghemoglobin species. The apparent resolution of Lbβ into three molecular species is believed to be due to differential oxidation and/or ligation states of heme, as previously observed in soybean leghemoglobins (Fuchsman and Appleby, 1979).

2D-PAGE (O'Farrell, 1975) of protein extracts from the cytoplasm of stem and root nodules confirmed the presence of only two leghemoglobin species in *Aeschynomene*, referred to as Lbα and Lbβ (data not shown). The component Lbβ occurs at about the same level in mature stem and root nodules, whereas component Lbα appears to be substantially more abundant in the stem tissue. In contrast to root nodules, since stem nodules contain active chloroplasts (see above), future studies will address the question of possible relationship between elevated oxygen levels and the expression of leghemoglobin Lbα. To this date it is not clear, however, if the components Lbα and Lbβ of *Aeschynomene* leghemoglobin represent products of two different genes, or if they derive from a single gene product due to post-translational modification(s) (compare Whittaker *et al.*, 1981). Whereas isolation and *in vitro* translation of 80 S-type polysomes from stem and root nodules, followed by immunoprecipitation of resulting products with an anti-leghemoglobin antibody, will allow us to answer the above question, molecular cloning and sequencing of genomic leghemoglobin clones from *A. scabra* (in progress) may provide a direct evidence regarding the structure, organization and expression of these symbiotic genes. Since leghemoglobin (Baulcombe and Verma, 1978; Sullivan *et al.*, 1981) and nodulins (Legocki and Verma, 1980) are plant gene products unique to the *Rhizobium*-legume symbiosis (see also Verma and Nadler, Chapter 3), their differential expression in stem and root nodules makes *Aeschynomene* a suitable system for studying the involvement of host genes and their induction in the symbiotic process.

VI. Characterization of Stem *Rhizobia*

Rhizobium strain BTAil was isolated from stem nodules of *A. indica* (Eaglesham and Szalay, 1983), and was shown to have a narrow host range, compatible only with some *Aeschynomene* species (Legocki *et al.*, 1983 a). A recent study by Stowers and Eaglesham (1983) indicated that this *Rhizobium* strain has physiological characteristics of both fast and slow-growing rhizobia, as classified in the literature (Allen and Allen, 1950; Jordan and Allen, 1974; Martinez-De Drets and Arias, 1972). Its generation time on glucose as a carbon source was found to be 3.2 h, on mannitol 4.0 h, on sucrose 4.8 h, and on arabinose 7.8 h (Stowers and Eaglesham, 1983). While the carbon nutrition patterns (Vincent, 1977) and colony morphology (Graham and Parker, 1964) of *Rhizobium* BTAil resembled those of slow-growing rhizobia, the relatively fast rate of growth and ability to utilize sucrose and lactose by this strain (Stowers and Eaglesham, 1983) were characteristic of the fast-growing *Rhizobium* strains. Thus, stem *Rhizobium* strain BTAil can be referred to as a "physiologically intermediate" *Rhizobium* (Stowers and Eaglesham, 1983).

Rhizobium ORS571, isolated from stem nodules of *Sesbania rostrata,* was recently shown to grow on atmospheric nitrogen as the sole nitrogen source (Dreyfus *et al.*, 1983). This strain is capable of fixing nitrogen *ex planta* in nitrogen-free media, however its growth was later shown to require nicotinic acid as both a vitamin and nitrogen source (Elmerich *et al.*, 1983). Under nitrogen-fixing conditions *Rhizobium* ORS571 required about ten times more nicotinic acid than when grown in the presence of ammonia.

Electrophoretic analysis of total DNA from stem *Rhizobium* BTAil using the procedures of Eckhardt (1978) and Kado and Liu (1981) indicated the presence of at least one large plasmid, about 150—160 kb (100—107 Md) in size (data not shown). This plasmid, designated pBTAil, does not appear to be lost at elevated temperatures up to 42° C (for comparison see Zurkowski and Lorkiewicz, 1978, and Kiss *et al.*, 1980). Since the aerial portion of *Aeschynomene* plants is generally exposed to temperatures higher than that of soil and root system, it is possible that stem *Rhizobium* is naturally more resistant to elevated temperatures than other rhizobia. Stem *Rhizobium* strain ORS 502 of *Sesbania rostrata* was recently shown to harbor a similar size plasmid (about 150 kb or 100 Md; Bogusz *et al.*, 1983). To this date it is not clear, however, if the symbiotic genes of stem rhizobia involved in nodulation (*nod*) and nitrogen fixation (*nif*) are located on this plasmid, other plasmid(s) (if any), or the chromosome.

The *nif*H, D and K genes of stem *Rhizobium* BTAil are adjacent (G. P. Nolan, personal communication), and thus resemble the polycistronic *nif* region organization of fast-growing rhizobia such as *R. meliloti* (Ruvkun and Ausubel, 1980). Whereas the *nif* KDH gene cluster was found in this strain on a 28 kb EcoRI fragment of its total DNA digest, a second copy of the *nif*H gene was localized on a 6 kb EcoRI fragment (R. P. Legocki, unpublished data). Furthermore, both promoter and N-terminal portions of the *nif*H gene, previously isolated from the 28 kb DNA clone (Legocki *et al.*, 1983 b), also showed hybridization to the 6 kb fragment. This observation suggested possible reiteration of *nif* genes in *Rhizobium* BTAil, as previously reported in *R. phaseoli* (Quinto *et al.*, 1982). A recent analysis of *nif* genes isolated from *Rhizobium* ORS 571 (Elmerich *et al.*, 1982) also suggested some reiteration of *nif* genes in this stem-nodulating strain (C. Elmerich, personal communication).

VII. Genetic Manipulation of Stem *Rhizobium*

The relatively fast growth of stem *Rhizobium* BTAil (Stowers and Eaglesham, 1983), along with its high nodulation (see above) and nitrogen fixation (270 μmoles/h/g dry nodule) capacities (Legocki *et al.*, 1983 a), suggest that this strain may be agriculturally important. Thus, a detailed study of its symbiotic genes involved in the processes of nodulation and nitrogen fixation, in comparison to those of the root-nodulating *Rhizobium* strains, may contribute much to our understanding of symbiotic nitrogen fixation.

Rhizobium genes can be divided into two classes: constitutive type, expression of which occurs in both free-living and bacteroid states, and symbiotically regulated, expressed only upon symbiosis with the host plant. With the exception of the structural genes encoding nitrogenase (*nif*), our knowledge of the structure and specific role of symbiotically regulated genes of *Rhizobium* remains very incomplete. Since the expression of at least some procaryotic genes appears to be determined at the level of their promoters (for review see Rosenberg and Court, 1979), isolation of specific *Rhizobium* promoters and comparative measurement of their activities in free-living rhizobia and in bacteroids could significantly improve our understanding of the genetic involvement of rhizobia in the symbiotic process. To accomplish this, it was essential to develop a generalized *Rhizobium* expression vector, in which the activity of a given *Rhizobium* promoter could be measured by a simple quantitative assay both in free-living rhizobia and in the nodule tissue. The *lacZ* gene of *E. coli*, encoding the enzyme β-galactosidase, was found to be a suitable expression marker for monitoring promoter activity in stem *Rhizobium* BTAil, and possibly other *Rhizobium* strains.

A *Rhizobium* expression vector, pREV1000, was recently constructed and tested in stem *Rhizobium* BTAil (Legocki and Szalay, 1983). It contains a *mob* (*ori*T) sequence used to mobilize pREV1000 into *Rhizobium* by conjugation, a fragment of stem *Rhizobium* chromosome to mediate homologous recombination, a kanamycin resistance gene for use as a transformation marker, and a *nif*H promoter-*lac*Z gene fusion. The *nif*H (nitrogenase reductase) promoter was previously isolated from stem *Rhizobium* BTAil and sequenced (Legocki *et al.*, 1984). Using an approach similar to that reported by Ruvkun and Ausubel (1981) and Ruvkun *et al.* (1982), the *nif*H promoter-*lac*Z gene fusion of pREV1000 was stably integrated into the genome of stem *Rhizobium* BTAil by homologous recombination (Legocki and Szalay, 1983). Measurements of β-galactosidase activity in stem nodules formed by rhizobia with and without the *nif*H promoter-*lac*Z fusion, in the presence and absence of supplied nitrogen, have shown that a symbiotically regulated *nif* promoter of *Rhizobium* can be used for the expression of foreign genes in bacteroids (Legocki and Szalay, 1983). The engineered strain BTAil1000, containing the *nif*H promoter-*lac*Z gene fusion, grows on X-gal[1] indicator plates as white (uninduced) colonies under aerobic conditions. Following a 15—24 h incubation under anaerobic conditions (97% N_2 — 3%O_2 v/v), the strain shows a substantial increase in β-galactosidase activity (blue colonies). Since anaerobic conditions were found to induce the expression of nitrogenase genes in some free-living *Rhizobium* strains (Keister, 1975; Kurz and LaRue, 1975; McComb *et al.*, 1975), it appears that the expression of the *lac*Z gene in *Rhizobium* BTAi1000 is controlled symbiotically by the *nif*H promoter.

Using the *lac*Z gene as an expression marker and its fusion to a given promoter-containing DNA fragment, both constitutive and symbiotic promoters of stem *Rhizobium* BTAil can be isolated and characterized. The ability to monitor β-galactosidase activity throughout the development of stem nodules may facilitate identification of *Rhizobium* promoters involved in the symbiotic process.

VIII. Future Outlook

Applicability of the stem-nodulating legumes to agriculture may range from insignificant to extremely useful, and can not be fully assessed on the basis of the laboratory data alone. The water-tolerance of *Aeschynomene* and *Sesbania* plants, nevertheless, along with their ability to nodulate either roots or stems or both under appropriate conditions, appear to be valuable characteristics in view of agricultural practices. In contrast to root-nodulating legumes, both *Aeschynomene* and *Sesbania* plants can be

[1] X-gal: 5-bromo-4-chloro-3-indolyl-β-D-galactopyranoside. Hydrolysis of X-gal by β-galactosidase produces a blue pigment.

propagated from branch cuttings, and their nodule mass can be substantially increased by repeated inoculations of stem with rhizobia.

Symbiotic association of rhizobia with stem-nodulating legumes may be considered a model system for studying the processes of nodulation and nitrogen fixation. Considering the morphological differences between root and stem nodules, e.g. the presence of chloroplasts in the latter, a comparative study of the expression of symbiotic genes in the two types of nodules may lead to a better understanding of the genetic interaction between legumes and rhizobia. The observed differences between leghemoglobins from stem and root nodules in *Aeschynomene scabra* indicates that some symbiotic genes of the host may be differentially expressed in the same plant. A similar study with respect to other nodule-specific plant gene products (nodulins), whose role in root-nodulating legumes remains largely unknown, could further contribute to our understanding of the symbiotic process (see also Verma and Nadler, Chapter 3).

Since stem nodules of *Neptunia oleracea* were found on adventitious roots, and since nodulation on stems of *Aeschynomene* and *Sesbania* may also initiate from adventitious root sites (see above), these nodules could in fact be designated as "aerial root nodules". Whether or not their appearance on the aerial portions of these legumes was a consequence of an ecological adaptation to waterlogged conditions in which roots could not nodulate, rather than a primary target of evolution, can not be answered at this time. The development of *Rhizobium* mutants capable of nodulating either stems or roots, but not both, as well as the isolation of isogenic lines of *Aeschynomene* and/or *Sesbania* with a similar deficiency, should be of great importance to our studies of the nodulation process in legumes.

IX. References

Allen, E. K., Allen, O. N., 1950: Biochemical and symbiotic properties of the rhizobia. Bacteriol. Rev. **14,** 273—330.

Allen, O. N., Allen, E. K., 1981: In: The Leguminosae, A Source Book of Characteristics, Uses, and Nodulation. Madison: The University of Wisconsin Press, 812 pp.

Appleby, C. A., Nicola, N. A., Hurrell, J. G. R., Leach, S. J., 1975: Characterization and improved separation of soybean leghemoglobins. Biochemistry **14,** 4444—4450.

Baulcombe, D., Verma, D. P. S., 1978: Preparation of a complementary DNA for leghemoglobin and direct demonstration that leghemoglobin is encoded by the soybean genome. Nucl. Acids Res. **5,** 4141—4153.

Bogusz, D., Delajudie, P., Dreyfus, B., 1983: Transposons, plasmids and bacteriophages as tools in genetic analysis of *Rhizobium* from *Sesbania rostrata*. In: Proceedings of the First Internat. Symp. Molec. Genetics of the Bacteria-Plant Interaction, Pühler, A. (ed.). Berlin – Heidelberg – New York – Tokyo: Springer (in press).

Burns, R. C., Hardy, R. W. F., 1975: Description and classification of diazotrophs. In: Nitrogen Fixation in Bacteria and Higher Plants, Kleinzeller, A., Whitmann, H. G. (eds.), 14—38, Berlin – Heidelberg – New York: Springer.

Davis, B. J., 1964: Disc electrophoresis, method and application to human serum proteins. Annals New York Acad. Sci. **121,** 404—427.

Dreyfus, B. L., Dommergues, Y. R., 1981: Stem nodules on the tropical legume *Sesbania rostrata*. In: Current Perspectives in Nitrogen Fixation, Gibson, A. H., Newton, W. E. (eds.), p. 471, Canberra: Australian Academy of Sciences.

Dreyfus, B. L., Elmerich, C., Dommergues, Y. R., 1983: Free-living *Rhizobium* strain able to grow on N_2 as the sole nitrogen source. Appl. Environmental Microbiol. **45,** 711—713.

Eaglesham, A. R. J., Szalay, A. A., 1983: Aerial stem nodules on *Aeschynomene* spp. Plant Sci. Lett. **29,** 265—272.

Eckhardt, T., 1978: A rapid method for the identification of plasmid deoxyribonucleic acid in bacteria. Plasmid **1,** 584—588.

Ellfolk, N., 1972: Leghaemoglobin, a plant haemoglobin. Endeavour **31,** 139—142.

Elmerich, C., Dreyfus, B. L., Reysset, G., Aubert, J.-P., 1982: Genetic analysis of nitrogen fixation in a tropical fast-growing *Rhizobium*. EMBO J. **1,** 449—503.

Elmerich, C., Dreyfus, B., Aubert, J.-P., 1983: Submitted (FEMS).

Fuchsman, W. H., Appleby, C. A., 1979: Separation and determination of the relative concentrations of the homogeneous components of soybean leghemoglobin by isoelectric focusing. Biochim. Biophys. Acta **579,** 314—324.

Graham, P. H., Parker, C. A., 1964: Diagnostic features in the characterization of the root-nodule bacteria of legumes. Plant and Soil **20,** 383—396.

Hagerup, O., 1982: En hygrofil baelgplante (*Aeschynomene* aspera L.) med bakterieknolde paa staengelen. Dan. Bot. Ark. **15,** 1—9.

Johnson, M. D., Allen, O. N., 1952: Cultural reactions of rhizobia with special reference to strains isolated from *Sesbania* species. Antonie van Leeuwenhoek J. Microbiol. Serol. **18,** 1—12.

Jordan, D. C., Allen, O. N., 1974: *Rhizobiaceae*. In: Bergey's Manual of Determinative Bacteriology, Buchanan, R. E., Gibbons, N. E. (eds.), 261—267, Baltimore: The Williams and Wilkins Co.

Kado, C. I., Liu, S. T., 1981: Rapid procedure for determination and isolation of large and small plasmids. J. Bacteriol. **145,** 1365—1373.

Keister, D. L., 1975: Acetylene reduction by pure cultures of rhizobia. J. Bacteriol. **123,** 1265—1268.

Kingsbury, J. M., 1964: In: Poisonous Plants of the United States and Canada. Englewood Cliffs, N. J.: Prentice-Hall. 626 pp.

Kiss, G. B., Dobo, K., Dusha, I., Breznovits, A., Orosz, L., Vincze E., Kondorosi, A., 1980: Isolation and characterization of an R-prime plasmid from *Rhizobium meliloti*. J. Bacteriol. **141,** 121—128.

Kurz, W. G. W., LaRue, T. A., 1975: Nitrogenase activity in rhizobia in absence of plant host. Nature (London) **256,** 407—408.

Laemmli, U. K., 1970: Cleavage of structural proteins during the assembly of the head of bacteriophasity. Nature New Biol. **227,** 680—685.

Legocki, R. P., Verma, D. P. S., 1979: A nodule-specific plant protein (nodulin-35) from soybean. Science **205,** 190—193.

Legocki, R. P., Verma, D. P. S., 1980: Identification of "nodule-specific" host proteins (nodulins) involved in the development of *Rhizobium*-legume symbiosis. Cell **20,** 153—163.

Legocki, R. P., Eaglesham, A. R. J., Szalay, A. A., 1983a: Stem nodulation in *Aeschynomene:* a model system for bacterium-plant interactions. In: Proceedings of the First Internat. Symp. Molec. Genetics of the Bacteria-Plant Interaction, Pühler, A. (ed.), 210—219, Berlin - Heidelberg - New York - Tokyo: Springer.

Legocki, R. P., Yun, A., Szalay, A. A., 1984: Manuscript in preparation.
Legocki, R. P., Szalay, A. A., 1984: Expression of Beta-galactosidase controlled by a nitrogenase promotor in stem nodules of *Aeschynomene scabra*. Manuscript submitted (PNAS).
Marsh, C. D., Clawson, A. B., 1920: Daubentonia longifolia (coffee bean), a poisonous plant. J. Agric. Res. **20,** 507—513.
Martinez-De Drets, G., Arias, A., 1972: Enzymatic basis for differentiation of *Rhizobium* into fast- and slow-growing groups. J. Bacteriol. **109,** 467—470.
McComb, J. A., Elliott, J., Dilworth, M. J., 1975: Acetylene reduction by *Rhizobium* in pure culture. Nature (London) **256,** 409—410.
Muenscher, W. C., 1939: In: Poisonous Plants of the United States. Rural Science Series. New York: Macmillan, 266 pp.
O'Farrell, P. H., 1975: High resolution two-dimensional electrophoresis of proteins. J. Biol. Chem. **250,** 4007—4021.
Quinto, C., de la Vega, H., Flores, M., Fernandez, L., Ballado, T., Soberon, G., Palacios, R., 1982: Reiteration of nitrogen fixation gene sequences in *Rhizobium phaseoli.* Nature **299,** 724—726.
Rosenberg, M., Court, D., 1979: Regulatory sequences involved in the promotion and termination of RNA transcription. Ann. Rev. Genet. **13,** 319—353.
Ruvkun, G. B., Ausubel, F. M., 1980: Interspecies homology of nitrogenase genes. Proc. Natl. Acad. Sci. U. S. A. **77,** 191—195.
Ruvkun, G. B., Ausubel, F. M., 1981: A general method for site-directed mutagenesis in procaryotes. Nature **289,** 85—88.
Ruvkun, G. B., Sundaresan, V., Ausubel, F. M., 1982: Directed transposon Tn5 mutagenesis and complementation analysis of *Rhizobium meliloti* symbiotic nitrogen fixation genes. Cell **28,** 551—559.
Schaede, R., 1940: Die Knöllchen der adventiven Wasserwurzeln von *Neptunia oleracea* und ihre Bakteriensymbiose. Planta **31,** 1—21.
Stowers, M. D., Eaglesham, A. R. J., 1983: A stem-nodulating *Rhizobium* with physiological characteristics of both fast and slow growers. J. Gen. Microbiol. (in press).
Sullivan, D., Brisson, N., Goodchild, B., Verma, D. P. S., Thomas, D. Y., 1981: Molecular cloning and organization of two leghaemoglobin genomic sequences of soybean. Nature **289,** 516—518.
Teale, F. W. J., 1959: Cleavage of the haem-protein link by acid methylethylketone. Biochim. Biophys. Acta **35,** 543.
Tsien, H. C., Dreyfus, B. L., Schmidt, E. L., 1983: Morphogenesis of stem nodules of *Sesbania rostrata.* In: Abstracts of the 9th North American *Rhizobium* Conference, Ithaca, N. Y., P6.
Verboom, W. C., 1966: The grassland communities of Barotseland. Trop. Agric. (Trinidad) **43,** 107—115.
Vincent, J. M., 1977: *Rhizobium*: a general microbiology. In: A Treatise on Dinitrogen Fixation, Sect. III: Biology, Hardy, R. W. F., Silver, W. S. (eds.), 277—366, New York: Wiley.
Whittaker, R. G., Lennox, S., Appleby, C. A., 1981: Relationship of the minor soybean leghemoglobins d_1, d_2, d_3 to the major leghemoglobins c_1, c_2 and c_3. Biochem. Int. **3,** 117—124.
Yatazawa, M., Yoshida, S., 1979: Stem nodules in *Aeschynomene indica* and their capacity of nitrogen fixation. Physiol. Plant. **45,** 293—295.
Zurkowski, W., Lorkiewicz, Z., 1978: Effective method for the isolation of non-nodulating mutants of *Rhizobium trifolii.* Genet. Res. (Camb.) **32,** 311—314.

Section III.
Plant Tumor Induction

Chapter 10

Induction of Cell Proliferation by *Agrobacterium tumefaciens* and *A. Rhizogenes:* A Parasite's Point of View

Jacques Tempé, Annik Petit, and Stephen K. Farrand[1]

Groupe de Recherche sur les Interactions entre Microorganismes et Plantes, Institut de Microbiologie, Bâtiment 409, Université de Paris-Sud, F-91405 Orsay, France

With 2 Figures

Contents

I. Introduction

The crown gall and the hairy root diseases which affect dicotyledonous plants are caused by the pathogenic soil bacteria *Agrobacterium tumefaciens* and *A. rhizogenes*. In these organisms genes responsible for pathogenicity are borne on large plasmids (200—400 kb) called respectively Ti plasmids (Tumor inducing) and Ri plasmids (Root inducing) (Zaenen *et al.*, 1974; Van Larebeke *et al.*, 1974, 1975; Watson *et al.*, 1975; White and Nester, 1980a; Costantino *et al.*, 1981; Chilton *et al.*, 1982; Petit *et al.*, 1983). Both diseases afflict crop plants and produce tumorous or rooty overgrowths which usually develop on the roots or at the crown of the plant, or more rarely on stems. In the laboratory, inoculation of a supsension of virulent bacteria produces typical symptoms (Fig. 1). The molecular basis for pathogenicity is the transfer, integration, and expression of a segment of Ti

[1] Present address: Department of Microbiology, Loyola University Stritch School of Medicine, 2160 South First Avenue, Maywood, IL 60153, U. S. A.

or Ri plasmid DNA into the nuclear genome of the host cells (Chilton *et al.*, 1977, 1980, 1982; Schell *et al.*, 1979; Lemmers *et al.*, 1980; Thomashow *et al.*, 1980; Willmitzer *et al.*, 1980, 1982; Spano *et al.*, 1982; White *et al.*, 1982). This segment, called T-DNA (Transferred DNA) carries several genes which confer upon the plant cell a specific phenotype (Garfinkel *et al.*, 1981; Willmitzer *et al.*, 1982; Leemans *et al.*, 1982, see also chapter 11).

The purpose of this paper is not to describe, once again, what is known about the crown gall and the hairy root systems, but to discuss, in the light of our current knowledge, a model describing the interactions between *Agrobacterium,* its plasmids and the host plant. According to this model, which is known as the *opine concept* (Petit *et al.*, 1978; Tempé *et al.*, 1978, 1979; Guyon *et al.*, 1980) or the *genetic colonization theory* (Schell *et al.*, 1979) the overgrowths elicited by pathogenic strains of *Agrobacterium* are ecological niches in which a favorable environment is responsible for the propagation of the pathogen.

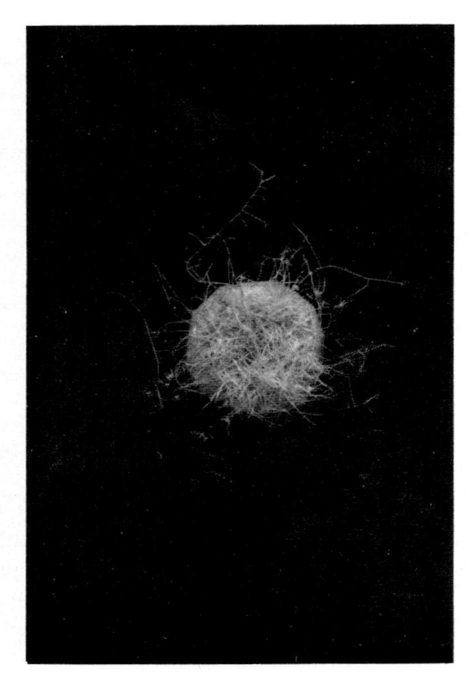

Fig. 1. Proliferations induced by members of the genus *Agrobacterium*
Left: A typical unorganized crown gall tumor formed following inoculation of *A. tumefaciens* strain B6 on the cut apical surface of *Datura tatula*. Photograph taken 6 weeks after inoculation
Right: Hairy root on a cut carrot disc. Photograph taken 8 weeks after inoculation of *A. rhizogenes* strain NCPPB 2659

II. Crown Gall Tumor Cell Phenotypes

The properties of crown gall cells were first studied by E. F. Smith (1916) whose cytological observations did not shed much light on this problem. At about the same time, C. O. Jensen (1910, 1918), who had previously been successful in grafting cancerous overgrowths in mice, was able to graft sugar beet crown gall tissue on healthy beets where it developed as a typical crown gall tumor. On the basis of these results and their similarities with those he had observed in the murine system he concluded that crown gall cells were cancerous, however his work was ignored. With the advent of plant tissue culture in the late thirties, it became possible to establish axenic crown gall tissue cultures. Of considerable significance, the crown gall tissues were shown to keep their tumorous properties indefinitely, (White and Braun, 1942). In addition to demonstrating that crown gall tumors are plant cancers, *in vitro* culture of crown gall cells and tissues has been, and still is, one of the most successful techniques used in this field.

In the mid-fifties crown gall and other types of plant tissues were first compared at the biochemical level. Lioret (1956) and Morel (1956) reported the presence of unusual compounds in crown gall tissues. Thus were isolated and identified, lysopine (Biemann *et al.*, 1960), octopine (Ménagé and Morel, 1964), octopinic acid (Ménagé and Morel, 1965) and nopaline (Goldmann *et al.*, 1969) (Table 1). Although the claim that these compounds were specific markers for crown gall cells was challenged (Seitz and Hochster, 1964; Johnson *et. al.*, 1974; Wendt-Gallitelli and Dobrigkeit, 1973; Ackermann *et al.*, 1973; Lippincott *et al.*, 1978) there is currently no doubt that their presence is characteristic for crown gall cells (Bomhoff, 1974; Holderbach and Beiderbeck, 1976; Kemp, 1976; Baldwin and Gresshof, 1978; Scott *et al.*, 1979). Furthermore, new compounds have been added to this list: histopine (Kemp, 1977), nopalinic acid (Firmin and Fenwick, 1977; Kemp, 1978), agropine (Firmin and Fenwick, 1978; Guyon *et al.*, 1980), agrocinopines (Ellis and Murphy, 1981) and three more opines related to agropine (Dahl *et al.*, In preparation) (Table 1).

The observation that the nature of the crown gall specific compounds is specified by the genome of the bacterial strain that incited the tumor (Goldmann *et al.*, 1968, Petit *et al.*, 1970) led to the hypothesis that bacterial DNA, transferred to the plant cell during crown gall induction, specifies the synthesis of these new compounds (Petit *et al.*, 1970). The discovery of T-DNA in the plant cell has subsequently confirmed these speculations (Chilton *et al.*, 1977; Schell *et al.*, 1979). Analysis of the expression of T-DNA in crown gall cells has permitted the establishment of functional maps of this segment of the Ti plasmid. The best characterized of these, the T-DNA of the octopine Ti plasmid (Fig. 2) is composed of two regions. The expression of genes in the oncogenic (ONC) region confers upon the cell its tumorous phenotype. Five genes have been identified by transcription and mutational analysis (Willmitzer *et al.*, 1982; Leemans *et al.*, 1982). These appear to negatively control the developmental pattern of plant cells: inactivation of one or the other leads to an altered tumor morphology

Table 1. Chemical Structures of Opines

octopine family	nopaline family	agropine family	agrocinopine family
NH_2 $CNH(CH_2)_3CHCOOH$ NH | NH | $CH_3CHCOOH$ octopine	NH_2 $CNH(CH_2)_3CHCOOH$ NH | NH | $HOOC(CH_2)_2CHCOOH$ nopaline	CH_2 $HOH_2C(CHOH)_3CH$ NH | | O $CH(CH_2)_2CONH_2$ CO agropine	
$NH_2(CH_2)_4CHCOOH$ | NH | $CH_3CHCOOH$ lysopine	$NH_2(CH_2)_3CHCOOH$ | NH | $HOOC(CH_2)_2CHCOOH$ nopalinic acid	$HOH_2(CHOH)_4CH_2$ | NH | $HOOCCH(CH_2)_2CONH_2$ mannopine	
$NH_2(CH_2)_3CHCOOH$ | NH | $CH_3CHCOOH$ octopinic acid		$HOH_2C(CHOH)_4CH_2$ | NH | $HOOCCH(CH_2)_2COOH$ mannopinic acid	
$HC{=}CCH_2CHCOOH$ | | | HN N NH CH | $CH_3CHCOOH$ histopine		$HOH_2C(CHOH)_4CH_2$ | N—CO | | $HOOCCH$ CH_2 CH_2 agropinic acid	

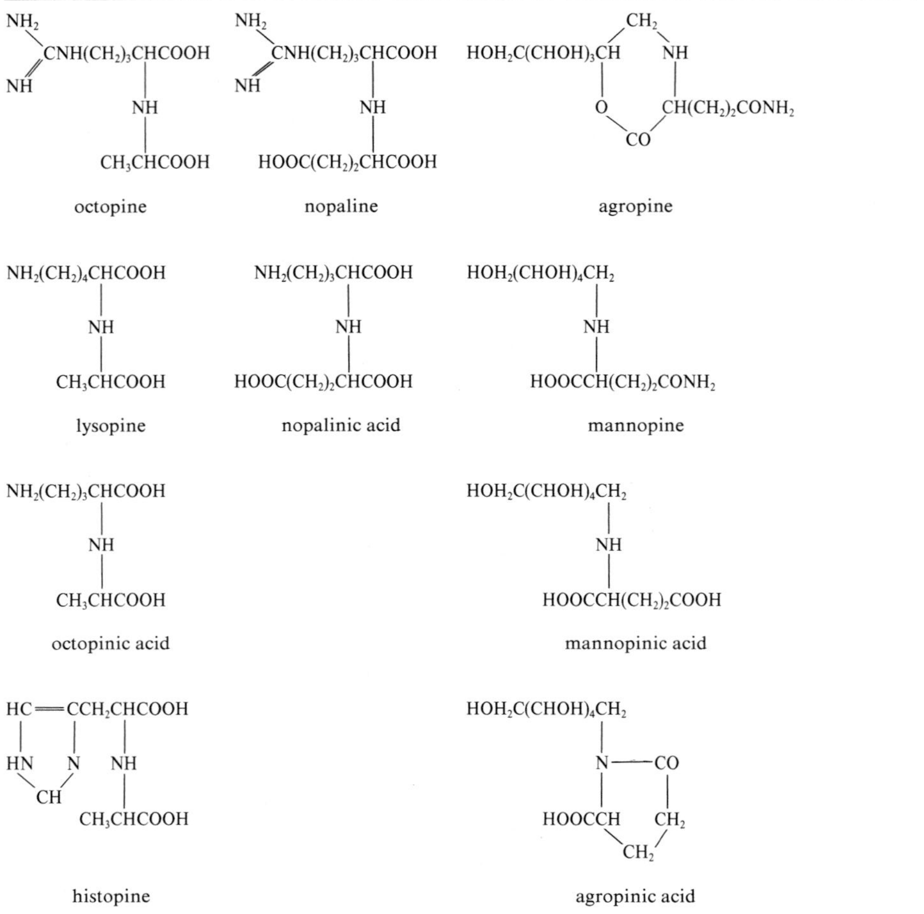

Data from: Biemann *et al.* (1960), Ménagé and Morel (1964, 1965), Goldmann *et al.* (1969), Firmin and Fenwick (1977), Kemp (1977), Coxon *et al.* (1980), Tate *et al.* (1982), Petit *et al.* (1983).

resulting in the production of organogenic tumors. The OPS (Opine Synthesis) region, adjacent to the ONC region, carries genes that code for enzymes which catalyse the synthesis of opines of the octopine family (octopine, lysopine, octopinic acid and histopine) (Schröder *et al.*, 1981; Murai and Kemp, 1982) and, presumable, the agropine family (agropine, mannopine, mannopinic acid, and agropinic acid) (Ellis, personal communication).

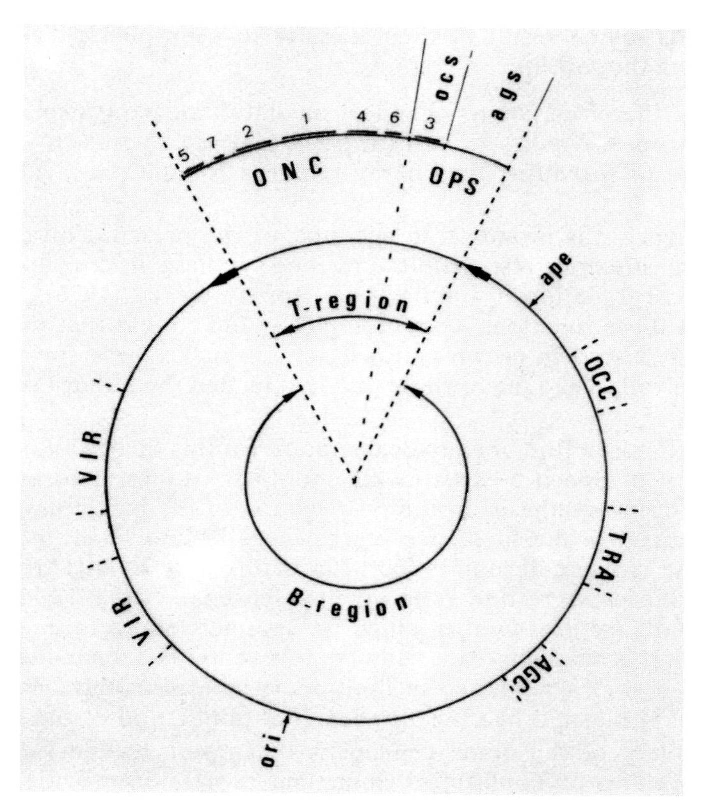

Fig. 2. Genetic organization of a typical octopine-type Ti plasmid. The T-DNA region has been expanded to show the position of ONC, with its six tumor transcripts and OPS which encodes octopine (ocs) and agropine (ags) synthesis in the tumors. The two arrows bracketing the T-DNA region indicate the direct repeats. The B-region, which never appears in the genomes of the cells of developed tumors, encodes non T-DNA trans acting virulence functions (VIR), conjugal transfer (TRA) and bacteriophage AP1 exclusion (ape) as well as octopine (OCC) and agropine AGC catabolism. Ori designates the site for the origin of Ti plasmid replication in the bacterium. Positions of markers are aproximate and distances are not to scale. Adapted from Willmitzer *et al.* (1982) and others

III. Plasmid Logic and Crown Gall Cell Phenotype

According to the opine concept, uncontrolled cell proliferation which results in the crown gall tumor or the hairy root hyperplasia serves to build an ecological niche where:

- bacterial strains harboring the pathogenic Ti or Ri plasmids find selective growth substrates (opines) which favor their multiplication, and,
- conjugation inducers (opines) trigger the conjugative activity of the Ti

plasmids which result in their transfer to saprophytic *Agrobacterium* strains in the soil.

Opines, therefore, act as chemical mediators of parasitism. Their synthesis constitutes a redirection of the plant photosynthetic activity whereby it is geared to providing the energy required for the propagation of the pathogen.

This theory was proposed to account for the presence on the Ti plasmids of the functions responsible for opine synthesis in crown gall tumors and opine degradation in the bacteria (Bomhoff *et al.*, 1976). It was proposed that these functions were biologically linked and that the rationale for opine synthesis in crown gall cells was to make these specific growth substrates available to the bacteria that had incited the tumor (Tempé *et al.*, 1979).

It is difficult to find unequivocal support for this theory in the literature on crown gall. Specific experiments should be designed to evaluate the impact of opine synthesis on the propagation of the bacterium. Paradoxically, in nurseries the incidence of crown gall tumors may be dramatic, whilst some crown gall tumors found in nature are virtually free of bacteria. Thus Jensen's grafting experiments (Jensen, 1910, 1918) were made with spontaneous beet tumors which we assume were fortuitously aseptic. However in this case where the tumors are essentially axenic one could still argue that opines synthesized in the tumors will ultimately become available to the Ti plasmid bearing agrobacterial population of the soil.

A corollary of the opine concept is that opine related functions are essential features of Ti plasmids since their expression provides the selective pressure required for their maintenance. If this is true, such functions should be present on every Ti plasmid. This could be examined by searching for opines in the "null type" tumors for which no opines have been described. The value of this type of approach was demonstrated when the agropine family of opines were shown to be present in tumors incited by one such null type strain (Guyon *et al.*, 1980, and unpublished results from this laboratory).

Further support for the theory came from the discovery of a new class of opines in nopaline- and in agropine-type tumors (Ellis and Murphy, 1981). The rationale leading to the discovery of these opines was circuitous. The existence of a suicide function, sensitivity to agrocin 84, on the plasmids responsible for the developement of nopaline-type tumors could easily be rationalized if the gene(s) responsible for agrocin sensitivity were part of a catabolic system for yet undiscovered opines. This was shown to be precisely the case when substances sharing the properties of other opines were discovered in these tumors. In addition, these opines, called agrocinopines (Ellis and Murphy, 1981) induced high levels of sensitivity to agrocin 84 in the bacterial strains. This sensitivity is due to a transport system for agrocinopines which also accepts agrocin 84 as a substrate (Murphy *et al.*, 1981). Agrocinopines are specific to the crown gall tissues induced by sensitive *Agrobacterium* strains. They are specifically catabolized by such strains and their catabolism is encoded by the Ti-plasmid. Finally, agrocinopines were found to be inducers of Ti-plasmid conjugative activity. It is also significant to note that the rationale leading to the discovery of agrocinopines was fullfilled.

The opine theory also affords an explanation, at the molecular biology level, for the observation that in crown gall tumors pathogenicity is efficiently transferred to non-pathogenic agrobacteria (Kerr, 1969, 1971). This transfer is due to the conjugative activity of the Ti plasmids which is inducible by opines. Such transfer results in spreading Ti plasmids into the saprophytic *Agrobacterium* population of the soil. The newly formed virulent strains are now able to induce tumors on plants grown in this soil.

The success of the opine concept as a theory of host parasite relations is largely due to the fact that the predictions it allowed workers to make turned out to be correct.

This is strong evidence that the opine concept describes, with sufficient accuracy, what happens in nature. The maintenance and propagation of the bacterial plasmids that govern the interactions between *Agrobacterium tumefaciens* strains and the plants are largely due to selective factors — the opines — which are produced as a result of these interactions. Are there other systems for which this may apply? The finding in hairy root tissues of specific catabolic growth substrates for the Ri plasmids of *A. rhizogenes,* constitutes an extension of the opine concept (Tepfer and Tempé, 1981; Petit *et al.*, 1983). There are two main types of Ri plasmids, the agropine type, which, in terms of its opines is similar to the agropine type Ti plasmid (opines of the agropine family; agrocinopine), and the mannopine type (opines of the agropine family except agropine; agrocinopine).

In general, the opine systems in *A. rhizogenes* strains are not as simple as those in *A. tumefaciens* strains. For instance, the Ti plasmids of *A. tumefaciens* carry all the functions necessary for the degradation of opines whose synthesis they induce in the tumors they elicit. In contrast, some of the catabolic functions of the agropine strains are not carried by the Ri plasmids, but by another plasmid, which can, interestingly enough, cointegrate with the virulence plasmid (Petit *et al.*, 1983). In addition, induction of the conjugative activity of the Ri plasmids by opines or any other compound has not been observed. These plasmids appear to be transferred constitutively at rather high frequencies.

IV. Mechanism Underlying the Opine Concept

The molecular basis for crown gall tumorigenesis and for hairy root hyperplasia is the transfer, integration and expression of Ti or Ri plasmid genes into plant cells. The tumor cells are redirected so that their activity results in the creation of an ecological niche for the pathogen. According to the opine theory the value of the system, for the parasite, will be dependent on the amount of selective factor produced and, finally, on the total number of plant cells carrying T-DNA copies. A system in which the transformed cells are dividing continuously will ensure an increasing supply of opines to the growing population of *Agrobacterium* cells carrying the plasmids. In other words the role of the ONC or HAIRY ROOT genes on the T-DNA serves the purpose of amplifying those encoding opine synthesis. This amplification of the T-DNA results in a secondary amplification of the pathogen in the environment of the crown gall tumor or the hairy root proliferations.

The discussion above points to the role of opines as selective agents for the propagation of a parasitic organism. At first glance it might be apparent that the bacterium itself is the primary parasite and the system is tuned towards supplying agrobacteria with a unique ecologial niche. Thus opines, provided by transformed plant cells, supply the bacteria with uniquely utilizable carbon sources and, therefore exert a strong selective force favoring their development. In an interesting side adaptation, strains like K84, which are non-pathogenic but catabolize opines are capable through production of a highly specific antibiotic, agrocin 84, of killing pathogenic strains. Owing to their opine catabolic functions, such strains would be able to colonize the niches created and occupied by their victims.

However, there exist two other possibilities worthy of discussion. These are interesting for their predictive value about the system and for the experimental directions to which they point.

The first is that the true parasite is not the bacterium, but rather the Ti or Ri plasmid itself. The bacterium then becomes a host allowing for the propagation and dissemination of the parasite.

This model fits well with the opine concept. The selective force exerted by opines becomes directed towards the propagation, amplification and dissemination of the Ti or Ri plasmids. In this context, it is easy to see how the chemical environment created in the crown gall tumor by the opine synthesis functions serves these purposes. Firstly, opine production allows bacterial cell division and therefore propagation of the pathogenic plasmids. Secondly, induction of conjugative activity by opines results in dissemination of the plasmid into the soil saprophytic agrobacterial population. Both of these processes result in increasing the frequency and the total number of molecules of the pathogenic plasmid within the agrobacterial population.

One can envisage that Ti plasmids have evolved two interrelated systems for ensuring their survival. The first is the ONC/OPS-containing translocatable element which allows the formation of an ecological niche producing a specifically utilisable nutrient source. What is more, T-DNA amplification, through the mechanism of uncontrolled plant cell proliferation ensures adequate supplies of these unique nutrient sources. The second system is comprised of the OPC (opine catabolic) function and the opine controlled conjugation system. With these two functions the plasmids take advantage of the availability of opines to increase their number as well as the diversity of the genetic background in which they find themselves.

The model allows for several testable predictions. First, all Ti, or Ri plasmids must encode opine catabolic traits and T-DNA borne opine synthetic functions. This has, in the main, proven to be true. However, there exist a series of *A. rhizogenes* isolates that induce development of roots in which no opine-like compounds have yet been isolated. Nor do these bacterial strains appear to catabolize any previously characterized opines (M. Ryter, personal communication). While this might seem to constitute a flaw in the argument, one must recall that similar "null type" agrobacteria were eventully shown to produce specific opines. Continuing work on these novel *A. rhizogenes* isolates may well lead to the discovery of hitherto unknown classes of opines.

The second prediction is that the catabolic and synthetic systems should be associated. This is clearly the case for all the Ti plasmids so far studied. However, as described above, certain *A. rhizogenes* isolates have opine catabolic functions split between two plasmids. This peculiarity does not constitute a real exception to the model since it has been observed that these two plasmids readily cointegrate and dissociate in their bacterial host (White and Nester, 1980; Petit *et al.*, 1983).

Finally, the model predicts that there may exist more diversity amongst the chromosomal backgrounds of the bacterial host then amongst the pathogenic plasmids themselves. Although there has been no systematic study along these lines, two facts are clear. Firstly, Ti plasmids from many isolates fall into only a few major classes based on phenotypic expression, incompatibility functions, and nucleotide sequence homology (Currier and Nester, 1976; Hooykaas *et al.*, 1980). In fact the Ti plasmids which have been studied to date shown a rather high degree of functional homology and similarity of genetic organization. Secondly, there is a rather large amount of chromosomal divergence amongst the agrobacteria. Thus, Kerr and coworkers describe three biovars, characterized by biochemical and physiological traits (Keane *et al.*, 1970; Kerr and Panagopoulos, 1977), and hybridization and numerical taxonomic evidence exists to show a rather wide diversity within the genus *Agrobacterium* (De Ley *et al.*, 1966; De Ley, 1974; Kersters *et al.*, 1973; Currier and Nester, 1976; White, 1972).

The previous discussion builds the hypothesis that the Ti or Ri plasmid, in its relationship with the bacterium and the plant, is the protagonist of the system. The opines then, act as the mediators by which the parasitic plasmid completes its life cycle. In this model the T-DNA is a sacrificial DNA element, completely subservient to the Ti or Ri plasmid which carries it. Its integration into the plant cell's genome constitutes a genetic dead end with its sole purpose being the transformation of plant cells into opine factories.

There is a least one other possibility open to discussion: namely T-DNA itself is the central protagonist in the crown gall or hairy root systems. The issue then becomes the preservation and dissemination of the T-DNA. The Ti and Ri plasmids become natural cloning vectors and the bacteria can be considered the hosts for propagation.

The interest of this hypothesis is that, not only does it propose the new concept of a DNA sequence being parasitic — an extreme case of selfish

DNA — but also that, as a working hypothesis it allows us to approach the problem of the evolution of T-DNA and of Ti and Ri plasmids.

If this model is correct then one may hypothesize that, like the oncogenes of retroviruses appearing to be of chromosomal origin, the T-DNA ONC or HAIRY ROOT genes have evolved from plant genes. Although we have no idea as to how these plant genes may have integrated into *Agrobacterium* plasmids we can imagine this to be a rather rare event, and difficult to observe. On the other hand if T-DNA from transformed tissues can move back from the plant cell and integrate into a bacterial plasmid, we should be able to detect this event by using engineered T-DNAs in which a selectable bacterial gene has been spliced. Such modified T-DNAs have been constructed by various workers (Herrera-Estrella *et al.*, 1983; Bevan *et al.*, 1983).

If, as proposed, T-DNA had its origin as plant genes, how did it evolve? It is not difficult to imagine that, since ONC genes, and probably HAIRY ROOT genes, interfere with normal plant cell development patterns, they have evolved from genes that control plant cell development. It is interesting here to draw attention to the analogy with the animal oncogenic retrovirus simian sarcoma virus for which it was recently demonstrated that the oncogene encodes a protein similar to or identical with platelet derived growth factor, a mitogenic polypeptide that stimulates growth of fibroblastic or neuroglial cells (Weiss, 1983). However, we are here in a different situation: most T-DNAs do not show homology with normal plant DNAs, whereas many animal oncogenes are homologous to normal cell genes. The only exception is the T-DNA of the agropine-type strains 15834 or 1855 of *A. rhizogenes* where homology with plant DNA has been observed reproducibly (White *et al.*, 1983; Spano *et al.*, 1982). Since T-DNA must have evolved in its bacterial host, it is possible that homology with the original DNA sequence was lost rather rapidly, due to the high frequency of DNA replication in bacteria. On the other hand, while the sequences have diverged, the functions encoded by ONC or HAIRY ROOT genes and which control cell proliferation resulting in crown gall tumor or hairy root hyperplasia must obvioulsy have been preserved during this evolution. Actually, loss of some of the ONC functions could be responsible for altered tumor phenotype as seen in strains inducing teratomas or hairy root. It is known that mutations in specific genes of the ONC region of the octopine-type Ti plasmid do result in teratomaceous or rooty tumors (Garfinkel *et al.*, 1981; Ooms *et al.*, 1981; Leemans *et al.*, 1982). Despite the divergence of the sequences of T-DNA and its hypothetical normal plant counterpart one could attempt to demonstrate the correctness of this hypothesis by looking in normal plant cells for proteins that would have compositional homology with the polypeptides encoded by T-DNA. The homology between agropine type T-DNA and normal plant DNA could be the Ariane's thread to follow in this investigation.

However, even if our hypotheses can be shown to be correct, we are left with the problem of the origin of the OPS genes. It is difficult to envisage that both OPS and OPC genes, which are not necessarily contiguous on the

Ti plasmids, have been acquired simultaneously by these plasmids. Another possibility would be that Ti or Ri plasmids have evolved from plasmids with an ancestral T-DNA encoding only opine synthesis, and without ONC or HAIRY ROOT genes. These primitive plasmids would have carried transformation functions responsible for integration of the OPS genes into the plant cell and opines produced by the transformed cells would have been degraded by OPC functions of these primitive plasmids. Such a transformation, leading to no cell proliferation would remain unnoticed since no macroscopic symptoms would be associated with it. However, if such opine inducing plasmids (Oi plasmids) do exist in nature, they could be easily screened for since they must carry opine catabolic functions. It may be relevant here to recall that several plasmids that encode opine catabolism and share rather extensive homology with Ti plasmids are been found in non pathogenic strains of *Agrobacterium* (Merlo and Nester, 1977). Some support for this hypothesis can be derived from the observation that the T-DNA of the octopine Ti plasmid consists of two subunits Tl and Tr which are apparently capable of independent integration into the plant genome (Willmitzer *et al.*, 1982). The fact that in the agropine-type Ri plasmid the T-DNA is composed of two separate regions (P. Costantino, J. Schell, personal communications) is also an argument in favour of this hypothesis. As previously proposed, acquisition of genes controlling cell proliferation by the T-DNA could be a significant step in the evolution of the parasite, be it the Ti of Ri plasmids or T-DNA.

V. Further Extensions of the Opine Concept

One may wonder whether the same kind of logic apply to other systems where plasmids govern plant bacteria interactions. More precisely one might ask whether these interactions result in the production of selective factors which favor the progagation of these plasmids.

Although such selective factors may not necessarily be opines, that is substrates for plasmid coded catabolic pathways, this remains an interesting possibility, since specific nutrients can exert a very powerful selective effect. We have begun to investigate the possibility that opine-like compounds exist in other systems and have found in alfalfa root nodules compounds that share the attributes of opines (Tempé *et al.*, 1982). That is they are specific markers for nodules incited by the *Rhizobium* strain studied and they are selective and specific growth substrates for it also. However, opine-like compounds have not yet been detected in nodules incited by other type of *Rhizobia*.

Further work on this system should give more insight on the status of these compounds and on the origin of the genetic information in which their production is encoded.

VI. Conclusion

From the beginning, the work on opines has proved successful in pointing the direction for elucidating the mechanism of crown gall and hairy root proliferations. As the molecular biology of these two diseases becomes clearer, the opine concept may, as we have discussed, generate new directions for research concerning questions on the origin and evolution of T-DNA and also of Ti and Ri plasmids. In addition, it might be short sighted to consider crown gall or hairy root with their associated nutritional *raison d'être* as isolated or unique relationships between a host, and a parasite or a symbiont. Thus it may be worthwile to take into account our knowledge of these systems when approaching the study of other systems of interaction.

Acknowledgements

The work from this laboratory which is described here was supported by grants form Centre National de la Recherche Scientifique, Institut National de la Recherche Agronomique, Ministère de l'Industrie et de la Recherche and Commission des Communautés Européennes. S. K. F. was recipient of a fellowship from Ministère de l'Industrie et de la Recherche and grants CA 19402 from the National Cancer Institute and 82-CRCR-1-1092 from the US Department of Agriculture.

VII. References

Ackermann, M. M., Teller, G., Hirth, L., 1973: Présence de lysopine libre dans des cultures de tissus de diverses origines. C. R. Acad. Sci. Ser. D. (Paris) **277,** 573—576.

Baldwin, J., Gresshof, P., 1978: On the question of the presence of octopine in normal plant cells and crown gall tumors: Use of a rapid biochemical assay for quantifying octopine in plant tissue extracts. Z. Pflanzenphysiol. **90,** 89—94.

Bevan, M., W., Falvell, R. B., Chilton, M. D., 1983: A chimaeric antibiotic resistance gene as a selectable marker for plant cell transformation. Nature **304,** 184—187.

Biemann, K., Lioret, C., Asselineau, J., Lederer, E., Polonski, J., 1960: Sur la structure chimique de la lysopine, nouvel acide aminé isolé des tissus de crown gall. Bull. Soc. Chim. Fr. **17,** 979—991.

Bomhoff, G. H., Klapwijk, P. M., Kester, H. C., Schilperoort, R. A., Hernalsteens, J. P., Schell, J., 1976: Octopine and nopaline synthesis and breakdown genetically controlled by a plasmid of *Agrobacterium tumefaciens.* Mol. Gen. Genet. **145,** 177—181.

Chilton, M. D., Drummond, M. H., Merlo, D. J., Sciaky, D., Montoya, A. L., Gordon, M. P., Nester, E. W., 1977: Stable incorporation of plasmid DNA into higher plant cells: the molecular basis of crown gall tumorigenesis. Cell **11,** 263—271.

Chilton, M. D., Saiki, R. K., Yadav, N., Gordon, M. P., Quétier, F., 1980: T-DNA from *Agrobacterium* Ti plasmid is in the nuclear DNA fraction of crown gall tumor cells. Proc. Natl. Acad. Sci. U. S. A. **77,** 4060—4064.

Chilton, M. D., Tepfer, D. A., Petit, A., David, C., Casse-Delbart, F., Tempé, J., 1982: *Agrobacterium rhizogenes* inserts T-DNA into the genomes of the host-plant root cells. Nature (London) **295,** 432—434.

Costantino, P., Mauro, M. L., Micheli, G., Risuelo, G., Hooykaas, P. J. J., Schilperoort, R. A., 1981: Fingerprinting and sequence homology of plasmids from different virulent strains of *Agrobacterium rhizogenes.* Plasmid **5,** 170—182.

Currier, T. C., Nester, E. W., 1976: Evidence for diverse types of large plasmids in tumor-inducing strains of *Agrobacterium.* J. Bacteriol. **124,** 157—165.

De Ley, J., 1974: Phylogeny of procaryotes. Taxon **23,** 291—300.

De Ley, J., Bernaerts, M., Rassel, A., Guilmot, A., 1966: Approach to an improved taxonomy of the genus *Agrobacterium.* J. Gen. Microbiol. **43,** 7—17.

Ellis, J. G., Murphy, P. J., 1981: Four new opines from crown gall tumours — their detection and properties. Mol. Gen. Genet. **181,** 36—43.

Firmin, J. L., Fenwick, R. G., 1977: N^2-(1-3-dicarboxypropyl) ornithine in crown gall tumours. Phytochemistry **16,** 761—762.

Firmin, J. L., Fenwick, R. G., 1978: Agropine — a major new plasmid-determined metabolite in crown gall tumours. Nature (London) **276,** 842—844.

Garfinkel, D. J., Simpson, R. B., Ream, L. W., White, F. F., Gordon, M. P., Nester, E. W., 1981: Genetic analysis of crown gall: fine structure map of the T-DNA by site-directed mutagenesis. Cell **27,** 143—153.

Goldmann, A., Tempé, J., Morel, G., 1968: Quelques particularités de diverses souches d'*Agrobacterium tumefaciens.* C. R. Soc. Biol. **162,** 630—631.

Goldmann, A., Thomas, D. W., Morel, G., 1969: Sur la structure de la nopaline, métabolite anormal de certaines tumeurs de crown gall. C. R. Acad. Sci. Paris, Sér, D. **268,** 852—854.

Guyon, P., Chilton, M. D., Petit, A., Tempé, J., 1980: Agropine in "null type" crown gall tumours: evidence for the generality of the opine concept. Proc. Natl. Acad. Sci. U. S. A. **77,** 2693—2697.

Herrera-Estrella, L., Depicker, A., Van Montagu, M., Schell, J., 1983: Chimeric genes are transferred and expressed in plant cells using a Ti plasmid-derived vector. Nature (London) **303,** 209—213.

Holderbach, E. Beiderbeck, R.: Octopinegehalt in normalen und Tumorgeweben einiger höherer Pflanzen. Phytochemistry **15,** 955—956 (1976).

Hooykaas, P. J. J., Den Dulk-Ras, H., Ooms, G., Schilperoort, R. A., 1980: Interactions between octopine and nopaline plasmids in *Agrobacterium tumefaciens.* J. Bacteriol. **143,** 1295—1306.

Jensen, C. O., 1910: Von echten Geschwülsten bei Pflanzen. Deuxième Conférence Internationale pour l'Etude du Cancer, Rapport, Paris, 1910, 234—254.

Jensen, C. O., 1918: Undersøgelser vedrørende nogle svulstlignende dannelser hos planter. Den Kgl. Veterinær- og Landbohøjskoles Aarsskrift 1918, 91—143.

Johnson, R., Guderian, R. H., Eden, F., Chilton, M. D., Gordon, M. P., Nester, E. W., 1974: Detection and quantitation of octopine in normal plant tissue and in crown gall tumours. Proc. Natl. Acad. Sci. U. S. A. **71,** 536—539.

Keane, P. J., Kerr, A., New, P. B., 1970: Crown gall of stone fruit. II. Identification and nomenclature of *Agrobacterium* isolates. Austr. J. Biol. Sci. **23,** 585—595.

Kemp, J. D., 1976: Octopine as a marker for the induction of tumorous growth by *Agrobacterium tumefaciens* strain B6. Biochem. Biophys. Res. Commun. **69,** 862—868.

Kemp, J. D., 1977: A new amino acid derivative present in crown gall tumor tissue. Biochem. Biophys. Res. Commun. **74,** 862—868 (1977).

Kemp, J. D., 1978: *In vivo* synthesis of crown gall-specific *Agrobacterium tumefaciens* directed derivatives of basic amino acids. Plant Physiol. **62,** 26—30.

Kerr, A., 1969: Transfer of virulence between isolates of *Agrobacterium.* Nature (London) **223,** 1175—1176.

Kerr, A., 1971: Acquisition of virulence by non-pathogenic isolates of *Agrobacterium.* Physiol. Plant Pathol. **1,** 241—246.

Kerr, A., Panagopoulos, C. G., 1977: Biotypes of *Agrobacterium radiobacter* var. *tumefaciens* and their biological control. Phytopath. Z. **90,** 172—179.

Kersters, K., De Ley, J., Sneath, P. H. A., Sackin, M., 1973: Numerical taxonomic analysis of *Agrobacterium.* J. Gen. Microbiol. **78,** 227—239.

Leemans, J., Deblaere, R., Willmitzer, L., De Greve, H., Hernalsteens, J. P., Van Montagu, M., Schell, J., 1982: Genetic identification of functions of TL-DNA transcripts in octopine crown galls. EMBO J. **1,** 147—152.

Lemmers, M., De Beuckeleer, M., Holtsers, M., Zambrysky, P., Hernalsteens, J. P., Van Montagu, M., Schell, J., 1980: Internal organization, boundaries and integration of Ti plasmid in nopaline crown gall tumors. J. Mol. Biol. **144,** 353—376.

Lemmers, M., Holsters, M., Engler, G., Van Montagu, M., Leemans, J., De Greve, H., Hernalsteens, J. P., Willmitzer, L., Otten, L., Schröder, J., Schell, J., 1981: Le plasmide Ti, vecteur potentiel pour la modification génétique des plantes. C. R. Acad. Agric. (Paris) **67,** 1052—1065.

Lioret, C., 1956: Sur la mise en évidence d'un acide aminé non identifié particulier aux tissus de crown gall. Bull. Soc. Fr. Physiol. Vég. **2,** 76.

Lippincott, J. A., Chi-Cheng Chang, Creaser-Pence, V. R., Birnberg, P. R., Rao, S. S., Margot, J. B., Whatley, M. H., Lippincott, B. B., 1978: Genetic determinants governing enhancement of tumor initiation by avirulent agrobacteria, *Agrobacterium*-Host adherence and octopine synthesis. Proc. 4th Int. Conf. Plant Pathog. Bact. 189—197.

Ménagé, A., Morel, G., 1964: Sur la présence d'octopine dans les tissus de crown gall. C. R. Acad. Sci. Paris **259,** 4795—4796.

Ménagé, A., Morel, G., 1965: Sur la présence d'un acide aminé nouveau dans le tissu de crown gall. C. R. Soc. Biol. Ses Fil. **159,** 561—562.

Merlo, D. J., Nester, E. W., 1977: Plasmids in avirulent strains of *Agrobacterium.* J. Bacteriol. **129,** 76—80.

Morel, G., 1956: Métabolisme de l'arginine par les tissus de crown gall de topinambour. Bull. Soc. Fr. Physiol. vég. **2,** 75.

Murai, N., Kemp, J. D., 1982: Octopine synthase m-RNA isolated from sunflower crown gall callus is homologous to the Ti-plasmid of *Agrobacterium tumefaciens.* Proc. Natl. Acad. Sci. U. S. A. **79,** 86—90.

Murphy, P. J., Tate, M. E., Kerr, A., 1981: Substituents at N^2 and C-5′ control selective uptake and toxicity of the adenine nucleotide bacteriocin, Agrocin 84 in *Agrobacterium.* Eur. J. Biochem. **115,** 539—543.

Ooms, G., Hooykaas, P. J., Moolenaar, G., Schilperoort, R. A.: Crown gall plant tumors of abnormal morphology induced by *Agrobacterium tumefaciens* carrying mutated octopine Ti plasmids: analysis of T-DNA functions. Gene **14,** 33—50.

Petit, A., David, C., Dahl, G. A., Ellis, J. G., Guyon, P., Casse-Delbart, F., Tempé, J., 1983: Further extension of the opine concept: Plasmids in *Agrobacterium rhizogenes* cooperate for opine degradation. Mol. Gen. Genet. **190,** 204, 214.

Petit, A., Delhaye, S., Tempé, J., Morel, G., 1970: Recherches sur les guanidines des tissus de crown gall. Mise en évidence d'une relation biochimique spécifique entre les souches d'*Agrobacterium tumefaciens* et les tumeurs qu'elles induisent. Physiol. Vég. **8,** 205—213.

Petit, A., Dessaux, Y., Tempé, J., 1978: The biological significance of opines. I. A study of opine catabolism by *Agrobacterium tumefaciens*. Proc. 4th Inter. Conf. Plant Pathog. Bact., M. Ridé, ed., INRA — Angers, 143—152.

Schell, J., Van Montagu, M., De Beuckeleer, M., De Block, M., Depicker, A., De Wilde, M., Engler, G., Genetello, C., Hernalsteens, J. P., Holsters, M., Seurinck, J., Silva, B., Van Vliet, F., Villaroel, R., 1979: Interaction and DNA transfer between *Agrobacterium tumefaciens,* the Ti-plasmid and the plant host. Proc. Royal Soc. London, Ser. B, **204,** 251-266.

Scott, I. M., Firmin, J. L., Butcher, D. N., Searles, L. M., Sogeke, A. K., Eagles, J., March, J. F., Self, R., Fenwick, G. R., 1979: Analysis of a range of crown gall and normal plant tissue for Ti plasmid-determined compounds. Mol. Gen. Genet. **176,** 57—65.

Schröder, J., Schröder, G., Huisman, H., Schilperoort, R. A., Schell, J., 1981: The m-RNA for lysopine dehydrogenase in plant tumor cells is complementary to a Ti-plasmid fragment. FEBS Lett. **129,** 166—138.

Seitz, E. W., Hochster, R. M., 1964: Lysopine in normal and in crown gall tumour tissue of tomato and tobacco. Can. J. Bot. **42,** 999—1004.

Smith, E. F., 1916: Studies on the crown gall of plants. Its relation to human cancer. J. Cancer Res. **1,** 231—309.

Spano, L., Pomponi, M., Costantino, P., Van Slogteren, G. M. S., Tempé, J., 1982: Identification of T-DNA in the root-inducing plasmid of the agropine type *Agrobacterium rhizogenes* 1855. Plant Mol. Biol. **1,** 291—300.

Tate, M. E., Ellis, J. G., Kerr, A., Tempé, J., Murray, K. E., Shaw, K. J.:, 1982: Agropine: a revised structure. Carbohydrate Research **104,** 105—120.

Tempé, J., Estrade, C., Petit, A., 1978: The biological significance of opines. II. The conjugative activity of the Ti-plasmids of *Agrobacterium tumefaciens*. Proc. 4th Inter. Conf. Plant Pathog. Bact., M. Ridé, ed., INRA — Angers, 153—160.

Tempé, J., Guyon, P., Tepfer, D., Petit, A., 1979: The role of opines in the ecology of the Ti-plasmids of *Agrobacterium*. In: Plasmids of Medical, Environmental, and Commercial Importance, Timmis, K. N., Pühler, A. eds., Amsterdam: Elsevier/North Holland Biomedical Press, 353—363.

Tepfer, D. A., Tempé, J., 1981: Production d'agropine par des racines formées sous l'action d'*Agrobacterium rhizogenes*. C. R. Acad. Sci. Paris, Ser. III, **292,** 153—156.

Thomashow, M. F., Nutter, R., Montoya, A. L., Gordon, M. P., Nester, E. W., 1980: Integration and organization of Ti plasmid sequences in crown gall tumours. Cell **19,** 729—739.

Van Larebeke, N., Engler, G., Holsters, M., Van den Elsacker, S., Zaenen, I., Schilperoort, R. A., Schell, J., 1974: Large plasmid in *Agrobacterium tumefaciens* essential for crown gall-inducing ability. Nature (London) **252,** 169—170.

Van Larebeke, N., Genetello, C., Schell, J., Schilperoort, R. A., Hemans, A. K., Hernalsteens, J. P., Van Montagu, M., 1975: Acquisition of tumor-inducing ability by non-oncogenic bacteria as a result of plasmid transfer. Nature (London) **255,** 742—743.

Watson, B., Currier, T. C., Gordon, M. P., Chilton, M. D., Nester, E. W., 1975: Plasmid required for virulence of *Agrobacterium tumefaciens*. J. Bacteriol. **123,** 255—264.

Weiss, R., 1983: Oncogenes and growth factors. Nature (London) **304,** 12.
Wendt-Gallitelli, M. F., Dobrigkeit, I., 1973: Investigations implying the invalidity of octopine as a marker for transformation by *Agrobacterium tumefaciens.* Z. Naturforsch., C. Biosci. **28 C,** 768—771.
White, F. F., Garfinkel, D. J., Huffman, G. A., Gordon, M. P., Nester, E. W., 1983: Sequences homologous to *Agrobacterium rhizogenes* T-DNA in the genomes of uninfected plants. Nature **301,** 348—350.
White, F. F., Ghidossi, G., Gordon, M. P., Nester, E. W., 1982: Tumor induction by *Agrobacterium rhizogenes* involves the transfer of plasmid DNA to the plant genome. Proc. Natl. Acad. Sci. U. S. A. **79,** 3193—3197.
White, F. F., Nester, E. W., 1980: Hairy root: Plasmid encodes virulence traits in *Agrobacterium rhizogenes.* J. Bacteriol. **141,** 1134—1141.
White, L. O., 1972: The taxonomy of the crown gall organism *Agrobacterium tumefaciens* and its relationship to Rhizobia and other Agrobacteria. J. Gen. Microbiol. **72,** 565—574.
White, P. R., Braun, A. C., 1942: A cancerous neoplasm of plants. Autonomous bacteria-free crown gall tissue. Cancer Res. **2,** 597—617.
Willmitzer, L., De Beuckeleer, M., Lemmers, M., Van Montagu, M., Schell, J., 1980: The Ti-plasmid derived T-DNA is present in the nucleus and absent from plastids of plant crown gall cells. Nature (London) **287,** 359—361.
Willmitzer, L., Simons, G., Schell, J., 1982a: The TL-DNA in octopine crown gall tumours codes for seven well defined polyadenylated transcripts. EMBO Journal **1,** 139—146.
Zaenen, I., Van Larebeke, N., Teuchy, H., Van Montagu, M., Schell, J., 1974: Supercoiled circular DNA in crown gall inducing *Agrobacterium* strains. J. Mol. Biol. **86,** 109—127.

Chapter 11

Gene Organization of the Ti-Plasmid

Jacques Hille, André Hoekema, Paul Hooykaas, and Rob Schilperoort

Department of Plant Molecular Biology, University of Leiden, Biochemistry Building, Wassenaarseweg 64, NL-2333 Al Leiden, The Netherlands

With 4 Figures

Contents

I. Introduction

A. Crown Gall

The phenomenon tumor formation on plants has intrigued biologists for a long time. More than 75 years ago, it was reported that a particular type of plant tumor, known as crown gall, was caused by bacteria (Smith and Townsend, 1907). The present taxonomic name of this soil bacterium is *Agrobacterium tumefaciens*. Crown gall is characterized by unlimited cell proliferations, which may lead to regression or even to death of the plant. The disease and its causative agent have been studied intensively, first to uncover concepts underlying the tumorous state, secondly because it is the cause of losses of economically important plants in Europe (Kerr and Panagopoulos, 1977), North-America (Kennedy and Alcorn, 1980), and Australia (New and Kerr, 1972), and thirdly, more recently, to develop plant gene vectors.

The crown gall plant tumor is an example of neoplastic transformation. This was shown by aseptically growing crown gall cells on a defined medium, on which normal plant cells could not grow (Braun and White, 1943). Braun (1956) demonstrated that this difference was due to the requirement for the exogenously added phytohormones auxin and cytokinin to normal plant cells. This led to the concept of a tumor inducing principle (TIP) present in *A. tumefaciens* and responsible for the neoplastic transformation of plant cells (Braun, 1947). Crown gall cells in addition have acquired the capacity to synthesize unusual aminoacid derivatives that have been given the familiy name opines. These compounds are tumor-specific (Petit *et al.*, 1970).

B. *Agrobacterium tumefaciens*

Agrobacterium belongs to the bacterial family of *Rhizobiaceae* (Bergey's Manual. Allen and Holding, 1974). This soil bacterium has been isolated from all over the world. It is a Gramnegative rod of (0.6—1.0) × (1—3) μm with 1 to 4 flagellae. Its optimal growth temperature is under aerobic conditions at 29° C. Agrobacteria can be distinguished from most other bacteria by their ability to induce tumors on dicotyledonous plants, although also non-tumorigenic agrobacteria have been isolated from soil. Wounding of a plant is essential for *A. tumefaciens* to be able to induce tumors. The bacteria do not invade the healthy plant cells, but rather attach to their cell walls. Selective media have been developed for the isolation of agrobacteria (New and Kerr, 1971; Kerr and Panagopoulos, 1977), and bacteriophages have been isolated for many agrobacterium strains (Hooykaas, 1979).

C. *Ti-Plasmids*

In search for the nature of TIP, thought to be present in tumorigenic *A. tumefaciens* strains, Kerr (1969, 1971) found that the tumor inducing capacity is transferred from one *Agrobacterium* strain to another, by inoculating a plant with a mixture of both strains. After a number of weeks, agrobacteria were recovered from the tumor and it appeared that non-tumorigenic agrobacteria had been converted into tumorigenic ones. If certain tumorigenic agrobacterium strains are cultured at higher temperatures (37° C), the tumorigenic property is lost (Hamilton and Fall, 1971). These data suggested a role for plasmids in tumor induction. All tumorigenic *Agrobacterium* strains were shown to contain a large plasmid, ranging in size from 120 to 150 megadaltons, which was absent in non-tumorigenic strains (Van Larebeke *et al.*, 1974). The final proof that this large plasmid is essential for the tumor-inducing capacity of *A. tumefaciens* came from experiments in which this plasmid was transferred from tumorigenic strains to non-tumorigenic plasmid free strains, converting the latter into tumorigenic strains (Van Larebeke *et al.*, 1975; Watson *et al.*, 1975). This type of plasmid was therefore named tumor-inducing (Ti) plasmid.

Several types of crown gall tumors are known depending on the *Agrobacterium* strain used to induce the tumor. The tumors were shown to contain tumor-specific nitrogenous compounds that were given the general name of opines. The three most studied opines are octopine (Ménagé and Morel, 1964), nopaline (Goldmann *et al.*, 1969) and agropine (Firmin and Fenwick, 1978). Tumors containing octopine also synthesize agropine, but not nopaline, while tumors having nopaline do neither show the synthesis of octopine nor agropine. Because of this difference in tumor compounds the respective *A. tumefaciens* strains are named octopine and nopaline strains. A third class of strains consists of bacteria which induce tumors without octopine or nopaline and therefore were called null-types strains. However, later on it was shown that their tumors contain agropine. For this reason this class of *A. tumefaciens* strains is now called agropine strains. Members of these classes of agrobacteria have the capacity to catabolize specifically the opines that are present in the type of tumor they induce. On certain plants, tumors can be distinguished also by their morphology (Schilperoort *et al.*, 1975; Hooykaas *et al.*, 1977). On *Kalanchoë daigremontiana,* tumors that contain octopine have a rough surface surrounded by many adventitious roots. Nopaline tumors have a smooth surface, produce only roots at the bottom of the tumor, and sometimes give rise to leaf like structures. Agropine tumors have a rough surface without (or a few) roots at the bottom. If an octopine producing *Agrobacterium* strain is cured of its Ti-plasmid, the resulting non-tumorigenic strain can no longer utilize octopine. Reintroduction of the Ti-plasmid in such a strain restores the ability to consume octopine as well as tumor inducing capacity. Introduction of a Ti-plasmid from an octopine agrobacterium strain into a Ti-plasmid cured nopaline agrobacterium strain, results in transconjugants which are able to induce octopine containing tumors and which can catabolize octopine. This indicates that the Ti-plasmid determines both the type of opine synthesized in the tumor and the catabolism of this compound by the bacterium (Bomhoff *et al.*, 1976; Kerr *et al.*, 1977). Therefore, also Ti-plasmids have been classified according to the main type of opine specified: octopine, nopaline and agropine (formerly null-type) Ti-plasmids (see Table 1).

Search for gene transfer from *A. tumefaciens* to the plant tumor cell was undertaken, and it was in 1977 that it was definitely proven that a piece of Ti-plasmid DNA was present in crown gall tumor cells (Chilton *et al.*, 1977), which was at least partly transcribed into RNA (Drummond *et al.*, 1977; Ledeboer, 1978). This segment is called T-region as part of the Ti-plasmid, and T-DNA as part of the plant tumor cell.

Table 1. Opines found in Tumors induced by *A. tumefaciens* strains containing different Ti-plasmids

type of Ti-plasmid	opines found in tumor
octopine	octopine octopinic acid histopine lysopine agropine mannopine
nopaline	nopaline nopalinic acid agrocinopine A agrocinopine B
agropine	agropine agrocinopine C agrocinopine D

II. Molecular Genetics of Ti-Plasmids

A. DNA Homology Among Ti-Plasmids

Ti-plasmids in different *A. tumefaciens* strains are not identical. DNA-DNA hybridization experiments have shown a varying amount of homology among them. Between various octopine Ti-plasmids at least 74% sequence homology exists, while this is 28—95% for various nopaline Ti-plasmids (Currier and Nester, 1976). Sequence homology between octopine and nopaline Ti-plasmids is 6—36% (Currier and Nester, 1976; Drummond and Chilton, 1978). Some octopine *A. tumefaciens* strains have a limited host range. They only induce tumors on a few plant species (Panagopoulos and Psallidas, 1973). In these strains it has been demonstrated that host range is determined by the Ti-plasmid (Loper and Kado, 1979; Thomashow *et al.*, 1980[c]). Sequence homology among these limited host range octopine Ti-plasmids amounts to about 64%, and with wide host range octopine and nopaline Ti-plasmids ranges in between 6—15% (Thomashow *et al.*, 1981). The homology between agropine type Ti-plasmids and octopine or nopaline Ti-plasmids is less clear; values between 11 and 62% have been reported (Currier and Nester, 1976; Drummond and Chilton, 1978).

Another similar plant disease, hairy root disease was shown to be caused by the bacterium *A. rhizogenes* (Riker, 1930). In analogy to *A. tumefaciens* a plasmid was found to be essential for the root induction and called Ri-plasmid (Moore *et al.*, 1979; White and Nester, 1980[a]). Between different Ri-plasmids strong homology is found (Costantino *et al.*, 1981), but only limited homology is observed with wide host range octopine and nopaline Ti-plasmids (White and Nester, 1980[b]; Risuelo *et al.*, 1982).

Restriction endonuclease digestion of purified Ti-plasmid DNA followed by agarose gel electrophoresis results in a pattern of DNA fragments with different sizes. For some Ti-plasmids, DNA restriction fragments obtained with various restriction endonucleases have been aligned on a circular physical map. Such a map has been constructed for the octopine Ti-plasmid pTiAch 5 (almost identical to pTiB 6) for the restriction endonucleases *Sma*I, *Hpa*I (Chilton *et al.*, 1978[a]), *Bam*HI, *Eco*RI, *Hind*III, *Kpn*I, *Xba*I (De Vos *et al.*, 1981), *Xho*I, *Sst*I, and *Sal*I (Knauf and Nester, 1982). In case of the nopaline Ti-plasmid pTi-C58 a physical map has been constructed for the enzymes *Bam*HI, *Eco*RI, *Hind*III, *Hpa*I, *Kpn*I, *Sma*I, and *Xba*I (De Picker *et al.*, 1980). Maps for an agropine type Ti-plasmid or for an Ri-plasmid have not yet been published.

Homologies among Ti-plasmids can now be visualized on the physical map. In Fig. 1 homologies are indicated on a map of the wide host range octopine Ti-plasmid pTiAch 5, compared with the nopaline Ti-plasmid pTiC 58. Extensive homology exists in mainly four areas, one of which is also homologous to Ri-plasmids. The segment of Ti-plasmid DNA, from octopine and nopaline Ti-plasmids, that is present as T-DNA in plant tumor cells, has an homologous part in common. This part of the T-region is strongly homologous in all wide host range Ti-plasmids (Chilton *et al.*, 1978[b]; De Picker *et al.*, 1978; Hepburn and Hindley, 1979), and consequently is highly conserved. It is called the common sequence and is assumed to be essential for oncogenicity. However, limited host range Ti-plasmids and Ri-plasmids have little homology to this common sequence (White and Nester, 1980[b]; Thomashow *et al.*, 1981; Risuleo *et al.*, 1982).

B. Genetic Map of a Ti-Plasmid

Functions that have to be ascribed to the Ti-plasmid were identified by comparing properties of various strains of *Agrobacterium* with and without a Ti-plasmid. The following functions were found: *1.* crown gall tumor induction, *2.* synthesis and catabolism of opines, *3.* conjugative transfer, *4.* incompatibility and Ti-plasmid replication, *5.* host range, *6.* agrocin sensitivity (in the case of nopaline Ti-plasmids only), *7.* phage AP-1 exclusion, *8.* arginine and ornithine catabolism.

Mutations in the Ti-plasmid were obtained in order to establish that these functions indeed are coded for by plasmid genes. These mutations were obtained either by insertion of transposons or by isolation of deletions. Restriction enzyme analysis of these mutated Ti-plasmids allowed the assignment of a phenotypic alteration to a physical map position, leading to a genetic map (Hernalsteens *et al.*, 1978; Koekman *et al.*, 1979; Garfinkel and Nester, 1980; Holsters *et al.*, 1980; Ooms *et al.*, 1980; De Greve *et al.*, 1981; see Fig. 1). In this way it was found that two regions on the Ti-plasmid are essential for tumor induction: the T-region which is transferred to plant tumor cells, and the Vir-region, which is expressed inside the bacterium (see section II F).

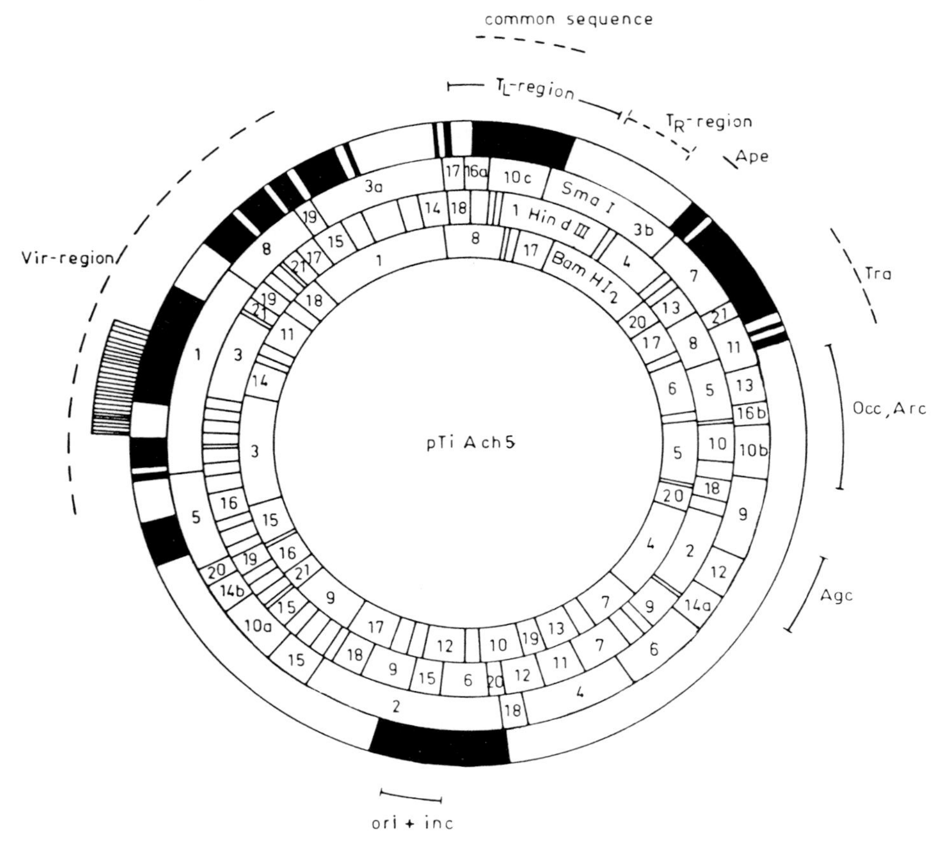

Fig. 1. Physical and genetic map of an octopine Ti-plasmid
A physical map of the octopine Ti-plasmid pTiAch5 is shown for the restriction endonucleases *Bam*HI, *Hind*III and *Sma*I. The black boxes in the outer circle represent regions of homology between this plasmid and the nopaline plasmid pTiC58. The segment of strong homology with an Ri-plasmid is indicated by a hatched box. The location of genetic determinants has been indicated: Tra, conjugative transfer; Ape, phage Ap-1 exclusion; ori + inc, origin of Ti-plasmid replication and incompatibility functions; Arc, arginine catabolism; Occ, octopine catabolism; Agc, agropine catabolism

Agrocin sensitivity

Some *Agrobacterium* strains produce bacteriocins, among which the best known is agrocin 84 (Kerr, 1972; Vervliet *et al.*, 1975). Ti-plasmid cured *A. tumefaciens* strains are resistant to agrocin 84, as are octopine Ti-plasmid containing strains. Nopaline *A. tumefaciens* strains, however, are sensitive to this agrocin. Sensitivity to agrocin 84 appeared to be correlated with the presence of the nopaline Ti-plasmid. More specifically, the nopaline Ti-plasmid encoded transport system for the newly discovered opines agrocinopine A and B is used by this agrocin (Ellis and Murphy, 1981; Murphy *et al.*, 1981). Agrocin 84 has been applied for practical purposes. If a non-tumorigenic agrocin 84 producing *A. tumefaciens* strain is present in the rhizosphere of a plant, the plant becomes protected against tumorigenic nopaline *A. tumefaciens* strains, which are killed by the agrocin (Kerr, 1980).

Phage AP-1 Exclusion

Phage AP-1 was found to form plaques on agropine Ti- and Ri-plasmid containing *Agrobacterium* strains as well as on Ti-plasmid cured strains, but not on octopine and most nopaline Ti-plasmid containing strains (Hooykaas *et al.*, 1977; Van Larebeke *et al.*, 1977; Hooykaas, 1979). The position of this genetic trait was determined by testing octopine Ti-plasmid deletion mutants for phage AP-1 exclusion (Koekman *et al.*, 1979; Ooms *et al.*, 1982).

C. Catabolic Functions and Conjugative Transfer

Catabolism of Opines

Crown gall tumors appear to be an ecological niche for tumorigenic agrobacteria, since opines forced to be produced in the tumor are catabolized specifically by the inciting bacterium. This led Petit *et al.* (1978) to propose the opine concept, stressing opines as mediators of parasitism. The genes for catabolism of these compounds are Ti-plasmid borne (Bomhoff *et al.*, 1976). Two enzymes are involved in the bacterial breakdown of octopine and nopaline, a permease and a membrane bound oxidase. The genes coding for these enzymes are coordinately controlled by a repressor (Klapwijk *et al.*, 1978; Klapwijk and Schilperoort, 1979; Petit and Tempé, 1978). The operon can be induced by octopine in the case of an octopine Ti-plasmid, and by nopaline for a nopaline Ti-plasmid (for reviews see Klapwijk and Schilperoort, 1982; Tempé and Petit, 1982). Both octopine and nopaline Ti-plasmids encode the degradation of arginine and ornithine, which are the break-down products of octopine and nopaline. These genes are also controlled by the regulator genes for opine catabolism (Ellis *et al.*, 1979). The genes for octopine and arginine catabolism have been cloned, thus facilitating accurate mapping on the Ti-plasmid (Knauf and Nester, 1982). In the following sections we will mainly concentrate on the wide host range octopine Ti-plasmid.

Plasmid Transfer

Currently, Ti-plasmid transfer can be accomplished by four different methods. The first method is performed with isolated Ti-plasmid DNA which is used to transform *A. tumefaciens* via the freeze-thawe method (Holsters *et al.*, 1978). Agrobacteria having received the Ti-plasmid can be selected for by their ability to catabolize either octopine or nopaline. An indicator medium has been developed to distinguish octopine utilizing strains from non-utilizing strains (Hooykaas *et al.*, 1979). The other three methods are based on conjugative transfer: transfer of the Ti-plasmid via mobilization by R-plasmids and per se in *in planta* or *ex planta* crosses. Indications showing that Ti-plasmids are conjugative plasmids themselves came from *in planta* crosses. In a developing tumor the Ti-plasmid is transferred between agrobacteria (Kerr, 1969). On a synthetic medium, however, initially this conjugative behavior could not be repeated. Ti-plasmid transfer could be accomplished at low frequency on a synthetic medium by mobilization with an R-plasmid (Chilton *et al.*, 1976; Hooykaas *et al.*, 1977). If a transposable element is present on both the mobilizing and mobilized plasmid, thus creating homology between the plasmids, the efficiency of transfer is increased (Hooykaas *et al.*, 1980[a]). In the presence of octopine on minimal medium Ti-plasmid transfer is observed, in its absence no such transfer occurs (Genetello *et al.*, 1977; Kerr *et al.*, 1977). Octopine is an inducer of both *occ*-genes and *tra*-genes, and it has been shown that these operons are coordinately regulated and negatively controlled by the same repressor (Klapwijk *et al.*, 1978; Klapwijk and Schilperoort, 1979; Petit and Tempé, 1978). More or less analogous situations have been reported for nopaline and agropine Ti-plasmids, in which respectively the opines agrocinopine A and B, and agrocinopine C and D act as inducers of coordinately regulated operons (Ellis *et al.*, 1982[a]; 1982[b]).

D. Replication and Incompatibility

Octopine and nopaline Ti-plasmids are incompatible, which means that they are unable to coexist stably as independent replicons within one cell. Therefore they have been classified as belonging to the same incompatibility group *inc* Rh-1. Limited host range Ti-plasmids also belong to this group (Hooykaas *et al.*, 1980b). Agropine Ti-plasmids form a different group (*inc* Rh-2), and Ri-plasmids a third group (*inc* Rh-3) (Hooykaas, 1979; Costantino *et al.*, 1980). Plasmids from different incompatibility groups can stably be maintained within one cell. It is assumed that replication and incompatibility are conservative traits, and thus might reflect evolutionary relatedness. This predicts that incompatible plasmids are more related than compatible plasmids.

The replicator and incompatibility region of the octopine Ti-plasmid is confined to a 3 M dalton segment, within *Hpa*I fragment 11 of pTiB6 (Koekman *et al.*, 1980). Further characterization of this region restricted the origin of replication to a 1.0 M dalton segment and revealed a region responsible for plasmid stability and copy number control (Koekman *et al.*, 1982). This 3 M dalton segment was shown to be largely homologous between octopine and nopaline Ti-plasmids (see Fig. 1), as could be expected from incompatibility studies.

Mutants have been described with transposons inserted into the replicator region. Such mutants turned out to be strongly attenuated in tumor induction (De Greve *et al.*, 1981; Ooms *et al.*, 1981; Koekman *et al.*, 1982). Moreover, it was difficult to isolate these plasmids, although genetically they were stably maintained. This raised the possibility that the replicator region is somehow involved in the tumor induction process. However, the Ti-plasmid replicator region has been deleted from a stable R:Ti cointegrate plasmid by *in vivo* cloning (Hille *et al.*, 1982). Strains harboring that R-prime plasmid normally induce tumors indicating that the Ti-replicator region has no essential function in the tumor induction process.

E. The T-Region

After the first successful proof that a segment of the Ti-plasmid is present in crown gall tumor cells, which has been obtained by using DNA renaturation experiments with Ti-plasmid fragments (Chilton *et al.*, 1977), a number of detailed studies (using Southern blot hybridization) have established and further extended this observation. The T-DNA was shown to be located in the nucleus in which it was covalently joined to plant genomic DNA (Thomashow *et al.*, 1980[b]; Chilton *et al.*, 1980; Willmitzer *et al.*, 1980). Some tumor lines, induced by octopine *A. tumefaciens* strains, appeared to contain two Ti-plasmid segments. In addition to T_L-DNA, always present in octopine tumor tissues in one or a few copies and containing the common sequence, a second fragment called T_R-DNA is sometimes present, either linked to T_L-DNA or independently integrated into the plant genome (Thomashow *et al.*, 1980[a]). The T_R-region is located close and to the right of the T_L-region on the physical map of the Ti-plasmid and often exists in a high copy number per tumor cell. Agrobacteria, harboring a Ti-plasmid lacking the T_R-region, but with an intact T_L-region, can normally induce tumors (Ooms *et al.*, 1982). If, however, the T_L-region on the Ti-plasmid is deleted, agrobacteria can no longer induce tumors (Koekman *et al.*, 1979), establishing that T_L-DNA is responsible for the tumorous state of plant cells.

Whether T-DNA is transcribed in plant tumor cells was assayed by using ^{32}P labelled tumor RNA as well as normal plant RNA in Southern blot hybridization experiments with Ti-plasmid DNA fragments. Almost the complete T-DNA was transcribed, but not all areas were transcribed with equal frequency (Drummond *et al.*, 1977; Ledeboer, 1978; Gurley *et al.*, 1979). Under the applied conditions normal plant RNA did not hybridize to the T-DNA. No differences in hybridization pattern were observed with nuclear RNA, polyA$^-$ and polyA$^+$ tumor RNA, showing that no extensive processing of these transcripts occurs. In tumor cells transcription of T-DNA appears to be inhibited by low concentrations of α-amanitin, indicating that T-DNA is transcribed by plant RNA polymerase II (Willmitzer *et al.*, 1981). Using northern blot hybridization, number, size and location of transcripts on T-DNA have been determined (Gelvin *et al.*, 1982; Murai and Kemp, 1982; Willmitzer *et al.*, 1982). From the T_L-DNA 8 transcripts appeared to be derived, and from T_R-DNA 4 transcripts (see Fig. 2). In addition, the direction of transcription was determined for most of the different transcripts and it was proven that initiation and termination of T-DNA transcripts occurs within the T-DNA, indicating that the

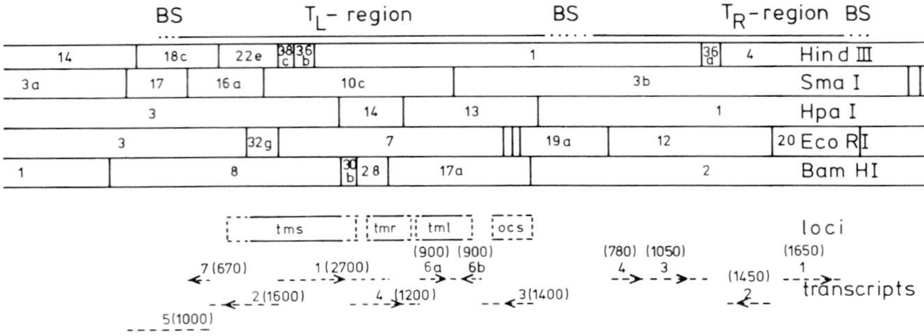

Fig. 2. T-region of the octopine Ti-plasmid pTiAch5
The position of the different plant-tumor transcripts has been indicated by wavy lines and the direction of transcription by an arrow. Transcripts from T_L- and T_R-DNA have been separately numbered. B.S. stands for border sequence. For further details see the text

prokaryotic T-region contains regulatory sequences recognizable by eukaryotic cells.

One transcript, present in octopine tumor tissues, was investigated in more detail. This 1400 nucleotides long transcript was translated *in vitro* into a protein of about M. W. 40,000 dalton, which reacted with antibodies raised against octopine synthase (Schröder *et al.*, 1981; Schröder and Schröder, 1982). Translation was inhibited by a "cap" analogue, again stressing the eukaryotic character of the transcript.

Mutations in the T-Region

Transposon insertions in the T-region allowed the positioning of a number of loci. Insertions at the left part of the T_L-region of the octopine Ti-plasmid resulted in strains that induce tumors with an increased capacity to stimulate the development of shoots, whereas mutations in the middle of the T_L-region resulted in increased root production when tested on *Kalanchoë* or tobacco plants (Ooms *et al.*, 1980; 1981; Holsters *et al.*, 1980; Garfinkel and Nester, 1980). Tumor induction by a rooter mutant was strongly stimulated and rather normal by addition of a cytokinin, and in the case of a shooter mutant the same was observed for addition of an auxin. It was therefore postulated that T-DNA influences the auxin and cytokinin activities in transformed plant cells (Ooms *et al.*, 1981). More recently, site-directed mutagenesis was applied to the T-region (Garfinkel *et al.*, 1981; Matzke *et al.*, 1981; Leemans *et al.*, 1982; Hille *et al.*, 1983[a]). Procedures for site-directed mutagenesis make use of a clone, consisting of an *E. coli* vector plasmid and the segment of the T-region that has to be mutated. A mutation is introduced into this segment by *in vitro* or *in vivo* manipulation. Finally, through *in vivo* recombination the mutation is introduced into the Ti-plasmid of *A. tumefaciens*. A large number of transposon insertion mutants in the T-region were isolated, mapped and examined for tumorigenicity. With a maximum interval of 400 basepairs, insertions were mapped in the T_L-region, revealing 4 separated loci (Garfinkel *et al.*, 1981). These loci were called *tmr, tms, tml,* and *ocs,* which stands for tumor morphology roots, shoots, large, and octopine synthase respectively. These genetic loci on T-DNA could be correlated with the transcription map. Two transcripts are involved in the *tms* or auxin locus and one in the *tmr* or cytokinin locus (see Fig. 2). No single insertions in these regions were found which caused complete loss of oncogenicity on all plant species.

Several octopine T-DNA transcripts have their homologous counterparts in nopaline tumors. This could be expected because of the presence of a common sequence in the T-region of octopine and nopaline Ti-plasmids. The transcripts 5, 2, 4, 6a, 6b, and presumably, were shown to be encoded by both octopine and nopaline T-DNA (Willmitzer *et al.*, 1982; 1983). This indicates that the loci *tmr* and *tml* may be fully conserved. Obviously, the *ocs*-locus did not have homology to the nopaline T-region.

Recently, sequence data became available for some parts of the T-region. The complete nucleotide sequence of the genes for the crown gall specific enzymes octopine synthase and nopaline synthase have been determined (De Picker *et al.*, 1982; De Greve *et al.*, 1983). Moreover, the *tmr*-locus on the octopine Ti-plasmid, coding for transcript 4, has been sequenced (Heidekamp *et al.*, 1983). These three genes appear to contain eukaryotic regulatory sequences (see Fig. 3). There are no indications for introns; all three genes have one large continuous open reading frame.

Mutations in both the *tmr* and the *tms* locus give rise to bacteria that are no longer able to induce tumors (Leemans *et al.*, 1982; Hille *et al.*, 1983[b, c]; Ream *et al.*, 1983). This suggests that either the oncogenes are inactivated

```
OCS        AGTTAAAGG-35-TATTTAA-23-CCAAAC-22-ATG
           |            |              |
           -76          -32            +1

NOS        GGTCACTAT-43-CATAAAT-18-AGAGTC-31-ATG
           |            |              |
           -79          -27            +1

tmr        AATGAATTT-42-TATAAAA-20-CTGCAA-7-ATG
           |            |              |
           -80          -29            +1

             C                A A
consensus  GG CAATCT-     -TATA A
             T                T T
             -70/-80       -34/-26
```

Fig. 3. Regulatory DNA sequences of three T-DNA transcripts
Strategic sequences from the 5′-region of the transcripts for octopine synthase, nopaline synthase and the *tmr* locus (transcript 4) are shown. The start position for transcription is indicated by +1. Consensus sequence is as described by Breathnach and Chambon (1981). Sequences at the 3′-end are not shown, but also have a eukaryotic character e.g. like polyA signals

but T-DNA is transferred or that the T-region from such strains is not transferred to the plant cells. Several lines of evidence show that these non-oncogenic bacteria do transfer T-DNA. Wound tissue of a tobacco plant, inocculated with a *tmr tms* double mutant, appears to contain octopine synthase activity. This is a function located on T_L-DNA, and therefore can only be expressed if T-DNA is indeed transferred to plant cells (Leemans *et al.*, 1982; Hille *et al.*, 1983[b, c]). Moreover, mixed infections with two non-oncogenic *tmr tms* mutants, in which the *tms* locus is mutated at the positions of different transcripts (1 respectively 2) in the mutants, leads to a rooting tumor phenotype, as if only a *tmr* mutant was used to induce the tumor. This indicates that both T-DNAs have been transferred to the plant cells and that complementation of oncogenes within a single locus (*tms*) does occur (Hille *et al.*, 1983[b, c]). These data indicate that oncogenic functions can be separated from the functions necessary for the transfer and expression of T-DNA loci.

T-Region Border Sequences

Transfer of T-DNA and its integration into plant cells is a largely unknown process. In general, T-DNA is a discrete entity, suggesting that T-region border sequences are important in the tumor induction process. Deletion of only the right border of the T_L-region has no effect on tumor induction (Leemans *et al.*, 1982; Hille *et al.*, 1983[c]), neither has deletion of the entire T_R-region. But if the T_R-region and the right border of the T_L-region is removed, tumor induction is strongly attenuated (Koekman *et al.*, 1979; Ooms *et al.*, 1982). This suggests that the righthand T_L-region border sequence has a function, but can be replaced by a different border sequence (right border of the T_R-region). In tumors induced by a strain harboring a deletion over the right T_L-border, always agropine is found, which is a function on T_R-DNA. This suggests that in the absence of the T_L-right hand border, the right hand border of T_R-DNA can be recognized, leading to the integration of both T_L- and T_R-DNA.

Transformation of plant protoplasts with naked Ti-plasmid DNA can result in tumorous plant cells. T-DNA analysis of such DNA transformants revealed the presence of stretches of T-DNA that are different from normal (Krens *et al.*, 1982). These results show that border sequences might not be essential for the process of T-DNA integration in the host genome, but might function as recognition signals inside the bacterium for transfer of a distinct DNA segment.

F. The Vir-Region

Transposon mutagenesis of Ti-plasmids has revealed the existence of a region, outside the T-region, which is essential for the bacterium to be tumorogenic (Garfinkel and Nester, 1980; Holsters *et al.*, 1980; Ooms *et al.*, 1980; De Greve *et al.*, 1981). This segment, which is called virulence or Vir-region, was shown to be strongly homologous between octopine and nopaline Ti-plasmids (Drummond and Chilton, 1978; Hepburn and Hindley,

1979). Moreover, the homologies in this region were precisely mapped between the octopine Ti-plasmid pTiAch5 and the nopaline Ti-plasmid pTiC58 by heteroduplex analysis. This revealed large areas of extensive homology that were interrupted by small areas of non-homology (Engler *et al.*, 1981). This region was also shown to be homologous to agropine Ti-plasmids (Drummond and Chilton, 1978), to Ri-plasmids (White and Nester, 1980[b]; Risuleo *et al.*, 1982), and partly to Sym-plasmids of *Rhizobium* species (Prakash *et al.*, 1982). Thus, sequences in this area seem to be conserved among related bacteria, and are essential for *A. tumefaciens* to be tumorigenic.

In search for DNA sequences present in the plant tumor cells which are homologous to the Ti-plasmid, Vir-region fragments have never been detected (Chilton *et al.*, 1977; Tomashow *et al.*, 1980[a]). This does not exclude, however, that the Vir-region is present in tumor cells during early stages of transformation, but is lost at later stage. The question then arises: Should the Vir-region only function in the plant cell, or only in the bacterium, or does it have to function in both bacterium and plant cell? How can we distinguish between these various possibilities? If it has to be expressed in the bacterium, then mutations in the Vir-region must be complementable *in trans* by the corresponding wild-type sequence. In addition, at the level of DNA sequences prokaryotic regulatory signals for transcription and translation are quite different from their eukaryotic counterparts, and their determination will thus give the final answer.

In order to perform genetic complementation studies in *A. tumefaciens* the entire Vir-region as well as parts of it have been cloned *in vivo* on the wide host range R-plasmid R772. These R-prime plasmids are compatible with Ti-plasmids. The various R-prime plasmids are introduced into non-tumorigenic mutants carrying a transposon insertion in the Vir-region, and the bacteria tested for tumor induction. Tumor inducing capacity of the mutants appears to be restored by the appropriate wildtype fragment introduced by the R-prime, indicating that complementation in fact occurs (Hille and Schilperoort, 1981; Hille *et al.*, 1982). These results show that most likely this region is expressed inside the bacterium. The second important region in tumorigenicity has been named the virulence (Vir)-region. This has been done in order to distinguish it from the T-region, which carries *onc*-genes, that have regulatory sequences recognizable by eukaryotic cells and are known to be expressed in the plant tumor cell.

Using a wide host range cosmid vector Klee *et al.* (1982) have cloned parts of the Vir-region *in vitro,* and have used the clones for complementation of *vir* mutations. With certain mutants it is observed that complementation with cosmid clones did not restore virulence while the same mutants are complemented with R-primes (Hille *et al.*, 1982; Hille *et al.*, 1983[d]). A difference between both studies is the size of Vir-region fragments used in the complementation experiments; far larger fragments are present in the R-prime plasmids. In case of complex regulation of *vir* genes the size of

Vir-region fragments used to complement mutations might explain differences in results.

Recently, we have isolated an avirulent mutant with a transposon Tn1831 insertion in the Vir-region, for which virulence could not be restored by complementation *in trans* via R-prime plasmids [LBA1505 (pAL1505); Hooykaas and Schilperoort, 1983]. This observation suggested that certain functions on the Vir-region might act *in cis* indeed, as was assumed earlier (Klee *et al.*, 1982). However, when a compatible R-prime plasmid, carrying almost the entire Ti-plasmid but lacking its replicator region, was introduced into this strain the transconjugant also appeared to be non-tumorigenic (Hille *et al.*, 1983[d]). In contrast to this result it was found that other avirulent *Agrobacterium* strains, with or without a Ti-plasmid, harboring this R-prime, have the capacity to induce full size tumors. This indicates that the mutated Ti::Tn1831 plasmid abolishes the tumor inducing capacity of an independent replicating tumor-inducing plasmid, which shows that the mutation acts *in trans* and has a trans-dominant effect.

If there exist *cis*-functions that are involved in the tumor induction process, located on the Vir-region, then a physical separation of Vir- and T-region on different compatible replicons within one bacterial cell, is expected to result in an *Agrobacterium* strain which does not have a tumor inducing capacity. Various explanations are possible if the bacterium is avirulent indeed, but a positive result, i. e. the induction of a tumor, would make it most likely that all Vir-functions act *in trans*. The latter has actually been observed. In the experiment an *Agrobacterium* strain was used in which the Vir-region was located on a pTi-derivative lacking the T-region, while the T-region was cloned on the R-plasmid R772. Agrobacteria, harboring these two compatible plasmids in one cell induce tumors normally on all plant species tested (Hoekema *et al.*, 1983[a]), which favors the view that *vir* genes act *in trans*. Moreover, it was found that also the nopaline Vir-region, the Vir-region of a limited host range Ti-plasmid, and that of the *A. rhizogenes* could facilitate the transfer of the octopine T-region to plant cells.

Structure of "Vir-Operon"

By using Vir-region clones on cosmids in complementation studies with transposon mutagenized Ti-plasmids, 5 areas of transcriptional activity have been mapped in the Vir-region, called *vir*-A, -B, -C, -D, -E (Iyer *et al.*, 1982; Klee *et al.*, 1983). Within 2 of these areas different operons were found: 2 in *vir*-A and 6 in *vir*-B, although there is some discrepancy in nomenclature between the two publications. One of the *vir*-A operons is called *vir*-B by the other. All of these operons, except for one, are within segments of homology between octopine and nopaline Ti-plasmids (see Fig. 4). The *vir*-C operon, however, only partially maps within the homologous sequences. We observed that *vir*-C actually consists of 2 operons, one having homology and the other non-homologous between octopine and nopaline Ti-plasmids. Therefore, we have suggested to refer to the homolo-

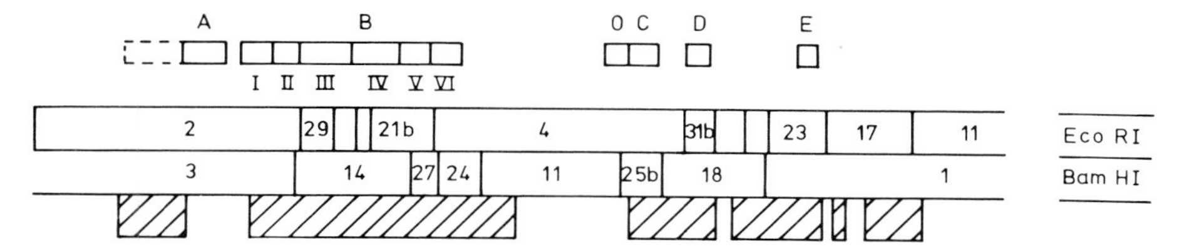

Fig. 4. Operon structure of the Vir-region
A physical map of the Vir-region of the octopine Ti-plasmid pTiAch5 is shown for the restriction endonucleases *Eco*RI and *Bam*HI. Homology to the nopaline Ti-plasmid pTiC58 is indicated by hatched boxes. Vir-operon structure is visualized by open boxes and named according to Iyer *et al.* (1982). Eleven operons have been indicated, and for one of them, *vir*-C, the direction of transcription is known (from left to right)

gous part as the *vir*-C operon and to the non-homologous part as *vir*-O, standing for octopine specific (Hille *et al.*, 1983[d]). The mutation in plasmid pAl 1505, described as dominant *in trans*, is also located in the non-homologous region. It has been shown that the nopaline Vir-region can transfer the octopine T-region to plant cells (Hoekema *et al.*, 1983[b]), while strains carrying a cointegrate, between the wild-type nopaline Ti-plasmid and the described octopine Ti-plasmid pAL 1505 are still avirulent (Hooykaas and Schilperoort, 1983). In the latter case, a wild-type nopaline Vir-region is present, and normally should be able to transfer the nopaline and the octopine T-region to plant cells. The result obtained therefore indicates that the mutation in plasmid pAL 1505 not only abolishes virulence of octopine Ti-plasmids, but also of nopaline Ti-plasmids. Currently, there is no explanation for this peculiar behavior of this mutation. Neither is it obvious why there are octopine Ti-plasmid specific *vir*-mutations, while the nopaline Vir-region can substitute for it. Mutations in *vir*-C cause complete loss of virulence, but mutations in *vir*-O do not. Mutations in *vir*-O have a host plant dependent effect. On tomato, almost no tumor formation is observed, while tumors on sunflower and *N. rustica* are apparently normal. A comparable host plant dependent effect has been reported for mutations in *vir*-A and *vir*-E regions. Mutations in between the several loci do not have effects on virulence. In summary, 11 operons have now been mapped, clustered in 5 areas, on the Vir-region of the octopine Ti-plasmid.

Regulatory sequences for the *vir*-C operon have been studied in some more detail (Hille *et al.*, 1983[d]). Part of the *vir*-C operon has been cloned on the promoter-probe *galK E. coli* vector pK0—1, on which introduced promoter activity can be assayed (via gene fusion) by activity of the enzyme galactokinase (McKenney *et al.*, 1981). Subsequently, this Vir-C fragment was mutagenized by insertions of the transposon Tn5, and analyzed for promoter activity. In this way, the promoter of *vir*-C, functional in *E. coli*, was localized in a fragment of 500 base pairs in size. This fragment was sequenced, revealing three different promoterlike sequences (−10 and

−35 boxes) which could be used by the *vir*-C operon. Downstream sequencing revealed an ATG triplet, which is preceded at a proper distance by a Shine and Dalgarno consensus — like sequence. Since the *vir*-C operon contains prokaryotic regulatory signals this proves, as discussed earlier, that *vir*-C has to function inside the bacterium for virulence.

Since substantial evidence is being gained for the expression of *vir*-genes inside the bacterium and that at least 11 *vir*-operons are involved, the question rises: What is their actual role in the tumor induction process? A tentative possibility is that one of the functions might be the recognition of T-region border sequences. Moreover, transfer of the T-region to plant cells might also be *vir*-specified. These assumptions would predict that one or more of the *vir*-gene products might be a DNA binding protein. In this regard it is noteworthy that recent observations suggest that only T-region border sequences are already sufficient for transfer into plant cells of DNA placed in between these borders (Hoekema *et al.*, 1983[b]). Vir-functions could also be required for bacterial conditioning of plant cells, e. g. to bring these cells in a competent state for transformation. It has been shown that production of the phytohormone trans-zeatin by *A. tumefaciens* is Ti-plasmid determined (Regier and Morris, 1982), as has also been demonstrated for indole-acetic acid production (Liu *et al.*, 1982). Clearly, further research is necessary to gain a better understanding of the first steps in the tumor induction process.

III. General Conclusion

For plant tumor induction by *A. tumefaciens* only 2 regions of the Ti-plasmid appeared to be essential, the T-region and the Vir-region. All other functions could be deleted without loss of tumor-inducing capacity. These regions are more or less required by the bacterium to profit from the tumor which it induces. T-region and Vir-region can be physically separated with maintenance of the tumor inducing capacity of the agrobacteria. Oncogenic functions on T-DNA can be suppressed without loss of T-DNA transfer, integration and expression into plant cells. Sequences on the T-region can be recognized by eukaryotic cells, as was expected, since they are expressed in the plant tumor cell. This is in contrast to a second set of genes located on the Vir-region, which are involved in the tumor induction process. All of these *vir* genes were shown to be complementable *in trans,* and for one of them it was established that the operon contains prokaryotic regulatory sequences. Therefore, both prokaryotic and eukaryotic functions, specified by the Ti-plasmid of *A. tumefaciens*, are involved in the induction of a tumor on a (eukaryotic) plant.

Acknowledgements

We gratefully acknowledge the help of all our colleagues of the MOLBAS research group. This work was supported in part by the Netherlands Foundation of Chemical Research (SON) and the Netherlands Foundation for Fundamental Biological Research (BION) with financial aid form the Netherlands Organization for Advancement of Pure Scientific Research (ZWO).

IV. References

Bomhoff, G., Klapwijk, P. M., Kester, H. C. M., Schilperoort, R. A., Hernalsteens, J. P., Schell, J., 1976: Octopine and nopaline synthesis and breakdown genetically controlled by a plasmid of *Agrobacterium tumefaciens*. Mol. Gen. Genet. **145,** 177—181.

Braun, A. C., 1943: Studies on tumor inception in the crown-gall disease. Am. J. Bot. **30,** 674—677.

Braun, A. C., 1947: Thermal studies on the factors responsible for tumor initiation in crown gall. Am. J. Bot. **34,** 234—240.

Braun, A. C., 1956: The activation of two growth-substance systems accompanying the conversion of normal to tumor cells in crown gall Cancer Res. **16,** 53—56.

Braun, A. C., White, P. R., 1943: Bacteriological sterility of tissues derived from secondary crown gall tumors. Phytopathology **33,** 85—100.

Breathnach, R., Chambon, P., 1981: Organization and expression of eukaryotic split genes coding for proteins. Ann. Rev. Biochem. **50,** 349—383.

Chilton, M.-D., Farrand, S. K., Levin, R. L., Nester, E. W., 1976: RP4 promotion of a large *Agrobacterium* plasmid which confers virulence. Genetics **83,** 609—618.

Chilton, M.-D., Drummond, M. H., Merlo, D. J., Sciaky, D., Montoya, A. L., Gordon, M. P., Nester, E. W., 1977: Stable incorporation of plasmid DNA into higher plant cells: the molecular basis of crown gall tumorigenesis. Cell **11,** 263—271.

Chilton, M.-D., Montoya, A. L., Merlo, D. J., Drummond, M. H., Nutter, R., Gordon, M. P., Nester, E. W., 1978[a]: Restriction endonuclease mapping of a plasmid that confers oncogenicity upon *Agrobacterium tumefaciens* strain B6-806. Plasmid **1,** 254—269.

Chilton, M.-D., Drummond, M. H., Merlo, D. J., Sciaky, D., 1978[b]: Highly conserved DNA of Ti-plasmids overlaps T-DNA maintained in plant tumors. Nature **275,** 147—149.

Chilton, M.-D., Saiki, R. K., Yadav, N., Gordon, M. P., Quetier, F., 1980: T-DNA from *Agrobacterium* Ti plasmid is in the nuclear DNA fraction of crown gall tumor cells. Proc. Natl. Acad. Sci. U. S. A. **77,** 4060—4064.

Costantino, P., Hooykaas, P. J. J., Den Dulk-Ras, H., Schilperoort, R. A., 1980: Tumor formation and rhizogenicity of *Agrobacterium rhizogenes* carrying Ti plasmids. Gene **11,** 79—87.

Costantino, P., Mauro, M. L., Micheli, G., Risuleo, G., Hooykaas, P. J. J., Schilperoort, R. A., 1981: Fingerprinting and sequence homology of plasmids from different virulent strains of *Agrobacterium rhizogenes*. Plasmid **5,** 170—182.

Currier, T. C., Nester, E. W., 1976: Evidence for diverse types of large plasmids in tumor-inducing strains of *Agrobacterium*. J. Bacteriol. **126,** 157—165.

De Greve, H., Decraemer, H., Seurinck, J., Van Montagu, M., Schell, J., 1981: The functional organization of the octopine *Agrobacterium tumefaciens* plasmid pTiB6S3. Plasmid **6,** 235—248.

De Greve, H., Dhaese, P., Seurinck, J., Lemmers, M., Van Montagu, M., Schell, J., 1983: Nucleotide sequence and transcript map of the *Agrobacterium tumefaciens* Ti-plasmid encoded octopine synthase gene. J. Mol. Appl. Genet. **2,** 499—511.

De Picker, A. Van Montagu, M., Schell, J., 1978: Homologous DNA sequences in different Ti-plasmids are essential for oncogenicity. Nature **275,** 150—153.

De Picker, A., De Wilde, M., De Vos, G., De Vos, R., Van Montagu, M., Schell, J., 1980: Molecular cloning of overlapping segments of the nopaline Ti-plasmid pTiC58 as a means to restriction endonuclease mapping. Plasmid **3,** 193—211.

De Picker, A., Stachel, S., Dhaese, P., Zambryski, P., Goodman, H. M., 1982: Nopaline synthase: transcript mapping and DNA sequence. J. Mol. Appl. Genet. **1,** 561—573.

De Vos, G., De Beuckeleer, M., Van Montagu, M., Schell, J., 1981: Restriction endonuclease mapping of the octopine tumor-inducing plasmid pTiAch5 of *Agrobacterium tumefaciens.* Plasmid **6,** 249—253.

Drummond, M. H., Chilton, M.-D., 1978: Tumor-inducing (Ti) plasmids of *Agrobacterium* share extensive regions of DNA homology. J. Bacteriol. **136,** 1178—1183.

Drummond, M. H., Gordon, M. P., Nester, E. W., Chilton, M.-D., 1977: Foreign DNA of bacterial plasmid origin is transcribed in crown gall tumours. Nature **269,** 535—536.

Ellis, J. G., Murphy, P. J. 1981: Four new opines from crown gall tumours — their detection and properties. Mol. Gen. Genet. **181,** 36—43.

Ellis, J. G., Kerr, A., Tempé, J., Petit, A., 1979: Arginine catabolism: a new function of both octopine and nopaline Ti-plasmids of *Agrobacterium.* Mol. Gen. Genet. **173,** 263—269.

Ellis, J. G., Kerr, A., Petit, A., Tempé, J., 1982[a]: Conjugal transfer of nopaline and agropine Ti-plasmids. The role of agrocinopines. Mol. Gen. Genet. **186,** 269—274.

Ellis, J. G., Murphy, P. J., Kerr, A., 1982[b]: Isolation and properties of transfer regulatory mutants of the nopaline Ti-plasmid pTiC58. Mol. Gen. Genet. **186,** 275—281.

Engler, G., De Picker, A., Maenhout, R., Villarroel, R., Van Montagu, M., Schell, J. (1981). Physical mapping of DNA base sequence homologies between an octopine and a nopaline Ti-plasmid of *Agrobacterium tumefaciens.* J. Mol. Biol. **152,** 183—208.

Firmin, J. L., Fenwick, G. R., 1978: Agropine: a major new plasmid-determined metabolite in crown gall tumours. Nature **276,** 842—844.

Garfinkel, D. J., Nester, E. W., 1980: *Agrobacterium tumefaciens* mutants affected in crown gall tumorigenesis and octopine catabolism. J. Bacteriol. **144,** 732—743.

Garfinkel, D. J., Simpson, R. B., Ream, L. W., White, F. F., Gordon, M. P., Nester, E. W., 1981: Genetic analysis of crown gall: fine structure map of the T-DNA by site-directed mutagenesis. Cell **27,** 143—153.

Gelvin, S. B., Thomashow, M. F., McPherson, J. C., Gordon, M. P., Nester, E. W., 1982: Sizes and map positions of several plasmid DNA-encoded transcripts in octopine-type crown gall tumors. Proc. Natl. Acad. Sci. U. S. A. **79,** 76—80.

Genetello, C., Van Larebeke, N., Holsters, M., De Picker, A., Van Montagu, M.,

Schell, J., 1977: The Ti-plasmids of *Agrobacterium* as conjugative plasmids. Nature **265,** 561—563.

Goldmann, A., Thomas, D. W., Morel, G., 1969: Sur la structure de la nopaline, métabolite anormale de certaines tumeurs de crown gall. C. R. Acad. Sci. Paris **268,** 852—854.

Gurley, W. B., Kemp, J. D., Albert, M. J., Sutton, D. W., Callis, J., 1979: Transcription of Ti-plasmid derived sequences in the octopine type crown gall tumor lines. Proc. Natl. Acad. Sci. U. S. A. **76,** 2828—2832.

Hamilton, R. H., Fall, M. Z., 1971: The loss of tumor-initiating ability in *Agrobacterium tumefaciens* by incubation at high temperature. Experientia **27,** 229—230.

Heidekamp, F., Dirkse, W. G., Hille, J., Van Ormondt, H., Schilperoort, R. A., De Groot, B., 1983: Nucleotide sequence of the *Agrobacterium tumefaciens* octopine Ti-plasmid encoded *tmr* gene. Nucl. Ac. Res., **11,** 6211—6223.

Hepburn, A. G., Hindley, J., 1979: Regions of DNA sequence homology between an octopine and a nopaline Ti-plasmid of *Agrobacterium tumefaciens.* Mol. Gen. Genet. **169,** 163—172.

Hernalsteens, J. P., De Greve, H., Van Montagu, M., Schell, J., 1978: Mutagenesis by insertion of the drug resistance transposon Tn7 applied to the Ti plasmid of *Agrobacterium tumefaciens.* Plasmid **1,** 218—225.

Hille, J., Schilperoort, R., 1981: The use of transposons to introduce well-defined deletions in plasmids: possibilities for *in vivo* cloning. Plasmid **6,** 151—154.

Hille, J., Klasen, I., Schilperoort, R., 1982: Construction and application of R-prime plasmids, carrying different segments of an octopine Ti-plasmid from *Agrobacterium tumefaciens* for complementation of vir genes. Plasmid **7,** 107—118.

Hille, J., Van Kan, J., Klasen, I., Schilperoort, R., 1983[a]: Site-directed mutagenesis in *Escherichia coli* of a stable R772: Ti cointegrate plasmid from *Agrobacterium tumefaciens.* J. Bacteriol. **154,** 693—701.

Hille, J., Van Kan, J., Schilperoort, R. 1983[b]: Onc- and vir-genes located on the Ti-plasmid of *Agrobacterium tumefaciens.* In: Proceedings of molecular genetics of the bacteria-plant interaction. Pühler, A. (ed.), pp. 223—220, Berlin-Heidelberg-New York: Springer.

Hille, J., Wullems, G., Schilperoort, R., 1983[c]: Non-oncogenic T-region mutants of *Agrobacterium tumefaciens* do transfer T-DNA into plant cells. Pl. Mol. Biol., **2,** 155—163.

Hille, J., *et al.*, 1983[d]: in preparation.

Hoekema, A., Hirsch, P. R., Hooykaas, P. J. J., Schilperoort, R. A., 1983[a]: A binary plant vector strategy based on separation of the Vir- and T-region of the *Agrobacterium tumefaciens* Ti-plasmid. Nature **303,** 179—180.

Hoekema, A. Van Haaren, M. J. J., Hille, J., Hoge, J. H. C., Hooykaas, P. J. J., Krens, F. A., Wullems, G. J., Schilperoort, R. A., 1983[b]: *Agrobacterium tumefaciens* and its Ti-plasmid as tools in transformation of plant cells. In: Plant Molecular Biology. Goldberg, R. B. (ed.), New York: Alan R. Liss.

Holsters, M., De Waele, D., De Picker, A., Messens, E., Van Montagu, M., Schell, J., 1978: Transfection and transformation of *Agrobacterium tumefaciens.* Mol. Gen. Genet. **163,** 181—187.

Holsters, M., Silva, B., Van Vliet, F., Genetello, C., De Block, M., Dhaese, P., De Picker, A., Inzé, D., Engler, G., Vilarroel, R., Van Montagu, M., Schell, J., 1980: The functional organization of the nopaline *A. tumefaciens* plasmid pTiC58. Plasmid **3,** 212—230.

Hooykaas, P. J. J., 1979: The role of plasmid determined functions in the interactions of *Rhizobiaceae* with plant cells. A genetic approach, 168 pp. Thesis, Leiden, The Netherlands.

Hooykaas, P. J. J., Schilperoort, R. A., 1983: The molecular genetics of crown gall tumorigenesis. In: Advances in Genetics. Scandalios, J. G. (ed.), New York: Academic Press.

Hooykaas, P. J. J., Klapwijk, P. M., Nuti, M. P., Schilperoort, R. A., Rörsch, A., 1977: Transfer of the *Agrobacterium tumefaciens* Ti-plasmid to avirulent agrobacteria and to Rhizobium *ex planta*. J. Gen. Microbiol. **98,** 477—484.

Hooykaas, P. J. J., Roobol, C., Schilperoort, R. A., 1979: Regulation of the transfer of Ti-plasmids of *Agrobacterium tumefaciens*. J. Gen. Microbiol. **110,** 99—109.

Hooykaas, P. J. J., Den Dulk-Ras, H., Schilperoort, R. A., 1980[a]: Molecular mechanism of Ti-plasmid mobilization by R-plasmids. Isolation of Ti-plasmids with transposon-insertions in *Agrobacterium tumefaciens*. Plasmid **4,** 65—74.

Hooykaas, P. J. J., Den Dulk-Ras, H., Ooms, G., Schilperoort, R. A., 1980[b]: Interactions between octopine and nopaline plasmids in *Agrobacterium tumefaciens*. J. Bacteriol. **143,** 1295—1306.

Iyer, V. N., Klee, H. J., Nester, E. W. 1982: Units of genetic expression in the virulence region of a plant tumor-inducing plasmid of *Agrobacterium tumefaciens*. Mol. Gen. Genet. **188,** 418—424.

Kennedy, B. W., Alcorn, S. M., 1980: Estimates of U. S. A. crop losses to prokaryotic plant pathogens. Plant Dis. **64,** 674—676.

Kerr, A. 1969: Transfer of virulence between isolates of *Agrobacterium*. Nature **223,** 1175—1176.

Kerr, A., 1971: Acquisition of virulence by non-pathogenic isolates of *Agrobacterium radiobacter*. Physiol. Plant Pathol. **1,** 241—246.

Kerr, A., 1972: Biological control of crown gall: seed inoculation. J. appl. Bact. **35,** 493—497.

Kerr, A., 1980: Biological control of crown gall through production of agrocin 84. Plant Dis. **64,** 25—30.

Kerr, A., Panagopoulos, C. G., 1977: Biotypes of *Agrobacterium radiobacter* var. *tumefaciens* and their biological control. Phytopath. Z. **90,** 172—179.

Kerr, A., Manigault, P., Tempé, J., 1977: Transfer of virulence *in vivo* and *in vitro* in *Agrobacterium*. Nature **265,** 560—561.

Klapwijk, P. M., 1979: The role of octopine in conjugation and plant tumour formation by *Agrobacterium tumefaciens*. Studies on tumour-inducing plasmids. Thesis, Leiden, The Netherlands, 128 pp.

Klapwijk, P. M., Schilperoort, R. A., 1979: Negative control of octopine degradation and transfer genes of octopine Ti-plasmids in *Agrobacterium tumefaciens*. J. Bacteriol. **139,** 424—431.

Klapwijk, P. M., Schilperoort, R. A., 1982: Genetic determination of octopine degradation. In: Molecular Biology of Plant Tumors. Kahl, G., Schell, J. (eds.), pp. 475—495. New York: Academic Press.

Klapwijk, P. M., Scheulderman, T., Schilperoort, R. A., 1978: Coordinated regulation of octopine degradation and conjugative transfer of Ti-plasmids in *Agrobacterium tumefaciens*: evidence for a common regulatory gene and separate operons. J. Bacteriol. **136,** 775—785.

Klee, H. J., Gordon, M. P., Nester, E. W., 1982: Complementation analysis of *Agrobacterium tumefaciens* Ti-plasmid mutations affecting oncogenicity. J. Bacteriol. **150,** 327—331.

Klee, H. J., White, F. F., Iyer, V. N., Gordon, M. P., Nester, E. W., 1983: Muta-

tional analysis of the virulence region of an *Agrobacterium tumefaciens* Ti-plasmid. J. Bacteriol. **153,** 878—883.
Knauf, V. C., Nester, E. W., 1982: Wide host range cloning vectors: a cosmid clone bank of an Agrobacterium Ti-plasmid. Plasmid **8,** 45—54.
Koekman, B. P., Ooms, G., Klapwijk, P. M., Schilperoort, R. A., 1979: Genetic map of an octopine Ti-plasmid. Plasmid **2,** 347—357.
Koekman, B. P., Hooykaas, P. J. J., Schilperoort, R. A., 1980: Localization of the replication control region on the physical map of the octopine Ti-plasmid. Plasmid **4,** 184—195.
Koekman, B. P., Hooykaas, P. J. J., Schilperoort, R. A., 1982: A functional map of the replicator region of the octopine Ti-plasmid. Plasmid **7,** 119—132.
Krens, F. A., Molendijk, L., Wullems, G. J., Schilperoort, R. A., 1982: *In vitro* transformation of plant protoplasts with Ti-plasmid DNA. Nature **296,** 72—74.
Ledeboer, A. M., 1978: Large plasmids in *Rhizobiaceae*. Studies on the transcription of the tumour inducing plasmid from *Agrobacterium tumefaciens* in sterile crown gall tumour cells, pp. 180, Thesis, University of Leiden, The Netherlands.
Leemans, J., Deblaere, R., Willmitzer, L., De Greve, H., Hernalsteens, J. P., Van Montague, M., Schell, J., 1982: Genetic identification of fuctions of T_L-DNA transcripts in octopine crown galls. EMBO J. **1,** 147—152.
Liu, S. T., Perry, K. L., Schardl, C. L., Kado, C. I., 1982: *Agrobacterium* Ti-plasmid indoleacetic acid gene is required for crown gall oncogenesis. Proc. Natl. Acad. Sci. U. S. A. **79,** 2812—2816.
Loper, J. E., Kado, C. I., 1979: Host range conferred by the virulence-specifying plasmid of *Agrobacterium tumefaciens*. J. Bacteriol. **139,** 591—596.
Matzke, A. J. M., Chilton, M.-D., 1981: Site-specific insertion of genes into T-DNA of the *Agrobacterium* tumor-inducing plasmid: An approach to genetic engineering of higher plant cells. J. Mol. Appl. Genet. **1,** 39—49.
McKenney, K., Shimatake, H., Court, D., Schmeissner, U., Brady, C., Rosenberg, M., 1981: A system to study promoter and terminator signals recognized by *E. coli* RNA polymerase. In: Gene amplification and analysis. Vol. 2, pp. 383—415. Chrikjian, J. G., Papas, T. S. (eds.). Elsevier/North Holland.
Ménagé, A., Morel, G., 1964: Sur la présence d'octopine dans les tissus de crown-gall. C. R. Acad. Sci (Paris) **259 D,** 4795—4796.
Moore, L., Warren, G., Strobel, G., 1979: Involvement of a plasmid in the hairy root-disease of plants caused by *Agrobacterium rhizogenes*. Plasmid **2,** 617—626.
Murai, N., Kemp, J. D., 1982: T-DNA of pTi 15955 from *Agrobacterium tumefaciens* is transcribed into a minimum of seven polyadenylated RNAs in a sunflower crown gall tumor. Nucl. Ac. Res. **10,** 1679—1689.
Murphy, P. J., Tate, M. E., Kerr, A., 1981: Substituents at N^6 and C-5′ control selective uptake and toxicity of the adenine-nucleotide bacteriocin, agrocin 84, in agrobacteria. Eu. J. Biochem. **115,** 539—543.
New, P. B., Kerr, A., 1971: A selective medium for *Agrobacterium radiobacter* biotype 2. J. appl. Bact. **34,** 233—236.
New, P. B., Kerr, A., 1972: Biological control of crown gall: field observations and glasshouse experiments. J. appl. Bact. **35,** 279—287.
Ooms, G., Klapwijk, P. M., Poulis, J. A., Schilperoort, R. A., 1980: Characterization of Tn904 insertions in octopine Ti-plasmid mutants of *Agrobacterium tumefaciens*. J. Bacteriol. **144,** 82—91.
Ooms, G., Hooykaas, P. J. J., Moolenaar, G., Schilperoort, R. A., 1981: Crown gall plant tumors of abnormal morphology, induced by *Agrobacterium tumefaciens*

carrying mutated octopine Ti-plasmids; analysis of T-DNA functions. Gene **14,** 33—50.
Ooms, G., Hooykaas, P. J. J., Van Veen, R. J. M., Van Beelen, P., Regensburg-Tuïnk, A. J. G., Schilperoort, R. A., 1982: Octopine Ti-plasmid deletion mutants of *Agrobacterium tumefaciens* with emphasis on the right side of the T-region. Plasmid **7,** 15—29.
Panagopoulos, C. G., Psallidas, P. G., 1973: Characteristics of Greek isolates of *Agrobacterium tumefaciens* (Smith & Townsend) Conn. J. appl. Bact. **36,** 233—240.
Petit, A., Tempé, J., 1978: Isolation of *Agrobacterium* Ti-plasmid regulatory mutants. Mol. Gen. Genet. **167,** 147—155.
Petit, A., Delhaye, S., Tempé, J., Morel, G., 1970: Recherches sur les guanidines des tissus de crown gall. Mise en évidence d'une relation biochimique spécifique entre les souches d'*Agrobacterium* et les tumeurs qu'elles induisent. Physiol. Veg. **8,** 205—213.
Petit, A., Tempé, J., Kerr, A., Holsters, M., Van Montagu, M., Schell, J., 1978: Substrate induction of conjugative activity of *Agrobacterium tumefaciens* Ti-plasmids. Nature **271,** 570—572.
Prakash, R. K., Schilperoort, R. A, 1982: Relationship between nif plasmids of fast-growng *Rhizobium* species and Ti-plasmids of *Agrobacterium tumefaciens.* J. Bacteriol. **149,** 1129—1134.
Ream, L. W., Gordon, M. P., Nester, E. W., 1983: Multiple mutations in the T-region of the *Agrobacterium tumefaciens* tumor-inducing plasmid. Proc. Natl. Acad. Sci. U. S. A. **80,** 1660—1664.
Regier, D. A., Morris, R. O., 1982: Secretion of *trans*-zeatin by *Agrobacterium tumefaciens*: a function determined by the nopaline Ti-plasmid. Biochem. Biophys. Res. Commun. **104,** 1560—1566.
Riker, A. J., 1930: Studies on infectious hairy root of nursery apple trees. J. Agr. Res. **41,** 507—540.
Risuleo, G., Battistoni, P., Costantino, P., 1982: Regions of homology between tumorigenic plasmids from *Agrobacterium rhizogenes* and *Agrobacterium tumefaciens.* Plasmid **7,** 45—51.
Schilperoort, R. A., Kester, H. C. M., Klapwijk, P. M., Rörsch, A., Schell, J., 1975: Plant tumors induced by *A. tumefaciens.* A genetic approach. In: Semaine d'étude agriculture et hygiène des plantes. Martens, P. (ed.), pp. 25—35, Gemblouse, Belgium.
Schröder, G., Schröder, J., 1982: Hybridization selection and translation of T-DNA encoded mRNAs from octopine tumors. Mol. Gen. Genet. **185,** 51—55.
Schröder, J., Schröder, G., Huisman, H., Schilperoort, R. A., Schell, J., 1981: The mRNA for lysopine dehydrogenase in plant tumor cells is complementary to a Ti-plasmid fragment FEBS Lett. **129,** 166—168.
Smith, E. F., Townsend, C. O., 1907: A plant tumor of bacterial origin. Science **25,** 671—673.
Tempé, J., Petit, A., 1982: Opine utilization by *Agrobacterium.* In: Molecular Biology of Plant Tumors. Kahl, G., Schell, J. (eds.), pp. 451—459, New York: Academic Press.
Thomashow, M. F., Nutter, R., Montoya, A. L., Gordon, M. P., Nester, E. W., 1980[a]: Integration and organization of Ti plasmid sequences in crown gall tumors. Cell **19,** 729—739.
Thomashow, M. F., Nutter, R., Postle, K., Chilton, M.-D., Blattner, F. R., Powel, A., Gordon, M. P., Nester, E. W., 1980[b]: Recombination between higher plant

DNA and the Ti-plasmid of *Agrobacterium tumefaciens*. Proc. Natl. Acad. Sci. U. S. A. **77,** 6448—6452.

Thomashow, M. F., Panagopoulos, C. G., Gordon, M. P., Nester E. W., 1980[c]: Host range of *Agrobacterium tumefaciens* is determined by the Ti plasmid. Nature **283,** 794—796.

Thomashow, M. F., Knauf, V. C., Nester, E. W., 1981: The relationship between the limited and wide host range octopine-type Ti plasmids of *Agrobacterium tumefaciens*. J. Bacteriol. **146,** 484—493.

Van Larebeke, N., Engler, G., Holsters, M., Van den Elsacker, S., Zaenen, I., Schilperoort, R. A., Schell, J., 1974: Large plasmid in *Agrobacterium tumefaciens* essential for crown-gall inducing ability. Nature **252,** 169—170.

Van Larebeke, N., Genetello, C., Schell, J., Schilperoort, R. A., Hermans, A. K., Hernalsteens, J. P., Van Montagu, M., 1975: Acquisition of tumour-inducing ability by non-oncogenic agrobacteria as a result of plasmid transfer. Nature **255,** 742—743.

Van Larebeke, N., Genetello, G., Hernalsteens, J. P., De Picker, A., Zaenen, I., Messens, E., Van Montagu, M., Schell, J., 1977: Transfer of Ti plasmids between Agrobacterium strains by mobilisation with the conjugative plasmid RP4. Mol. Gen. Genet. **152,** 119—124.

Vervliet, G., Holsters, M., Teuchy, H., Van Montagu, M., Schell, J., 1975: Characterization of different plaque-forming and defective temperate phages in *Agrobacterium* strains.

Watson, B., Currier, T. C., Gordon, M. P., Chilton, M.-D., Nester, E. W., 1975: Plasmid required for virulence of *Agrobacterium tumefaciens*. J. Bacteriol. **123,** 255—264.

White, F. F., Nester, E. W., 1980[a]: Hairy root: plasmid encodes virulence traits in *Agrobacterium rhizogenes*. J. Bacteriol. **141,** 1134—1141.

White, F. F., Nester, E. W., 1980[b]: Relationship of plasmids responsible for hairy root and crown gall tumorigenicity. J. Bacteriol. **144,** 710—720.

Willmitzer, L., De Beuckeleer, M., Lemmers, M., Van Montagu, M., Schell, J., 1980: DNA from Ti plasmid present in nucleus and absent from plastids of crown gall plant cells. Nature **287,** 359—361.

Willmitzer, L., Schmalenbach, W., Schell, J., 1981: Transcription of T-DNA in octopine and nopaline crown gall tumours is inhibited by low concentrations of α-amanitin. Nucl. Ac. Res. **9,** 4801—4812.

Willmitzer, L., Simons, G., Schell, J., 1982: The T_L-DNA in octopine crown-gall tumours codes for seven well-defined polyadenylated transcripts. EMBO J. 139—146.

Willmitzer, L., Dhaese, P., Schreier, P. H., Schmalenbach, W., Van Montagu, M., Schell, J., 1983: Size, location and polarity of T-DNA encoded transcripts in nopaline crown gall tumors; common transcripts in octopine and nopaline tumors. Cell **32,** 1045—1056.

Chapter 12

Phytohormone-Mediated Tumorigenesis by Plant Pathogenic Bacteria

C. I. Kado

Davis Crown Gall Group, Department of Plant Pathology, University of California, Davis, CA 95616, U. S. A.

With 1 Figure

Contents

I. Introduction

There are a number of plant pathogens that cause visible morphological distortions of the plants they infect. Past studies on the effects of these particular pathogen-host interactions have been mainly descriptive and the precise mechanisms which cause these distortion were not well understood. Many of the early efforts have been to examine the diseased plant at the site of infection, and few workers have emphasized the detailed study of the pathogen itself. As stressed previously (Kado, 1976), the rewards will be great when we gain a comprehensive knowledge of the pathogen of interest. This is clearly evidenced by the studies of *Agrobacterium tumefaciens,* the results of which have yielded a wealth of new information not only in plant pathology, but also in molecular biology and molecular genetics. Furthermore, the discovery that genetic information is naturally transferred from a prokaryote to an eukaryote during the interaction sets a new precedence in biology.

The plant distorting pathogens *Agrobacterium tumefaciens, A. rhizogenes, Pseudomonas savastanoi,* and *Corynebacterium fascians* are currently amenable to genetical studies. Each of these bacteria have disease causing mechanisms that culminate in the morphological distortions of their respective host plants. Recent molecular and genetic studies, as presented in this chapter, have provided substantial information on the involvement of phytohormones induced by the pathogen. Evidence accumulated in these studies support many of the early suggestions that phytohormones such as indole-3-acetic acid (IAA) and the cytokinins might be responsible for the disease response.

II. Crown Gall

The tumorous overgrowth caused by *A. tumefaciens* is best known as crown gall. The mechanism by which this bacterium induces these tumors in healthy plants is unique. All virulent strains of *A. tumefaciens* harbor a large plasmid that is responsible for conferring the tumorigenic phenotype. This plasmid of approximately 220 kb, designated the Ti plasmid (for tumor-inducing), may differ in size and genetic properties. However, all virulent plasmids contain a section of DNA that is detected in the genome of the crown gall tumor cell. This specific DNA segment, defined as the T-DNA, may be stably maintained in the plant cells in single or multiple copies. This results in the manifestation of transformed cells which form the crown gall tumor. Detailed early studies indicated that when these tumors were freed of bacteria they could be maintained as callus tissues on culture medium in the absence of IAA and cytokinin, in contrast to normal calli which require these growth regulators (Braun, 1956). Analyses of normal and crown gall tumor calli revealed two-fold or greater levels of IAA in the hormone autotrophic tumor callus lines (Table 1). The mechanism causing the hyperauxinic state remains obscure. In some cases the levels of IAA in tobacco tumor calli return to normal levels, suggesting that the hyperauxinic state is unnecessary for continued autonomous growth of the crown gall tumors (Atsumi and Hayashi, 1978; Nakajima *et al.,* 1979; Pengelly and Meins, 1982). Recent studies have shed some light on the possible role of the Ti plasmid and its T-DNA in this phenomenon.

A. IAA Synthesis in Crown-Gall Tumor Cells

Crown gall tumor cells have long been recognized for their phytohormone autotrophy. One plausible explanation for this is that the levels of IAA, either in the bound or free form, are higher in these cells than in comparable normal cells (Table 1). However, the concentrations of IAA can vary depending on the phase of growth of the tumor calli (Mousdale, 1981,

Table 1. Hyperauxinic State of Crown Gall Tumors

Host Plant	*A. tumefaciens* strain used	Level of indole-3-acetic acid in crown gall tissue relative to normal tissue	Reference
Lycopersicum esculentum L. cv. Bonnie Best	unspecified	5 ×	Link and Eggers (1941)
Scorsonera sp.	unspecified	23.9—40.5 ×	Kulescha and Gautheret (1948)
Helianthus annuus L. cv. Giant Russian	H13	12 ×	Henderson and Bonner (1952)
L. esculentum	unspecified	8—12 ×	Dye *et al.* (1961)
Impatiens balsamia L.	unspecified	100 ×	Bouillenne and Gaspar (1970)
H. annuus L. cv. Giant Russian	ATCC4452	1—14.4 ×	Atsumi and Hayashi (1978); Atsumi (1980)
Nicotiana tabacum L. cv. Hick-2	unspecified	300—500 ×	Nakajima *et al.* (1979)
N. tabacum cv. White Burley	unspecified	2.5 ×	Chirek (1979)
H. annuus L. cv. Borowski Prazkowany	unspecified	3 ×	Chirek (1979)
Cantharanthus roseus L.	B6	16.2 ×	Weiler (1981)
H. annuus L. cv. Russian Giant	B6	30 × [a]	Mousdale (1981)
Beta vulgaris L.	B6	2.8 ×	Weiler and Spanier (1981)
L. esculentum L.	B6	1.5 ×	Weiler and Spanier (1981)
Blitum virgatum L.	T37	2.3 ×	Weiler and Spanier (1981)
N. tabacum L. TC4096	A6	3 ×	Weiler and Spanier (1981)
N. tabacum L. cv. Hicks-2	unspecified	750 ×	Nakajima *et al.* (1981)
N. rustica L.	B6	5 ×	Mousdale (1982)
H. annuus L. cv. Giant Russian	B6	7.7 ×	Mousdale (1982)

[a] Where applicable the data were estimated from graphical data which were presented as quantitative values based on weight equivalents obtained from either biological or chemical assays.

Table 1. Continued

Host Plant	*A. tumefaciens* strain used	Level of indole-3-acetic acid in crown gall tissue relative to normal tissue	Reference
Kalanchoë daigremontiana L.	B6	5 ×	Mousdale (1982)
Parthenocissus tricuspidata (Sieb. and Zucc.) Planch.	B6	6 ×	Mousdale (1982)
Daucus carota L. cv. Chantenay	B6	17 ×	Mousdale (1982)
N. tabacum L. cv. Turkish	T37	53.4 ×	Pengelly and Meins (1982)
N. tabacum L. cv. Wisconsin 38	A6—2	290 ×	Amasino and Miller (1982)
N. tabacum L. cv. Wisconsin 38	B6—2	350 ×	Amasino and Miller (1982)
N. tabacum L. cv. Wisconsin 38	542	180 ×	Amasino and Miller (1982)
N. tabacum L. cv. Wisconsin 38	T37	8—10 ×	Amasino and Miller (1982)
N. tabacum L. cv. Wisconsin 38	C58	8—10 ×	Amasino and Miller (1982)
N. tabacum L. cv. Wisconsin 38	EU6	8—10 ×	Amasino and Miller (1982)
N. tabacum L. cv. Wisconsin 38	A6—1	34 ×	Amasino and Miller (1982)
N. tabacum L. cv. Wisconsin 38	B6—1	29 ×	Amasino and Miller (1982)

1982; Nakajima *et al.*, 1981). Furthermore, these variations, together with variations in cytokinin levels, may be responsible for the differences in crown gall tumor morphology (e. g. *Nicotiana tabacum* L. var. Wisconsin-38, Amasino and Miller, 1982). At least three different tumor morphologies have been recognized: friable unorganized, compact unorganized and teratoma types. The friable unorganized types of *N. tabacum* tumor calli contain about ten-fold greater concentrations of IAA than compact unorganized tumor calli and about twenty to thirty-fold greater concentrations of IAA than teratoma calli (Amasino and Miller, 1982). It is clear that plant cells transformed by *A. tumefaciens* display a hyperauxinic state, but it remains unclear whether this condition itself is the factor responsible for auxin autotrophy (Atsumi and Hayashi, 1978; Nakajima *et al.*, 1979; Pengelly and Meins, 1982). There is a close correlation between autotrophy and the production of growth factors in genetic tumors of certain interspecific *Nicotiana* hybrids (Schaeffer and Smith, 1963; Cheng, 1972). It is interesting to note that the growth of crown gall cells is not enhanced by supplying either auxin or cytokinin to the growth medium (Matsumoto *et al.*, 1975).

The role of the T-DNA that is maintained in the plant genome, either as a single copy or in multiple copies, in eliciting hormone production and activity is also unclear. There are several possibilities to explain how the intergrated T-DNA may function. (1) The mere insertion of the T-DNA may result in the inactivation of genetic determinants that are usually responsible for repressing or regulating plant hormone production genes under normal conditions. (2) The T-DNA itself contains genetic determinants that redirect the normal regulation of hormone production. (3) The T-DNA may carry phytohormone production genes of its own.

Accumulating evidence suggests that the T-DNA contains genetic information for the regulation of phytohormone production in crown gall tumors. In a parallel case with genetic tumors of interspecific *Nicotiana* hybrids, the tumors seem to be caused by altered gene regulation rather than by gene mutation (Smith, 1972). Braun (1958) had earlier proposed that tumor autonomy results from the activation of normally repressed biosynthetic capacities of the plant cell. Although the precise functional roles of the T-DNA genes remain to be identified, transposon mutagenesis of this region on octopine Ti plasmids (e.g., pTIB6806, pTiAch5, pTiB6S3, and pTiA6NC) causes alterations in tumor morphology (Garfinkel and Nester, 1980; Ooms *et al.*, 1981; De Greve *et al.*, 1981; Garfinkel *et al.*, 1981). Similarly, transposon insertions in the T-DNA of nopaline Ti plasmids (e. g., pTiT37, pTiC58) cause morphological changes in the development of the crown gall tumor (Bevan and Chilton, 1982; Van Montagu and Schell, 1982; Joos *et al.*, 1983). Mapping of these insertion mutations in, for example, pTiA6NC has revealed three distinct loci each of which affects crown gall tumor morphology differently (Garfinkel *et al.*, 1981). Each transposon insertion within a 1.25 kilobase region, designated *tml,* caused the bacteria to incite tumors two to three times larger than normal. A second set of insertions contained within a 1 kilobase region (designated *tms*) caused tumors with prolific root growth. The third set of insertions distributed over a 3.1 kilobase region (designated *tms*) caused tumors with shoots growing from the pTiA6NC tumor callus. With the T-DNA of pTiB6S3, replacement of a 0.7 Md *Eco*RI fragment by a 5.8 Md fragment of the IncW plasmid pSa was observed to cause small, greenish tumors that were covered with numerous shoots (Leemans *et al.*, 1981). Thus, the insertion of foreign genetic elements such as transposons and R plasmid fragments into particular regions of the T-DNA affect the developmental organization properties of crown gall tumors. Although the mechanism remains unclear, the present evidence strongly suggests that phytohormone imbalances are in part responsible for the developmental alterations. Clearly the persistent high level of IAA is indicative that such imbalances exist in crown gall tumor. The absence of high levels of IAA may result in the decrease of tumor growth and therefore the size of the tumor, or in tumor morphology changes manifested macroscopically by shoot and root formations. One means of testing the possibility that higher levels of IAA are needed for tumor production is to supplement the T-DNA mutants with IAA and/or cytokinin.

Ooms *et al.* (1981), observing that mutations in the T-DNA of pTiAch5 affected the morphology of crown gall tumors, noted that one of these mutations was caused by the insertion of an IS element identified as IS*60*. The *Agrobacterium* harboring this mutant plasmid (pAL108) incited tumors that were smaller than those caused by the wild-type strain LBA4001(pTiAch5). Interestingly, attenuated crown gall tumors initiated by this mutant and a second one obtained by Tn1831 insertion nearby (ca. 1.7 kb) were stimulated to grow to "normal" size by adding naphthalene acetic acid (NAA) to tomato plants inoculated with these mutants. These results are reminiscent of the early experiments of Brown and Gardner (1936), Locke *et al.* (1938), Braun and Laskaris (1942), Thomas and Riker (1948), and Klein and Link (1952). In these early experiments, the attenuated *A. tumefaciens* strain A66, which induces very small, slow growing, infrequent tumors, was able to induce tumors of normal size when the inoculated tomato plants were treated with either IAA or NAA at either the wound site or at the decapitated surface of the stem. Recent analysis of strain A66 revealed a naturally existing 2.7 kb insertion in the T-DNA region of the pTiA6 plasmid that apparently resulted in the attenuated condition of this strain and enabled strain A66 to cause shoot-forming tumors (Binns *et al.*, 1982). Thus, it appears that IAA or an IAA induced product within the inoculated plant is essential for restoring strain A66 to full virulence.

Analyses were made by Leemans *et al.* (1982) using deletion mutants of the T-DNA of pTiB6S3. In this case, 2,4-dichlorophenoxyacetic acid (2,4D) was incorporated into agar medium on which inoculated carrot slices were placed. Restoration to wild-type levels of virulence were observed with mutants containing deletions within *Bam*H1 fragment 8 of the pTiB6S3 T-DNA.

To determine if regions outside the T-DNA were also responsive to exogenously applied IAA, a series of transposon insertion mutants of pTiC58 were tested on several host plant species (Kao *et al.*, 1982). Analysis of these mutants and a deletion mutant revealed that mutations in the T-DNA, located near the left terminus of *Eco*RI fragment 1 (HindIII fragments 19 and 41) of pTiC58, resulted in decreased virulence of strain C58 typified by slow-growing tumors in some plant species but *not* in others. Plants that responded poorly to these T-DNA mutants could respond normally by the exogenous application of IAA or NAA, thereby resulting in the restoration of the natural host range of strain C58. Kao *et al.* (1982) postulated that host specifying functions are located in the T-DNA and perhaps elsewhere on the pTiC58 plasmid, and that these functions are

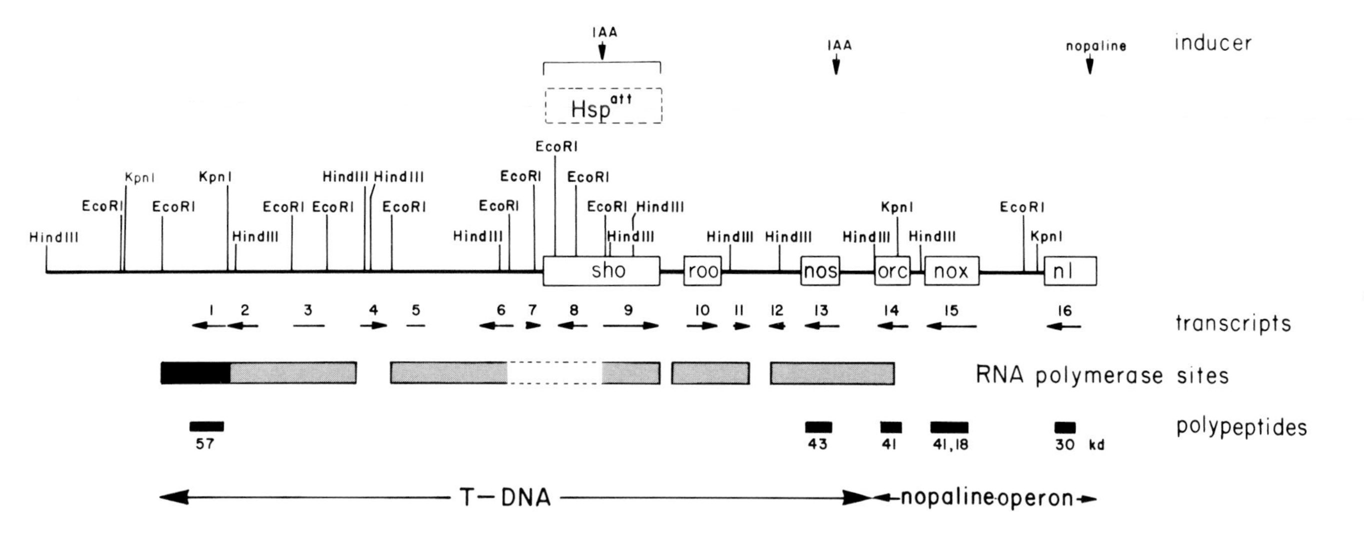

Fig. 1. Restriction endonuclease, transcriptional and translational-product map of that part of the pTiC58 plasmid of *Agrobacterium tumefaciens* C58 which contains the T-DNA and the nopaline operon. Portions of the T-DNA inducible with indole-3-acetic acid (indicated as IAA) are in the shoot-organization region (sho) and the nopaline synthase region (nos). The root-organization region (roo) is activated by cytokinin. The nopaline operon is induced by nopaline, which leads to the induction of the 30 kilodalton nopaline binding protein nl. The production of nopaline oxidase is directed by the nox region and an unidentified ornithine degradative enzyme maps in the region designated orc (Schardl and Kado, 1983). Transcript locations were identified by compiling transcript map information described for TiT37 (Bevan and Chilton, 1982) and for TiC58 (Joos *et al.*, 1983) and by DNA sequence and transcript run-off experiments (R. C. Tait, H. Rempel, and C. I. Kado, manuscript submitted). The polypeptide maps were obtained from a *E. coli* 1592 mini-cell analysis of cloned fragments bearing specific deletions or Tn5 insertions (Schardl and Kado, 1983; M. Hagiya, R. C. Tait, and C. I. Kado submitted for publication). Purified *A. tumefaciens* C58 RNA polymerase was employed in polymerase 'walking' experiments to deduce sites of specific binding (R. C. Tait, T. J. Close, R. L. Rodriguez, and C. I. Kado, manuscript submitted)

regulated by IAA. It was recently shown by Joos *et al.* (1983) that insertion and deletion mutants mapping in the same area responded to NAA treatment of potato discs. Since this region of the T-DNA contains two RNA transcripts in pTiC58 induced tumors, it was postulated that the products of both genes are critical for tumor formation but that tumors can be restored by the addition of NAA. Akiyoshi *et al.* (1983) recently reported decreased levels of IAA in primary crown gall tumors caused by pTiA6 plasmids containing mutations in the *tms* region of the T-DNA. It was not reported if the function of this region can be chemically complemented with IAA. However, since the *tms* region confers shoot organization like that of the *sho* region (for shoot organization) of the T-DNA of pTiC58 (Fig. 1), it is quite likely that the *tms* region will respond to IAA treatment.

The present evidence clearly shows that certain T-DNA genes respond to IAA (via treated plants) and are somehow involved in auxin metabolism (also cytokinin metabolism as discussed below). Whether or not IAA directly influences T-DNA genes was unclear until it was recently shown that IAA can switch on the synthesis of polypeptides encoded by the T-DNA in *Escherichia coli* mini-cells (Tait *et al.,* 1982) (Fig. 1). That certain attenuated T-DNA mutants of either octopine (pTiA6, pTiAch5, pTiB6S3) or nopaline (pTiC58) Ti plasmids respond to IAA treatment indicates a general requirement of *A. tumefaciens* for IAA for the expression of full virulence. This requirement is met by the endogenous IAA (or its immediate precursors) of the developing host plant and may explain why one observes variation of tumor size depending on the host (Kao *et al.,* 1982) and on the overall 'tumorigenic competence' of the plant (for review see Beardsley, 1972). Because free IAA concentrations vary between plant parts and at various stages of plant growth, the T-DNA genes may be developmentally regulated by IAA and cytokinin. The observation that the T-DNA is actively transcribed in the crown gall tumor cell and weakly transcribed in the *Agrobacterium* cell (Gelvin *et al.,* 1981) and the recent observation that IAA can switch-on T-DNA genes in bacteria, as judged by the synthesis of specific polypeptides including nopaline synthase (Tait *et al.,* 1982), supports this hypothesis.

B. IAA Synthesis by A. tumefaciens

Early studies on the nature of the tumor-inducing principle as studied in *A. tumefaciens* had implicated various bacterial constituents including IAA (for review see Kado, 1976). Because IAA was well recognized as a powerful growth substance in plants, several workers analyzed various *A. tumefaciens* strains for IAA. It was soon learned that *A. tumefaciens* indeed secrets IAA into the culture fluid (Table 2). *A. tumefaciens* cells grown in medium containing L-tryptophan produce copious quantities of IAA. One pathway for the synthesis of IAA is through the conversion of tryptophan to indolepyruvic acid, which is converted to indoleacetaldehyde and then to IAA (Kaper and Veldstra, 1958). The enzymes of this pathway are tryptophan aminotransferase (transaminase), indole-3-pyruvate decarboxylase and

Table 2. Synthesis of indole-3-acetic acid by *Agrobacterium tumefaciens* strains

Bacterial strain	Assay method	Level of IAA relative to control	Reference
Unspecified	lanolin paste assay	detected	Brown and Gardner (1936)
Unspecified	lanolin paste assay	4×	Link *et al.* (1937)
Unspecified	lanolin paste assay	detected	Link and Wilcox (1937)
Unspecified	colorimetric	synthesized	Berthelot and Amoureux (1938)
B.t., Ra, A-Da	lanolin paste assay	detected	Dame (1938)
A6	*Avena* coleoptile (ACT)	4×	Locke *et al.* (1938)
E III.9.6.1	paper chromatography	synthesized 72%	Kaper and Veldstra (1958)
Unspecified	*Avena* coleoptile test	44×	Vyas and Jain (1973)
ATV	colorimetric	1.2×	Beltra *et al.* (1978)
C58	Gas-liquid chromatography, high performance thinlayer chromatography	5—10×	Liu and Kado (1979)
B6	Radioimmunoassay	1.4×	Weiler and Spanier (1981)
Octopine strains 15955, A277, B6, 1058	Gas liquid chromatography	2—3×	Lobanok *et al.* (1982)
Nopaline strains 8934, 8628, C58	Gas-liquid chromatography	2—3×	Lobanok *et al.* (1982a, 1982b)

indole-3-acetalaldehyde oxidase respectively (Sembdner *et al.,* 1980). Evidence in support of this pathway comes from the presence of tryptophan transaminase activity (Atsumi, 1980). This enzyme is encoded by a chromosomal gene rather than a Ti plasmid gene of *A. tumefaciens* (Liu *et al.,* 1982) and has been purified 58-fold (Sukanya and Vaidyanathan, 1979). The enzyme converts L-tryptophan to indole-3-pyruvic acid, a product which is readily isolated from these cells and culture fluid (Liu, *et al.,* 1982). The indole-3-pyruvate decarboxylase and indole-3-acetaldehyde oxidase have not yet been identified as products encoded by *Agrobacterium* genes. However, comparative analysis made of virulent and avirulent Ti plasmid-free strains showed a distinct difference in the quantity of IAA produced (Liu and Kado, 1979; Weiler and Spanier, 1981; Lobanok *et al.,* 1982a, 1982b). The extent of this difference was small when cells of these strains were grown in medium containing tryptophan, but the difference was quite striking and significant when the medium contained tyrosine. When the Ti plasmid was transferred to the pTi-free strain NT1, the amount of IAA synthesized increased to the amount normally synthesized by the parental wild-type strain (Liu and Kado, 1979). These results suggested that the Ti plasmid was needed for elevated IAA production. Whether the elevated production of IAA was the result of indole-3-pyruvate decarboxylase and/or indole-3-acetaldehyde oxidase encoded by the Ti plasmid genes has not been demonstrated owing to the labile nature of the substrates involved. However, Liu *et al.* (1982) have shown the pTiC58 plasmid contains a gene(s) necessary for IAA production. A plasmid-free mutant defective in tryptophan transaminase activity and therefore carrying a mutation in the *iaaC* gene, was used to test whether the pTi plasmid indeed contained genetic determinants that encoded IAA producing functions. When nopaline or octopine Ti plasmids were inserted by transformation into the avirulent *iaaC* mutant, the resulting transformants produced increased levels of IAA and virulence was restored. It was then demonstrated that the Ti plasmid did not encode tryptophan transaminase. One gene responsible for IAA production was located in the virulence *(vir)* region of the TiC58 plasmid by transposon mutagenesis. When this *iaaP* mutant Ti plasmid was introduced into the *iaaC* mutant, no increase in IAA was observed and virulence was not restored. It is clear from these experiments that the Ti plasmid encodes functions essential for elevated IAA production and virulence. That the *iaaP* gene is located within a region that was defined by transposon mutagenesis as the virulence region, indicates IAA is one of the essential components of *Agrobacterium* virulence.

C. Cytokinin Synthesis in Crown Gall Tumors

Crown gall tumors synthesize not only auxin, put also high levels of cytokinins. As discussed, when untransformed plant cells are placed in tissue culture, an exogenous source of both IAA and cytokinin is necessary for sustained growth on tissue culture medium. As the concentration of IAA in the medium is increased, the concentration of cytokinin required for growth is decreased (Einset, 1977). Einset (1980) argued that because of this phenomenon in tobacco calli, the finding of high relative cytokinin levels in crown galls is not sufficient evidence, in itself, to account for the capability of these tumor tissues to grow on nutrient medium devoid of cytokinin. As shown in Table 3 it is clear that, like IAA concentrations, the cytokinin levels are elevated in crown gall tissues in most cases. There are of course variations in these concentrations depending on the plant, its age, and stage of tumor growth, all of which reflect the physiological state of the host plant. Because the levels of cytokinin and IAA are high in crown gall tissues, any imbalances between the levels of these growth factors will obviously result in morphological deviations from the usually observed unorganized crown gall tumor. Indeed, addition of cytokinin can promote

Table 3. Cytokinin in Crown Gall Tumor Tissue

Plant species	*A. tumefaciens* strain	Cytokinin[a] levels relative to control		Reference
		trans-zeatin riboside	isopentenyl-adenosine	
Parthenocissus tricuspidata (Sieb. and Zucc.) Planch	unspecified	nd	nd	Tegley *et al.* (1971)
Vinca rosea L.	A6	nd	nd	Peterson and Miller (1976)
N. tabacum cv. Hicks-2	unspecified	40—80 ×	nd?	Nakajima *et al.* (1979)
N. tabacum L. cv. Wisconsin-38	C58	8 ×	nd	Einset (1980)
N. tabacum L. cv. Wisconsin-38	27	220 ×	nd	Einset (1980)
N. tabacum L. cv. Wisconsin-38	86	1620 ×	nd	Einset (1980)
N. tabacum. L. cv. Wisconsin-38	CGIC	40 ×	nd	Einset (1980)
N. tabacum L. cv. Wisconsin-38	AT4	450 ×	nd	Einset (1980)
N. tabacum L. cv. Hicks-2	unspecified	15 ×	nd	Nakajima *et al.* (1979)
Euphorbia lathyris L.	B6	181 ×	1.7 ×	Weiler and Spanier (1981)
E. myrsinites Brot.		11.5 ×	1.4 ×	Weiler and Spanier (1981)
Papaver rhoeas L.	B6	1.9 ×	1.2 ×	Weiler and Spanier (1981)
Kalanchoë blossfeldiana cv. Poelln (Hamet & Perr) Bathie	B6	1.3 ×	1.3 ×	Weiler and Spanier (1981)

[a] Isopentenyl adenosine is N^6-(Δ^2-isopentenyl)adenosine.
trans-zeatin riboside is 6-(4-hydroxy-3-methylbut-2-enylamino)-9-β-D-ribofuranosyl purine.
The levels of cytokinin were based on quantitative differences between normal and tumor tissues where applicable. The quantitative units were either reported as cytokinin equivalents, from bioassays, or on a weight equivalent basis from chromatographic analyses.

Table 3. Continued

Plant species	*A. tumefaciens* strain	Cytokinin[a] levels relative to control *trans*-zeatin riboside	isopentenyladenosine	Reference
K. daigremontiana	B6	8.4 ×	2.9 ×	Weiler and Spanier (1981)
Phaseolus vulgaris L. cv. Troketta	B6	20.9 ×	nd	Weiler and Spanier (1981)
Clarkia rubicunda (Lindl) Lewis and Lewis	B6	32.7 ×	1.6 ×	Weiler and Spanier (1981)
Apium graveolens cv. secalinum Alef.	B6	1.5 ×	1.2 ×	Weiler and Spanier (1981)
Arabis turrita L.	B6	1.3 ×	1.7 ×	Weiler and Spanier (1981)
Sinapis alba L.	B6	190.3 ×	4.0 ×	Weiler and Spanier (1981)
Tropaeolum majus L.	B6	40.6 ×	1.2 ×	Weiler and Spanier (1981)
Agrostemma githago L.	B6		– 0.9 ×	Weiler and Spanier (1981)
Tetragonia tetragonoides (Pall.) Kuntze	B6	0.1 ×	1.7 ×	Weiler and Spanier (1981)
Portulaca oleracea L. spp. sativa (Haw.) Celak	B6	– 0.8 ×	– 0.8 ×	Weiler and Spanier (1981)
Basella alba L.	B6	– 0.6 ×	1.1 ×	Weiler and Spanier (1981)
Beta vulgaris L. cv. alba DV (var. crassa (Alef.) Wittm.)	B6	1.8 ×	2.4 ×	Weiler and Spanier (1981)
Amaranthus candatus L.	B6	– 0.5 ×	1.0 ×	Weiler and Spanier (1981)
Catharanthus roseus L.	B6	83.3 ×	3.6 ×	Weiler and Spanier (1981)
Hyoscyamus niger L.	B6	18.5 ×	6.8 ×	Weiler and Spanier (1981)
Petunia violaceae Lindl.	B6	– 0.9 ×	1.1 ×	Weiler and Spanier (1981)
Digitalis lanata L.	B6	1.8 ×	1.5 ×	Weiler and Spanier (1981)

Table 3. Continued

Plant species	*A. tumefaciens* strain	Cytokinin[a] levels relative to control *trans*-zeatin riboside	isopentenyladenosine	Reference
Hyssopus officinalis L.	B6	16.4 ×	1.2 ×	Weiler and Spanier (1981)
Leonurus cardiaca L.	B6	1.4 ×	1.0 ×	Weiler and Spanier (1981)
Ocimum basilicum L.	B6	−0.9 ×	1.5 ×	Weiler and Spanier (1981)
Calendula officinalis L.	B6	59.5 ×	4.3 ×	Weiler and Spanier (1981)
Zea may L.	—	0	1.4 ×	Weiler and Spanier (1981)
N. tabacum L. cv. Wisconsin-38	A6—2	230 ×	nd	Amasino and Miller (1982)
N. tabacum L. cv. Wisconsin-38	B6—2	23 ×	nd	Amasino and Miller (1982)
N. tabacum L. cv. Wisconsin-38	542	14 ×	nd	Amasino and Miller (1982)
N. tabacum L. cv. Wisconsin-38	T37	24 ×	nd	Amasino and Miller (1982)
N. tabacum L. cv. Wisconsin-38	C58	14 ×	nd	Amasino and Miller (1982)
N. tabacum L. cv. Wisconsin-38	EU6	27 ×	nd	Amasino and Miller (1982)
N. tabacum L. cv. Wisconsin-38	A6—1	200 ×	nd	Amasino and Miller (1982)
N. tabacum L. cv. Wisconsin-38	B6—1	150 ×	nd	Amasino and Miller (1982)
N. tabacum L. cv. Wisconsin-38	A6—3	170 ×	nd	Amasino and Miller (1982)
N. tabacum L. cv. Xanthi-nc	A6NC	49.5 ×	1.2 ×	Akiyoshi *et al.* (1983)

the formation of shoots in cultured crown gall tissues of tobacco (Skoog and Miller, 1957; Sacristan and Melchers, 1969).

The predominant cytokinin in plants is 6-(4-hydroxy-3-methyl-but-2-enylamino)-9-β-D-ribofuranosylpurine (Hall *et al.*, 1967), which is assigned the trivial name zeatin riboside and abbreviated as io^{6}A. As shown in Table 3, zeatin riboside levels are generally higher in crown gall tumor tissues than in comparable untransformed tissues. Because cytokinin levels and IAA levels are high in crown gall tissue cultures that originated from cloned tissue lines, it can be argued that the crown gall cell is the site of synthesis of these growth factors and does not represent a "sink" or storage site where cytokinins and IAA have accumulated as might be the case with crown galls on plants. Thus, the high levels of free zeatin riboside and also of glucosyl ribosylzeatin (Peterson and Miller, 1977; Morris, 1977) in crown gall tumor cells can be explained as a result of the interaction of the Ti plasmid with the plant cell and the intergration of a portion of the plasmid (T-DNA) in the nuclear DNA. Because the T-DNA appears to be inserted in multiple sites in the plant genome (Lemmers *et al.*, 1980; Thomashow *et al.*, 1980a) and in some cases as tandem repeats (Lemmers *et al.*, 1980), and because there are no unique sites of T-DNA integration in plant DNA (Thomashow *et al.*, 1980a, 1980b; Yadav *et al.*, 1980), the possible mechanism that the T-DNA abolishes a host genetic regulatory element responsible for cytokinin synthesis appears the least plausible explanation for the elevated levels of cytokinin. Instead, it appears more likely that the T-DNA itself either contains genetic determinants for regulating cellular cytokinin concentrations, or encodes cytokinin synthesis directly.

Like IAA, cytokinin, when sprayed onto the inoculated plant in the form of kinetin (6-furfurylaminopurine), was able to restore full virulence of a mutant containing a transposon insertion in the T-DNA of pTiAch5 (Ooms *et al.*, 1981). The mutant did not respond to IAA treatment. The lesion in the T-DNA mapped approximately 4.6 kb downstream from the region that responds to IAA. These results were supported by the direct analyses of crown-gall tumors induced by T-DNA mutants of *A. tumefaciens* A348 (Akiyoshi *et al.*, 1983). Tn5 insertions in the root-promoting region greatly lowered *trans*-ribosylzeatin content in the tumors, whereas mutations in the shoot-promoting region greatly enhanced *trans*-ribosylzeatin content. Beiderbeck (1973) showed that kinetin promoted the growth of tumorlike proliferations on *Kalanchoë daigremontiana* leaves that were inoculated with *A. rhizogenes*, a pathogen which usually causes root proliferation. All of the evidence suggests that crown-gall tumor morphology is tightly regulated by cytokinins. Although further analyses are required to define the limits of the regions for IAA and cytokinin regulation, the evidence seems clear that separate loci in the T-DNA are involved. The view that the Ti plasmid encodes for the synthesis of cytokinins is supported by the fact that the Ti plasmid will substantially increase isopentenyladenine production when introduced into *Rhizobium leguminosarum* (Wang *et al.*, 1982b).

D. Cytokinin Synthesis by A. tumefaciens

A. tumefaciens strains have been analyzed for the production of cytokinins. Cytokinin activity was first observed in culture extracts of virulent strains A6 and B6 but not in an avirulent mutant M39 (Romanow *et al.*, 1969). The cytokinin in the culture filtrate of strain B6 was identified as 6-(methyl-2-butenylamino)purine (2iP) (Upper *et al.*, 1970). Analysis of culture extracts of the same strain of *A. tumefaciens* showed that N^6-(Δ^2-isopentenyl)adenine was also present (Hahn *et al.*, 1976). Messens and Claeys (1978) measured the cytokinins 6-(3-methyl-2-butenylamino)purine (i^6Ade) and its 4-hydroxy derivatives *cis*- and *trans*-zeatin (*c*-io^{6}Ade and *t*-io^{6}Ade) in culture extracts. Interestingly, *trans*-zeatin was detected only in wild-type virulent strains C58 and Ach5 but not in their Ti plasmid-free counterparts C58C1 and Ach5C1. These observations were supported by the analyses of culture extracts of virulent strain C58 and its plasmid-free derivative NT1 (McCloskey *et al.*, 1980). Both *cis*- and *trans*-zeatin ribosides were identified in only the culture extract of strain C58. A second cytokinin N^6-isopentenyladenine was identified in culture extracts of strain C58 (Sonoki *et al.*, 1981). Both *cis*- and *trans*-zeatin ribosides were found in the tRNA of strain IIBNV6, which contained 6-(methylbut-2-enylamino)-9-β-D-riborfuranosylpurine (iPA) and 6-(3-methylbut-2-enylamino)-2-methylthio-9-β-D-ribofuranosyl purine (ms-iPA) (Chapman *et al.*, 1976). Cherayil and Lipsett (1977) similarly found 2-methylthio-N^6-(4-hydroxy-3-methylbut-2-enyl)adenosine (ms^2io^6A) in the tRNA of *A. tumefaciens* 15955. Chapman, *et al.* (1976) estimated that *trans*-zeatin comprised about 30% of the total cytokinin-active nucleosides in IIBNV6 tRNA. In strain C58 this level was much lower (5—10% of the total) and a fourth unidentified active component was present. Further analysis of strain C58 showed that the culture filtrate itself contained zeatin, of which 85% was *trans*-zeatin, 15% *cis*-zeatin, and 2-methylthio-ribosylzeatin (Kaiss-Chapman and Morris, 1977).

E. A Mechanism for the Role of IAA and Cytokinin in Crown Gall Formation

The transformation of the normal plant cell to a neoplastic crown gall cell involves the interaction between *A. tumefaciens* and target cells in a wounded site. Sometime during this interaction, ranging from a few hours to a few days, either a portion or the entire Ti plasmid is transferred into these target cells. When isolated from the crown gall tumor these cells are autotrophic for IAA and cytokinin and contain only the T-DNA sequences in their nuclei. Because the *iaa*P gene is located in the *vir*-region of the pTiC58 plasmid and because a mutation in this gene renders the bacterium avirulent (Liu *et al.*, 1982), it is reasonable to consider that the *vir*-region directs IAA production as an essential component of virulence maintenance function. If this is the case, then it seems difficult to explain the hyperauxinic and elevated cytokinin state of these tumor cells. However, the model of Liu *et al.* (1982) which suggests that more than merely the T-DNA of the Ti plasmid is transferred to the target cells might be applicable. The *vir*-region containing the *iaa*P gene functions in the plant to produce IAA to abnormally high levels. High levels of IAA may be sufficient to induce the expression of IAA production genes normally repressed in the plant cell. Once the plant IAA production genes are switched-on they stably maintain production of IAA, the concentrations of which can vary with the age and growth stage of the tumor (Nakajima *et al.*, 1981). Although the mechanism for the persistent production of IAA in crown gall cells remains to be understood, there is the possibility that auxin autotrophy is the result of positive feedback loops in which IAA either triggers its own biosynthesis or inhibits its degradation. In fact, Meins (1975) has observed a positive-feedback relationship between the treatment of crown gall teratoma cells at 35° C and subsequent changes in tissue phenotype, namely that tissues treated without added IAA became less autotrophic and tissues treated with IAA became more autotrophic. This is supported by the experiments of Syono and Furuya (1974) who showed that tobacco calli can be converted to auxin autotrophy by exposure to either low or high IAA or NAA concentration depending on the cell line used. Also Cheng (1972) has shown that IAA induces its own production in cultured explants of pith from a *Nicotiana glauca* X *langsdorfii* hybrid. A parallel positive-feedback loop mechanism appears to be operating for cytokinin production in habituated tobacco cells (Meins and Binns, 1978; Meins *et al.* 1980). Thus, the need for the *iaa*P and possibly cytokinin production genes in the *Vir*-region of the Ti plasmid is transitory once the endogenous IAA production mechanism is switched-on. Hence, the Ti plasmid segments containing these hormone genes are naturally eliminated during the DNA processing of the transferred Ti plasmid and the T-DNA represents the vestigial detectable element that remains responsive to endogenous IAA and cytokinins in established tissue culture lines.

III. Olive Knot

The disease known as olive knot, observed already in biblical times, is characterized by the formation of tumor-like growths in olive (*Olea europaea* L.), oleander (*Nerium oleander* L.) and privet (*Liqustrum vulgare* L.) (Savastano, 1886). The causal agent of olive knot is the gram-negative rod-shaped bacterium *Pseudomonas savastanoi* (or *P. syringae pv. savastanoi*). Like *A. tumefaciens,* wounds on the host plant are primary sites of infection. Beltra (1959) reported that *P. savastanoi* produced IAA when grown in a medium containing tryptophan and that higher amounts of IAA were present in tumor tissues than in normal counterparts. That this pathogen produces IAA in the presence of tryptophan was confirmed by Wilson and Magie (1963). The biosynthetic pathway for the conversion of tryptophan to IAA is different from that of *A. tumefaciens.* Indoleacetamide seems to be an intermediate in the pathway leading to IAA (Magie *et al.,* 1963) and this contention was verified by the isolation of an oxidative carboxylase (Kosuge, *et al.,* 1966). Mutants of *P. savastanoi,* which lacked tryptophan oxidative carboxylase activity, failed to accumulate IAA in the culture medium. These mutants were unable to incite gall formation on oleander; and since they grew as well as IAA-producing strains in culture medium, the loss of tumorigenicity was not attributed to a lack of bacterial growth at the inoculation site (Smidt and Kosuge, 1978).

The oleander strain of *P. savastanoi* 2009 harbors four plasmids of which one, a 34 mdal plasmid, was cured by acridine orange treatment.

Derivatives lacking this plasmid were defective in IAA production (Comai and Kosuge, 1980). That this plasmid, designated pIAA1, contained the genetic determinants for IAA production was demonstrated by transferring pIAA1 to competent IAA defective mutants by co-transformation with pRSF1010 and selecting for pRSF1010 markers such as streptomycin resistance. Analysis of a transformant showed that it had regained IAA production. Concomitantly two enzyme activities of the IAA synthesis pathway, tryptophan 2-monooxygenase and indoleacetamide hydrolase were restored. A 2.75 kb *Eco*RI restriction fragment containing the tryptophan 2-monooxygenase locus of pIAA1 was cloned into *E. coli* using pRSF1010 as the cloning vector (Comai and Kosuge, 1982). Interestingly, tryptophan 2-monooxygenase activity was efficiently expressed in *E. coli* SK1592 and thus permitted further genetic analysis of this locus (e.g., transposon mutagenesis). It should be noted that since the screening of clones was carried out using the tryptophan analogue, α-methyl tryptophan, a possible explanation for the expression in the heterologous background is that a new "promoter" was created. Such transcription-initiating signals have appeared in *Salmonella typhimurium* mutants derived from 5-methyltryptophan selection (Callahan and Balbinder, 1970). Further analyses of a number of *P. savastanoi* strains revealed that although these strains harbored plasmids, their plasmids contained no DNA sequences that were homologous to pIAA1 (Comai *et al.*, 1982). Only a 73 kb plasmid present in two strains showed homology. Comai *et al.* (1982) have observed that all virulent oleander strains examined bare the IAA genes on a plasmid. On the other hand, all olive and privet strains studied do not contain such plasmid genes, thus implicating a chromosomal origin of analogous genes that function in IAA production. Although not yet examined, the possibility that one or more IAA genes are flanked by chromosomal elements, IS elements, or resolvase sites (res) would allow the interchange of such genes between resident plasmids and the chromosome. An additional possibility that megaplasmids play a role has not been resolved.

Whether of plasmid or chromosomal origin, it is evident that the precise role of the IAA genes on pIAA plasmids in the virulence of *P. savastanoi* remains to be elucidated.

Besides IAA production, *P. savastanoi* also seems to induce cytokinins since cytokinin-like activity has been detected in culture extracts (Surico *et al.*, 1975). The production of IAA and cytokinin by this organism would indeed provide an explanation for the manifestation of tumorous growths during colonization of wound sites on the host plant.

So far, there is no evidence that a genetic determinant of *P. savastanoi* is introduced into the plant host cell as is the case with *A. tumefaciens*. Southern blot hybridization analyses showed no homologies between the T-DNA, *vir*-region DNA, and pIAA plasmid DNA (S. T Liu *et al.*, unpublished). Both organisms invade the plant host tissues and cause hypertrophy, hyperplasia and distorted differentiations. Nevertheless, *P. savastanoi* is very well adapted for survival in the galls.

IV. Fasciation and Witches' Broom

The growth distorting disease known as fasciation and witches' broom is caused by *Corynebacterium fascians*. This disease attacks sweet peas and other plants and is characterized by release of apical dominance and profuse outgrowths of lateral buds which often give the appearance of a witches' broom. Leaves are smaller than normal and in later stages of the disease the internodes of the shoots and sometimes also those of the main stem become abnormally thick. These symptoms can be mimicked by treatment of young plants with kinetin alone or in combination with auxins, suggesting that *C. fascians* may induce cytokinins (Samuels, 1961; Thimann and Sachs, 1966). Indeed, a number of cytokinins have been identified in the culture extracts of *C. fascians*. They are: 6-(3-methyl-2-butenylamino)purine (i^6Ade), 6-(4-hydroxy-3-methyl-*cis*-2-butenylamino)purine (*c*io^{6}Ade), 6-(4-hydroxy-3-methyl*cis*-2-butenylamino)-2-methylthiopurine (ms^2*c*-io^{6}Ade), 6-(4-hydroxy3-methyl-*trans*-2-butenylamino)purine (*t*-io^6 Ade), 6-(4-hydroxy-3-methyl2-butenylamino)-9-β-D-ribofuranosylpurine (io^{6}A) and 6-(3-methyl-2-butenylamino)-9-β-D-ribofuranosylpurine (i^6A) (Klämbt *et al.,* 1966; Helgeson and Leonard, 1966; Scarbrough *et al.*, 1973; Armstrong *et al.,* 1976; Rathbone and Hall, 1972). *C. fascians* also contains cytokinins such as ribosyl-*cis*-zeatin as a moiety of tRNA (Matsubara *et al.*, 1968; Einset and Skoog, 1977; Murai *et al.,* 1980). Analysis of virulent and avirulent strains of *C. fascians* might yield information on the possible correlation of cytokinin production with virulence. Such a study was conducted by Murai *et al.* (1980) who observed that avirulent strains and strains attenuated in virulence released very little cytokinin in the culture medium and contained somewhat lower cytokinin-active ribonucleosides in their tRNA. In contrast, the virulent strain expressed copious amounts of detectable cytokinin activity. Examination of four pathogenic strains revealed the presence of plasmids of approximately 100 megadaltons. A weakly pathogenic strain (Cfl5) contained only a small plasmid, which was also present in a moderately pathogenic strain (Cfl). The avirulent strain (Cf16) contained no detectable plasmid. These results suggest that a plasmid may be involved in virulence expression. However, many different plant pathogenic *Corynebacterium* species harbor one or more plasmids ranging from 23 to 77 megadaltons (Gross *et al.,* 1979). Examination of three *C. fascians* strains showed that they contained a plasmid of 77 megadaltons. Recently, the studies of Lawson *et al.* (1982) on the plasmid content of ten different *C. fascians* strains, which included the same strains used by Murai *et al.* (1980), showed that a 78 megadalton plasmid was harbored in all strains, agreeing with the results of Gross *et al.* (1979). Three of the strains that were examined were avirulent and three others displayed intermediate virulence. The rapid plasmid detection procedure used by Lawson *et al.* (1982) easily resolves plasmids of 100 megadaltons (Kado and Liu, 1981). Lawson *et al.* (1982) thus concluded that the large plasmid observed by Murai *et al.* (1980) was probably the same as those observed by them and by Gross *et al.* (1979). Since the 78 megadalton plasmid is harbored in both virulent and avirulent strains of *C. fascians*, and since no other plasmids were detected (unless megaplasmids of > 500 megadaltons are present), the genetic determinants for conferring virulence on this organism seem to occupy a chromosomal location. It is clear that detailed genetic analyses are necessary to understand the basis of virulence of *C. fascians*. The use of mapping by transposon and deletion mutagenesis would be reasonable if a plasmid role in virulence of *C. fascians* is clearly demonstrated. For chromosomal analyses these procedures might be too tedious.

V. Concluding Remarks

The most notable of host-parasite relationships are the plant diseases characterized by unusual growth. *A. tumefaciens, A. rhizogenes, P. savastanoi*, and *C. fascians* are prime examples of plant pathogens that divert the normal growth differentiating machinery of the host plant to unique and often bizarre phenotypes at and near the site of infection. Because differentiation in plants involves phytohormones, of which IAA and cytokinins are most prominent, it has been the natural strategy to explore the possibility that these pathogens themselves produce these growth regulators. Indeed, this seems to be the case for these bacteria. That some of the genetic determinants responsible for the production of these phytohormones are encoded on plasmids is a fortunate circumstance, for it has greatly increased our understanding of their role in pathogenesis. Clearly, one of the main contributing factors in pathogenesis is the ability of the pathogenic bacterium to synthesize and induce production of growth hormones. It is apparent that these bacteria require this capacity for successful survival in particular hosts. It is also clear that more than these growth hormones are necessary for infection. For successful incessant colonization *A. tumefaciens* makes use of unusual amino acids, organic acids and phosphorylated sugar compounds, collectively known as 'opines', and of enzymes for phytohormones that are synthesized under the direction of T-DNA genes. The high levels of IAA and probably cytokinins synthesized by *P. savastanoi* may be a means to divert food substances into tumorous cells or a signal to the tumorous cells themselves to serve as the food releasing reservoir. The use of cytokinins by *C. fascians* for its survival in nature can be explained in similar terms. In all of these cases the importance of the phytohormone requirement is clear. The original source of the genes that encode these growth substances remains unresolved. That these genes are of plant origin is debatable. The possibility that these genes in the plant may be of microbial origin also exists. Whatever the case may be, detailed studies of these genes and their relation to growth abnormalities on plants will certainly lead to an understanding of the molecular basis of phytohormone action.

Acknowledgements

This work was supported in part by grant MV-102 from the American Cancer Society and NIH grant CA-11526 from the National Cancer Institute, DHHS.

VI. References

Akiyoshi, D. E., Morris, R. O., Hinz, R., Mischke, B. S., Kosuge, T., Garfinkel, D. J., Gordon, M. P., Nester, E. W., 1983: Cytokinin/auxin balance in crown gall tumors is regulated by specific loci in the T-DNA. Proc. Nat. Acad. Sci. U. S. A. **80,** 407—411.

Amasino, R. M., Miller, C. O., 1982: Hormonal control of tobacco crown gall tumor morphology. Plant Physiol. **69,** 389—392.

Atsumi, S., 1980: Relation between auxin autotrophy and tryptophan content in sunflower crown gall cells in culture. Plant Cell Physiol. **21,** 1031—1039.

Atsumi, S., Hayashi, T., 1978: The relationship between auxin concentration, auxin protection and auxin destruction in crown gall cells cultured in vitro. Plant Cell Physiol. **19,** 1391—1397.

Armstrong, D. J., Scarbrough, E., Skoog, F., Cole, D. L., Leonard, N. J., 1976: Cytokinins in *Corynebacterium fascians* cultures. Isolation and identification of 6-(4-hydroxy-3-methyl-*cis*-2-butenylamino)-2-methylthiopurine. Plant Physiol. **58,** 749—752.

Beardsley, R. E., 1972: The inception phase in the crown gall disease. In: Braun, A. C. (ed.), Progress in Experimental Tumor Research: Plant Tumor Research. Vol. **15,** 1—75. Basel - München - Paris - London - New York - Sydney: S. Karger.

Beiderbeck, R., 1973: Wurzelinduktion an Blättern von *Kalanchoë daigremontiana* durch *Agrobacterium rhizogenes* und der Einfluß von Kinetin auf diesen Prozeß. Z. Pflanzenphysiol. **68,** 460—467.

Beltra, R., 1959: Bacterial origin of β-indoleacetic acid in vegetative tumors. Rev. latinoam. Microbiol. **2,** 23—32.

Beltra, R., Sanchez-Serrano, J. J., Serrada, J., 1978: Relationship between plasmids and plant tumorigenesis. Proc. 4th Int. Conf. Plant Path. Bact. (Angers), Station de Pathologie Végétale of Phytobactériologie (ed.), I. N. R. A., pp. 199—205.

Berthelot, A., Amoureaux, G., 1938: Sur la formation d'acide indol-3-acétique dans l'action de *Bacterium tumefaciens* sur le tryptophane. Compt. Rend. **206,** 537—540.

Bevan, M. W., Chilton, M.-D., 1982: Multiple transcripts of T-DNA detected in nopaline crown gall tumors. J. Mol. Appl. Genet. **1,** 539—546.

Binns, A. N., Sciaky, D., Wood, H. N., 1982: Variation in hormone autonomy and regenerative potential of cells transformed by strain A66 of *Agrobacterium tumefaciens.* Cell **31,** 605—612.

Bouillenne, C., Gaspar, T., 1970: Auxin catabolism and inhibitors in normal and crown gall tissues of *Impatiens balsamina.* Can. J. Bot. **48,** 1150—1163.

Braun, A. C., 1956: The activation of two growth-substance systems accompanying the conversion of normal to tumor cells in crown gall. Cancer Res. **16,** 53—56.

Braun, A. C., 1958: A physiological basis for autonomous growth of the crown-gall tumor cell. Proc. Nat. Acad. Sci. U. S. A. **44,** 344—349.

Braun, A. C., Laskaris, T., 1942: Tumor formation by attenuated crown gall bacteria in the presence of growth promoting substances. Proc. Nat. Acad. Sci. U. S. A. **28,** 468—477.

Brown, N., Gardner, F. E., 1936: Galls produced by plant hormones, including a hormone extracted from *Bacterium tumefaciens.* Phytopath. **26,** 708—713.

Callahan, R., Balbinder, E., 1970: Tryptophan operon: structural gene mutation creating a "promoter" and leading to 5-methyltryptophan dependence. Science **168,** 1586—1589.

Chapman, R. W., Morris, R. O., Zaerr, J. B., 1976: Occurrence of trans-ribosyl-zeatin in *Agrobacterium tumefaciens* tRNA. Nature **262,** 153—154.

Cherayil, J. D., Lipsett, M. N., 1977: Zeatin ribonucleosides in the transfer ribonucleic acid of *Rhizobium leguminosarum, Agrobacterium tumefaciens, Corynebacterium fascians* and *Erwinia amylovora.* J. Bacteriol. **131,** 741—744.

Cheng, T. Y., 1972: Induction of indoleacetic acid synthetase in tobacco pith explants. Plant Physiol. **50,** 723—727.

Chirek, Z., 1979: Comparison of auxin activity in tumorus and normal callus cultures from sunflower *Helianthus annuus* and tobacco *Nicotiana tabacum* plants. Acta Soc. Bot. Pol. **48,** 47—54.

Comai, L., Surico, G., Kosuge, T., 1982: Relation of plasmid DNA to indoleacetic acid production in different strains of *Pseudomonas syringae* pv. savastanoi. J. Gen. Microbiol. **128,** 2157—2163.

Comai, L., Kosuge, T., 1982: Cloning and characterization of *iaa*M, a virulence determinant of *Pseudomonas savastanoi.* J. Bacteriol. **149,** 40—46.

Comai, L., Kosuge, T., 1980: Involvement of plasmid deoxyribonucleic acid in indoleacetic acid synthesis in *Pseudomonas savastanoi.* J. Bacteriol. **143,** 950—957.

Dame, F., 1938: *Pseudomonas tumefaciens* (Sm. et Towns.) Stev., Der Erreger des Wurzelkropfes in seiner Beziehung zur Wirtspflanze. Zentralblatt für Bakt. Parasit. Infektionskrankheit. II. Abt. **98,** 385—429.

De Greve, H., Decraemer, H., Seurinck, J., Van Montagu, M., Schell, J., 1981: The functional organization of the octopine *Agrobacterium tumefaciens* plasmid pTiB6S3. Plasmid **6,** 235—248.

Dye, M. H., Clarke, G., Wain, R. L., 1961: Investigations on the auxins in tomato crown-gall tissue. Proc. Royal Soc. (London) **B155,** 478—492.

Einset, J. W., 1977: Two effects of cytokinin on the auxin requirement of tobacco callus cultures. Plant Physiol. **59,** 45—47.

Einset, J. W., 1980: Cytokinins in tobacco crown gall tumors. Biochem. Biophys. Res. Commun. **93,** 510—515.

Einset, J. W., Skoog, F., 1977: Isolation and identification of ribosyl-*cis*-zeatin from transfer RNA of *Corynebacterium fascians.* Biochem. Biophys. Res. Commun. **79,** 1117—1121.

Garfinkel, D., Nester, E. W., 1980: *Agrobacterium tumefaciens* mutants affected in crown gall tumorigenesis and octopine catabolism. J. Bacteriol. **144,** 732—743.

Garfinkel, D. J., Simpson, R. B., Ream, L. W., White, F. F., Gordon, M. P., Nester, E. W., 1981: Genetic analysis of crown gall: fine structure map of the T-DNA by site-directed mutagenesis. Cell **27,** 143—153.

Gelvin, S. B., Gordon, M. P., Nester, E. W., Aronson, A. I., 1981: Transcription of the Agrobacterium Ti plasmid in the bacterium and in crown gall tumors. Plasmid **6,** 17—29.

Gross, D. C., Vidaver, A. K., Keralis, M. B., 1979: Indigenous plasmids from phytopathogenic *Corynebacterium* species. J. Gen. Microbiol. **115,** 479—489.

Hall, R. H., Csonka, L., David, H., McLennan, B., 1967: Cytokinins in the soluble RNA of plant tissues. Science **156,** 69—71.

Hahn, H., Heitmann, I., Blumbach, M., 1976: Cytokinins: production and biogenesis of N^6-(Δ^2-isopentenyl)adenine in cultures of *Agrobacterium tumefaciens* strain B6. Z. Pflanzenphysiol. **79,** 143—153.

Helgeson, J. P., Leonard, N. J., 1966: Cytokinins: identification of compounds isolated from *Corynebacterium fascians.* Proc. Natl. Acad. Sci. U. S. A. **56,** 60—63.

Henderson, J. H. M., Bonner, J., 1952: Auxin metabolism in normal and crown-gall tissue of sunflower. Amer. J. Bot. **39,** 444—451.

Joos, H., Inze, D., Caplan, A., Sormann, M., Van Montagu, M., Schell, J., 1983: Genetic analysis of T-DNA transcripts in nopaline crown galls. Cell **32,** 1057—1067.

Kado, C. I., 1976: The tumor-inducing substance of *Agrobacterium tumefaciens.* Ann. Rev. Phytopath. **14,** 265—308.

Kado, C. I., Liu, S.-T., 1981: Rapid procedure for detection and isolation of large and small plasmids. J. Bacteriol. **145,** 1365—1373.

Kaiss-Chapman, R. W., Morris, R. O., 1977: Trans-zeatin in culture filtrates of *Agrobacterium tumefaciens.* Biochim. Biophys. Res. Commun. **76,** 453—459.

Kaper, J. M., Veldstra, H., 1958: On the metabolism of the tryptophan by *Agrobacterium tumefaciens.* Biochem. Biophys. Acta **30,** 402—420.

Kao, J. C., Perry, K. L., Kado, C. I., 1982: Indoleacetic acid complementation and its relation to host range specifying genes on the Ti plasmid of *Agrobacterium tumefaciens.* Mol. Gen. Genet. **188,** 425—432.

Klämbt, D., Thies, G., Skoog, F., 1966: Isolation of cytokinins from *Corynebacterium fascians.* Proc. Natl. Acad. Sci. U. S. A. **56,** 52—59.

Klein, R. M., Link, G. K. K., 1952: Studies on the metabolism of plant neoplasms, V. auxin as a promoting agent in the transformation of normal to crown-gall tumor cells. Proc. Natl. Acad. Sci. U. S. A. **38,** 1066—1072.

Kosuge, T., Heskett, M. G., Wilson, E. E., 1966: Microbial synthesis and degradation of indole-3-acetic acid I. The conversion of L-tryptophan to indole-3-acetamide by an enzyme system from *Pseudomonas savastanoi.* J. Biol. Chem. **241,** 3738—3744.

Kulescha, Z., Gautheret, R., 1948: Sur l'élaboration des substances de croissance par 3 types de cultures de tissus de *Scorsonere:* cultures normales, cultures de crown-gall et cultures accoutumées à l'hétéroauxine. Comptes Rendus des Séances de l'Académie des Sciences (Paris) **227,** 292—294.

Lawson, E. N., Gantotti, B. V., Starr, M. P., 1982: A 78-megadalton plasmid occurs in avirulent strains as well as virulent strains of *Corynebacterium fascians.* Curr. Microbiol. **7,** 327—332.

Leemans, J., Shaw, Ch., Deblaere, R., De Greve, H., Hernalsteens, J. P., Maes, M., Van Montagu, M., Schell, J., 1981: Site-specific mutagenesis of *Agrobacterium* Ti plasmids and transfer of genes to plants. J. Mol. Appl. Gent. **1,** 149—164.

Leemans, J., Deblaere, R., Willmitzer, L., De Greve, H., Hernalsteens, J. P., Van Montagu, M., Schell, J., 1982: Genetic identification of functions of TL-DNA transcripts in octopine crown galls. EMBO. J. **1,** 147—152.

Lemmers, M., De Berickeleer, M., Hosters, M., Zambryski, P., Depicker, A., Hernalsteens, J. P., Van Montagu, M., Schell, J., 1980: Internal organization, boundaries and integration of Ti-plasmid DNA in nopaline crown gall tumours. J. Mol. Biol. **144,** 353—376.

Link, G. K. K., Eggers, V., 1941: Hyperauxiny in crown gall of tomato. Bot. Gaz. **103,** 87—106.

Link, G. K. K., Wilcox, H. W., 1937: Tumor production by hormones from *Phytomonas tumefaciens*. Science **86**, 126—127.

Link, G. K. K., Wilcox, H. W., Link, A. S., 1937: Responses of bean and tomato to *Phytomonas tumefaciens, P. tumefaciens* extracts, β-indoleacetic acid, and wounding. Bot. Gaz. **98**, 816—867.

Liu, S.-T., Kado, C. I., 1979: Indoleacetic acid production: a plasmid function of *Agrobacterium tumefaciens*. Biochem. Biophys. Res. Commun. **90**, 171—178.

Liu, S.-T., Perry, K. L., Schardl, C. L., Kado, C. I., 1982: *Agrobacterium* Ti plasmid indoleacetic acid gene is required for crown gall oncogenesis. Proc. Natl. Acad. Sci. U. S. A. **79**, 2812—2816.

Lobanok, E. V., Fomicheva, V. V., Kartel, N. A., 1982a: The synthesis of indolylacetic acid by oncogenic and nononcogenic strains of *Agrobacterium tumefaciens*. Doklady Akad. Nauk BSSR **26**, 565—566.

Lobanok, E. V., Fomicheva, V. V., Chernin, L. S., 1982b: Correlation between tumorigenicity and phytohormonal activity in *A. tumefaciens*. Conf. Metabolic Plasmids, Tallinn, Estonian SSR, USSR 19—23 October 1982 (abstr.), pp. 146—148.

Locke, S. B., Riker, A. J., Duggar, B. M., 1938: Growth substance and the development of crown gall. J. Agr. Res. **57**, 21—39.

Magie, A. R., Wilson, E. E., Kosuge, T., 1963: Indoleacetamide as an intermediate in the synthesis of indoleacetic acid in *Pseudomonas savastanoi*. Science **141**, 1281—1282.

Matsubara, S., Armstrong, D. J., Skoog, F., 1968: Cytokinins in tRNA of *Corynebacterium fascians*. Plant Physiol. **43**, 451—453.

Matsumoto, T., Okunishi, K., Nishida, K., Noguchi, N., 1975: Growth profiles of crown gall cells of tobacco in suspension culture. Agr. Biol. Chem. **39**, 485—490.

Meins, F., Jr., 1975: Temperature-sensitive expression of auxin-autotrophy by crown-gall teratoma cells of tobacco. Planta **122**, 1—9.

Meins, F., Jr., Binns, A. N., 1978: Epigenetic clonal variation in the requirement of plant cells for cytokinins. In: Subtelny, S., Sussex, I. M. (eds.), The clonal basis of development, pp. 185—201. New York - San Francisco - London: Academic Press.

Meins, F., Jr., Lutz, J., Binns, A. N., 1980: Variation in the competence of tobacco pith cells for cytokinin-habituation in culture. Differentiation **16**, 71—75.

Messens, E., Claeys, M., 1978: Ti-plasmid encoded production of cytokinins in culture media of *Agrobacterium tumefaciens*. Proc. 4th Int. Conf. Plant Path. Bact. (Angers). Station de Pathologie Végétale et Phytobactériologie (ed.), I. N. R. A., Beaucouze, Angers, pp. 169—175.

McCloskey, J. A., Hashizume, T., Basile, B., Ohno, Y., Sonoki, S., 1980: Occurrence and levels of *cis*- and *trans*-zeatin ribosides in the culture medium of a virulent strain of *Agrobacterium tumefaciens*. FEBS Lett. **111**, 181—183.

Morris, R. O., 1977: Mass spectroscopic identification of cytokinins. Gluosyl zeatin and glucosyl ribosylzeatin from *Vinca rosea* crown gall. Plant Physiol. **59**, 1029—1033.

Mousdale, D. M. A., 1981: Endogenous indolyl-3-acetic acid and pathogen induced plant growth disorders: distinction between hyperplasia and neoplastic development. Experientia **37**, 972—973.

Mousdale, D. M., 1982: Endogenous IAA and the growth of auxin dependent and auxin autotrophic crown gall plant tissue cultures. Biochem. Physiol. Pflanz. **177**, 9—17.

Murai, N., Skoog, F., Doyle, M. E., Hanson, R. S., 1980: Relationships between cytokinin production presence of plasmids, and fasciation caused by strains of *Corynebacterium fascians*. Proc. Nat. Acad. Sci., U. S. A. **77,** 619—623.

Nakajima, H., Yokota, T., Matsumoto, T., Noguchi, M., Takahashi, N., 1979: Relationship between hormone content and autonomy in various autonomous tobacco *Nicotiana tabacum* cells cultured in suspension. Plant Cell Physiol. **20,** 1489—1500.

Nakajima, H., Yokota, T., Takahashi, N., Matsumoto, T., Noguchi, M., 1981: Changes in endogenous ribosyl-trans zeatin and IAA levels in relation to the proliferation of tobacco *Nicotiana tabacum* cultivar Hicks-2 crown galls cells. Plant Cell Physiol. **22,** 1405—1410.

Ooms, G., Hooykaas, P. J. J., Moolenaar, G., Schilperoort, R. A., 1981: Crown gall plant tumors of abnormal morphology, induced by *Agrobacterium tumefaciens* carry mutated octopine Ti plasmids; analysis of T-DNA functions. Gene **14,** 33—50.

Pengelly, W. L., Meins, F., Jr., 1982: The relationship of IAA content and growth of crown gall tumor tissues of tobacco *Nicotiana tabacum* cultivar Turkish in culture. Differentiation **21,** 27—31.

Peterson, J. B., Miller, C. O., 1976: Cytokinins in *Vinca rosea* L. crown gall tumor tissue as influenced by compounds containing reduced nitrogen. Plant Physiol. **57,** 393—399.

Peterson, J. B., Miller, C. O., 1977: Glucosyl zeatin and glubosyl ribosylzeatin from *Vinca rosea* L. crown gall tissue. Plant Physiol. **59,** 1026—1028.

Rathbone, M. P., Hall, R. H., 1972: Concerning the presence of the cytokinin, N^6-(Δ-isopentenyl)adenine, in cultures of *Corynebacterium fascians*. Planta **103,** 93—102.

Regier, D. A., Morris, R. O., 1982: Secretion of trans-zeatin by *Agrobacterium tumefaciens:* a function determined by the nopaline Ti plasmid. Biochem. Biophys. Res. Commun. **104,** 1560—1566.

Romanow, I., Chalvignac, M. A., Pochon, J., 1969: Recherches sur la production d'une substance cytokinique par *Agrobacterium tumefaciens.* (Smith et Town) Conn. Ann. L'Institute Pasteur **117,** 58—63.

Sacristán, M. D., Melchers, G., 1969: The caryological analysis of plants regenerated from tumorous and other callus cultures of tobacco. Mol. Gen. Genet. **105,** 317—333.

Samuels, R. M., 1961: Bacterial-induced fasciation in *Pisum sativum* var. Alaska. Ph. D. Thesis, Indiana University, Bloomington, 119 pp.

Savastano, L., 1886: Les maladies de l'olivier: hyperplasies et tumeurs. Compt. Rend. Acad. Agr. **103,** 1278.

Scarbrough, E., Armstrong, D. J., Skoog, F., Frihart, C. R., Leonard, N. J., 1973: Isolation of *cis*-zeatin from *Corynebacterium fascians* cultures. Proc. Natl. Acad. Sci. U. S. A. **70,** 3825—3829.

Schaeffer, G. W., Smith, H. H., 1963: Auxin-kinetin interaction in tissue cultures of *Nicotiana* species and tumor conditioned hybrids. Plant Physiol. **38,** 291—297.

Schardl, C. L., Kado, C. I., 1983: A functional map of the nopaline catabolism genes on the Ti plasmid of *Agrobacterium tumefaciens* C58. Mol. Gen. Genet. **191,** 10—16.

Skoog, F., Miller, C. O., 1957: Chemical regulation of growth and organ formation in plant tissues cultured in vitro. Symp. Soc. Exptl. Biol. **11,** 118—131.

Smidt, M., Kosuge, T., 1978: The role of indole-3-acetic acid accumulation by alpha methyl tryptophan-resistant mutants of *Pseudomonas savastanoi* in gall formation on oleanders. Physiol. Pl. Path. **13,** 203—214.

Sembdner, G., Gross, D., Liebisch, H.-W., Schneider, G., 1980: Biosynthesis and metabolism of plant hormones. Encyclop. Pl. Physiol. **9,** 281—444.

Smith, H. H., 1972: Plant genetic tumors. Prog. Exp. Res. **15,** 138—164.

Sonoki, S., Ohno, Y., Sugiyama, T., Iizuka, M., Hashizume, T., 1981: Identification of a cytokinin, N^6-isopentenyladenine, produced by plant pathogenic bacterium, *Agrobacterium tumefaciens.* J. Chem. Soc. Japan: Nippon Kagaku Kaisi **5,** 899—901.

Sukanya, N. K., Vaidyanathan, C. S., 1979: Tryptophan phenyl pyruvate amino transferase of *Agrobacterium tumefaciens:* purification and general properties of the enzyme. J. Indian Inst. Sci. Sect. C. Biol. Sci. **61,** 51—62.

Surico, G., Sparapano, L., Lerario, P., Durbin, R. D., R. D., Iacobellis, N., 1975: Cytokinin-like activity in extracts from culture filtrates of *Pseudomonas savastanoi.* Experientia **31,** 929—930.

Syono, K., Furuya, T., 1974: Induction of auxin-non requiring tobacco calluses and its reversal by treatments with auxins. Plant Cell Physiol. **15,** 7—17.

Tait, R. C., Close, T. J. Hagiya, M., Lundquist, R. C., Kado, C. I., 1982: Construction of cloning vectors from the Inc W plasmid pSa and their use in analysis of crown gall tumor formation. In: Genetic Engineering in Eukaryotes (P. F. Lurquin and A. Kleinhofs, eds.) pp. 111—123. New York: Plenum Publishing Company.

Tegley, J. R., Witham, F. H., Krasnuk, M., 1971: Chromatographic analysis of a cytokinin from tissue cultures of crown-gall. Plant Physiol. **47,** 581—585.

Thomas, J. E., Riker, A. J., 1948: The effects of representative plant growth substances upon attenuated-bacterial substances upon attenuated bacterial crown galls. Phytopath. **38 :** 26. (abstr.).

Thomashow, M. F., Nutter, R. Montoya, A., Gordon, M. P., Nester, E. W., 1980a: Integration and organization of Ti plasmid sequences in crown gall tumors. Cell **19,** 729—739.

Thomashow, M. F., Nutter, R., Postle, K., Chilton, M.-D., Blattner, F. R., Powell, A., Gordon, M. P., Nester, E. W., 1980b: Recombination between higher plant DNA and the Ti plasmid of *Agrobacterium tumefaciens.* Proc. Nat. Acad. Sci. U. S. A. **77,** 6448—6452.

Thimann, K. V., Sachs, T., 1966: The role of cytokinins in the "fasciation" disease caused by *Corynebacterium fascians.* Amer. J. Bot. **53,** 731—739.

Upper, C. D., Helgeson, J. P., Kemp, J. D., Schmidt, C. J., 1970: Gas-liquid chromatographic isolation of cytokinins from natural sources. 6-(3-methyl-2-butenylamino)purine from *Agrobacterium tumefaciens.* Plant Physiol. **45,** 543—547.

Van Montagu, M., Schell, J., 1982: The Ti plasmids of Agrobacterium. Curr. Top. Microbiol. Immunol. **96,** 237—254.

Vyas, K. M., Jain, S. K., 1973: Production of auxins by microorganisms. Hindustan Antibiotics Bull. **16,** 20—21.

Wang, T. L., Wood, E. A., Brewin, N. J., 1982: Growth regulators, *Rhizobium* and nodulation in peas. Indole-3-acetic acid from the culture medium of nodulating and non-nodulating strains of *R. leguminosarum.* Planta **155,** 345—349.

Wang, T. L., Wood, E. A., Brewin, N. J., 1982b: Growth regulators, *Rhizobium* and nodulation in peas. The cytokinin content of a wild-type and a Ti-plasmid-containing strain of *R. leguminosarum.* Planta **155,** 350—355.

Weiler, E. C., 1981: Radioimmunoassay for pmol-quantities of indole-3-acetic acid

for use with highly stable ^{125}I and ^{3}H derivatives as radiotracers. Planta **153,** 319—325.
Weiler, E. W., Spanier, K., 1981: Phytohormones in the formation of crown gall tumors. Planta **153,** 326—337.
Went, F. W., Thimann, K. V., 1937: Phytohormones. New York: Macmillan Company, 294 pp.
Wilson, E. E., Magie, A. R., 1963: Physiological, serological, and pathological evidence that Pseudomonas tonelliana is identical with Pseudomonas savastanoi. Phytopath. **53,** 653—659.
Yadav, N. S., Postle, K., Saiki, R. K., Thomashow, M. F., Chilton, M.-D., 1980: T-DNA of a crown gall teratoma is covalently joined to host plant DNA. Nature (London) **287,** 458—461.

Section IV.
Plant Pathogens and Defence Mechanisms

Chapter 13

Genetic and Biochemical Basis of Virulence in Plant Pathogens

N. J. Panopoulos[1], J. D. Walton[2], and D. K. Willis[1]

[1]Department of Plant Pathology, University of California, Berkeley, Calif., and [2]Department of Plant Breeding, Cornell University, Ithaca, N.Y., U.S.A.

Contents

I. Introduction

Higher plants are routinely exposed to microorganisms, both above and below the ground. Yet, each plant is susceptible to infection by only a handful of them. Indeed, as with animals and microorganisms, susceptibility of plants to infectious agents is an exception in nature, while resistance is the rule. The special properties which enable microorganisms to inflict economically or biologically significant damage on living plants are a main theme of research in plant pathology. Equally interesting, and at least as important in scope, are the properties which make plants susceptible to infection, the biochemical mechanisms by which the cellular damage is sustained, and the nature of plant resistance to microorganisms with a demonstrated pathogenic potential on other plants often within the same narrow taxonomic group (species, genus, etc.).

The potential importance of this basic knowledge to plant protection and disease management is generally recognized. However, with few exceptions, plant-pathogen systems have not gained great appeal among basic cell biologists at large. One important reason for this is the unwelcome complexity of interactive processes that are simultaneously controlled by two different genetic and cellular systems. With the increasing interest in more complex biological phenomena, the new molecular-genetic tools now available, and the hopes of biotechnological exploitation of genome manipulation, greater contributions by scientists from outside the field of plant pathology should be forthcoming.

The biochemistry and genetics of virulence in plant pathogens is a very broad subject. This article, by necessity, is only a partial incursion into a few aspects of microbial pathogenicity, and, as such, can only offer a glimpse of the diversity of biological phenomena and systems, and the research opportunities they provide.

II. Genetic Control of Host-Pathogen Interactions

Pathogenicity or virulence of plant pathogens and susceptibility or resistance of plants to infection are probably controlled by a constellation of genes in each organism. The most thoroughly investigated pathogens, from a genetic point of view, are obligate parasitic fungi of major crop plants where systematic analysis of natural genetic variability in the pathogen and the host led to the foundation of the gene-for-gene hypothesis regarding race/cultivar specificity. Systematic studies in other systems are few by comparison. Molecular genetics is assuming an increasingly important role in the analysis of virulence and pathogenicity in bacteria. By far the most advanced studies thus far have been with the *Agrobacterium tumefaciens* virulence system (Chapters 10, 11), while studies with other bacterial pathogens are following a similar trend. In this and other sections of this chapter, we discuss representative examples of various plant pathogen systems. Due to space limitations, viral pathogens are not covered and the reader is referred to the several comprehensive reviews on the subject (Fraenkel-Conrat and Wagner, 1977).

A. The Gene-for-Gene Hypothesis for Race/Cultivar Specificity

The gene-for-gene hypothesis of cultivar/race interactions was proposed by Flor (1955) as a result of his genetic analysis of the flax *(Linum usitatissimum)*/rust *(Melampsora lini)* system. His analysis revealed 26 single genes for resistance in flax which were counteracted by an equal number of corresponding virulence genes in flax rust. The flax resistance genes occurred as multiple alleles at five loci while the virulence genes of flax rust were non-allelic and defined 26 separate loci. Flor found that resistance was inherited as a dominant trait in the host, while virulence was recessive to avirulence in the pathogen. In diploid genomes, there are nine genetically unique combinations of specific cultivar resistance (*R1*)/susceptibility (*r1*) genes and pathogen race avirulence (*P1*)/virulence (*p1*) genes that result in only two types of phenotypic expressions, i. e. either compatible or incompatible reactions. The incompatible phenotype is specified only by *R/P* allelic pairs and is epistatic to the compatible interaction which other allelic pairs (e.g., *R/p, r/P, r/p*) permit. An important aspect of this genetic relationship is that *R/r* genes in the host cultivar can only be detected by the presence of corresponding race specific *P/p* genes in the pathogen. Conversely, the phenotypic detection of race specific *P/p* genes in the pathogen is only possible in the context of *R/r* genes in the host. Because of epistasis, the recognition of *R* and *P* genes also requires recombinational separation of individual *R/p* gene pairs in cultivar/race combinations.

Gene-for-gene interactions have been demonstrated or suggested from the genetic analysis of many plant/pathogen systems that include fungal, bacterial, viral, insect, and nematode pathogens (Day, 1974; Ellingboe, 1976; Flor, 1971; Keen, 1982; Mills and Gonzalez, 1982; Sequeira, 1979; Sidhu, 1975; Van der Plank, 1982).

The genetics of race/cultivar interactions have been used to construct several contrasting models concerning the basis of specificity. One view, shared by Van der Plank (1982), suggests that specificity in race/cultivar interactions resides in the compatible reaction. However, a more prevailing view (e. g., Ellingboe, 1976) considers the dominance of host resistance and pathogen avirulence as an indication that specificity lies in the incompatible reaction. In this view, compatibility in race/cultivar systems is a nonspecific reaction and a mere consequence of the lack or failure of incompatibility. This concept finds experimental support from the analysis of temperature sensitive host/pathogen interactions, the multiple allelism within *R* genes, and mutational changes affecting race specificity. Changes from an incompatible to a compatible reaction as a result of temperature shifts from low to high are common (e. g., Ellingboe, 1976; Holliday *et al.*, 1981; Vanderplank, 1982). When these can be attributed to individual gene pairs, only the *R/P* (not *R/p, r/P* or *r/p*) pair shows temperature sensitivity. This suggests that incompatibility rather than compatibility is the specifically determined response. In allelic series sytems only one allele at the host *R* locus leads to incompatibility against a pathogen race carrying a *P* gene at a functionally correspondent locus. Other alleles at the *R* locus permit a compatible reaction to develop with the same pathogen strains, although each specifies incompatibility against strains carrying corresponding dominant alleles at different *P* loci (Ellingboe, 1976; Moseman, 1966). There is good evidence for true allelism rather than pseudo-allelism in some cases, although this cannot be said for most. Finally, mutations to virulence and susceptibility attributable to $P \rightarrow p$ and $R \rightarrow r$ type transitions, respectively, are frequent while the converse ($p \rightarrow P$, $R \rightarrow r$) has not been reported (Gabriel *et al.*, 1982; Kirosawa, 1982; McIntosh, 1977; Simmons, 1979). Direct support for the above concept has been provided by the cloning of a P gene from *Pseudomonas syringae* pv. *glycinea* (Staskawicz, Dahlbeck, Keen, in preparation).

The gene-for-gene hypothesis predicts the patterns of interaction between *R* and *P* gene alleles in a simple manner, but offers limited insight into the biochemical and physiological mechanisms determining specificity. Nevertheless, it provides a basis for building provisional biochemical models or excluding others. For example, the apparent residence of specificity in the incompatible reaction has led to the hypothesis that race specific avirulence and cultivar specific resistance involve as a key recognitional step the formation of a heterodimer between the primary products of functionally correspondent *P* and *R* genes which results in an incompatible interaction (Ellingboe, 1982). Keen (1982), on the other hand, postulates that the *R* genes encode or determine structural components of specific receptors, probably proteins or glycoproteins, located on the plant cell surface; the *P* genes are similarly thought to encode specific glycosyltransferases that catalyze the addition of unique sugar residues to extracellular or surface-bound pathogen glycoproteins which interact with the plant receptors leading to induction of phytoalexin biosynthetic pathways. Recognition between such elicitors and their receptors may also be a key step in the activation of other host defense systems, such as the hyparsensitive reaction (see section VI), which restrict pathogen growth and development. Finally, elicitor-suppressor models (Bushnell and Rowell, 1981; Heath, 1981 a, 1981 b) propose that specificity is determined by the production by the parasite of race- or species-specific suppressors that prevent the pathogen elicitors from triggering resistant responses in the host.

The above models are based on the genetic analysis of the best understood and most genetically malleable plant/parasite interactions. Exceptions to both the dominance of cultivar resistance/race avirulence and the one-to-one correspondece of *R* and *P* genes have been described. A single *R* gene, *Rp3*, in dent corn (*Zea mays*) apparently conditions resistance to two races, 901 aba and 933 a, of the rust fungus, *Puccinia sorghi* (Hooker and Saxena, 1967, 1971); however, only in the *Rp3/Rp3* homozygote is this resistance fully expressed. The heterozygote *Rp3/rp3* is resistant to race 901 aba but sensitive to 933 a, and the homozygote *rp3/rp3* is sensitive to both races. The authors investigated the possibility that two closely linked genes were responsible for the resistant phenotype of the *Rp3* gene. However, in a test cross involving 4800 progeny, no segregation of the dominant and recessive phenotypes was detected (Hooker and Saxena, 1971). It was concluded that if *Rp3* consisted of two genes, they were linked by less than 0.02 map units. The authors also suggested the alternative hypothesis, namely a gene dosage effect, to explain the reaction differential. Although two copies of the *Rp3* gene give resistance to both *P. sorghi* races, a single copy of *Rp3* does not provide enough gene product to establish a resistance threshold with respect to race 933 a, the more aggressive of the two races.

Another example of two *P* genes interacting with a single *R* gene has been described in the flax/flax rust system (Lawrence *et al.*, 1981). Two allelic gene pairs in *M. lini* consisting, of a gene pair for avirulence/virulence (designated *A/a*) and an inhibitor gene pair (*I/i*), were found to interact in determining pathogenicity on cultivars of *L. usitatissimum* carrying the *M1, L1, L7,* or *L10* resistance genes. When the inhibitor gene was present as a homozygous recessive (*i/i*), then *A*-gene specific avirulence was dominant (i. e., the genotypes *ii Aa* and *ii AA* give incompatible interactions) on hosts containing the *M1* resistance gene. However, in the presence of a dominant inhibitor allele (*I/I* or *I/i*), this avirulence was not expressed. A similar relationship with different inhibitor alleles was detected with *P* gene pairs specific for the *L1, L7,* or *L10* resistance genes.

Recessive genes for resistance to disease are found in many host/parasite interactions. For example, such single gene resistance is found in rice toward bacterial blight (Mew and Khush, 1981), in legumes to several fungal and viral pathogens (Meiners, 1981), in barley to powdery mildew (Jorgensen, 1971), and toward nematodes (Bingefors, 1971). Three single recessive genes provide resistance to stem rust of oats in a manner that is suggestive of a gene-for-gene interaction (McKenzie *et al.*, 1971). In certain proposed models of gene-for-gene interactions, these deviations from the most common genetic pattern of cultivar/race specificity have been interpreted as gene dosage effects leading to incomplete dominance of *R* genes or the presence of inhibitor genes (i.e., regulatory genes) which influence the expression of *R* alleles.

In addition to these apparent exceptions to Flor's gene-for-gene concept, other factors concerning host-pathogen incompatibility shoud be noted. Resistance to disease is not an all-or-none phenomenon; disease reactions comprise a gradient of symptoms which are often arbitrarily classified as either resistant (incompatible) or susceptible (compatible). In the flax/flax rust system, Flor (1971) recognized five infection types, of which types 0, 1, and 2 he considered resistant, while types 3 and 4 he considered compatible. Daly (1976) considers this classification scheme an oversimplification and suggests that such consistent differences in the degree of incompatibility need to be addressed in any model concerning the biochemistry of gene-for-gene interactions.

The apparent deviations from the basic tenants of the gene-for-gene concept only underscore the need for the isolation of *R* and *P* gene products in order to gain an understanding of the interactions of these products at a molecular level. The date, the products of single *R* and *P* genes have not been identified. It is hoped that the application of the tools of molecular biology will soon lead to the cloning of *P* genes from other bacterial phytopathogens which are the most amenable to such manipulations, as well as from fungi. This will allow the genetic analysis of the dominance/recessiveness of race specificity genes and other aspects of *P-R* gene interactions. It will also permit the identification of specific *P* gene, and possibly *R* gene, products.

B. Genetic Analysis in Other Systems

The work of Ellingboe and Gabriel (1977), Gabriel and Ellingboe (1979), and Gabriel *et al.* (1982) illustrates a classical mutational analysis which has been applied to many fungal and bacterial pathogens. They used chemical mutagenesis and obtained conditional (temperature sensitive) as well as nonconditional mutants in three different fungi. Three classes of mutants were obtained in *Colletotrichum lindemuthianum,* a facultative saprophyte with no known perfect stage; those that were temperature sensitive (*ts*) with respect to both growth at 28 C in fully supplemented agar media (therefore, not *ts* auxotrophs) and production of expanding lesions in the host which were thus presumed to be affected in basic growth or ontogenic functions; those that were *ts* for growth as above, but not for lesion formation in the host; and those that were *ts* for lesion development but not for growth *in vitro* and which appeared, therefore, to be affected in genes whose functions are necessary for parasitism/pathogenesis. In *Phyllostricta maydis,* another imperfect fungus, two other types of mutants were obtained: those which were *ts* for growth on agar media, but cold-sensitive (*cs*) for lesion formation (i. e., formed lesions at 28 but not at 22 C); and those which grew normally on agar media at both temperatures and produced lesions only at 28 C but not at 22 C. Assuming that the *ts* and *cs* phenotypes result from thermal inactivation and "configurational freezing", respectively, of primary gene products, these mutants can be presumed to be affected in genes whose products are required for an incompatible relationship, similar to race/cultivar specificity genes.

In the obligate parasite *Erysiphe graminis* f. sp. *tritici,* mutants were obtained which gave infection type 4 (virulent) on host lines carrying single dominant resistance *Pm1a* or *Pm4a,* but retained the incompatible parental reaction type on host lines carrying other *Pm* genes. Normally, an infection type 0 (avirulence) is seen with the parental strain of the fungus. The results were consistent with the hypothesis that *Pm1a/P1a* and *Pm4a/P4a* host-pathogen gene pairs specified incombatibility. Another class of mutants were temperature sensitive, i. e., infected plants normally at 20 C but failed to do so at 25 C. Their infection type (at 20 C) on host ines with *Pm* genes were unaffected; they were thought to be affected in genes crucial to fungal growth and development.

Several considerations emerge from these studies. First, a variety of mutant types, including some unexpected ones, can be obtained and rationalized. Second, in obligate parasites, nonconditional mutants affected in genes needed for parasitism will be lethal and are therefore impossible to study at present. Third, studies of this type are the only ones possible in imperfect fungi unless a parasexual cycle or nonconventional genetic tools are available.

Mutations to resistance in barley induced by X-rays illustrate several other aspects of genetic control of host-pathogen interactions. Single mutations conferring resistance to *Erysiphe graminis* f. sp. *hordei* were obtained, all mapping at a single locus (*m1-o*) on chromosome no. 4 of barley. This location is distinct from other loci conferring natural mildew resistance, which map either at a different site on the same chromosome or at several loci on chromosome no. 5 (Jorgensen, 1974, 1977). The *ml-o* mutations conferred resistance against all strains of powdery mildew and were genetically recessive, unlike the majority of naturally occurring mildew resistance genes which are genetically dominant or semi-dominant (Jorgensen, 1971). The wild type allele at this locus specifies race non-specific susceptibility. Its products and, by implication, those of the corresponding gene(s) in the pathogen, may be required for the establishment of basic compatibility between host and parasite (basic susceptibility and basic pathogenicity, respectively). At present, there is no information as to what steps in the infection process these types of genes control and what their products might be.

The *ml-o* mutants form spontaneous chlorotic/necrotic flecking ("autogenic necrosis") on the leaves reminiscent of disease lesion mimics in other plants (Gracen, 1982; Marchetti *et al.*, 1983; Walbot *et al.*, 1983). The often striking similarity of such symptoms has led to speculation about underlying similarities in mechanisms. In one case (Langford, 1948), the expression of a recessive gene for autogenic necrosis was dependent on the presence of a dominant gene for disease resistance. The biochemical basis of disease mimic symptoms is as yet unexplored. Although there is no general assurance that disease mimic syndromes share common steps and components with disease syndromes, there is also no a priori reason to doubt such possibilities.

Pseudomonas solanacearum, a vascular pathogen of solanaceous plants, provides an example where biochemical and genetic investigations into the virulence system and other aspects of bacteria-plant interaction have been extensively investigated. It is discussed here in some detail to illustrate the complexity of such analysis, and some recent progress through molecular genetics.

Kelman (1954) described a type of spontaneous mutation in this bacterium which leads to loss of extracellular polysaccharide (EPS) production (nonfluidal colony morphology on tetrazolium media), loss of virulence and flagella function (Kelman and Hruschka, 1973), inability to produce polygalacturonase and pectate lyase, and decreased levels of cellulase (Cx) and several other polysaccharidases (Ofuya and Wood, 1981 a). Fluctuation analysis estimates indicate a mutation frequency of 10^{-5} to 10^{-6} per cell (Staskawicz, pers. commun.), and reversion with respect to colony fluidity has not been observed. One report (Ofuya and Wood, 1981 b), which needs to be confirmed, claims restoration of the polysaccharidase activities in one mutant following transformation with plasmid DNA from the wild type parent. However, repeated attempts to associate plasmid loss with the Kelman variation have been unsuccessful (Currier and Morgan, 1981; Zischek *et al.,* 1981).

Another class of avirulent mutants, designated Acr, different from the type described by Kelman (1954) were obtained by Message *et al.* (1979) after selection for resistance to acridine orange. These mutants retained their fluidal colony phenotype (EPS^+) but showed other pleiotropic differences from their wild type parent: loss of virulence towards tomato; inability to induce hypersensitive reaction (HR) on tobacco; loss of ability of heat-killed cells to protect against HR induced on tobacco by the wild type strain (they still protected against HR caused by an avirulent derivative described as "rough" with respect to its lipopolysaccharide); methionine requirement; and production of a brownish pigment. Avirulence on tomato may have been due to methionine auxotrophy (this was not tested), as Met^- mutants of another strain of this bacterium were avirulent or partially virulent on this host, altough fully virulent in tobacco (Coplin *et al.*, 1974). However, methionine auxotrophy apparently was not the reason for their HR^- phenotype.

Whatley *et al.* (1980) described two groups of strains among avirulent EPS^- variants of strain K60. Group I mutants failed to induce HR on tobacco and had "rough" type (incomplete) lipopolysaccharide (LPS) lacking "O" type antigens normally characteristic of bacteria with "smooth" (complete) LPS, while group II mutants were rough with respect to LPS composition and induced HR on tobacco. Several other mutants obtained recently (Sequeira, pers. commun.) also have rough LPS and were unable to induce HR. LPS by itself does not induce HR in this system but protects against HR induced by avirulent derivatives (Graham *et al.*, 1977). Both "rough" and "smooth" LPS have this ability. Although the relationship between components and recognition mechanisms involved in HR induction and in protection against HR is not entirely clear. The working hypothesis is that LPS may be necessary for initial host-bacterium recognition in a process that leads to induction of HR. The O antigen of strains with smooth LPS, while not required for protection activity, may mask the portion of LPS (lipid A) believed to be recognized by a host component, presumably lectin(s) (Sequeira and Graham, 1977).

Staskawicz *et al.* (1983 and pers. commun.) recently obtained nonfluidal avirulent mutants by transposon (Tn*5*) mutagenesis. These mutants represent a genetically better defined class although they have not yet been biochemically characterized. Two insertions were detected in these mutants: one, by IS*50* (the transposable terminal repeat of Tn*5*), located in a 1.5 kb target fragment common in all mutants; plus an additional Tn*5* insertion located in a larger restriction fragment, which differed in size for each mutant. The Tn*5* insertions were apparently unrelated to the nonfluidal phenotype observed, as apparently precise excision of IS*50* in one mutant without detectable change in the Tn*5* target fragment restored both fluidal colony appearance and virulence (wilting). The 1.5 kb fragment was apparently chromosomal, as it did not hybridize with plasmids of *P. solanacearum* strains (Boucher, pers. commun. to B. J. Staskawicz). Site-directed transposon mutagenesis and marker exchange (homogenotization) experiments (Staskawicz, pers. commun.) further proved that a gene(s) on this fragment controls virulence and fluidal colony type (EPS production?). Other evidence (Boucher, pers. commun; cited in Staskawicz *et al.*, 1983) suggests that several other loci may also control these properties.

The unique aspects of the *Agrobacterium tumefaciens* virulence system are discussed elsewhere in this volume (Chapter 10, 11). Here we discuss only some data pertinent to host specificity, as this pathogen provides the opportunity to investigate the biochemical genetics of host range across wide taxonomic boundaries.

The host range of *A. tumefaciens* among dicotyledonous hosts is primarily a property of the pTi plasmids (Kao *et al.*, 1982; Knauf *et al.*, 1982; Loper and Kado, 1979; Thomashow *et al.*, 1980). In the majority of cases, chromosomally isogenic strains that differ in their pTi plasmids express different host ranges and pairs of different strains carrying the same pTi plasmid have the same host range. In one recent study (Knauf *et al.*, 1982), pTi^+ transconjugants of an avirulent (pTi^-) derivative of the broad host range wild type strain C-58 displayed five different host specificity patterns, which were characteristic of the natural isolates from which the pTi's originated. Certain deletion and insertion mutations that map on pTi lead to host range alterations (Holsters *et al.*, 1980; Kao *et al.*, 1982; Matzke and Chilton, 1981). However, host range appears also to be controlled by non-pTi, presumably chromosomal genes (Garfinkel and Nester, 1980; Hamada and Farrand, 1980; Knauf *et al.*, 1982). Furthermore, plant target systems also determine host-species as well as host-tissue specificity of *Agrobacterium*. For instance, certain mutants, designated *tmr* because they induce root-forming tumors on Kalanchoe stems, were avirulent on Kalanchoe leaves (F. White, pers. commun.), while others, designated *hsp* (Kao *et al.*, 1982), were tumorigenic on some hosts but non-tumorigenic on others. Host specificity mutations on pTi plasmid map both within as well as outside the T-DNA region, ca. 75—77 kb to the right of T-DNA (Kao *et al.*, 1982; see also Chapter 11). Some mutants carrying a small (1 kb) deletion within T-DNA were complemented for virulence on tomato by indoleacetic or naphthalenacetic acid. As the authors suggested, the T-DNA region may encode IAA-dependent function(s) for virulence that may also be linked to host specificity. Restoration of these properties may result from IAA-dependent expression of T-region genes in certain plant cell types but not in others. This suggestion was based on the ability of IAA to induce the synthesis of proteins in *E. coli* minicells carrying a cloned T-DNA fragment that contained the region deleted in the hormone-conditional mutants (Tait *et al.*, 1982). It is not certain whether expression was mediated by promoters on the fragment or the vector. It is interesting to note that IAA can substitute for cAMP in the expression of certain CAP-dependent operons in *E. coli* (Kline *et al.*, 1980).

One as yet unexplained aspect of crown gall infection, important to the genetic manipulation of plant genomes is the inability of *A. tumefaciens* to infect monocotyledonous hosts which include plants of major agronomic importance. The mutational alterations of host specificity reported thus far represent reduction of host range. We are not aware of any successful host range extension, and there are no obvious clues at present whether such extension is possible.

Epiphytic bacteria, such as *Pseudomonas syringae* and *Erwinia herbicola* are incitants of frost injury to major agricultural crops (Lindow *et al.*, 1978). This is due to the ice nucleating activity (Ina) of these bacteria, i. e., the ability to catalyze the liquid-to-solid state transition of water molecules at temperatures below 0 C. A role for Ina in infection by *P. s. syringae* has also been suggested based on circumstantial evidence. Frost injury has often been reported to be a predisposing factor for infection of some plants (e. g., pear, apricot, peach) and is either required for or favors disease development. For instance, in spite of the fact that significant populations of bacteria are present on plant surfaces, disease often occurs only after the ambient temperature drops to or below freezing (Panagopoulos and Crosse, 1964). Presumably, localized frost injury causes the release of nutrients from damaged cells and water congestion of tissues which facilitate bacterial multiplication and access to internal tissues. Thus, Ina may be considered a frost-conditional virulence factor in this bacterium.

Little is known about the nature of the substances responsible for Ina. The property is associated with the bacterial outer membrane and one or more proteins appear to be associated directly or indirectly with its expression (Sprang and Lindow, 1981). The genes responsible for Ina were recently cloned (Orser *et al.*, 1982; Orser *et al.*, submitted). They are clustered in a small DNA fragment (ca. 4—5 kb), apparently on the chromosome. The cloned fragments conferred Ina to *E. coli* and restored the property in Ina$^-$ mutants of the DNA source strain. These studies should facilitate the identification of the primary product of the genes involved and the study of other biochemical aspects of Ina. Mutants deleted in the *ina* genes are also attractive as potential biocontrol agents of frost damage because they can be expected to be genetically stable (nonreverting) Ina$^-$ and to be free of silent mutations, often induced by chemical mutagens, which may adversely affect competitive fitness.

The biochemical genetics of another virulence system that is well understood is *Pseudomonas syringae* pv. *savastoni* where a plant growth hormone is required for virulence. The cloning of *iaaM,* a gene encoding tryptophane monooxygenase (Comai and Kosuge, 1982; Comai *et al.*, 1982) and other aspects of growth hormone production and virulence are covered in Chapter 12 of this book.

III. Toxins as Virulence Factors

The idea that pathogenic organisms produce chemicals with the ability to perturb the normal functioning of their hosts dates back to the earliest days of research on diseases. Microorganisms, including plant pathogenic bacteria and fungi, produce a vast variety of chemicals with biological activity against plants and other organisms.

Many toxins produced by plant pathogens have been studied for their possible roles in pathogenesis. Phytotoxins have never been implicated as necessary pathogenicity factors in any diseases caused by obligate parasites such as the rusts and powdery mildews. This follows logically from the consideration that successful obligate parasites must keep their host cells alive in order to survive and reproduce. However, it has been proposed that resistance against an obligate parasite may be due to hypersensitive response of the host to a "toxin" produced by the pathogen (Jones and Deverall, 1978; Litzenberger, 1949).

There are two groups of phytotoxins: those that are nonspecific (i. e., have host ranges that are wider than those of the pathogens that produce them) and those that are host-specific with respect to plant species, variety, or genotype, like their producer pathogens. Nonspecific phytotoxins often also affect animals and other microorganisms (see Stoessl, 1981; and Mitchell, 1981). Antimicrobial activity has sometimes been exploited in developing sensitive microbiological assays for these toxins and in the identification of their target systems (Backman and DeVay, 1971; Owens *et al.*, 1968; Staskawicz and Panopoulos, 1979).

Some toxins are neither precisely host-specific nor totally nonspecific. Tentoxin from *Alternaria tenuis* inhibits photophosphorylation in certain plants scattered throughout the plant kingdom (Durbin and Uchytill, 1977). The toxin from *Alternaria altenata* f. sp. *lycopersici* deviates from the strict correspondence beetween sensitivity to toxin and susceptibility to the fungus required for a toxin to be called host-specific in an interesting way: whereas resistance in tomato (*Lycopersicon esculentum*) to the pathogen is controlled by a single genetic locus and is completely dominant to susceptibility, heterozygous tomato plants have intermediate sensitivity to the toxin (Gilchrist and Grogan, 1976).

In several cases, nonspecific toxins can reproduce some of the characteristic symptoms of the particular disease, e. g., wilting caused by both *Fusicoccum amygdali* and fusicoccin (Ballio *et al.*, 1964), chlorosis caused by both *Pseudomonas syringae* pv. *phaseolicola* and phaseolotoxin (Staskawicz and Panopoulos, 1979), or chlorosis induced by the fungus *Alternaria tenuis* and its toxin, tentoxin (Fulton *et al.*, 1965). However, only few nonspecific toxins have actually been shown to contribute to pathogenesis by the producing organism in nature. In several cases, a lack of correlation between an organism's ability to produce a particular nonspecific toxin and its ability to be a pathogen has been shown (Holenstein, 1982; Kinoshita *et al.*, 1972; and see Durbin and Steele, 1979). When nonspecific toxins are involved in pathogenicity at all, and are not just fortuitous secondary metabolites, they appear to contribute to increased virulence of the pathogen (see Panopoulos and Staskawicz, 1981), perhaps enabling it to colonize host tissue more readily or to extend its host or tissue range. It has recently been proposed that the nonspecific toxin phaseolotoxin, from *Pseudomonas phaseolicola,* acts not only as a classical chlorotic toxin against the host, *Phaseolus vulgaris,* but also as an inhibitor of hypersensitive resistance (Ganamanickam and Patil, 1977), similar to the *Phytophthora infestans* race-specific suppressors of hypersensitive resistance in potato (Doke *et al.*, 1979).

A host-specific toxin can easily account for the specificity of the host-pathogen interaction at the molecular level. All of the approximately fifteen host-specific toxins known to date are synthesized by fungi, mainly by species and races in the genera *Helminthosporium* (now known as *Bipolaris*) and *Alternaria.* The importance of host-specific toxins in pathogenesis has been well documented mainly by genetic studies. Isolates or mutants of *Bipolaris* species that lack ability to produce their respective host-specific toxins are avirulent or only weakly pathogenic; in genetic crosses of the fungi, pathogenicity or increased virulence always segregates with toxin production; and in genetic crosses of the host, plant sensitivity to the toxin segregates with susceptibility to the fungus.

A distinction has been made between toxins that are pathogenicity factors (ones that dictate the qualitative state of disease or no disease) or that are virulence factors (ones that contribute to the quantitative level of disease). Host-specific toxins have been considered to be pathogenicity factors by some and virulence factors by others (e.g., see Yoder, 1980, 1983; Wheeler, 1981). HMT toxin of *B. maydis* race T is clearly a virulence factor since *B. maydis* race O, which does not produce HMT-toxin, is still a pathogen but exhibits no increased virulence on maize with Texas male-sterile cytoplasm. Even victorin, the classic example of a "primary determinant of disease", might be more properly called a virulence rather than a pathogenicity factor since the fungus is capable of infecting plants that are insensitive to victorin (Meehan and Murphy, 1946). The host-specific toxins produced by *Alternaria* have been considered to be virulence factors that allow the normally weakly pathogenic *Alternaria* species to inflict more severe damage on plants that are sensitive to the particular toxin produced (Nishimura *et al.*, 1979).

What is the advantage to a fungus of producting a host-specific instead of a nonspecific toxin? It would appear that insofar as the host-range of a host-specific toxin and its producing fungus are identical, if the fungus could produce a toxin with a wider host range its range as a pathogen might also be greater. It is possible that host-specific toxins are flukes of nature and of our own efforts to artificially alter the genotypes of our crop plants, and perhaps nonspecific toxins will eventually be seen as more universal molecular agents of virulence. Another possibility is that host-specific toxins are less likely to be lethal or deleterious to the producing organism than nonspecific toxins and, therefore, less likely to require that the producer organism co-evolves or already possesses a toxin immunity system (Panopoulos and Staskawicz, 1981). An interesting example of producer organism immunity to its own toxin is the apparent duplication of the target enzyme (ornithine carbomoyltransferase) in the bacterial pathogen *P. s. phaseolicola* whose phytotoxin is a universal inhibitor of the enzyme from bacterial, plant, and mammalian cells. One of the two forms of this enzyme is insensitive to the toxin and its production is regulated by temperature in the same way as production of the toxin (Staskawicz *et al.*, 1980). A different mechanism for insensitivity to tabtoxin in *P. s. tabaci* has been suggested (Durbin, 1982). In this case, neither the tabtoxins nor tabtoxinine beta-lactam, the active moiety against glutamine synthetase, are taken up by the producer bacterium.

Knowledge of the chemical structures of toxins is important for at least two reasons. First, it can give clues about their modes of action and about possible mechanisms of host resistance and susceptibility (e. g., Walton and Earle, 1983). Second, it is essential for elucidating the biochemical pathways by which pathogens synthesize toxins, which is necessary for studies on biochemical genetics of toxin production and on the evolution of vilurence factors in nature. The evidence to date indicates that neither phytotoxicity nor hostspecificity are restricted to a particular class of chemicals, and therefore that bacteria and fungi that produce toxins are not characterized by specialization in any particular biochemical pathway.

The structures of hundreds of nonspecific fungal and three bacterial phytotoxins are known (see Mitchell, 1981; Stoessel, 1981, for most recent reviews). There has been a good deal of research on structural aspects of host-specific toxins in the last few years, but still only a few such toxins are known. The host-specific phytotoxins from *Bipolaris maydis,* called HMT-toxin, are a family of linear C_{37} to C_{43} polyketides (see Kono *et al.,* 1981). *Bipolaris sacchari* synthesizes HS-toxin, also known as helminthosporoside, a sequiterpine with four beta-galactofuranose residues attached (Macko *et al.,* 1982). The structure of HC-toxin from *Bipolaris zeicola (Helminthosporium carbonum)* race 1 was proposed, corrected, and confirmed by the work of several laboratories in 1982. It is a cyclic tetrapeptide containing L- and D-alanine, D-proline, and an unusual alpha-amino acid containing a terminal epoxide (Kawai *et al.,* 1983; Walton *et al.,* 1982). The structure of victorin from *B. victoriae* is still unknown, although at least three laboratories are now working on it. The most recent structural work suggests that it is a terpenoid-peptide (Pringle and Braun, 1958). *Alternaria mali* produces a group of three closely related cyclic tetradepsipeptide toxins (Ueno *et al.,* 1975). The host-specific toxins produced by *Alternaria alternata* f. sp. *lycopersici* have been recently identified as esters of 1,2,3-propane carboxylic acid and 1-aminodimethylheptadecapentol (Bottini and Gilchrist, 1981; Bottini *et al.*, 1981). AM-toxins I and II (see Rich, 1981) and HC toxin (D. R. Rich *et al.*, unpublished) have been synthesized.

Several genetic studies with the various species of phytopathogenic *Bipolaris* (perfect stage *Cochliobolus*) host-specific toxins have concluded that host-specific toxin production is essential for pathogenicity because toxin production and pathogenicity always segregated together in sexual crosses (Lim and Hooker, 1971; Scheffer *et al.*, 1967; Yoder and Gracen, 1975). These and other studies have also come to the surprising conclusion that production of host-specific toxins in *Bipolaris* is controlled by single but different Mendelian genes. Progeny from crosses between *B. victoriae* and *B. zeicola* (*C. carbonum*) race 1 segregated 1:1:1:1 for production of just victorin or just HC-toxin, neither toxin, or both toxins (Scheffer *et al.*, 1967). Crosses between races O and T of *B. maydis* (*C. heterostrophus*) have suggested but usually been unable to conclusively demonstrate the involvement of more than one gene in HMT-toxin synthesis (Lim and Hooker, 1971; Panopoulos and Staskawicz, 1981; Yoder and Gracen, 1975). Intraspecific crosses among isolates of *B. zeicola* that did (= race 1) or did not (= race 2) produce HC-toxin showed 1:1 segregation for toxin production (Nelson and Ullstrup, 1961; Scheffer *et al.*, 1967).

These results could have several explanations. First, there may have been insufficient numbers of crosses made and analyzed. In some experiments only one isolate of each race or species was used as a parent. The crosses beteen *B. victoriae* and *B. zeicola* race 1 which indicated that control of production of the toxins was under the control of single but different genetic loci was based on an analysis of less than 150 progeny from three crosses using a total of three parents (Scheffer *et al.*, 1967). These experiments are difficult because interspecific crosses within *Bipolaris* are usually sterile. However, many hundreds of crosses have been made between *B. maydis* race T (which produces HMT-toxin) and *B. maydis* race O (which does not produce HMT-toxin) (Lim and Hooker, 1971; Yoder, 1980; Yoder and Gracen, 1975). The conclusion that only gene controlled pathogenicity (later discovered to be due to HC-toxin production) in *B. zeicola* race 1 was based on analysis of 973 ascospore isolates from crosses involving 22 different parents (Nelson and Ullstrup, 1961). Second, perhaps all other genes involved in toxin biosynthesis are critical for survival of the fungus, even when grown *in vitro,* and thus mutations in any gene except the one committed to toxin production are lethal. This explanation assumes that host-specific toxins differ by only a single enzymatic step from some other fungal product or intermediate. It has been suggested that perhaps HMT-toxin is an intermediate in some as yet unknown biochemical pathway (Yoder, 1980), but it is difficult to imagine that the cyclic peptide toxins (e. g., HC-toxin, AM-toxin, and tentoxin) are single-step byproducts of some pathway necessary for survival *in vitro.* There are precedents, however, for complex antibiotics being completely synthesized by as few as two polypeptides. Gramicidin S, a cyclic decapeptide, is synthesized by *Bacillus brevis* from L-amino acids by a multienzyme composed of only two subunits each of which is apparently a single polypeptide. One subunit racemizes L-phenylalanine to D-phenylalanine, while the other larger subunit synthesizes and cyclizes the decapeptide (Koischwitz and Kleinkauf, 1976; Vater and Kleinkauf, 1976).

The question of whether toxins may play a role outside the pathogen-plant interactive phase, such as in normal cellular metabolism or in the saprophytic phase is largely a speculative subject. As mentioned earlier in this section, nonspecific toxins often inhibit microorganisms, although few are universal antibiotics. Antimicrobial action may help the pathogens minimize competition by soil-resident or plant-associated microflora (Durbin, 1982). There is virually no direct evidence that toxins play a role in cellular metabolism of the producer organisms. However, it is interesting in this respect that indole-acetic acid switches on protein synthesis in *E. coli* minicells carrying a cloned fragment of T-DNA from *A. tumefaciens* (Tait *et al.,* 1982) and that IAA and certain other indoles will substitute for cAMP in the induction of certain catabolite sensitive operons in *E. coli* (Kline *et al.,* 1980). Production of IAA is fairly widespread among plant pathogens, but its role in disease is established only in *Pseudomonas syringae* pv. *savastanoi* and *A. tumefaciens* (see Chapter 12 of this book). One wonders whether the hormone plays a similar role in other diseases or in other aspects of pathogen biology unrelated to disease.

Research on the biosynthesis of chemical virulence factors such as host-specific and nonspecific toxins could potentially tell us a great deal about the ability of plant pathogenic fungi to adapt to new ecological niches created by our agronomic practices. In particular, studies on *in vitro* biosynthesis of host-specific toxins may tell us why most host-specific toxins are made by species of *Alternaria* and *Bipolaris*. Do these fungi possess an extraordinary ability to synthesize all types of secondary metabolites, which are then screened by natural selection? Do the genes controlling their secondary metabolite pathways have an unusually high mutation rate? Or can these fungi somehow obtain from other fungi by legitimate or illegitimate genetic means the ability to synthesize particular toxins?

Bipolaris and *Pseudomonas* would be preferred organisms for studies on *in vitro* biosynthesis of toxins. One reason is that they can be manipulated genetically and cloning approaches are either available or within reach. Naturally-occurring nontoxigenic (Tox^-) isolates are frequently reported in these organisms (e.g., Leach *et al.*, 1982; see Panopoulos and Staskawicz, 1981) and chemical or transposon-mediated mutagenesis has successfully yielded Tox^- mutants in *P. s. syringae, P. s. phaseolicola,* and *P. s. tabaci* (Durbin, 1982; Lam and Strobel, pers. commun.; Panopoulos and Staskawicz, 1981). The study of extraribosomal peptide biosynthesis in bacteria and fungi (e. g., Yurioka and Winnick, 1966) also provides good guidance in biochemical studies of peptide toxins.

Genetic approaches could help answer the important question of how toxigenic traits originate and evolve, what the constraints of such evolution may be and whether toxin biosynthesis is the result of a newly acquired enzymatic capability or of an acquired block in a normal pathway. There are many examples of biochemical inermediates that are normally found at low concentrations in cells and become toxic when regulation of their biosynthesis is disrupted. Studies with induced heterokaryons between toxin producing and nonproducing isolates of *Bipolaris maydis* race T indicate that toxin production is a dominant trait (Leach *et al.*, 1982). This is consistent with HMT-toxin being an end-product of a biochemical pathway rather than an intermediate which accumulates only when the normal biochemical pathway is blocked. From consideration of chemical structures, genetic dominance of toxin production seems likely for the peptide toxins, but not necessarily for the terpenoid composite toxins such as HS-toxin or victorin.

Claims that accessory genetic elements (e. g., bacterial plasmids) are involved in toxigenesis have not been substantiated. The only well-established examples of plasmid-controlled production of low molecular weight virulence factors are those involving plant growth hormones in *P. s. savastanoi* and *A. tumefaciens* (see Panopoulos and Staskawicz, 1981, and Chapter 12 of this book).

IV. Cutinases and Pectinases

The cell well system of plant cells constitutes a complex network of homo- and heteropolymers, each with distinct monomer composition, physical properties and chemical linkages. Whereas it is an exceedingly complex enzymological substrate and physical barrier to penetration, fungal pathogens penetrate through the entire system repeatedly during colonization of the host tissues. Fungal hyphae and bacteria growing in the conductive elements of the vascular system or intercellularly between parenchyma cells also cause selective dissolution of cell wall polymers. Combinations of these actions are characteristic of many diseases.

The role of various cell wall degrading enzymes in these events has been extensively investigated. Since the subject has been reviewed extensively (Bateman and Millar, 1966; Bateman and Basham, 1976; Collmer *et al.*, 1982; Cooper, 1977; Kolattukudy, 1980a, 1980b, 1981), only certain aspects regarding cutinases and pectinases are briefly discussed here.

Cutin, a polyester of hydroxy and epoxy fatty acids, is a main constituent of the cuticle, which forms the outermost layer of the aerial surfaces of plants (Holloway, 1982a, 1982b). Many plant pathogenic fungi and a streptomycete (*S. scabies*) produce cutinolytic enzymes as extracellular glycoproteins (3.5 to 6% carbohydrate), 22,000 to 26,000 daltons in molecular weight, with active serine residues (serine esterases) and similar amino acid compositions (see Kolattukudy, 1980a, 1980b). Indirect evidence from histopathological (McKeen, 1974), enzymological and other studies (reviewed by Van Den Ende and Linskens, 1974), and more direct evidence from studies by Kolattukudy's group support the notion that cutinolytic enzymes are required for fungal penetration. The cutinases of *Fusarium solani* f. sp. *pisi,* a cortical pathogen on pea, and *Colletotrichum gloeosprodioides,* a pathogen on papaya fruit, are immunologically detectable at infectionsites on the host surface (Dickman *et al.*, 1982; Shaykh *et al.*, 1977). Antibodies prepared against purified cutinase and chemical inhibitors of the enzyme prevented infection of (nonwounded) tissues. Supplementation of a cutinase-deficient mutant of *F. solani* f. sp. *pisi* with cutinase, along with pectolytic enzymes for which it was also deficient, restored virulence levels to those observed in wounded stems where the cuticle barrier presumably was broken (Koller *et al.*, 1982). It appears, therefore, that cutinase is a significant virulence factor in these fungi. The gene for cutinase was recently cloned from this fungus (Kolattukudy, pers. commun.).

Pectic substances interconnect the plant cell wall hemicellulose polymers in which cellulose microfibrils are enmeshed. They are also the major constituent of the middle lamella, the region which joins adjacent cells (Darvil *et al.*, 1980; Keegstra *et al.*, 1973). Dissolution of this layer is the basis for the tissue maceration characteristic of soft rot and other diseases.

The pectolytic enzymes found among plant pathogenic fungi and bacteria include one or more of the following types (McMillan and Sherman, 1974; Rombouts and Pilnik, 1972): pectin methylesterase (PME), catalyzing methoxylation; polygalacturonases (PG) and transeliminases (lysases, PL), which cleave the α-1,4 glycosidic linkages of pectic polymers by hydrolysis and beta-elimination, respectively; and oligo-as well as digalacturonic acid lyases (OGL, DGL) which are intracellular enzymes and degrade pectic oligomers to their constituent monomer. Exo-acting pectin methylesterases or pectin lyases have not been described (Rombouts and Pilnik, 1972). Molecular weights between 32,000 and 42,000 daltons have been reported for several fungal and bacterial endopectic enzymes (Brookhouser, 1974; Cervone *et al.*, 1977, 1978; Gardner and Kado, 1976; Hancock, 1976; Marciano *et al.*, 1982) and 65,000 to 67,000 for a polygalacturonase of *Fusarium roseum* "Avenaceum" (Mullen and Bateman, 1975). A model for the degradation and assimilation of host pectic substances by the combined action of various components of the pectic enzyme complex in *Erwinia carotovora* has been proposed (Stack *et al.*, 1980).

In both fungi and bacteria multiple isoenzymic forms, especially for endopectic enzymes, are common, often within a given strain (e. g., Hancock, 1976; Marciano *et al.*, 1982; Pupillo *et al.*, 1976; Quantik *et al.*, 1983; Stack *et al.*, 1980). These are distinguished by their isoelectric points, molecular weights, or both. The relationship between these isoenzymes is not clearly understood. In one instance (Cervone *et al.*, 1978), size polymorphism of polygalacturonase isoforms was attributed to protein-linked carbohydrate residues. Both intra- and extracellular forms may be polymorphic with respect to pI (Stack *et al.*, 1980). Nothing is known about the genetic basis of these polymorphisms.

Most of our knowledge regarding pectic enzymes as virulence components comes from enzymological studies, with a smaller but increasing contribution from genetic analysis (Bateman and Basham, 1976; Collmer *et al.*, 1982). The weight of experimental data supports the involvement of these enzymes in disease, particularly in soft rots (Chatterjee and Starr, 1977; Collmer *et al.*, 1982). However, the role of individual components of the pectic enzyme complex is unsettled, even for soft-rot causing pathogens. Given enzymes or isoenzyme forms are often produced in culture but not *in planta,* or vice versa, and the biochemical or physiological parameters determining the production spectra *in vitro* and *in planta* are not fully understood. Plant tissues also contain inhibitors which are often specific for individual pectolytic enzymes (Albesheim and Anderson, 1971; Fielding, 1981; Fisher *et al.*, 1973; Gobel and Bock, 1978; Hoffman and Turner, 1982). Although there is no direct evidence that these inhibitors are involved in disease resistance, they, by interferring with enzyme recovery, complicate the demonstration of enzyme production *in planta.*

Enzyme deficient mutants of soft rot bacteria have been isolated through both chemical and molecular (transposon) mutagenesis in several laboratories (Beraha and Garber, 1971; Collmer and Bateman, 1981; Guimaraes, 1977; Thurn *et al.*, 1982). Those which have been characterized appear to be either pleiotropic (Beraha and Garber, 1971), regulatory (Collmer and Bateman, 1981), secretion deficient (Chatterjee, pers. commun.), or defective in cAMP biosynthesis which indirectly controls induction (Chatterjee and Starr, 1977; Mount, *et al.,* 1979). To our knowledge, single structural pectic enzyme gene mutants have not yet been described. The ultimate test of the role of these enzymes will be through the use of structural gene mutants. The successful cloning of polygalacturonic acid transeliminase-encoding genes from *Erwinia Chrysanthemi* (Keen, Dahlbeck, Staskawicz, Belser, in preparation) should assist in this as well as facilitate other studies on pectolytic enzymes.

Pectolytic enzymes have also been implicated in the vascular wilt syndrome. However, physiological evidence on this subject is contradictory (Keen and Erwin, 1971; Mussell, 1973). Studies with pectolytic enzyme deficient mutants of two fungi (Howell, 1976; Mann, 1962; McDonnell, 1958; Puhalla and Howell, 1975) suggest that the enzymes are not essential for vascular colonization although they may be required for the prior penetration of cortical tissues.

The extent to which pectolysis seems to be associated with different diseases also differs greatly. In soft rots tissue maceration is extensive but in other diseases degradation of pectic polymers is limited. Pectolytic and other cell wall degrading enzymes may be needed only for short periods in these cases (e. g., when the pathogen initially penetrates the cells and colonizes the tissues) and may be detrimental thereafter (e. g., in infections by obligate parasites). The recent finding (Northnagel *et al.*, 1983) that a galacturonic acid oligosaccharide from plant cell walls elicits phytoalexin production and that other pectic oligosaccharides may be involved in the induction of hypersensitive cell death (Yamazaki *et al.*, 1983; see Section IV) suggest additional, more intricate roles for these enzymes in disease processes.

V. Extracellular Macromolecules Implicated in Vascular Wilts

The idea that extracellular macromolecules may function as wilt inducing agents was promoted by the demonstration (Hodgson *et al.*, 1949) that synthetic polymers, such as polyethylene glycol and polyvinyl alcohol, fed to tomato cuttings induced wilt symptoms. A number of such macromolecules are capable of causing wilt in *in vitro* assays. For example, plants are susceptible to picomole levels of synthetic dextrans and their molecular weight is an important paramenter in wilt induction (Carpita *et al.*, 1979; Van Alfen, 1982; Van Alfen and Allard-Turner, 1979; Van Alfen and McMillan, 1982). In alfalfa, dextran molecules larger than 7×10^4 daltons in molecular weight reduce xylem conductivity, probably by blocking pit membranes between vessels, while those with lower molecular weight probably accumulate on the walls of leaf mesophyll cells, interferring with water movement from leaf veins into the mesophyl cells. The primary walls of plant cells exclude dextrans with molecular weights of 6.5×10^3 or greater (Carpita *et al.*, 1979). Such macromolecules can accumulate at apoplastic sites of water movement and interfere with normal water transport. Greater quantities may be required in this case than for xylem blockage by larger molecules. Thus, depending on their size, pathogen-produced macromolecules may block different sites in the plant's water conducting system. Associative properties of such molecules may also determine which sites are blocked and the quantities of macromolecule required for wilt induction. Ability to associate, rather than molecular weight *per se,* was considered the critical parameter for wilt induction by amylovorin (Sijam *et al.*, 1982). The unusual associative properties of cerato-ulmin (Takai and Richards, 1978) may also be important in this respect.

At present, the pathological significance of wilt inducing biogenic macromolecules *in planta* is in many cases considered presumptive (Pegg, 1981). One reason for this is the frequent inability to demonstrate their presence in vascular systems. Another is the possibility of artifacts in wilt bioassays (Van Alfen and McMillan, 1982). Furthermore, a plethora of alternative candidate components and mechanisms have been proposed in various wilt systems, particularly in fungi, among them low molecular weight wilt toxins, fungal biomass, enzymes releasing plant pectic polymers into the transpiration stream, auxins, and collapse of xylem parenchyma cells (see Pegg, 1981). However, the relative significance and contribution of these components to wilt have not been fully evaluated in each system.

Van Alfen (1982) provides an interesting discussion of wilt induction by bacterial EPS which illustrates the complexities of analysis in wilt systems. In alfalfa plants infected with *C. michiganeuse* pv. *insidiosum*, disruption of waterflow occurs at petiole junctions and leaflets, the two locations most susceptible to vascular plugging by macromolecules. Bacterial populations in these regions are four orders of magnitude lower than in the plant's crown. Therefore, disruption of waterflow was attributed to EPS accumulating at and plugging the pit membranes between adjacent xylem vessels in these sites rather than to the bacterial cells themselves. Infected xylem vessels are thought to be no longer functional in the transpiration stream due to gas embolisms. EPS produced within these vessels must move into functional vessels to cause water stress. Pit membrane pores are larger in alfalfa stems than in the petiole junctions. In the model proposed by Van Alfen (1982), the three molecular species of the bacterial EPS assume different importance in alfalfa wilt. A species larger than 5×10^6 daltons is too large to pass through pores of any pit membranes and presumably cannot exit from the vessel in which it is produced to play a role in wilt. The second largest (5×10^6 daltons) will pass through pit membrane pores in the stem but will be blocked at the petiole junctions. The third and small molecular species probably passes through both but accumulates in the primary walls of leaf cells, blocking water movement from veins to mesophyll cells.

VI. Hypersensitive Reaction, Phytoalexins, and Inhibitor Detoxification

A common response of platns to avirulent or heterologous pathogens is the so-called hypersensitive reaction (HR). It is typified by localized cell death at the site of infection and differs from the susceptible interaction with respect to time dynamics, being detectable as early as 30 minutes in the case of *Phytophthora infestans* (Tomiyama *et al.*, 1979) or within a few hours. In the case of bacterial pathogens, a need for physical contact between bacteria and plant cells has also been suggested based on the inability of cells embedded in agarose to induce HR (Stahl and Cook, 1979). A further requirement appears to be *de novo* synthesis of protein(s) in the pathogen as well as the host. Thus, heat- or otherwise killed cells do not trigger HR and transcription/translation inhibitors block the reaction if administered early after inoculation (Keen *et al.*, 1981; Klement, 1982; Meadows and Stahl, 1981; Sasser, 1982; Tomiyama *et al.*, 1979). These studies suggest that the pathogen metabolites which trigger the response *in vivo* may need to be synthesized *de novo* following an inductive signal. Alternatively, they may be constitutively present but must be released from a bound form by host or pathogen enzymes which themselves must be synthesized *de novo.* The protein synthesis requirement in the host cells suggests that either a component(s) of the plant target or receptor systems are not constitutively present or that *de novo* synthesis is required for the completion of cell-death specific events subsequent to be initial receptor-signal interaction.

HR is the first recognized visual manifestation of many, although not all resistance reactions. Its role in resistance and host-pathogen specificity have been subjects of continuing dispute (Heath, 1976). For obligate parasites, host cell death provides a reasonable explanation for the arrest of pathogen growth in resistant reactions. This is clearly inadequate for nonobligate pathogens which can still grow on dead cells.

There are some indications that HR is not a single phenomenon and that an earlier, as yet uncharacterized event(s) is the basis for resistance. For example, each host or nonhost resistance response to rusts differs from the others in the timing and the extent of cellular necrosis relative to the time of death of fungal haustoria or of inhibition of pathogen growth (Deverall and McLeod, 1980; Heath, 1976; Littlefield, 1973; Niks and Kuiper, 1983). Furthermore, in one case where host cell necrosis in wheat carrying the *Lr20* race-specific gene could be prevented by a photosystem II inhibitor, resistance to pathogen growth was still expressed (Campbell and Deverall, 1980).

The nature of pathogen-produced HR-triggering metabolites remains elusive. Early claims that such substances could be isolated from bacterial pathogens remain unconfirmed and efforts to obtain such substances from hypersensitively interacting tissues have been unsuccessful (see Sequeira, 1976). Cell wall glucans (Peters *et al.*, 1978), uncharacterized cell wall preparations (Currier, 1981), and a lipid fraction (Kuranz and Zacharius, 1981) from *Phytophthora infestans* have been reported to cause necrosis in potato. Similar suggestions have been made for cell wall glycoproteins from *Cladosporium fulvum* in tomato (DeWit and Kode, 1981), for lipopolysaccharides from *Pseudomonas solanacearum* (Whatley *et al.*, 1980) and for pectic lyase from *Erwinia rubrifaciens* in tobacco (Gardner and Kado, 1976). However, the physiological significance of these findings is uncertain. In the case of *E. rubrifaciens,* HR^- mutants which still produce pectic lyase *in vitro* were recently obtained (Kado, pers. commun.). Furthermore, endogenous elicitors may also be important in HR, as in phytoalexin induction (see below). Yamazaki *et al.* (1983) recently desribed a low molecular weight plant cell wall preparation capable of killing and preventing ^{14}C-leucine incorporation into acid precipitable polymers in sycamore cell suspension cultures.

In a recent paper, DeWit and Spickman (1982) reported that intercellular fluids from tomato leaves undergoing compatible interactions with *Gladosporium fulvum* elicited necrosis and varying degrees of chlorosis in tomato leaves in a manner unusually related to the race-cultivar specific reactions observed in this system. The fluids were inactive on cultivars from which they originated but induced necrosis/chlorosis on cultivars carrying *Cf* gene(s) for resistance but were either absent from the infected cultivar ore present but ineffective against the infecting pathogen race. Only cultivars carrying *Cf* genes specifying an extreme type of hypersensitive resistance responded with significant necrosis to fluid injection. The origin (i.e., pathogen or host) and chemical nature of these "necrosis elicitors" is not yet known, and their relevance to HR is uncertain at this point (e. g., they may be more properly regarded as toxins).

A molecular genetic approach to the study of HR would appear worthwhile at this stage and is entirely within reach in bacterial pathogens. Although serious genetic analysis of the HR-inducing ability has not been undertaken thus far, mutants unable to trigger the reaction (HR^-) have been repeadly obtained in bacteria (Guimaraes, 1977; Kado, pers. commun., Message *et al.*, 1979; Sequeira, pers. commun.; Whatley *et al.*, 1980).

Another important mechanism used by plants to restrict pathogen growth is the accumulation in response to infection of antimicrobial substances, termed phytoalexins. It is now established, although only in a few systems, that accumulation of phytoalexins in and around infection sites is sufficient, with respect to levels (less consistently) and timing, to account for inhibition of pathogen growth (Mansfield, 1982; Yoshikawa *et al.*, 1978 b). Genetic evidence also indicates that the phytoalexin-generating capacity of gamma-ray induced potato mutants correlates with degree of resistance to *Phytophthora infestans* (El-Sayed, 1977) and phytoalexin tolerance levels can correlate with the degree of virulence (see below). In other systems, a singular role of phytoalexins in pathogen growth inhibition is questionable (e.g., Bruegger and Keen, 1979; Holliday and Keen, 1982).

Research on phytoalexins as a disease resistance mechanism has been done mainly with non-obligate host/pathogen systems which show race/cultivar specificity but have not been extensively studied genetically. There is little evidence that phytoalexins have a role in resistance in any disease caused by an obligate parasite or that they *per se* determine race/cultivar specificity in well-established gene-for-gene systems (Keen, 1982). Since phytoalexins are the products of complex enzymatic pathways and, therefore, are presumably controlled by many genes, Ellingboe (1982) has argued that phytoalexin production does not determine specificity in race/cultivar systems which follow a one-to-one gene-for-gene relationship.

A number of physical treatments and chemical agents, both biotic and abiotic in origin, can bring about the induction of the phytoalexin synthesizing systems of plants. As in the case of HR, *de novo* mRNA and protein synthesis is required (Bailey, 1982; Yoshikawa *et al.*, 1978 a). Phytoalexin "elicitors" of pathogen origin include polypeptides, glycoproteins, poly- and obligosaccharides, lipoidal materials, and enzymes with endopeptidase (protease) and endopectic activity (Albersheim and Valent, 1978; Bailey,

1982; Cruickshank and Perrin, 1968; Keen, 1982; Lee and West, 1981). These compounds do not share a common chemistry, although polysaccharide elicitors generally have beta-1,3 glucan linkages with C-4 and C-6 branches. Phytoalexin elicitors of plant origin (endogenous elicitors) have also been described (Hahn *et al.*, 1981; Hargreaves and Bailey, 1978; Hargreaves and Selby, 1978). One of these is a galacturonic acid-rich obligosaccharide originating from plant cell walls (Northnagel *et al.*, 1983) and another is sucrose (Cooksey *et al.*, 1983). It has been suggested that biologically relevant elicitors are of plant origin and are produced or released during infection by the pathogen's cell wall degrading enzymes (Hahn *et al.*, 1981).

There have been several reports of host-specific phytoalexin elicitors activity of various pathogen extracts (Keen, 1975; Keen and Lengrand, 1980). Most elicitors, however, appear to be nonspecific, i. e., regardless of source they elicit phytoalexin production equally well in resistant and susceptible plants (Bailey, 1982; Mansfield, 1982).

The relation between phytoalexin synthesis and accumulation and HR is unclear (Ward and Stoessl, 1976). Although the two processes often occur together (e.g., Doke and Furuichi, 1982; Keen, 1982), activation of the phytoalexin synthesizing system can occur in the absence of cell death, for example, in cell suspension culture systems (Albersheim, 1977; Dixon *et al.*, 1983; Fett and Zaccharius, 1983; Hahlbrock *et al.*, 1981; Paxton, 1977). Furthermore, HR proceeds normally in the presence of inhibitors of phytoalexin biosynthesis such as glyphosate (Holliday and Keen, 1982). Thus, apparently neither of these processes necessarily results in the occurrence of the other.

Successful pathogens must either prevent/suppress the elicitation of HR and other active resistance response or must inactive or otherwise tolerate the presence of inhibitors in the host tissue. Doke and co-workers, among others, have found race-specific suppressors of phytoalexin biosynthesis and HR for example in the *Phytophthora infestans*/potato system (Doke and Furuichi, 1982; Doke and Tomiyama, 1980). Although hyphal cell wall fractions from all races of *P. infestans* could elicit phythioalexin accumulation and cell necrosis in aged potato tuber discs, only a hyphal fraction containing water soluble glucans from virulent *P. infestans* races could prevent phytoalexin accumulation in the corresponding sensitive potato genotype. The ability of extracellular invertase of this fungus to inhibit glyceolin accumulation in soy bean (Ziegler *et al.*, 1982) is interesting in view of the report (Cooksey *et al.*, 1983) that sucrose elicits phytoalexin production in other legumes.

At least twenty fungal, but so far no bacterial, pathogens are known to metabolize phytoalexins, usually to less toxic compounds (Van Etten *et al.*, 1982). The types of conversions which have been biochemically characterized, although rarely tested in cell-free systems, include oxygenation, hydration, carbonyl reduction, and ether cleavage. Both inducible and constitutive degradation systems have been found in different fungi. In some cases, phytoalexin metabolism may be a unique biological role of the enzymes involved, while in others merely a fortuitous function of relatively nonspecific enzymes having some other primary role in the biology of the organism. Nondegradative tolerance mechanisms of an unknown nature are also known. Where a combination of metabolism and tolerance is operative (e. g., *Nectria haematococca* MPVI and pisatin), the degradative component is essential for a high level of tolerance (Tegtmeier and Van Etten, 1982; Van Etten *et al.*, 1980). In some cases, there is direct evidence that pathogens do contact and metabolize phytoalexins *in planta* (Van Etten *et al.*, 1982).

A more direct approach to the possible importance of phytoalexin tolerance/detoxification in virulence has been genetic analysis. Van Etten's group (Tegtmeier and Van Etten, 1982; Van Etten *et al.*, 1980) has exploited the naturally occurring variation in pisatin sensitivity among interfertile isolates of *Nectria haematococca* (mating population group VI or MPVI). All isolates that could not demethylate pisatin (a conserion catalyzed by an inducible pisatin demethylase) were either nonpathogenic or of low virulence on pea. When different isolates were crossed, tolerance to the phytoalexin was always linked to pisatin demethylase, i. e., all ascospore progeny that were pisatin sensitive were low in virulence, and all highly or moderately virulent progeny were pisatin tolerant. Low virulence among progeny with high tolerance to and ability to demethylate pisatin was attributed to other independently segregating virulence traits. No combination of virulence genes was obtained that could give high virulence without pisatin demethylating ability. These studies, therefore, provide strong evidence that pisatin demethylase is an essential component for maximum virulence in this system.

An example of pathogen intolerance to a host inhibitor being the basis for avirulence has been investigated in a *Fusarium* species by Defago and Kern (1983) and Defago *et al.* (1983). Green tomato fruits, unlike other parts of the tomato plant, have high quantities of tomatine, a steroid glycoalkaloid toxic to many fungi *in vitro*. *Fusarium solani* mutants that were insensitive to tomatine caused a severe not in green tomatoes, while the wild type strain did not. The mutants were neither more nor less aggressive than the wild type strains on red tomatoes or pea plants, which did not contain high levels of tomatine. In crosses between two tomatine insensitive mutants and wild type *Nectria haematococca* (perfect stage of the fungus), tomatine insensitivity and pathogenicity on green tomatoes were 100% linked (no exceptions). The basis for tomatine insensitivity in the mutants did not appear to involve degradative mechanisms. Instead, the low sterol content observed in the mutants was thought to result in reduced complex formation between the glycoalkaloid and sterols in plasma membrane, the presumed target system in the fungus.

VII. Summary

The interactions between plants and their pathogens represent complex phenomena of interative biology. Partly because of this complexity their study has progressed relatively slowly by comparison with other areas of microbiology, genetics, plant and cell biology. At present our knowledge regarding disease mechanisms is incomplete and we are far from being able to provide a total description of a host-pathogen interaction even for the simplest pathogens, i.e., the viroids and viruses. Nevertheless, substantial progress has been made with respect to the nature and role of individual virulence components in various pathogens. This is one of the several areas where molecular-genetic approaches will undoubtedly make significant impact and help clarify present ambiguities. Other such areas include the development of molecular diagnostics and the study of disease processes or pathogen virulence traits whose biochemical nature is presently unknown or not obvious. The basis of host pathogen incompatibility and specificity at the race/cultivar and heterologous pathogen/non-host level, and the related task of identifying the gene products of pathogen virulence/host resistance genes are also major challenges in molecular plant pathology. Recent progress in the molecular manipulation of both microbial and plant systems raises hopes for the cloning of *P* and *R* genes in the near future. Manipulation of pathogens with respect to virulence, pathogenicity, and host range traits offers attractive possibilities for biological control of pathogens and weeds which will undoubtedly be explored. Better understanding of the molecular basis of pathogenicity and resistance wil hopefully suggest other novel approaches to plant diseases-control as well.

Acknowledgments

We thank all colleagues mentioned in personal communication citations in the text for sharing unpublished information with us and the office personnel of our departments for their prompt response to typing deadlines. The writing of this review was partially supported by Grant No. 79-59-2061-1-1241-1 from the United States Department of Agriculture/SEA Competitive Grants Office.

VIII. References

Albersheim, P., 1977: General discussion of fungal elicitors, chaired by Paxton, J. In: Solheim, B., Raa, J. (eds.): Cell wall biochemistry related to specificity in host-plant pathogen interactions, pp. 155—162. Tromso – Oslo – Bergen: Universitetsforlaget.

Albersheim, P., Anderson, A. J., 1971: Proteins from plant cell walls inhibit polygalacturonases secreted by plant pathogens. Proc. Nat. Acad. Sci. U. S. A. **68,** 1815—1819.

Albersheim, P., Valent, B. S., 1978: Host-pathogen interactions in plants. Plants, when exposed to oligosaccharides of fungal origin, defend themselves by accumulating antibiotics. J. Cell. Biol. **78,** 627—643.

Backman, P. A., DeVay, J. E., 1971: Studies on the mode of action and biogenesis of the phytotoxin syringomycin. Phyiol. Plant Pathol. **1,** 215—233.

Bailey, J. G., 1982: Mechanisms of phytoalexin accumulation. In: Bailey, J. A., Mansfield, J. W. (eds.): Phytoalexins, pp. 284—318. New York: Wiley.

Ballio, A., Chain, E. B., DeLeo, P., Erlanger, B. F., Mauri, M., Tonolo, A., 1964: Fusicoccin: a new wilting toxin produced by *Fusicoccum amygdali* Del. Nature **203,** 297.

Bateman, D. F., Basham, H. G., 1976: Degradation of plant cell walls and membranes by microbial enzymes. In: Heitefuss, R., Williams, P. H. (eds.), Encyclopedia of Plant Physiology, New Series, Vol. 4, Physiological Plant Pathology, pp. 317—355. Berlin - Heidelberg - New York: Springer.

Bateman, D. F., Millar, R. L., 1966: Pectic enzymes and tissue degradation. Annu. Rev. Phytopathol. **4,** 119—146.

Beraha, L., Garber, E. D., 1971: Avirulence and extracellular enzymes of *Erwinia carotovora.* Phytopathol. Z. **70,** 335—344.

Bingefors, S., 1971: Resistance to nematodes and the possible value of induced mutations. In: Mutation breeding for disease resistance, pp. 209—235. Int. Atomic Energy Agency, Vienna, 1971.

Bottini, A. T., Gilchrist, D. G., 1981: Phytotoxins. I. A l-aminodimethylheptadecapentol from *Alternaria alternata* f. sp. *lycopersici.* Tetra Lett. **22,** 2719—2722.

Bottini, A. T., Bowen, J. R., Gilchrist, D. G., 1981: Phytoxins. II. Chracterization of a phytotoxic fraction from *Alternaria alternata* f. sp. *lycopersici.* Tetra. Lett. **22,** 2723—2726.

Brookhouser, L. W., 1974: Characterization of endopolygalacturonase produced by *Rhizoctonia solani* during infection of cotton seedings and in response to host exudates. Ph. D. Thesis, University of California, Berkeley.

Brueger, B. B., Keen, N. T., 1979: Specific elicitors of glyceolin in the *Pseudomonas glycinea*-soybean host-parasite system. Physiol. Plant Pathol. **15,** 43—51.

Bushnell, W. R., Rowell, J. B, 1981: Suppressors of defense reactions: a model for roles in specificity. Phytopathology **71,** 1012—1014.

Campbell, G. K., Deverall, B. J., 1980: The effects of light and a photosynthetic inhibitor on the expression of the *Lr-20* gene for resistance to leaf rust. Physiol. Plant Pathol. **16,** 415—423.

Carpita, N., Sabularse, D., Montezinos, D., Delmar, D. P., 1979: Determination of pore size of cell walls of living plant cells. Science **205,** 1144—1147.

Cervone, F., Scala, A., Scala, F., 1978: Polygalacturonase from *Rhizoctonia fragariae:* further characterization of two isoenzymes and their action towards strawberry tissue. Physiol. Plant Pathol. **12,** 19—26.

Cervone, F., Scala, A., Foresti, M., Cacace, M. G., Noviello, G., 1977: Endopolygalacturonase from *Rhizoctonia fragariae:* purification and characterization of two isoenzymes. Biochim. Biophys. Acta **482,** 379—385.

Chatterjee, A. K., Starr, M. P., 1977: Donor strains of the soft-rot bacterium *Erwinia chrysanthemi* and conjugational transfer of the pectolytic capactity. J. Bacteriol. **132,** 862—869.

Collmer, A., Bateman, D. F., 1981: Impaired induction and self-catabolite repression of extracellular pectate lyase in *Erwinia chysanthemi* mutants deficient in oligogalacturonide lyase. Proc. Nat. Acad. Sci. U. S. A. **78,** 3920—3924. (Erratum: **78,** 7844.)

Collmer, A., Berman, P., Mount, M. S., 1982: Pectate lyase regulation and bacterial soft-rot pathogenesis. In: Mount, M. S., Lacy, G. H. (eds.), Phytopathogenic Procaryotes, pp. 395—422. New York: Academic Press.

Comai, L., Kosuge, T., 1982: Cloning and characterization of *iaaM,* a virulence determinant of *Pseudomonas sayastanoi.* J. Bacteriol. **149,** 40—46.

Comai, L., Surico, G., Kosuge, T, 1982: Relation of plasmid DNA to indole-acetic acid production in different strains of *Pseudomanos syringae* pv. *savastanoi.* J. Gen. Microbiol. **128,** 2157—2163.

Cooksey, C. J., Garratt, P. J., Dahiya, J. S., Strange, R. N., 1983: Sucrose: a constitutive elicitor of phytoalexin synthesis. Science **220,** 1398—1400.

Cooper, R. M., 1977: Regulation of synthesis of cell wall-degrading enzymes of plant pathogens. In: Solheim, B., Raa, J. (eds.): Cell wall biochemistry related to specificity in host-plant pathogen interactions (Proc. of a Symposium, Univ. of Tromso, Tromso, Norway, 1976), pp. 163—211. Oslo: Universitetsforlaget.

Coplin, D. L., Sequeira, L., Hanson, R. S., 1974: *Pseudomonas solanacearum:* virulence of biochemical mutants. Can. J. Microbiol. **20,** 519—529.

Cruickshank, I. A. M., Perrin, D. R., 1968: The isolation and partial characterization of Monilicolin A, a polypeptide with phaseolin-inducing activity from *Monilinia fructicola.* Life Sci. **7,** 449—458.

Currier, T. C., Morgan, M. K., 1981: Plasmids are not associated with formation of nonecapsulated variants of *Pseudomonas solanacearum.* In: Lozano, J. C. (tech. ed.), Proc. 5th Int. Conf. Plant Pathogenic Bacteria, Cali, Clumbia, 1981, pp. 420—426. Cali, Columbia: Centro International de Agricultura Tripical (CIAT).

Currier, W. W., 1981: Molecular controls in the resistance of potato to late blight. Trends Biochem. Sci. **6,** 191—194.

Daly, J. M., 1976: Some aspects of host-pathogen interactions. In: Heitefuss, R., Williams, P. H. (eds.): Physiological plant pathology, pp. 27—50. Berlin - Heidelberg - New York: Springer.

Darvil, A., McNeil, A. C., Albersheim, P., Delmer, D. P., 1980: The primary cell wall of flowering plants. In: Tolbert, N. E. (ed.): The plant cell, pp. 91—162. New York: Academic Press.

Day, P. R., 1974: Genetics of host-parasite interactions. 238 pp. San Francisco: W. H. Freeman.

Defago, G., Kern, H., 1983: Induction of *Fusarium solani* mutants insensitive to tomatine, their pathogenicity and aggressiveness to tomato fruits and pea plants. Physiol. Plant Pathol. **2,** 29—37.

Defago, G., Kern, H., Sedlar, L., 1983: Genetic analysis of tomatine insensitivity, sterol content and pathogenicity for green tomato fruits in mutants of *Fusarium solani.* Physiol. Plant Pathol. **22,** 39—43.

Deverall, B. J., McLeod, S., 1980: Responses of wheat cells around heat-inhibited rust mycelia and associated with the expression of the *Lr20, Sr6* and *Sr15* alleles for resistance. Physiol. Plant Pathol. **17,** 213—219.

DeWit, P. J. G. M., Kode, E., 1981: Further characterization of and cultivar specificity of glycoprotein elicitors from culture filtrates and cell walls of *Cladosporium fulvum (Fulvia fulva).* Physiol. Plant Pathol. **18,** 297—314.

DeWit, P. J. G. M., Spickman, G., 1982: Evidence for occurrence of race and cultivar-specific elicitors of necrosis in intercellular fluids of compatible interactions of *Cladosporium fulvum* and tomato. Physiol. Plant Pathol. **21,** 1—11.

Dickman, M. B., Patil, S. S., Kolattukudy, P. E., 1982: Purification, characterization, and role in infection of an extracellular cutinolytic enzyme from *Colletotrichum gloeosporioides* Penz. on *Carica papaya* L. Physiol. Plant Pathol. **20,** 333—347.

Dixon, R. A., Dey, D. M., Lawton, M. A., Lamb, C. J., 1983: Phytoalexin induction in french bean: intercellular transmission of elicitation in cell suspension cultures and hypocotyl sections of *Phaseolus vulgaris*. Plant Physiol. **71,** 251—256.

Doke, N., Garas, N. A., Kuc, J., 1979: Partial characterization and aspects of the mode of action of a hypersensitivity-inhibiting factor (HIF) isolated from *Phytophthora infestans*. Physiol. Plant Pathol. **15,** 127—140.

Doke, N., Tomiyama, K., 1980: Effect of hyphal wall components from *Phytophthora infestans* on protoplasts of potato tuber tissues. Physiol. Plant Pathol. **16,** 167—176.

Doke, N., Furuichi, N., 1982: Response of protoplasts to hyphal wall components in relation to resistance in potato to *Phytophthora infestans*. Physiol. Plant Pathol. **21,** 23—30.

Durbin, R. D., 1982: Toxins and pathogenesis. In: Mount, M. S., Lacy, G. H. (eds.), Phytopathogenic procaryotes, Vol. 1, pp. 423—441. New York: Academic Press.

Durbin, R. D., Steele, J. A., 1979: What are thous, O specificity? In: Daly, J. M., Uritani, I. (eds.): Recognition and specificity in plant host-parasite interactions, pp. 115—131. Tokyo: Japan Scientific Societies.

Durbin, R. D., Uchytil, T. F., 1977: A survey of plant insensitivity to tentoxin. Phytopathology **67,** 602—603.

Ellingboe, A. H., 1976: Genetics of host-parasite interactions. In: Heitefuss, R., Williams, P. H. (eds.): Encyclopedia of Plant Physiology, Vol. 4, Physiological Plant Pathology, pp. 761—778. Berlin - Heidelberg - New York: Springer.

Ellingboe, A. H., 1982: Genetic aspects of active defense. In: Wood, R. K. S. (ed.): Active defense mechanisms in plants, pp. 179—192. New York: Plenum Press.

Ellingboe, A. H., Gabriel, D. W., 1977: Induced conditional mutants for studying host-pathogen interactions. In: Induced mutations against plant diseases (Proc. of a Symposium organized by the International Atomic Energy Agency), pp. 35—46. Vienna: International Atomic Energy Agency.

El-Sayed, S. A., 1977: Phytoalexin-generating capacity in relation to late blight resistance in certain tomato mutants induced by gamma irradiation of seeds. In: Induced mutations against plant diseases (Proc. of a Symposium organized by the International Atomic Energy Agency), pp 265—274. Vienna: International Atomic Energy Agency.

Fett, W. F., Zacharius, R. M., 1983: Bacterial growth and phytoalexin elicitation in soybean cell suppression cultures inoculated with *Pseudomonas syringae* pathovars. Physiol. Plant Pathol. **22,** 151—172.

Fielding, A. H., 1981: Natural inhibitors of fungal polygalacturonases in infected fruit tissues. J. Gen. Microbiol. **123,** 377—381.

Fisher, M. L., Anderson, A. J., Albersheim, P., 1973: Host-pathogen interactions. VI. A single protein efficiently inhibits endopolygalacturonase secreted by *Colletotrichum lindermuthianum* and *Aspergillus niger*. Plant Physiol. **51,** 489—491.

Flor, H. H., 1955: Host-parasite interaction in flax rust — its genetics and other implications. Phytopathology **45,** 680—685.

Flor, H. H., 1971: Current status of the gene-for-gene concept. Annu. Rev. Phytopathol. **9,** 275—296.

Fraenkel-Conrat, H., Wagner, R. R. (eds.): 1977: Comprehensive Virology, Vol. 11. Regulation and genetics, plant viruses. New York — London: Academic Press.

Fulton, N. D., Bollenbacher, K., Templeton, G. E., 1965: A metabolite from *Alternaria tenuis* that inhibits chlorophyll production. Phytopathology **55,** 49—51.

Gabriel, D. W., Ellingboe, A. H., 1979: Mutations affecting virulence in *Phyllosticta maydis*. Can. J. Bot. **57,** 2639—2643.

Gabriel, D. W., Lister, N., Ellingboe, A. H., 1982: The induction and analysis of two classes of mutations affecting pathogenicity in an obligate parasite. Phytopathology **72,** 1026—1028.

Gardner, J. M., Kado, C. I., 1976: Polygalacturonic acid *trans*-eliminase in the osmotic shock fluid of *Ewinia rubrifaciens:* characterization of the purified enzyme and its effect on plant cells. J. Bacteriol. **127,** 451—460.

Garfinkel, D. J., Nester, E. W., 1980: *Agrobacterium tumefaciens* mutants affected in crown gall tumorigenesis and octopine catabolism. J. Bacteriol. **144,** 732—734.

Gilchrist, D. G., Grogan, R. G., 1976: Production and nature of a host-specific toxin from *Alternaria alternata* f. sp. *lycopersici*. Phytopathology **66,** 165—171.

Gnanamanickam, S. S., Patil, S. S., 1977: Phaseolotoxin suppresses bacterially induced hypersensitive reaction and phytoalexin synthesis in bean cultivars. Physiol. Plant Pathol. **10,** 169—179.

Gobel, H., Bock, W., 1978: Isolierung und Charakterisierung eines Pektinase-inhibitors aus grünen Bohnen (*Phaseolus vulgaris*). Die Nahrung **22,** 809—818.

Gracen, V. E., 1982: Role of genetics in etiological plant pathology. Annu. Rev. Phytopathol. **20,** 219—233.

Graham, T. L., Sequeira, L., Huang, T. S. R. 1977: Bacterial lipopolysaccharide as inducers of disease resistance in tobacco. Appl. Env. Microbiol. **34,** 424—432.

Guimaraes, W. V., 1977: Studies with plasmids in plant pathogenic bacteria. I. Conjugational plasmid transmission and plasmid promoted chromosome mobilization in *Pseudomonas phaseolicola* and *Erwinia chrysanthemi*. II. Effect of plasmids on physiology and pathogenicity of *Erwinia chrysanthemi*. Ph. D. Thesis, University of California, Berkeley.

Hahlbrock, K., Lamb, C. J., Purwin, C., Ebel, J., Fautz, E., Schafer, E., 1981: Rapid response of suspension-cultured parsley cells to the elicitor from *Phytophthora megasperma* var. *sojae*, induction of the enzymes of general phenylpropanoid metabolism. Plant Physiol. **67,** 768—773.

Hahn, M. G., Darvill, A. G., Albersheim, P., 1981: Host-pathogen interactions. XIX. The endogenous elicitor, a fragment of a plant cell wall polysaccharide that elicits phytoalexin accumulation in soybeans. Plant Physiol. **68,** 1161—1169.

Hamada, S. E., Farrand, S. K., 1980: Diversity among B6 strains of *Agrobacterium tumefaciens*. J. Bacteriol. **141,** 1127—1133.

Hancock, J. G., 1976: Multiple forms of endo-pectate lyase formed in culture and in infected squash hypocotyls by *Hypomyces solani* f. sp. *cucurbitae*. Phytopathology **66,** 40—45.

Hargreaves, J. A., Bailey, J. A., 1978: Phytoalexin production by hypocotyls of *Phaseolus vulgaris* in response to constitutive metabolites released by damaged bean cells. Physiol. Plant Pathol. **13,** 89—100.

Hargreaves, J. A., Selby, C., 1978: Phytoalexin formation in cell suspensions of *Phaseolus vulgaris* in response to an extract of bean hypocotyls. Phytochemistry **17,** 1099—1102.

Heath, M. C., 1976: Hypersensitivity: the cause or consequence of rust resistance? Phytopathology **66,** 935—936.

Heath, M. C., 1981 a: Resistance of plants to rust infection. Phytopathology **71,** 971—974.

Heath, M. C., 1981 b: A generalized concept of host-parasite specificity. Phytopathology **71,** 1121—1123.

Hodgson, R., Peterson, R. H., Riker, A. J., 1949: The toxicity of polysaccharides and other large molecules to tomato cuttings. Phytopathology **39,** 47—62.

Hoffman, R. M., Turner, J. G., 1982: Partial purification of proteins from pea leaflets that inhibit *Ascochyta pisi* endopolygalacturonase. Physiol. Plant Pathol. **20,** 173—187.

Holenstein, J. E., 1982: On the transmission of napthazarine production and pathogenesis in *Nectria hematococca.* Thesis, Confederate Technical University, Zurich.

Holliday, M., Keen, N. T., 1982: The role of phytoalexin in the resistance of soybean leaves to bacteria: effect of glyphosate on glyceolin accumulation. Phytopathology **72,** 1470—1474.

Holliday, M. J., Long, M., Keen, N. T., 1981: Manipulation of the temperature-sensitive interaction between soybean leaves and *Pseudomonas syringae* pv. *glycinea*-implications on the nature of deteminative events modulating hypersensitive resistance. Physiol. Plant Pathol. **19,** 206—216.

Holloway, P. J., 1982 a: The chemical constitution of plant cutins. In: Cutler, D. F., Alvin, K. L., Price, C. E. (eds.), The Plant Cuticle (Linnean Society Symposium Series No. 10), pp. 45—85. London: Academic Press.

Holloway, P. J., 1982 b: Structure and histochemistry of plant cuticular membranes: an overview. In: Cutler, D. F., Alvin, K. L., Price, C. E. (eds.), The Plant Cuticle (Linnean Society Symposium Series No. 10). London: Academic Press.

Holsters, M., Silva, B., Van Vleit, F., Genetello, C., DeBlock, M., Dhaese, P., Depicker, A., Inze, D., Engler, G., Villaroel, R., Van Montague, M., Schell, J., 1980: The functional organisation of the nopaline *A. tumefaciens* plasmid pTiC 58. Plasmid **3,** 212—230.

Hooker, A. L., Saxena, K. M. S., 1967: Apparent reversal of dominance of a gene in corn for resistance to *Puccinia sorghi.* Phytopathology **57,** 1372—1374.

Hooker, A. L., Saxena, K. M. S., 1971: Genetics of disease resistance in plants. Annu. Rev. Genetics **5,** 407—424.

Howell, C. R., 1976: Use of enzyme-deficient mutants of *Verticillium dahliae* to assess the importance of pectolytic enzymes in symptom expression of Verticillium wilt of cotton. Physiol. Plant Pathol. **9,** 279—283.

Jones, D. R., Deverall, B. J., 1978: The use of leaf transplants to study the cause of hypersensitivitry to leaf rust, *Puccinia recondita,* in wheat carrying the *Lr 20* gene. Physiol. Plant Pathol. **12,** 311—319.

Jorgensen, J. H., 1971: Comparison of induced mutant genes with spontaneous genes in barley conditioning resistance to powdery mildew. In: Mutation breeding to disease resistance (Proc. of a Panel, organized by the International Atomic Energy Agency, Vienna, 1970), pp. 117—124. Vienna: International Atomic Energy Agency.

Jorgensen, J. H., 1974: Induced mutations for powdery mildew resistance in barley — a progress report. In: Induced mutations for disease resistance to crop plants (Proc. of a Coordination Meeting organized by the International Atomic Energy Agency, Vienna, 1973), p. 67. Vienna: International Atomic Energy Agency.

Jorgensen, J. H., 1977: Location of the *ml-o* locus on barley chromosome 4. In: Induced mutations against plant diseases (Proc. of a Symposium organized by the International Atomic Energy Agency, Vienna, 1977), pp. 533—549. Vienna: International Atomic Energy Agency.

Kao, J. C., Perry, K. L., Kado, C. I., 1982: Indoleacetic acid complementation and its relation to host range specificity genes on the Ti plasmid of *Agrobacterium tumefaciens*. Mol. Gen. Genet. **188,** 425—432.

Kawai, M., Rich, D. H., Walton, J. D., 1983: The structure and conformation of HC-toxin. Biochem. Biophys. Res. Comm. **111,** 398—403.

Keegstra, K., Talmadge, K., Bauer, W. D., Albersheim, P., 1973: The structure of plant cell walls. III. A model of the walls of suspension-cultured sycamore cells based on the interconnections of the macromolecular components. Plant Physiol. **51,** 188—196.

Keen, N. T., 1975: Specific elicitors of plant phytoalexin production: determinants of race specificity in pathogens? Science **187,** 74—75.

Keen, N. T., 1982: Specific recognition in gene-for-gene host-parasite systems. In: Ingram, D. S., Williams P. H. (eds.), Advances in Plant Pathology, Vol. 1, pp. 35—82. New York: Academic Press.

Keen, N. T., Ersek, T., Long, M., Bruegger, B., Holliday, M., 1981: Inhibition of the hypersensitive reaction of soybean leaves to incompatible *Pseudomonas* spp. by blasticidin S, streptomycin or elevated temperature. Physiol. Plant Pathol. **18,** 325—337.

Keen, N. T., Erwin, D. C., 1971: Endopolygalacturonase: evidence against involvement in verticillium wilt of cotton. Phytopathology **61,** 698—203.

Keen, M. T., Lengrad, M., 1980: Surface glycoproteins: evidence that they may function as the race specific phytoalexin elicitor of *Phytophthora megasperma* f. sp. *glycinea*. Physiol. Plant Pathol. **17,** 175—182.

Kelman, A., 1954: The relationship of pathogenicity in *Pseudomonas solanacearum* to colony appearance on a tetrazolium agar medium. Phytopathology **44,** 693—695.

Kelman, A., Hruschka, J., 1973: The role of motility and aerotaxis in the selective increase of avirulent bacteria in still cultures of *Pseudomonas solanacearum*. J. Gen. Microbiol. **76,** 177—188.

Kinoshita, T., Renbutsu, Y., Khan, I. D., 1972: Distribution of tenuazonic acid production in the genus *Alternaria* and its pathological evaluation. Ann. Phytopath. Soc. Japan **38,** 397—404.

Kirosawa, S., 1982: Genetics and epidemiological modeling of breakdown of plant disease resistance. Ann. Rev. Phytophathol. **20,** 93—117.

Klement, Z., 1982: Hypersensitivity. In: Mount, M. S., Lacy, G. H. (eds.), Phytopathogenic Procaryotes, Vol. 2, pp. 150—177. New York: Academic Press.

Kline, E. L., Brown, C. S., Bankaitis, V., Montefiori, D. C., Craig, K., 1980: Metabolite gene regulation of the L-arabinose operon in *Escherichia coli* with indoleacetic acid and other indole derivatives. Proc. Nat. Acad. Sci. (U. S. A.) **77,** 1768—1772.

Knauf, V. C., Panagopoulos, C. G., Nester, E. W., 1982: Genetic factors controlling the host range of *Agrobacterium tumefaciens*. Phytopathology **72,** 1545—1549.

Koischwitz, H., Kleinkauf, H., 1976: Gramicidin S-synthetase: electrophoretic characterization of the multienzyme. Biochim. Biophys. Acta **429,** 1052—1061.

Kolattukudy, P. E., 1980a: Biopolyester membranes of plants: cutin and suberin. Science **208,** 990—1000.

Kolattukudy, P. E., 1980b: Cutin, suberin, and waxes. In: Stumpf, P. K., Conn, E. E. (eds.), The biochemistry of plants, Vol. 4, pp. 571—645. New York: Academic Press.

Kolattukudy, P. E., 1981: Structure, biosynthesis and degradation of cutin and suberin. Annu. Rev. Plant Physiol. **32,** 539—567.

Koller, W., Allan, C. R., Kolattukudy, P. E., 1982: Role of cutinase and cell wall degrading enzymes in infection of *Pisum sativum* by *Fusarium solani* f. sp. *pisi.* Physiol. Plant Pathol. **20,** 47—60.

Kono, Y., Knoche, H. W., Daly, J. M., 1981: Structure: fungal host specific. In: Durbin, R. D. (ed.), Toxins in plant disease, pp. 221—257. New York: Academic Press.

Kuranz, M. J., Zacharius, R. M., 1981: Hypersensitive response in potato tuber: elicitation by combination of non-eliciting components from *Phytophthora infestans.* Physiol. Plant Pathol. **18,** 67—77.

Langford, A. N., 1948: Autogenous necrosis in tomatoes immune from *Gladosporium fulvum* Cooke. Can. J. Res. **26,** 35—64.

Lawrence, G. J., Mayo, G. M. E., Shepard, K. W., 1981: Interactions between genes controlling pathogenicity in the flax rust fungus. Phytopathology **71,** 12—19.

Leach, J., Tegtmeier, K. J., Daly, J. M., Yoder, O. C., 1982: Dominance at the Toxl locus controlling T-toxin production by *Cochiobolus heterostrophus.* Physiol. Plant Pathol. **21,** 327—333.

Lee, S.-C., West, C. A., 1981: Polygalacturonase from *Rhizopus stolonifer,* an elicitor of casbene synthetase activity in castor bean (*Ricinus communis* (L.)) seedlings. Plant Physiol. **67,** 633—639.

Lim, S. M., Hooker, A. L., 1971: Southern corn leaf blight: genetic control of pathogenicity and toxin production in race T and race O of *Cochliobolus heterostrophus.* Genetics **69,** 115—117.

Lindow, S. E., Arny, D. C., Borchet, W. R., Upper, C. D., 1978: The role of bacterial ice nuclei in frost injury to sensitive plants. In: Li, P. H. (ed.), Plant cold hardiness and freezing stress, pp. 249—263. New York: Academic Press.

Littlefield, L. J., 1973: Histological evidence for diverse mechanisms of resistance to flax rust, *Melampsora lini* (Ehreng.) Lev. Physiol. Plant Pathol. **3,** 241—247.

Litzenberger, S. C., 1949: Nature of susceptibility to *Helminthosporium victoriae* and resistance to *Puccinia coronata* in Victoria oats. Phytopathology **39,** 300—318.

Loper, J. E., Kado, C. I., 1979: Host range conferred by the virulence specifying plasmid of *Agrobacterium tumefaciens.* J. Bacteriol. **139,** 591—596.

Macko, V., Acklin, W., Arigoni, D., 1982: Structure of host-specific toxins produced by *Helminthosporium sacchari.* Abstracts of Papers, 83rd Amer. Chem. Soc. Mtg., 3252.

Mann, B., 1962: Role of pectolytic enzymes in the Fusarium syndrome of tomato. Trans. Brit. Mycol. Soc. **45,** 169—178.

Mansfield, J. W., 1982: The role of phytoalexins in disease resistance. In: Bailey, J. A., Mansfield, J. W. (eds.), Phytoalexins, pp. 253—288. New York: Wiley.

Marchetti, M. A., Bollich, C. N., and Uecker, R. A., 1983: Spontaneous occurrence of the Sekiguichi lesion in two American rice lines: its induction inheritance and utilization. Phytopathology **73,** 603—606.

Marciano, P., DiLenna, P., Magro, P., 1982: Polygalacturonase isozymes produced by *Schlerotinia schlerotiorum in vivo* and *in vitro.* Physiol. Plant Pathol. **20,** 201—212.

Matzke, A. J. M., Chilton, M.-D., 1981: Site-specific insertion of genes into the T-DNA of the *Agrobacterium* tumor-inducing plasmid: an approach to genetic engineering of higher plant cells. J. Mol. Appl. Genet. **1,** 39—49.

McDonnell, K., 1958: Absence of pectolytic enzymes in a pathogenic strain of *Fusarium oxysporum* f. *lycopersici.* Nature **182,** 1025—1026.

McIntosh, R. A., 1977: Nature of induced mutations affecting disease resistance in wheat. In: Induced mutations against plant diseases (Proc. of a Symposium Organized by the International Atomic Energy Agency, Vienna, 1977), pp. 551—565. Vienna: International Atomic Energy Agency.

McKeen, W. B., 1974: Mode of penetration of epidermal cell walls of *vicia faba* by *Botrytis cinerea*. Phytopathology **64,** 461—467.

McKenzie, R. I. H., Martens, J. W., Green, G. J., 1971: The oat stem rust problem. In: mutation breeding for disease resistance, pp. 151—157. Vienna: International Atomic Energy Agency.

McMillan, J. D., Sherman, M. I., 1974: Pectic enzymes. In: Whitaker, J. R. (ed.), Food Related Enzymes, pp. 101—130. Advances in Chemistry Series, No. 136. American Chemical Society, Washington, D. C.

Meadows, M. F., Stahl, R. E., 1981: Different induction periods for hypersensitivity in peper to *Xanthomonas vesicatoria* determined with antimicrobial agents. Phytopathology **71,** 1024—1027.

Meehan, F., Murphy, H. C., 1946: A new *Helminthosporium* blight of oats. Science **104,** 413—414.

Meiners, J. P., 1981: Genetics of disease resistance in edible legumes. Annu. Rev. Phytopathol. **21,** 189—109.

Message, B., Biostard, P., Pitrat, M., Schmidt, J., Boucher, C., 1979: A new class of fluidal avirulent mutants of *Pseudomonas solanacearum* unable to induce a hypersensitive reaction. In: Proceedings of the IVth International Conference on Plant Pathogenic Bacteria, pp. 823—833. Angers, France, 1978: Station de Pathologie Végétale et Phytobacteriologie.

Mew, T. W., Kush, G. S., 1981: Breeding for bacterial blight resistance in rice. In: Lozano, J. C. (ed.), Proc. 5th Int. Conf. on Plant Pathogenic Bacteria, pp. 505—577. Centro Internacional de Agricultura Tropical (CIAT), Cali, Columbia.

Mills, D., Gonzalez, C. F., 1982: The evolution of pathogenesis and race specificity. In: Mount, M. S., Lacy, G. H. (eds.), Phytopathogenic Prokaryotes, pp. 77—119. New York: Academic Press.

Mitchell, R. E., 1981: Structure: bacterial. In: Durbin, R. D. (ed.), Toxins in plant disease, pp. 259—293. New York: Academic Press.

Moseman, J. G., 1966: Genetics of powdery mildews. Ann. Rev. Phytopathol. **4,** 269—290.

Mount, M. S., Berman, P. M., Mortlock, R. P., Hubbard, J. P., 1979: Regulation of endopolygalacturonate trans-eliminase in an adenosine 3′, 5′-cyclic monophosphate-deficient mutant of *Erwinia coratovora*. Phytopathology **69,** 117—120.

Mullen, J. M., Bateman, D. F., 1975: Polysaccharide degrading enzymes produced by *Fusarium roseum* "Avenaceum" in culture and during pathogenesis. Physiol. Plant Pathol. **6,** 233—246.

Mussell, H. W., 1973: Endopolygalacturonase: evidence for involvement in Verticillium wilt of cotton. Phytopathology **63,** 62—70.

Nelson, R. R., Ullstrup, A. J., 1961: The inheritance of pathogenicity in *C. carbonum*. Phytopathology **51,** 1—2.

Niks, R. E., Kuiper, H. J., 1983: Histology of the relation between minor and major genes for resistance of barley to leaf rust. Phytopathology **73,** 55—59.

Nishimura, S., Kohmoto, K., Otani, H., 1979: The role of host-specific toxins in saprophytic pathogens. In: Daly, J. N., Uritani, I. (eds.), Recognition and specificity in plant host-parasite interactions, pp. 133—146. Tokyo: Japan Scientific Societies Press.

Northnagel, E. A., McNeil, M., Albersheim, P., Dell, A., 1983: Host-pathogen interactions. XXII. A galacturonic acid oligosaccharide from plant cell walls elicits phytoalexins. Plant Physiol. **71,** 916—926.

Ofuya, C. O., Wood, R. K. S., 1981 a: Virulence of *Pseudomonas solanacearum* in relation to extracellular polysaccharides. In: Lozano, J. C. (tech. ed.), Proc. 5th Int. Conf. Plant Pathogenic Bacteria, Cali, Colombia, 1981, pp. 263—269. Centro International de Agricultura Tropical (CIAT), Cali, Colombia.

Ofuya, C. O., Wood, R. K. S., 1981 b: Cell wall degrading polysaccharidases and extrachromosomal DNA in wilts caused by *Pseudomonas solanacearum.* In: Lozano, J. C. (tech. ed.), Proc. 5th Int. Conf. Plant Pathogenic Bacteria, Cali, Colombia, 1981, pp. 270—279. Centro International de Agricultura Tropical (CIAT), Cali, Colombia.

Orser, C. O., Staskawicz, B. J., Panopoulos, N. J., Dahlbeck, D., Lindow, S. E., 1982: Cloning and expression of ice nucleation genes from *Pseudomonas syringae* and *Erwinia herbicola* in *Escherichia coli.* Phytopathology **72,** 1000 (abstr.).

Owens, L. D., Thomson, J. F., Pitcher, R. G., Williams, T., 1968: *Rhizobium* synthesized phytotoxin: an inhibitor of beta-cystathionase in *Salmonella typhimurium.* Biochim. Biophys. Acta **158,** 217—225.

Panagopoulos, C. G., Crosse, J. E., 1964: Frost injury as a predisposing factor in blossom blight of pear caused by *Pseudomonas syringae* van Hall. Nature **202,** 1352—1355.

Panopoulos, N. J., Staskawicz, B. J., 1981: Genetics of production. In: Durbin, R. D. (ed.), Toxins in plant disease, pp. 79—107. New York: Academic Press.

Paxton, J. D., 1977: *Phytophthora* inducers of soybean phytoalexin production. In: Solheim, B., Rad, J. (eds.), Cell Wall Biochemistry Related to Specificity in Host-Pathogen Interactions, pp. 147—154. (Proc. of a Symposium, University of Tromso, Norway, August 1976.) Tromso - Oslo - Bergen: Universitetsforlaget.

Pegg, G. F., 1981: Biochemistry and physiology of pathogenesis. In: Mace, M. E., Bell, A. A., Beckman, C. H. (eds.), Fungal Wilt Diseases of Plants, pp. 193—253. New York: Academic Press.

Peters, B. M., Cribbs, D. H., Stelzic, D. A., 1978: Agglutination of plant protoplasts by fungal cell glucans. Science **201,** 364—365.

Pringle, R. B., Braun, A. C. 1958: Constitution of the toxin of *Helminthosporium victoriae.* Nature **181,** 1205—1206.

Puhalla, J. E., Howell, C. R., 1975: Significance of endopolygalacturonase activity to symptom expression of Verticillium wilt in cotton, assessed by the use of mutants of *Verticillium dahliae.* Physiol. Plant Pathol. **7,** 147—152.

Pupillo, P., Mazzuchi, U., Pierini, G., 1976: Pectic lyase isozymes produced by *Erwinia chrysanthemi* Burkh. *et al.* in polypectate broth or in Dieffenbachia leaves. Physiol. Plant Pathol. **9,** 113—120.

Quantick, P., Cervone, F., Wood, R. K. S, 1983: Isoenzymes of a polygalacturonate trans-eliminase producted y *Erwinia atroseptica* in potato tissue and in liquid culture. Physiol. Plant Pathol. **22,** 77—86.

Rich, D. R., 1981: Chemical synthesis. In: Durbin, R. D. (ed.), Toxins in plant disease, pp. 79—107: New York: Academic Press.

Rombouts, F. M., Pilnik, W., 1972: Research on pectin depolymerases in the sixties, a literature review. Chem. Rubb. Comp. Crit. Rev. Fd. Technol. **3,** 1—26.

Sasser, M., 1982: Inhibition by antibacterial compounds of the hypersensitive reaction induced by *Pseudomonas pisi* in tobacco. Phytopathology **72,** 1513—1517.

Scheffer, R. P., Nelson, R. R., Ullstrup, A. J., 1967: Inheritance of toxin production and pathogenicity in *Cochliobolus carbonum* and *Cochliobolus victoriae*. Phytopathology **57,** 1288—1291.

Sequeira, L., 1976: Induction and suppression of the hypersensitive reaction by phytopathogenic bacteria; specific and non-specific components. In: Wood, R. K. S., Ganiti, A. (eds.), Specificity in Plant Disease, pp. 289—309. New York: Plenum Press.

Sequeira, L., Graham, T. L., 1977: Agglutination of avirulent strains of *Pseudomonas solanacearum* by potato lectin. Physiol. Plant Pathol. **11,** 43—54.

Sequeira, L., 1979: Recognition between hosts and parasites. In: Nickol, B. B. (ed.), Host-Parasite Interfaces, pp. 71—84. New York: Academic Press.

Shaykh, M., Soliday, C., Kolattukudy, P. E., 1977: Proof for the production of cutinase by *Fusarium solani* f. *pisi* during penetration into its host, *Pisum sativum*. Plant Physiol. **60,** 170—172.

Sidhu, G. S., 1975: Gene-for-gene relationships in plant parasitic systems. Sci. Prog. Oxf. **62,** 467—485.

Sijam, K., Karr, A. L., Goodman, R. N., 1982: Relationship of molecular weight and viscosity to wilt-inducing properties of *Erwinia amylovora*-EPS. Phytopathology **72,** 945 (abstr.).

Simmons, M. D., 1979: Modifications of host-parasite interactions through artificial mutagenesis. Ann. Rev. Phytopathol. **17,** 75—96.

Sprang, M. L., Lindow, S. E., 1981: Subcellular localization and partial characterization of ice nucleation activity of *Pseudomonas syringae* and *Erwinia herbicola*. Phytopathology **71,** 256.

Stack, J. P., Mount, M. S., Berman, P. M., Hubbard, J. P., 1980: Pectic enzyme complex from *Erwinia coratovora:* a model for degradation and assimilation of host pectic fractions. Phytopathology **70,** 267—272.

Stahl, R. E., Cook, A. A., 1979: Evidence that bacterial contact with plant cells is necessary for the hypersensitive reaction, but not the susceptible reaction. Physiol. Plant Pathol. **14,** 77—89.

Staskawicz, B. J., Panopoulos, N. J., 1979: A rapid and sensitive microbiological assay for phaseolotoxin. Phytopathology **69,** 663—666.

Staskawicz, B. J., Panopoulos, N. J., Hoogenraad, N. J., 1980: Phaseolotoxin-insensitive ornithine carbamyoltransferase of *Pseudomonas syringae* pv. *phaseolicola:* basis for immunity to phaseolotoxin. J. Bacteriol. **142,** 720—723.

Staskawicz, B. J., Dahlbeck, D., Miller, J., Damm, D., 1983: Molecular analysis of virulence genes in *Pseudomonas solanacearum*. In: Proc. of 1st Int. Conf. on the Molecular Genetics of the Bacteria-Plant Interaction, 1978, Bielefeld, Federal Republic of Germany. (In press.)

Stoessl, A., 1981: Structure and biogenetic relations: fungal non-host specific. In: Durbin, R. D. (ed.), Toxins in plant disease, pp. 109—219. New York: Academic Press.

Tait, R. C., Close, T. J., Hagiya, M., Lundquist, R. C., Kado, C. I., 1982: Construction of cloning vectors from the IncW plasmid pSa and their use in analysis of crown gall tumor formation: In: Lurquin, P. F., Kleinhofs, A. (eds.), Genetic Engineering in Eucaryotes. New York: Plenum Press.

Takai, S., Richards, W. C., 1978: Cerato-ulmin, a wilting toxin of *Ceratocystis ulmi:* isolation and some properties of cerato-ulmin from the culture of *C. ulmi*. Phytopathol. Z. **91,** 129—146.

Tegtmeier, J. D., Van Etten, H. D., 1982: The role of pisatin tolerance and degradation in the virulence of *Nectria haematococca*. A genetic analysis. Phytopathology **72,** 608—612.

Thomashow, M. F., Panagopoulos, C. G., Gordin, M. P., Nester, E. W., 1980: Host range of *Agrobacterium tumefaciens* is determined by the Ti plasmid. Nature **283,** 794—796.

Thurn, K. K., Tyrel, D. J., Chatterjee, A. K., 1982: Tn*5* induced mutations alter extracellular enzyme production in *Erwinia chrysanthemi.* Phytopathology **72,** 934 (abstr.).

Tomiyama, K., Doke, N., Nozue, M., Ishiguri, Y., 1979: The hypersensitive response of resistant plants. In: Daly, J. M., Uritami, I. (eds.), Recognition and specificity in plant host-parasite interactions, pp. 69—81. Tokyo: Japan Scientific Societies Press, and Baltimore: University Park Press.

Ueno, T., Kanakashima, T., Hayashi, Y., Fukami, H., 1975: Structures of AM-toxin I and II. Host-specific phytotoxic metabolites produced by *Alternaria mali.* Agr. Biol. Chem. **39,** 1115—1122.

Van Alfen, N. K., 1982: Wilts: concepts and mechanisms. In: Mount, M. S., Lacy, G. H. (eds.), Phytopathogenic Procaryotes, pp. 459—474. New York: Academic Press.

Van Alfen, N. K., Allard-Turner, V., 1979: Susceptibility of plants to vascular disruption by macromolecules. Plant Physiol. **63,** 1072—1075.

Van Alfen, N. K., McMillan, B. D., 1982: Macromolecular plant-wilting toxins: artifacts of the bioassay? Phytopathology **72,** 132—135.

Van den Ende, G., Linskens, H. F., 1974: Cutinolytic enzymes in relation to pathogenesis. Annu. Rev. Phytopathol. **12,** 247—258.

Van Etten, H. D., Matthews, D. E., Smith, D. A., 1982: Metabolism of phytoalexins. In: Bailey, J. A., Mansfield, J. W. (eds.), Phytoalexins, pp. 181—217. New York: Wiley.

Van Etten, H. D., Matthews, P. S., Tegtmeir, K. J., Dietert, M. F., Stein, J. I., 1980: The association of pisatin tolerance and demethylation with virulence on pea in *Nectria haematococca.* Physiol. Plant Pathol. **16,** 257—268.

Vanderplank, J. E., 1982: Host-pathogen interactions in plant disease, pp. 83—121. New York: Academic Press.

Vater, J., Kleinkauf, H., 1976: Gramicidin S-synthetase. A further characterization of phenylalanine racemase, the light enzyme of gramidicin S-synthetase. Biochim. Biophys. Acta **429,** 1062—1072.

Walbot, V., Hoisington, D. A., Neuffer, M. G., 1983: Disease lesion mimic mutations. In: Kosuge, T., Meredith, C. P., Hollaender, A., (eds.), Genetic Engineering of plants: an agricultural perspective, Proc. Symp., Aug. 1982: Davis, Calif., pp. 431—442. New York - London: Plenum Press.

Ward, E. W. B., Stoessl, A., 1976: On the question of "elicitors" or "inducers" of incompatible interactions between plants and fungal pathogens. Phytopathology **66,** 940—941.

Walton, J. D., Earle, E. D., 1983: The epoxide in HC-toxin is required for activity against susceptible maize. Physiol. Plant Pathol., in press.

Walton, J. D., Earle, E. D., Gibson, B. W., 1982: Purification and structure of the host-specific toxin from *Helminthosporium carbonum* race 1. Biochem. Biophys. Res. Comm. **107,** 785—794.

Whatley, H. M., Hunter, N., Cantrell, M. A., Hendrick, C., Keegstra, K., Sequeira, L., 1980: Lipopolysaccharide composition of the wilt pathogen *Pseudomonas solanacearum.* Plant Physiol. **65,** 557—559.

Wheeler, H., 1981: Role in pathogenesis. In: Durbin, R. D. (ed.), Toxins in Plant Disease, pp. 477—494. New York: Academic Press.

Yamazaki, N., Fry, S. C., Darvill, A. G., Albersheim, P., 1983: Host-pathogen inter-

actions. XXIV. Fragments isolated from suspension-cultured sycamore cell walls inhibit the ability of the cells to incorporate (^{14}C)-leucine into proteins. Plant Physiol. **72,** 864—869.
Yoder, O. C., 1980: Toxins in pathogenesis. Annu. Rev. Phytopathol. **18,** 103—129.
Yoder, C. O., 1983: Use of pathogen-produced toxins in genetic engineering of plants and pathogens. In: Kosuge, T., Meredith, C. P., Hollaender, A. (eds.), Genetic engineering of plants: an agricultural perspective. Proc. Symp., Aug. 1982, Davis, CA, pp. 335—353. New York - London: Plenum Press.
Yoder, O. C., Gracen, V. E., 1975: Segregation of pathogenicity types and host-specific toxin production in progenies of crosses between races T and O of *Helminthosporium maydis (Cochliobolus heterostrophus).* Phytopathology **65,** 273—276.
Yoshikawa, M., Yamauchi, K., Masago, H., 1978a: *De novo* messenger RNA and protein synthesis are required for phytoalexin-mediated disease resistance in soybean hypocotyls. Plant Physiol. **61,** 314—317.
Yoshikawa, M., Yamauchi, K., Masago, H., 1978b: Glyceolin: its role in restricting fungal growth in resistant soybean hypocotyls infected with *Phytophthora megasperma* var. *sojae.* Physiol. Plant Pathol. **12,** 73—82.
Yukioka, M., Winnick, T., 1966: Synthesis of malformin by an enzyme preparation from *Aspergillus niger.* J. Bacteriol. **91,** 2237—2244.
Ziegler, E., Pontzen, R., 1982: Specific inhibition of glucan-elicited glyceolin accumulation in soybeans by an extracellular mamman-glycoprotein of *Phytophthora megasperma* f. sp. *glycinea.* Physiol. Plant Pathol. **20,** 321—331.
Zischek, C., Tuyen, L. T. K., Boistard, P., Boucher, C., 1981: Presence of megaplasmids in *Pseudomonas solanacearum.* In: Lozano, J. C. (tech. ed.), Proc. 5th Int. Conf. Plant Pathogenic Bacteria, Cali, Colombia, 1981, pp. 427—437. Cali, Colombia: Centro International de Agricultura Tropical (CIAT).

Chapter 14

Defense Responses of Plants

C. A. Ryan

Institute of Biological Chemistry and Biochemistry/Biophysics Program, Washington State University, Pullman, WA 99164–6340, U.S.A.

Contents

Introduction

One of the most promising areas of research concerned with plants and their reaction to their environments is that of natural plant defense systems. Within these defense systems lies a variety of both constitutive and active defenses in which a broad spectrum of chemicals are involved. These chemicals are either present all the time as toxins of deterrents or are induced in response to signals that are released from the sites of pest attacks. In this latter category are two types of responses; (i) those that are localized, in which signals travel to only a relatively few cells within the vicinity of the attack site and, (ii) those that are distal response. In the latter case, signals travel throughout the plant to mobilize defense systems in cells many cm away. Many of the active defense responses can produce resistances that are very broad, and can include a variety of pathogens, including viruses, bacteria and fungi (see Kúc, 1982; Rhodes, 1980; Schultz, 1982). The biochemistry of the regulation of the induced defense compounds and the molecular biology of the signals and the genes that are regulated are generally not known.

Only a few plant systems are known in which well characterized proteins (inlcuding enzymes and/or other proteins) are expressed in response to external stimuli. The most studied systems to date are the heat shock proteins, storage proteins and lectins in developing corn, cereal and legume seeds, and the wound-induced or pest-induced hydoxyproline-rich glycoproteins, enzymes of the phytoalexin pathways, and proteinase inhibitors in leaves and seedlings. The regulation of synthesis of these proteins at the translational and transcriptional levels, is not well understood.

The synthesis of phytoalexins and proteinase inhibitors, occur in a number of plant species in response to pest attacks. The phytoalexins are considered to be antimicrobial agents (West, 1981) and the proteinase inhibitor proteins, antinutrients (Ryan, 1981). Both the phytoalexins and proteinase inhibitors can significantly reduce the nutritional quality of the tissues in which they accumulate. Recently, several reviews have been written concerning the phytoalexins (Stöessl, 1980; West, 1981) and the proteinase inhibitor proteins (Ryan, 1981; Laskowski and Kato, 1980) and their roles in plant protection will not be discussed here, except to say they are both considered potentially important classes of compounds among the large arsenal of protective agents that are known to occur in plants.

The accumulation of a considerable body of research concerning the biochemistry and physiology of these two systems over the past several years has reinforced their potentials for the study of the regulation of genes by pest attacks, using current DNA technology. In this paper these two systems will be reviewed with emphasis on the recent progress that has been made toward understanding the molecular basis of environmental regulation of the plant genome.

The processes of induction and accumulation of the phytoalexins and proteinase inhibitor proteins are, superficially, quite different. The phytoalexins are induced, or elicited, in cells surrounding the site of fungal or bacterial attack and follow, or accompany, a hypersensitive reaction of the plant in response to the attacks. The proteinase inhibitors, on the other hand, are induced throughout the plant by insect wounding (Green and Ryan, 1972) and can also be induced by compounds released by pectolytic enzymes of microorganisms during attacks (Bishop *et al.,* 1981). However, there are also considerable similarities between the two systems, considered particularly with respect to the nature of the kinds of signals that can initiate them. This, in turn, indirectly implies a similarity in the nature of the entire process of regulation of genes involved in these systems.

In the phytoalexin response, a substance, or substances, is (are) released from the attack site that diffuse(s) to the nearby cells to elicit synthesis of the antimicrobial phytoalexins (West, 1981). This elicitation involves the induction of several enzymes that synthesize the phytoalexins. The elicitors, or "messengers", appear to be of several types (Stöessl, 1980; West, 1981). The attacking organisms can secrete enzymes that break down the plant's cell wall to release pectic fragments that apparently diffuse from the infection site to surrounding viable cells to signal, or elicit, synthesis of the phytoalexins (Bruce and West, 1981; Hahn *et al.,* 1981). Alternatively, the plant contains β-glucanases (Boller, 1982; Verma *et al.,* 1982) and chitosanases (Boller *et al.,* 1983) that can fragment fungal cell walls and release polysaccharide fragments that are also effective elicitors of phytoalexins (West, 1981). Thus, at least some part of the sensing mechanism for defense appears to involve signals that are fragments that are released from the cell walls of either the plant or microorganisms during the earliest period of fungal attack.

In contrast to phytoalexin elicitation, proteinase inhibitor induction is systemic. Signals are released at the attack sites that travel throughout the plant to signal the induction of the proteinase inhibitor proteins in leaves. The same types of signals (pectic fragments) that elicit phytoalexin synthesis in nearby cells of soybeans and castor beans can also induce proteinase inhibitor synthesis in potato, tomato and alfalfa leaves (Bishop *et al.*, 1981; Brown and Ryan, in review). This indicates that a similar receptor system may be at work in signalling both phytoalexins and proteinase inhibitors. However, the size of the signals appears to be important. In the phytoalexin system only polysaccharides above about DP = 10 are active elicitors (Nothnagel *et al.*, 1983). In proteinase inhibitor induction in tomato plants, oligogalacturonic polymers below DP = 10 are still active, even down to the dimer (Bishop *et al.*, in review). Additionally, β-glucans and chitosans, fragments of the fungal cell wall that elicit phytoalexins, are also active inducers of proteinase inhibitors (Bruce and West, 1981; Walker-Simmons *et al.*, 1983).

These similarities in signals strongly support a hypothesis that a similar communication systems probably do exist in many plants (Ryan *et al.*, 1980). Similar signals (poly and oligosaccharide fragments of the cell walls of plants, microorganism and insects) may have evolved in various plant genera and species to trigger a wide variety of genes that are involved in plant defense. Knowledge about such mechanisms, and their possible breadth and significance in nature, could be fundamentally important in many aspects of fashioning plants to suit man's needs, and particularly to molecular biologists who are seeking ways to manipulate and regulate plant genes.

II. Phytoalexin Induction

Phytoalexins are chemical compounds that are produced by plants in response to microbes and which inhibit the growth of the attacking organisms (Stöessl, 1980; West, 1981). Their presence was recognized over 50 years ago, but they were identified and defined as "phytoalexins" in 1940 by Müller and Börger (1940). The phytoalexins are now included with a broad spectrum of chemicals calles "secondary plant compounds" that are thought to be important in plant interactions with their environment (Janzen, 1979). The phytoalexins are found among several chemical families including isoflavonoids, sesquiterpenes, diterpenes, polyacetylenes, dihydrophenarthrenes and stilbenes (Stöessl, 1980; West, 1981). Most of the phytoalexins reported to date are found in the legumes, although they are present in other plant families. Plants within any given family tend to produce phytoalexins that are chemically related to other phytoalexins found within that family. The most extensively studied phytoalexins found in the legumes are flavonoids, which are produced by members of this family in response to microbe attack (Stöessl, 1980).

The resistance response is initiated by signals, called elicitors, that can be either the cell wall fragments mentioned earlier, or any of a variety of chemicals or other agents (such as UV light) that can initiate the synthesis of phytoalexins in higher plant tissues (West, 1981). Many different chemicals have been found to trigger the process. However, the majority of elicitors appear to originate from the attacking organisms or from the plant itself and are classed as "biotic elicitors" (Hargreaves and Bailey, 1978). These biotic elicitors besides cell wall fragments, can also include peptides, enzymes, polysaccharides and lipids (West, 1981).

One of the major classes of phenolic compounds synthesized by plants is the flavonoids, which include isoflavoves, isoflavinones, isoflavans and ptercocarpins (Stöessel, 1980; West, 1981). These compounds have a variety of roles, such as formation of flower pigments, synthesis of phytoalexins and U.V.-protective compounds (Hahlbrock, 1981) as well as the antimicrobial phytoalexins. The initial steps in the synthesis of flavonoids is the formation of 4-coumoroyl-CoA from phenylalanine by the three enzymes; phenylalanine ammonia-lyase (PAL), cinnamate-4-hydrolase (C4H) and 4-coumoroyl-CoA ligase (4Cl). This pathway is called the phenylpropanoid pathway and can lead to a variety of structurally related flavonoids, including a number of phytoalexins.

The enzymes of the flavonoid pathway can be induced by elicitors. Much of the research to elucidate the regulation of the enzymes of flavonoid metabolism has been performed with parsley (Hahlbrock, 1981), where U.V. irradiation causes induction of about 16 enzymes leading to several flavonoid glycosides. An absolute requirement for U.V. light for their induction has been demonstrated, and their presence in parsley appears to be as protective agents against the excessive U.V. exposure rather than against pathogens. The system, however, is an excellent model to study the regulation of some of the enzymes leading to the synthesis of phytoalexins. A central pathway, from phenylalanine to 4-coumoroyl-CoA, as described above, provides an intermediate (4-coumoroyl-CoA) that is a key compound for three branch pathways leading to (a) lignin and associated compounds, (b) flavonoids and isoflavonoids and (c) coumorins (Hahlbrock and Grisebach, 1979). The enzymes of the central pathway have been isolated and studied in detail. However, with the exception of the first enzyme, chalcone synthase, leading to flavonoids and isoflavonoids, and the enzyme, UDP-apiose synthase, the enzymes of the three branches have not been isolated and characterized.

The enzymes involved in flavonoid and isoflavonoid synthesis have been grouped according to their positions in the pathway from phenylalanine to the flavonoids. The three initial enzymes of the pathway from phenylalanine to 4-coumoroyl-CoA, i. e., PAL, C4H, 4Cl, have been called the Group I enzymes (Hahlbrock, 1981). The Group II enzymes are those of the flavonoid branch from chalcone synthase, including about 13 enzymes (Hahlbrock and Grisebach, 1979).

Suspension cultures, particularly of parsley *(Petroselinum hortense)* and soybean (*Glycine max* L.) have provided model systems to study the pathways of phenylpropanoid metabolism that produce the phytoalexins as well as other flavonoids. In parsley, changes in both culture dilution and U.V. irradiation regulate the two groups of enzymes, Group I and Group II, separately. A dilution in suspension cells triggers a 5—50 fold increase in activity of the Group I enzymes for 12—15 hr (Hahlbrock and Grisebach, 1979). Group II enzymes are unaffected. U.V. irradiation induces both Group I and II enzymes that also peak at 12—15 hr (Hahlbrock and Grisebach, 1979). At certain growth stages of the cultures, U.V. irradiation causes Group I enzymes to reach maximal activities several hours before those of Group II. The lag period for Group I enzymes after the initial induction is about 2 hr and that for Group II somewhat larger. The activities of the two groups of enzymes are regulated together within groups, but each group is regulated independently.

Following U.V. irradiation, the changes in activities of the specific enzymes in each group, PAL and 4-coumerate ligase in Group I, and CS, and UDP Apiose synthase in Group II, were shown to result from changes in levels of mRNA (Schröder *et al.,* 1979; Gardiner *et al.,* 1980; Ragg *et al.,* 1981). The rates of appearance of mRNA for Group I and II enzymes were in agreement with rates of appearance of enzyme activities. Group I mRNA and enzyme activity preceeded those of Group II, again indicating that the mRNA and enzyme synthesis within each group was coordinately regulated and changes in enzyme activity was a function of mRNA present.

Although coordinately regulated within each group, recent evidence indicates that this coordination does not extend to the relative quantities of each enzyme synthesized. For example the ratio of translatable mRNAs for two enzymes of Group I, i.e. PAL and 4Cl, induced by irradiation, is 1.0, 10 hr after induction by U.V. light, whereas this ratio in cells induced by an elicitor is about 0.5 (Ragg *et al.,* 1981).

A crude elicitor, isolated from *Phytophtera megasperma* var. sojae, presumably containing cell wall fragments, was shown to induce PAL in parsley cell cultures (Hahlbrock *et al.,* 1981) with similar efficiencies as in cell cultures of *Glycine max* cv Harosoy 63, which is resistant to several lines of the fungi (Ebel *et al.,* 1976). The elicitor induced PAL activity and PAL mRNA in the parsley cells more rapidly than did U.V. light and there was a 2—4 fold increase in translatable mRNA for PAL. In parsley, however, only the Group I enzymes were induced whereas Group II enzymes were not. In fact, the presence of the elicitor inhibited the induction of Group II enzymes by U.V. light (Hahlbrock *et al.,* 1981). Apparently another pathway, that of furanocorumarin, rather than the isoflavonoid pathway, is induced (Hahlbrock *et al.,* 1983). The enzymes of this pathway have tentatively been assigned to the category called Group III.

On the other hand, in suspension cultures of French bean (*Phaseolus vulgaris* L.) a crude elicitor from the fungus *Colletotrichum lindimutheanum,* a pathogen of French beans, caused a rapid increases in PAL, a Group I enzyme, and chalcone synthase, a Group II enzyme (Lawton *et al.,* 1983 a, 1983 b). In both cases the increases in enzymes of either Groups I or II were dependent upon mRNA synthesis, as measured with an *in vitro* translation system (Lawton *et al.,* 1983 a).

Poly $(A)^+$ mRNA, isolated from U.V. irradiated parsley cell suspension cultures which contained both Group I and II enzymes, was used to obtain a library of cDNA clones (Kreuzaler *et al.,* 1983). A clone containing cDNA for chalcone synthase mRNA was identified and used to quantify mRNA levels in U.V. irradiated cells. Within 3 hr following U.V. irradiation chalcone synthase mRNA could be detected. Its levels increased for over 10 hr, then decreased thereafter. cDNAs for PAL and 4Cl have now been obtained from the cDNA library and used to quantify levels of both mRNAs following U.V. treatment of parsley cell cultures. mRNA to both enzymes was detected within 1 hr after elicitor treatment (Kuhn *et al.,* 1983).

The aquisition of cDNA probes opens the way toward the isolation and characterization of the genes that code for several of the enzymes whose expression is regulated by the elicitor or U.V. irradiation. The genes specifically regulated by elicitor may be of great value in exploring the promotor and enhancer regions of the genes and the mechanism by which the elicitor exerts its influence on gene expression. As mentioned earlier, it is possible that such information may relate to a common mechanism that also regulates other genes which respond to environmental stress inflicted by pest attacks, such as the proteinase inhibitor genes, which are discussed in the following section.

III. Proteinase Inhibitor Induction

Synthesis of two small proteinase inhibitor proteins is induced in plant leaves by the release of a systemic "wound hormone" from tissues in response to severe wounding (Ryan, 1979). The induction of two non-homologous gene products by wounding provides not only the opportunity to study hormone-regulated expression of specific nuclear genes in plant leaves, but provides different proteinase inhibitor genes to compare gene products that are both developmentally and environmentally regulated in plants. The wound-regulated expression of proteinase inhibitor genes is a unique system for the study of gene regulation in plants.

The proteinase inhibitors that are wound-induced in potato and tomato leaves are among a group of several small proteins originally found in potato tubers where they are under developmental control and stored in fairly high concentrations (Ryan *et al.,* 1977). These proteins are powerful inhibitors of the serine endopeptidases (trypsin, chymotrypsin and elastase). These endopeptidases are among the major digestive enzymes utilized by most animals and many microorganisms, but rarely by plants, for digesting food proteins (Ryan, 1981). Proteinase inhibitors, because of their broad, powerful inhibitory specificities toward animal and microorganism proteinase, and their high concentrations in plant storage organs, have long been thought to be protective agents against plant pests by interfering with protein digestion and thereby severely reducing the nutritional value of the tissues (Ryan, 1979, 1981).

The two inhibitor proteins, Inhibitor I and II were first discovered in potato tubers where they appear early during tuberization and continue to accumulate until the tubers are mature. The two inhibitors can account for over 10% of the soluble proteins of potato tubers, depending upon the variety (Ryan *et al.*, 1977).

The two proteinase inhibitors that accumulate in leaves of wounded tomato leaves have been isolated and characterized (Table I). Inhibitor I has a native molecular weight of 41,000 and is composed of subunits with molecular weights of 8,100 (Plunkett *et al.*, 1981). It is, therefore, a pentamer in its native state. Each subunit possesses an active site specific for chymotrypsin with a K_I of about 10^{-9}M. Inhibitor II has a molecular weight of about 23,000, is composed of two subunits, and strongly inhibits both trypsin and chymotrypsin with K_I values of about 10^{-8} and 10^{-7}M, respectively.

Table I. Biochemical Properties of tomato Inhibitors I and II

Property[1]	Inhibitor I	Inhibitor II
Molecular Weights		
Dissociated Monomer	8,100	12,300
Multimer (native)	41,000	23,000
Disulfide bonds (per monomer)	1	5
Reactive Sites (per monomer)	1	2
Specificites		
Reactive Site I	chromotrypsin-like enzymes	chromotrypsin-like enzymes
Reactive Site II	—	trypsin-like enzymes

[1] Plunkett *et al.*, 1982.

Messenger RNA has been prepared from leaves of wounded and unwounded tomato plants and only leaves of wounded plants contain translatable mRNAs specific for Inhibitors I and II (Nelson and Ryan, 1980a). The two inhibitors are synthesized as preproteins that are post-translationally cleaved and sequestered in the central vacuoles of the leaf cells after where they have half lives of days or weeks. Neither Inhibitor I nor II are glycoproteins and their mode of synthesis and transport into the vacuole is virtually unknown. Since both proteins have been shown to be translated *in vitro* in a reticulocyte lysate system as preinhibitors, 2000—3000 daltons larger than those synthesized and accumulated *in vivo* (Nelson and Ryan, 1980a) it appears that these precursors are important in the process of compartmentation of the inhibitors into the central vacuole, perhaps in the same manner as seed protein synthesis and compartmentation in the vacuole (Sengupta *et al.*, 1982).

When young tomato plants are wounded, by chewing insects or by a severe crushing of any type, the levels of total poly(A)$^+$ translatable mRNA for both proteinase Inhibitor I and Inhibitor II rise rapidly during the first four hours after wounding (Nelson and Ryan, 1980b). Translatable mRNAs for Inhibitor I and II are present at near maximum levels within four hours following a single wounding. The levels remain high until nine hours when they decrease to less than half their original levels. A second wounding of the leaves, nine hours after the initial wounding, doubled or tripled (this varied with groups of plants) the rate of accumulation of both Inhibitors I and II over those of once wounded plants (Nelson and Ryan, 1980b). This second wound also resulted in the maintenance of the mRNA levels already present at nine hours so that the decrease in translatable mRNA noted in singly wounded plants did not occur. The second wound apparently provides more mRNA when the plant's translational system is already operating at high efficiency.

IV. Molecular Cloning and Characterization of Proteinase Inhibitor Genes

The poly(A)$^+$ mRNA from leaves of wounded plants was cloned into the plasmid pUC9 and clones containing Inhibitor I and II cDNAs were identified (Graham *et al.*, 1983). Utilizing the cDNAs for Inhibitor I and II in Northern blots with tomato leaf mRNA from wounded plants, the molecular weights of the specific mRNAs coding for these two inhibitors was established. In Table II the lengths of the cDNAs and their respective mRNAs are given.

Table II. Size of Inhibitor I and II cDNAs and mRNAs[1]

Inhibitor	Size	
	dscDNA (base pairs)	mRNA (bases)
I	570	600
II	700	750

[1] Graham, J., Brown, W., Ryan, C. A., unpublished data.

The cDNAs are full length for their coding regions and they include a signal sequence for each inhibitor (Graham *et al.*, 1983). The cDNAs have been employed to select genomic clones from lambda phage libraries of both the potato and tomato genomes. With the cDNAs as probes for mRNA synthesis, and the genomic clones to study gene structure it may soon be possible to identify important regulatory mechanisms that control wound-induced transcription and translation of proteinase inhibitors in plants. With this information it should be possible, at the gene level, to begin to compare these mechanisms with those of the antimicrobial phytoalexin elicitation.

V. Summary

Studies of two pest induced systems, the insect-induced synthesis of proteinase inhibitors and the fungal-induced synthesis of phytoalexins, are progressing toward the same important goals of understanding the biochemistry and molecular biology of how plants respond to pest attack by activating specific genes to defend themselves.

Despite the apparent differences in the regulation of synthesis of phytoalexin and proteinase inhibitors, some striking similarities among these systems have emerged that suggest that at least some facets of a similar communication system may be common to each. The full understanding of such system may still lie far into the future but answers are being sought from two different approaches. One course of research is to define the signals that trigger these systems, both localized and systemic, and to discover the mechanism of transport, reception and their amplification at the cellular level and to unravel the sequence of events leading to gene expression. The second line of attack is to isolate the genes that are regulated and elucidate the promoter and enhancer regions and to learn how these regions are controlled by pest attacks. Both approaches have severe technical problems and experimental difficulties that must be overcome. These are new frontiers of plant science.

On one hand, some of the signals have been identified and the evidence suggests that cell wall fragments from both the pests and the host plant can be involved. With these factors, the receptor mechanism can now be studied.

At the molecular biological end, some progress has also been made. Isolation of mRNA from induced plants has provided cDNAs for several of the enzymes of the pathway of phytoalexin synthesis and for two proteinase inhibitors. These cDNAs have proven to be useful tools for studying the elicitor-induced regulation of expression of genes involved in phytoalexin synthesis and in isolating specific proteinase inhibitor genes from appropriate plant gene libraries. It is anticipated that studies of the genes for both the phytoalexin pathway enzymes and for proteinase inhibitors will not only provide important new information concerning the environmental regulation of the plant genome but will also provide useful genetic materials for future use in genetically engineering plants.

VI. References

Bishop, P. D., Makus, D., Pearce, G., Ryan, C. A., 1981: Proteinase inhibitor-inducing factor activity in tomato leaves resides in oligosaccharides enzymically released from cell walls. Proc. Natl. Acad. Sci. (U. S. A.) **78,** 3536—3540.

Bishop, P. D., Pearce, G., Bryant, J., Ryan, C. A., 1984: Isolation and characterization of the proteinase inhibitor induction factor and a PIIF-active oligosaccharide fragments, in review.

Boller, T., Gehri, A., Mauch, F., Vögeli, V., 1983: Chitinase in bean leaves: induction by ethylene, purification, properties and possible function. Planta **157,** 22—31.

Boller, T., 1982: Enzymatic equipment of plant vacuoles. Physical Vég. **20,** 247—257.

Brown, W., Ryan, C. A. (in review): Isolation and characterization of a wound-induced trypsin inhibitor from alfalfa leaves.

Bruce, R. J., West, C. A., 1981: The role of pectic fragments of the plant cell wall in elicitation by a fungal endopolygalacturonase. Plant Physiol. **69,** 1181—1188.

Ebel, J., Ayers, A. R., Albersheim, P., 1976: Host-pathogen interaction. XII. Response of suspension cultured soybean cells to elicitor isolated from *Phytophthora megasperma* var. sojae, a fungal pathogen of soybeans. Plant Physiol. **57,** 775—779.

Graham, J., Okita, T., Pearce, G., Merryweather, J., Titani, K., Walsh, K., Ryan, C. A., 1983: Molecular cloning and characterization of tomato leaf Inhibitors I and II. Fed. Proc. **42,** 1143.

Gardiner, S. E., Schröeder, J., Matern, I., Hammer, D., Hahlbrock, K., 1980: mRNA-dependent regulation of UDP-aprose synthase activity in irradiated plant cells. J. Biol. Chem. **225,** 10752—10757.

Green, T., Ryan, C. A., 1972: Wound-induced proteinase inhibitor in plant leaves; a possible defense mechanism against insects. Science **175,** 776—777.

Hahlbrock, K., Grisebach, H., 1979: Enzymatic controls in the biosynthesis of lignin and flavonoids. Ann. Rev. Plant Physiol. **30,** 105—130.,

Hahlbrock, K., 1981: Flavonoids. In: Stumpf, P. K., Conn, E. E. (eds.), The biochemistry of plants, pp. 425—456. New York: Academic Press.

Hahlbrock, K., Lamb, C., Purwin, C., Ebel, J., Fauty, E., Schäfer, E., 1981: Rapid response of suspension-cultured parsley cells to the elicitor from *Phytophthora megasperma* var. sojae. Plant Physiol. **67,** 768—771.

Hahlbrock, K., Boudet, A. M., Chappell, J., Kreuzaler, F., Kuhn, D. N., Ragg, H., 1983: Differential induction of mRNAs by light and elicitor in cultured plant cells. In: Proc. NATO Conference on Structure and Function of the Plant Genome, Ciferri, O., and Dure, L., eds., pp. 15—24.

Hahn, G. M., Darvill, A., Albersheim, P., 1981: Host pathogen intraction, XIX. The endogenous elicitor and fragment of a plant cell wall polysaccharide that elicits phytoalexin accumulation in soybeans. Plant Physiol. **68,** 1161—1169.

Hargreaves, I. A., Bailey, J. A., 1978: Phytoalexin production by hypocotyls of *Phaseolus vulgares* in response to constitutive metabolites released by damaged bean cells. Physiol. Plant Pathol. **13,** 89—100.

Janzen, D., 1979: New horizons in the biology of plant defenses, In: Herbivores: their interaction with secondary plant metabolites, pp. 331—351. New York: Academic Press.

Kreuzaler, F., Ragg, H., Fauty, E., Kuhn, D., Hahlbrock, K., 1983: U.V. irradiation of chalcone synthase in cell suspension cultures of *Petroselenum hortense*. Proc. Natl. Acad. Sci. U. S. A. **80,** 2591—2593.

Kúc, J., 1980: Plant immunization-mechanisms and practical implication. In: Active defense mechanisms in plants, pp. 157—178. New York: Plenum Press.

Kúc, J., 1982: Induced immunity to plant disease. BioScience **32,** 854—860.

Kuhn, D. N., Chappell, J., Hahlbrock, K., 1983: Identification and use of cDNAs of phenylalanine ammonia-lyase and 4-coumorate-CoA lyase mRNAs in studies of the induction of phytoalexin biosynthetic enzymes in cultured parsley cells. In: Proc. NATO Conference on "Structure and Function of the Plant Genome", Ciferri, O., and Dure, L., eds., pp. 329—336.

Lawton, M. A., Dixon, R. A., Hahlbrock, K., Lamb, C., 1983a: Elicitor induction of mRNA activity: Rapid effects of elicitor on phenylalanine ammonia-lyase and chalcone synthase mRNA in bean cells. Eur. J. Biochem. **130,** 131—139.

Lawton, M. A., Dixon, R. A., Hahlbrock, K., Lamb, C., 1983b: Rapid induction of the synthesis of phenylalanine ammonia-lyase and of chalcone synthase in elicitor-treated plant cells. Eur. J. Biochem. **129,** 593—601.

Laskowski, M., Jr., Kato, I., 1980: Protein inhibitors of proteinases. Ann. Rev. Biochem. **49,** 593—626.

Müller, K. O., Börger, H., 1940: Experimentelle Untersuchungen über die *Phytophthora*-Resistenz der Kartoffel. Arb. Biol. Reichsanst. **23,** 189—231.

Nelson, C., Ryan, C. A., 1980a: *In vitro* synthesis of pre-proteins or vacuolar compartmented proteinase inhibitors that accumulate in leaves of wounded tomato plants. Proc. Natl. Acad. Sci. U. S. A., **77,** 1975—1979.

Nelson, C., Ryan, C. A., 1980b: Temporal shifts in the apparent *in vivo* transcriptional efficiencies of tomato leaf proteinase Inhibitors I and II mRNAs following wounding. Biochem. Biophys. Res. Commun. **94,** 355—359.

Nothnagel, E. A., McNeil, M., Albersheim, P., Dell, A., 1983: Host-Pathogen interactions. XXII. A galacturonic acid oligosaccharide from plant cell walls elicits phytoalexins. Plant Physiol. **71,** 916—926.

Plunkett, G., Senear, D. F., Zuroske, G., Ryan, C. A., 1982: Proteinase Inhibitors I and II from leaves of wounded tomato plants: purification and properties. Arch. Biochem. Biophys. **213,** 463—472.

Ragg, H., Kuhn, D. N., Hahlbrock, H., 1981: Coordinated regulation of 4-coumorate-CoA ligase and phenylalanine ammonia-lyase mRNAs in cultured plant cells. J. Biol. Chem. **256,** 10061—10065.

Rhodes, D., 1979: Evolution of plant chemical defenses against herbivores. In: Herbivores, their interaction with secondary plant metabolites, pp. 3—54. New York: Academic Press.

Ryan, C. A., 1981: Proteinase inhibitors. In: Stumpf, P. K., Conn, E. E. (eds.), The biochemistry of plants, Vol. 6, pp. 351—370. New York: Academic Press.

Ryan, C. A., 1979: Proteinase inhibitors. In: Herbivores, their interaction with secondary metabolites, pp. 599—618. New York: Academic Press.

Ryan, C. A., 1973: Proteolytic enzymes and their inhibitors in plants. Ann. Rev. Plant Physiol. **24,** 173—196.

Ryan, C. A., Bishop, P., Pearce, G., Darvill, A. G., McNeil, M., Albersheim, P., 1981: A sycamore cell wall polysaccharide and a chemically related tomato leaf polysaccharide possess similar proteinase inhibitor-inducing activities. Plant Physiol., **68,** 616—618.

Ryan, C. A., Kuo, T., Pearce, G., Kunkel, R., 1977: Variability in the concentration of three heat-stable proteinase inhibitor proteins in potato tubers. Am. Potato J. **53,** 433—440.

Schröder, J., Kreuzaler, F., Schäfer, E., Hahlbrock, K., 1979: Concomitant induction of phenylalanine ammonia-lyase and flavone synthase mRNAs in irradiated plant cells. J. Biol. Chem. **254,** 57—65.

Schultz, J. C., 1983: Impact of variable plant defense chemistry on susceptibility of insects to natural enemies. In: Plant resistance to insects, Am. Chem. Soc., pp. 39—45.

Sengupta, C., Deluca, V., Bailey, D. C., Verma, D. P. S., 1981: Post-translational processing of 7 S and 15 S components of soybean storage proteins. Plant Mol. Biol. **1,** 19—34.

Stöessl, A., 1980: Phytoalexins — a biogenic perspective. Phytopath. Z. **99,** 251—272.

Verma, D. P. S., Kumar, V., Maclachlan, G., 1982: β-glucanases in higher plants: localization and potential functions. In: Cellulose and other natural polymer systems. Biogenesis, structure and degradation. Brown, M., Jr. (ed.), pp. 459—488. New York: Plenum Press.

Walker-Simmons, M., Ryan, C. A., 1983: Chitosans and pectic polysaccharides both induce the accumulation of the antifungal phytoalexin pisatin in pea pods and anti-nutrient proteinase inhibitors in tomato leaves. Biochem. Biophys. Res. Comm. **110,** 194—199.

West, C. A., 1981: Fungal elicitors of the phytoalexin response in higher plants. Naturwissenschaften **68,** 447—457.

Subject Index

A pagenumber with * refers to a whole chapter

Printed by Druckerei G. Grasl, A-2540 Bad Voeslau